Mixing
Chaos and Turbulence

NATO ASI Series

Advanced Science Institutes Series

A series presenting the results of activities sponsored by the NATO Science Committee, which aims at the dissemination of advanced scientific and technological knowledge, with a view to strengthening links between scientific communities.

The series is published by an international board of publishers in conjunction with the NATO Scientific Affairs Division

A	**Life Sciences**	Kluwer Academic/Plenum Publishers
B	**Physics**	New York, Boston, Dordrecht, London, Moscow
C	**Mathematical and Physical Sciences**	Kluwer Academic Publishers Dordrecht, Boston, and London
D	**Behavioral and Social Sciences**	
E	**Applied Sciences**	
F	**Computer and Systems Sciences**	Springer-Verlag
G	**Ecological Sciences**	Berlin, Heidelberg, New York, London,
H	**Cell Biology**	Paris, Tokyo, Hong Kong, and Barcelona
I	**Global Environmental Change**	

PARTNERSHIP SUB-SERIES

1. **Disarmament Technologies**	Kluwer Academic Publishers
2. **Environment**	Springer-Verlag
3. **High Technology**	Kluwer Academic Publishers
4. **Science and Technology Policy**	Kluwer Academic Publishers
5. **Computer Networking**	Kluwer Academic Publishers

The Partnership Sub-Series incorporates activities undertaken in collaboration with NATO's Cooperation Partners, the countries of the CIS and Central and Eastern Europe, in Priority Areas of concern to those countries.

Recent Volumes in this Series:

Volume 370—Supersymmetry and Trace Formulae: Chaos and Disorder
edited by Igor V. Lerner, Jonathan P. Keating, and David E. Khmelnitskii

Volume 371—The Gap Symmetry and Fluctuations in High-T_c Superconductors
edited by Julien Bok, Guy Deutscher, Davor Pavuna, and Stuart A. Wolf

Volume 372—Ultrafast Dynamics of Quantum Systems: Physical Processes and
Spectroscopic Techniques
edited by Baldassare Di Bartolo

Volume 373—Mixing: Chaos and Turbulence
edited by H. Chaté, E. Villermaux, and J.-M. Chomaz

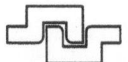

Series B: Physics

Mixing
Chaos and Turbulence

Edited by

H. Chaté

CEA-Service de Physique de l'Etat Condensé
Gif-sur-Yvette, France
and
LadHyX, Ecole Polytechnique
Palaiseau, France

E. Villermaux

LEGI-CNRS
Institute de Mécanique de Grenoble
Grenoble, France

and

J.-M. Chomaz

LadHyX, Ecole Polytechnique
Palaiseau, France

Springer Science+Business Media, LLC

Proceedings of a NATO Advanced Study Institute on
Mixing: Chaos and Turbulence,
held July 7–20, 1996
in Cargèse, France

NATO-PCO-DATA BASE

The electronic index to the NATO ASI Series provides full bibliographic references (with
keywords and/or abstracts) to about 50,000 contributions from international scientists published
in all sections of the NATO ASI Series. Access to the NATO-PCO-DATA BASE is possible via a
CD-ROM "NATO Science and Technology Disk" with user-friendly retrieval software in English,
French, and German (©WTV GmbH and DATAWARE Technologies, Inc. 1989). The CD-ROM
contains the AGARD Aerospace Database

The CD-ROM can be ordered through any member of the board of Publishers or through NATO-
PCO, Overijse, Belgium.

Additional material to this book can be downloaded from http://extra.springer.com.

ISBN 978-1-4613-7127-4   ISBN 978-1-4615-4697-9 (eBook)
DOI 10.1007/978-1-4615-4697-9

© 1999 Springer Science+Business Media New York
Originally published by Kluwer Academic / Plenum Publishers in 1999

10 9 8 7 6 5 4 3 2 1

Jacques Villermaux passed away a few months after this school. A chemical engineer and specialist in mixing, he was also an enthusiastic popularizer of Georg Bauer *De Re Metallica*, from which the engravings illustrating this volume are taken.

Preface

Why mixing?

–Because the subject, as shall become clear in the following pages, is ubiquitous and varied, with problems ranging from the fundamental to the very concrete, straddling the frontiers of many fields: a truly federative theme.

This, at least, was the spirit in which this Advanced Study Institute was planned, intending to bring together diverse communities sharing an interest in the problem of mixing, either as a theme of fundamental interest or as a necessary step in various applications.

The success of this lively school owed much to the quality of the speakers. It is reflected, we believe, in the quality of their contributions to this volume, which cover many facets of mixing. We thank them for their participation in our project, which was initially proposed — and since continually supported by Etienne Guyon, himself one of the organizers, ten years ago, of a school on "Disorder and Mixing" in Cargèse. We thank our sponsors, NATO, NSF, CNRS and DRET, and the staff of the Institut des Etudes Scientifiques de Cargèse, for making this school a success.

H. Chaté, E. Villermaux, and J.-M. Chomaz

Contents

Part IV: Statistical Methods and Mixing

Why Mixing?

EMMANUEL VILLERMAUX,[1] HUGUES CHATÉ,[2,3] AND JEAN-MARC CHOMAZ[3]

[1]*LEGI–CNRS, Institut de Mécanique de Grenoble, 38041 Grenoble, France*
[2]*CEA–Service de Physique de l'Etat Condensé, 91191 Gif-sur-Yvette, France*
[3]*LadHyX, Ecole Polytechnique, 91128 Palaiseau, France*

Mixing is a subject which suffers from the *Bourgeois Gentilhomme* complex. Like Monsieur Jourdain in Molière's play (1670), scientists, engineers, and indeed all of us often "do mixing without even knowing it" (just think about yourself trying to prepare mayonnaise...).

As an operation, it consists simply in putting together two or more initially segregated constituents and stirring, in order to attain uniformity, or a new product, or the complete disappearance of one of the constituents etc.; mixing thus is indeed at the cross-roads of many different areas of science. One often needs to mix elements in order to make a new product, an homogeneous blend, or to make possible a chemical reaction or an efficient combustion. One also needs to understand how nature mixes or has mixed to gain information on e.g., the size of a pollutant spot in a valley, the rate of destruction of ozone in the atmosphere, or the dynamics of the earth's mantle. More than a fascinating subject in itself, mixing is thus ubiquitous and a key process in many complex man-made, or natural operations.

It is no doubt *because* of its universality that mixing, as noted with regret by J. Ottino (1989) [17], "...does not enjoy the reputation of being a very scientific subject..." as is frequently the fate of interdisciplinary topics.

The dendritic nature of the subject matter is revealed by the different angles from which the various scientific communities attack the problem, according to their needs These approaches can be roughly grouped into three naturally overlapping categories, each of them of interest to different schools of thought:

- Geometry
- Kinetics
- Structures

Beyond the tools and attitudes developed by different scientific groups regarding the problem of mixing, one might appreciate that mixing, as suggested by common

Mixing: Chaos and Turbulence, edited by Chaté *et al.*
Kluwer Academic / Plenum Publishers, New York 1999.

sense, is the operation by which a system evolves from one state of simplicity (the initial segregation) to another state of simplicity (the complete uniformity). Between these two extremes, complex patterns emerge and die. Questions then naturally arise: how can the geometry of complex patterns be characterized, what is the clock, the time-scale of the process, and what are the structures involved in the flow?

1.   Geometry

Very early on, emphasis was placed on the geometry of the mixing zone in the combustion context. Indeed the involved, multiscale geometry of the interface which separates two streams being mixed is not only a spectacular facet of the process, but is sometimes at the core of the physical problem. Exothermic reactions in gases, or reactions with a fast kinetics in liquids confine the reaction zone to a thin region of space and the flame appears as a sharp interface for most scales in the flow. The total extent of the flame area dictates the propagation speed of the front in premixed reactants, as noted by Damköhler (1940) [9], and the net combustion rate in diffusion flames (see the celebrated experiments of Hawthorne et al. (1949) [13]. Knowing how and why the front is distorted by the underlying motions is thus crucial for predicting the flame extent.

The multiscale structure of the contour of a scalar blob immersed in a disordered flow was recognized by Welander (1955) [24], who suggested how a connection could be made between the internal structure of the underlying motions and the complexity of the blob shape. Welander even made reference to the Koch curve, long before fractals were popularized in this and other contexts by Mandelbrot (1975) [14].

The program, still very active today, for investigating the geometry of the scalar support, or the iso-concentration surfaces in flows was certainly already contained in Reynolds (1894) [20]. Reynolds was probably even more ambitious, suggesting that watching the dynamics of "coloured bands" in a flow was a route towards understanding the driving motions. We have not yet gone so far.

2.   Kinetics

Mixing is, in the strict sense, a transient process from initial segregation to ultimate homogeneity. This calls for the understanding of the kinetics and timescales of mixing processes.

Studies on dispersion address the problem of the growth rate of the radius of a tracer blob in a prescribed displacement field. In addition to pure molecular diffusion in the medium at rest, fluid motion usually enhances dispersion (Taylor 1921, 1953) [22, 23] and alters not only the diffusion law, by a renormalization of the diffusion coefficient, but also the structure of the law itself. Due to persistent ballistic movements and ever larger jumps in turbulent flows, dispersion laws exhibit, in the absence of traps or slow recirculating motions, a faster than linear growth of the mean squared radius of the blob (Prandtl 1925 [18], Richardson 1926 [21]). The presence of bypasses or dead-ends in complex geometries alter, in continuous flow systems (a river, a valley through which wind blows, an open chemical reactor), the residence time distribution of a tracer introduced at the inlet of the system. The novel features of the distribution are spikes at short times if a short circuit is present, and/or long tails caused by traps and slow motions in confined cavities. This was first described by Danckwerts (1953) [11].

The kinetics of mixing does not only refer to dispersion. Dispersion may result solely from a spatial reorganization of the quantity to be mixed, with no interpenetration with the substrate at the molecular level.

Mixing, as opposed to stirring, actually means homogeneisation at the smallest, i.e. diffusive, scales, and the appropriate quantity to define and measure the mixing time, namely the "intensity of segregation," was introduced by Danckwerts (1952) [10]. The mixing time (for instance in a tank stirred with an impeller, familiar in the chemical industry) is found to be solely determined by the large scale features of the flow (integral scale, root mean square velocity; see, e.g., Nagata (1975) [15]), regardless of the intimate structure of turbulence, provided the Reynolds number is large enough (typically larger than 10^4), in a wide variety of flows. This absence of a role for turbulence and its structure in the characteristic time of the inhomogeneities decay is an important point which has been somewhat overlooked. The intimate structure of the local rearrangements in the flow does, however, play a role when chemical reactions with a nonlinear kinetics, or consecutive-concurrent chemical reactions occur within the mixture (Danckwerts 1953 [11], Epstein 1990 [12]).

3. Structures

Disordered flows develop a broad hierarchy of scales of motion which convect and distort the scalar field. Notwithstanding the fact that mixing is, like the decay of the energy-containing eddies, a transient phenomenon, the problem of the interaction of the scalar field with the underlying turbulent flow has focused on hypothetical "stationary conditions". This approach parallels the Kolmogorov quasi-equilibrium picture of turbulence (Corrsin 1951 [7], Oboukhov 1949 [16], Batchelor 1959 [1], Batchelor et al. 1959 [2]) and it is in this assumed limit that the timescales of the stirring motions which distort the scalar field are all shorter than the global mixing time (i.e. the variance of the scalar fluctuations is stationary as seen on the timescale of these motions), allowing the possibility of resorting to cascade arguments, and spectral analysis.

It is customary, in turbulence, to oppose the statistical approaches, the cascade representation being one of those, to the approaches focusing on "structures" more or less coherent or permanent in the flow. This opposition is somewhat artificial. Modern statistical models of turbulence rely in fact on a description of the flow as a hierarchical structure in the real space, possibly fractal, the structure of the hierarchy inducing the statistics. The cascade picture, notably, is thus in a sense a "structural approach".

In this cascade picture, the mixing time is the time needed for the scalar to travel, by a process of successive reductions of scale imposed by the hierarchy of pre-existing scales of motion (eddies) in the flow, from its initial size to the dissipation scale, prescribed by the Reynolds and Schmidt numbers (Corrsin 1964) [8]. This description presents many shortcomings, the first one being its inconsistency with the observed fact that the mixing time depends linearly on the initial size of the blob to be mixed, instead of on its size raised to the power 2/3. If it is clear that mixing macroscopic objects implies a reduction of their transverse size and a multiplication of ever smaller scales for diffusion to act efficiently, there is no direct proof that a scalar blob should follow the "Kolmogorov cascade" *step by step* to be ultimately erased by molecular diffusion. Here is a sign that, in turbulence, even many basic properties escape our precise understanding; the future should therefore be rich in forthcoming works.

The will to reduce a complicated problem such as turbulence to a set of elementary objects containing, presumably, all of the desired information, is in keeping with a long tradition in physics and fluid mechanics. Coherent structures, filaments, worms and sheets are frequently-invoked paradigms. In the context of mixing, the discovery of large-scale vortical structures sustained by shear flows has prompted a surge of activity (Brown and Roshko 1974) [6]. Originating essentially from a Kelvin-Helmhotz type of instability, these intense, long-lived structures are serious candidates to compete with the cascade representation since they directly couple injection scales with dissipation scales, eliminating intermediary steps. Their existence alone is not, however, sufficient to ensure a rapid mixing. When formed by a given velocity difference across a shear layer, their size (equivalent to the Reynolds number) has to grow beyond a critical value to allow the onset of the fine-scales activity, thereby hastening the uniformisation within the layer. This is the so-called "mixing transition" (Breidenthal 1981) [5].

Little is known in detail about what occurs during and after this transition in mixing layers. It is, nevertheless, a known fact that the deformation tensor in turbulent flows bears, in the mean, two directions of stretching and one direction of compression as first analyzed by Betchov (1956) [4]. This property leads to an increase in the length and area of material lines and surfaces (Batchelor and Townsend 1956) [3]. The stretched sheet is thus likely to be the elementary brick of mixing, whether turbulent or not (Ranz 1979 [19], Ottino 1989 [17]).

4. Towards a new paradigm

Restricting our field of sight to the particular limit of mixing in homogeneous fluids and keeping the Pandora's box of mixing in complex, non-Newtonian fluids, multiphase flows, granular flows etc... carefully sealed, we may wonder what the minimal set of ingredients should be to reach a satisfactory description of the above three facets of the mixing phenomenon. Let us examine mixing patterns obtained in different flow conditions, going from laminar (Fig. 1) to strongly turbulent (Fig. 3), via transitional shear instabilities (Fig. 2).

Figure 1 shows a start-up vortex (a kind of a smoke ring), several turns after its formation. The fluid constitutive of the ring develops a chemical reaction with the ambient medium in which it is rolled up. This is a fast chemical reaction and the reaction-diffusion zone, very thin in this case, traces the spiraling interface between the two media. The ring is a non-turbulent object. It directly couples its own injection scale with the diffusive scale via the spiraling motion and the accretion of sheets of ambient fluid at each turn. However, the ring presents many scales, even a continuous spectrum of scales due to the continuous increase of the striation

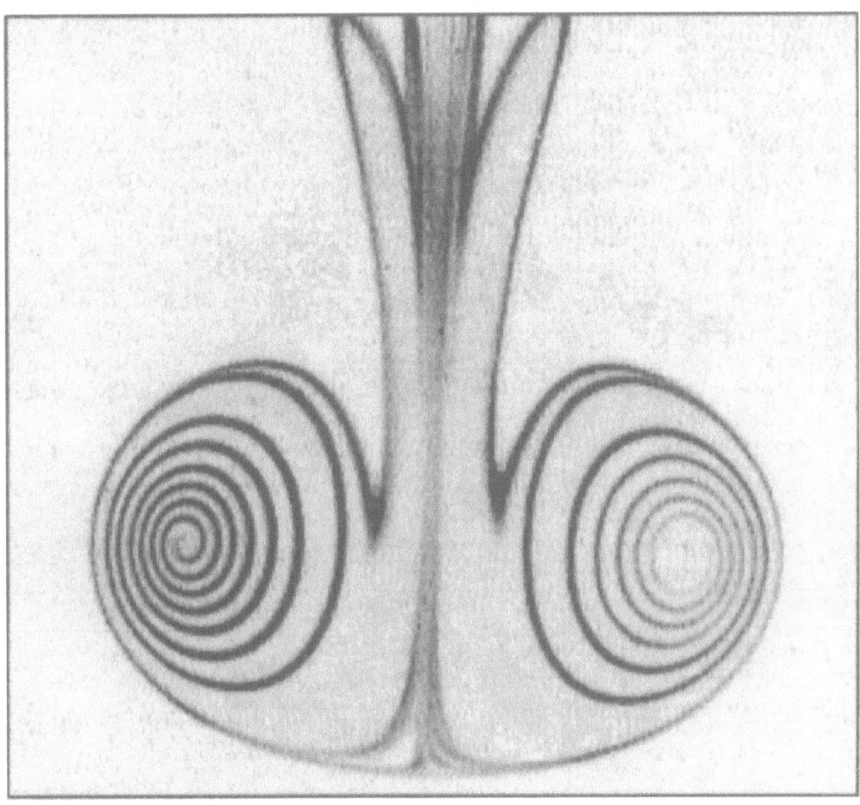

Fig. 1. — Start-up vortex. A fast chemical reaction between the ring and the ambient fluid traces the spiraling interface between the two media.

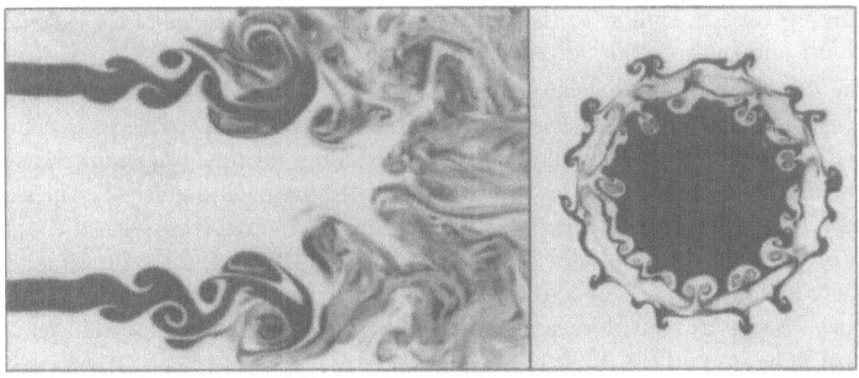

Fig. 2. — Cuts in the longitudinal and transverse directions of the flow of a coaxial jet showing the rolling-up structures resulting from the shear instability between the streams.

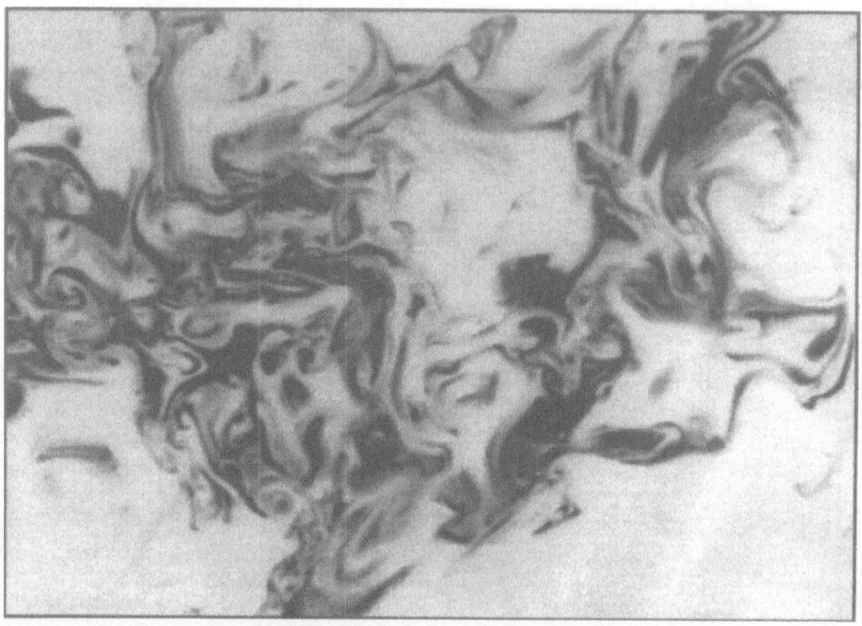

Fig. 3. — Blob of dye deposed in a large scale sustained turbulent flow converted into disjointed sheets which dilute in the surrounding medium.

thickness from the core of the spiral to its edge. This is a fractal whose covering dimension is close to 3/2. This is a multiscale, dissipative object; but where is the cascade?

Figure 2 represents cuts in the longitudinal and transverse directions of the flow of a coaxial jet. The inner jet is slower than the outer annular jet, and the pictures show the structures which result from the instability of the shear between the streams. One recognizes in the longitudinal structures formed from the shear instability the rolling-up vortices similar to those in Fig. 1. But the coherence of these structures gets lost and the resulting mixture finally looks like Fig. 3.

Figure 3 illustrates what happened to a small compact blob of scalar deposed in a large scale sustained turbulent flow. The blob has been progressively converted into disjointed sheets which dilute in the surrounding turbulent medium. Except for loose, diffuse sheets of dye, no obvious structures can be distinguished. The striking observation is nevertheless the broad fluctuations in the transverse size of the sheets, and in the concentration level they carry. Although captured at the same instant of time in the snapshot of Fig. 3, those sheets have not all experienced the same history. Some are still thick and dark, while others are already so thin that they almost fade away in the diluting medium.

What is the paradigm, the elementary structure of mixing in disordered flows? The coherent rolling-up vortex, the cascade, the stretched sheet? It seems that none of these caricatures is fully satisfactory and could give way to a new paradigm which would account for the parallel, distributed histories of cumulated stretchings experienced by the material elements in the course of time. This new paradigm could reconcile the three facets of mixing by unifying its geometrical, temporal, and structural aspects.

References

[1] Batchelor, G. K. 1959 Small-scale variation of convected quantities like temperature in a turbulent fluid. Part 1. General discussion and the case of small conductivity, *J. Fluid Mech.* **5**, 113–133.

[2] Batchelor, G. K., Howells, I. D. & Townsend, A. A. 1959 Small-scale variation of convected quantities like temperature in a turbulent fluid. Part 2. The case of large conductivity, *J. Fluid Mech.* **5**, 134–139.

[3] Batchelor, G. K. & Townsend, A. A. 1956 Turbulent diffusion. In Batchelor, G. K. and Davis, R. M. (eds), *Surveys in Mechanics*, Cambridge University Press, 352–399.

[4] Betchov, R. 1956 An inequality concerning the production of vorticity in isotropic turbulence. *J. Fluid Mech.* **1**, 497–504.

[5] Breidenthal, R. 1981 Structure in turbulent mixing layers and wakes using a chemical reaction. *J. Fluid Mech.* **109**, 1–24.

[6] Brown, G. L. and Roshko, A. 1974 On density effects and large scale structures in turbulent mixing layers. *J. Fluid Mech.* **64**, (4), 775–815.

[7] Corrsin, S. 1951 On the spectrum of isotropic temperature fluctuations in an isotropic turbulence. *J. Appl. Phys.* **22**, 469–473.

[8] Corrsin, S. 1964 The isotropic turbulent mixer: Part II. Arbitrary Schmidt number. *AIChE J.* **10**, (6), 870–877.

[9] Damköhler, D. 1940 Der Einfluss der Turbulenz auf die Flammengeschwindigkeit in Gasgemischen. *Z. f. Elektroch.* **46**, (11), 601–652.

[10] Danckwerts, P. V. 1952 The definition and measurement of some characteristic mixtures. *Appl. Sci. Res. A* **3**, 279–296.

[11] Danckwerts, P. V. 1953 Continuous flow systems. Distribution of residence times. *Chem. Eng. Sci.* **2**, 1–13.

[12] Epstein, I. R. 1990 Shaken, stirred—but not mixed. *Nature* **346**, 16–17.

[13] Hawthorne, W. R., Wendell D. S. & Hottel, H. C. 1949 Mixing and combustion in turbulent gas jets. In *Third Symposium on Combustion and Flame and Explosion Phenomena*, Baltimore, Williams & Wilkins, 266–288.

[14] Mandelbrot, B. B. 1975 On the geometry of homogeneous turbulence, with stress on the fractal dimension of the iso-surfaces of scalars. *J. Fluid Mech.* **72**, (2), 401–416.

[15] Nagata, S. 1975 *Mixing, principles and applications.* John Wiley & Sons, New York.

[16] Oboukhov, A. M. 1949 Structure of the temperature field in a turbulent flow. *Izv. Acad. Nauk SSSR, Geogr. i Geofiz* **13**, 58–69.

[17] Ottino, J. M. 1989 *The kinematics of mixing: stretching, chaos, and transport.* Cambridge University Press.

[18] Prandtl, L. 1925 Über die ausgebildete Turbulenz. *ZAMM* **5**, 136–139.

[19] Ranz, W. E. 1979 Application of a stretch model to mixing, diffusion and reaction in laminar and turbulent flows. *AIChE Journal* **25**, (1), 41–47.

[20] Reynolds, O. 1894 Study of fluid motion by means of coloured bands. *Nature* **50**, 161–164.

[21] Richardson, L. F. 1926 Atmospheric Diffusion shown on a distance-neighbour graph. *Proc. R. Soc. Lond. A* **110**, 709–737.

[22] Taylor, G. I. 1921 Diffusion by continuous movements. *Proc. Lond. Math. Soc.* **20**, 196–212.

[23] Taylor, G. I. 1953 Dispersion of a soluble matter in a solvent flowing slowly through a tube. *Proc. R. Soc. Lond. A* **218**, 44–59.

[24] Welander, P. 1955 Studies on the general development of motion in a two-dimensional, ideal fluid. *Tellus* **7**, (2), 141–156.

Part I

History, Lexicon, and Basics

Why Mixing?

Emmanuel Villermaux,[1] Hugues Chaté,[2,3] and Jean-Marc Chomaz[3]

[1] LEGI–CNRS, Institut de Mécanique de Grenoble, 38041 Grenoble, France
[2] CEA–Service de Physique de l'Etat Condensé, 91191 Gif-sur-Yvette, France
[3] LadHyX, Ecole Polytechnique, 91128 Palaiseau, France

Mixing is a subject which suffers from the *Bourgeois Gentilhomme* complex. Like Monsieur Jourdain in Molière's play (1670), scientists, engineers, and indeed all of us often "do mixing without even knowing it" (just think about yourself trying to prepare mayonnaise...).

As an operation, it consists simply in putting together two or more initially segregated constituents and stirring, in order to attain uniformity, or a new product, or the complete disappearance of one of the constituents etc.; mixing thus is indeed at the cross-roads of many different areas of science. One often needs to mix elements in order to make a new product, an homogeneous blend, or to make possible a chemical reaction or an efficient combustion. One also needs to understand how nature mixes or has mixed to gain information on e.g., the size of a pollutant spot in a valley, the rate of destruction of ozone in the atmosphere, or the dynamics of the earth's mantle. More than a fascinating subject in itself, mixing is thus ubiquitous and a key process in many complex man-made, or natural operations.

It is no doubt *because* of its universality that mixing, as noted with regret by J. Ottino (1989) [17], "...does not enjoy the reputation of being a very scientific subject..." as is frequently the fate of interdisciplinary topics.

The dendritic nature of the subject matter is revealed by the different angles from which the various scientific communities attack the problem, according to their needs These approaches can be roughly grouped into three naturally overlapping categories, each of them of interest to different schools of thought:

- Geometry
- Kinetics
- Structures

Beyond the tools and attitudes developed by different scientific groups regarding the problem of mixing, one might appreciate that mixing, as suggested by common

Mixing: Chaos and Turbulence, edited by Chaté *et al.*
Kluwer Academic / Plenum Publishers, New York 1999.

methods for instance). This is why one has to dig in history to recover the thread of thought to guide our present research.

Our paper is organized in three parts: the first, written by Etienne Guyon, is on mixing, the second and the third, written by Marie Farge, are on fully-developed turbulence. We will restrict ourselves to incompressible Newtonian fluids and fully-developed (i.e. highly nonlinear) turbulent flows, without addressing the question of the transition towards a turbulent state.

1.   Mixing

1.1.   Philosophical journey

Mixing is present everywhere. It is observed in a large number of natural phenomena involving fluid as well as solid granular substances. It has been used throughout the history of mankind in a wide range of applications. Moreover, it often takes place at the scale of direct observation. Thus we should not be surprised that it has been studied in science and philosophy from the beginning of humanity. In the classical Greek philosophy in particular, it is found as a subject of direct investigation as well as metaphorically. On the one hand Democritus and Epicurus introduced atomism and the notion of random motion of atoms leading to mixing and separation. On the other hand Plato, Aristotle and their followers made constant reference to the pure and the mixed in the reflections on the self as well as on the social and the political. Rather than following an historical approach, which would lead us finally to contemporary philosophers, let us consider a few basic concepts related to mixing which have been analysed throughout the twenty five centuries of philosophy. Mixing is often described in terms of oppositions such as: pure/impure, mix/disperse, near/away, simple (*stoicheia*)/complex, elemental/composite or reversible/irreversible. In doing so, we follow Heraclitus who claimed, starting from similar pairs of opposites, that the association of the contraries is a necessary condition in order to reach the unique. This dialectics has been the object of many philosophical studies which we will briefly review.

The dialectics of *'Pure and impure'* has been recently analysed by the contemporary philosopher Vladimir Jankélévitch in a beautifully written essay. One of the basic questions of philosophy is indeed that of unity and purity, which leads to the definition of the self. Purification is the result of distillation, of an elimination of foreign elements. But this operation repeated to the extreme makes it impossible to define the pure which would have lost all its added ingredients. In some philosophies the mere reference to the pure is enough to render it impure. Pure is for Plato the essence of matter. In the quest for an organized universe it is of the same perfect nature as the classical polyhedron shapes and it relates to rationality. The impure and composite is the irrational and thus depreciated. The symbolism of water as described by G. Bachelard in *'L'eau et les rêves'* associates water with symbols of purification. However, water is seldom pure. It accepts many elements by dissolution. For the stoicians, a single drop of wine in the ocean makes it alcoholized. We also know that dust particles or material heterogeneities are necessary to mark and follow the flow of water. We are indeed condemned to a mixed life. We make full use of impurities throughout our life to characterize, by contrast, what could be pure, just as the crystallographer uses the topology of defects to characterize the symmetry group of a perfect crystal. Life in society itself is impure. In Plato's 'Politics', couple and family are compared to the interwoven threads (another form of mixed texture) which associate dissimilar beings in a permanent combination. This is further applied to threads of social life, which combine and hold contraries. There is a clear need to go beyond these contradictions. Rather than trying to define impure and composite structures starting from association of pure elements, it is more realistic to start from impure and see purity as an ultimate renunciation to impure.

Dual approaches tend to oppose and discriminate: good and bad, black and white. The separation of the wheat from the chaff or that of the saved from the damned in the last judgement leaves little chance for an intermediate permanent state. Many natural processes operate similar discriminations: oil separates naturally from vinegar, active transport in biological membranes leads to an increase of ion concentration gradients. Granular flows of large and small grains paradoxically also lead to segregation. In all the above dual operations, we obtain purity by separation. However, most of the time we are faced with a plural approach in which gradations are present on a multiplicity of scales. Part of the popular success of fractal concepts comes clearly from the fact that it has filled a gap in our unconscious perception of the infinite obtained by iteration over an infinite range of scales. "Multiple" accepts impure and diversity through this pluralist view. Complexity which has recently become fashionable recently aims at a description of composite multiple-scale structures. It has a long tradition in philosophy in its attempt of a definition of the pure and unique in the world of the multiple.

Philosophers, as well as scientists, search for a unifying mechanism which would hold together the diverse elements of a whole. Bergson often uses the example of an orange colour. It should be considered as pure because our perception of it is pure despite the fact that it could have been obtained by amalgamation of several colours. Thus, from this perceptual point of view, the decomposition of orange into elementary colours is artificial. The unity of our being, even that of body and soul, is to be found throughout the diversity of its elements. In the science of composite materials in particular, we are aiming for such global descriptions which lead to a new kind of purity, that of homogenized structures. The mixture of dielectric grains is replaced by an equivalent and homogeneous composite from the early works of Clausius and Mossotti in the mid 18th century. Another example of homogenization is given by Albert Einstein's calculation of the viscosity of a dilute suspension which gives it the status of a pure liquid slightly more viscous than the suspending one. We propose to employ the term "small disorder" to describe mixtures to which homogenization treatments can be applied. Fractal structures which could only be homogeneized at the upper scale of similarity correspond to a situation of "large disorder".

We should not, however, overlook the scale of the original heterogeneities which have been averaged out in the homogeneisation process. They are essential, in particular when we deal with chemical reaction. For stoicians, the distinction between *fusion* and *juxtaposition* was as crucial as it is today for a physical chemist. The operation of melting is a way to discriminate between the two kinds of composite structures. In the melting of an alloy, the temperature varies continuously during heating. On the other hand, in a composite material temperature shows steps corresponding to the melting of the individual grains of the various phases. The original nature of the elements has been lost forever. In such a juxtaposition process, the individual elements can be separated again in principle. The stoician view is not far from the distinction, amply discussed throughout the Cargèse school, between *macromixing*, obtained in stirring, and *micromixing* which is the ultimate step resulting from diffusion and which is necessary for a chemical reaction to take place. Between fusion and juxtaposition so defined, a third type of mixing is introduced, *true mixing (krasis)*, that of a drop of wine in the ocean, in which the elements are coextensive and present down to the smallest scale while retaining their identity and being, in principle, separable again. For the stoicians, this last kind of mixing is that of the body and the soul.

The pre-scientific (from the atomic physicists to Descartes and Newton) images of reactivity resulting from mixing and close associations of elements are varied: mixing is seen via images of branching, fingering, interweaving, with only loose connection with observations. In this view, the reaction occurs through attachments of elements induced by their various shapes, local glue when elements are in contact or, again, by involving forces acting at a distance. Only at very small distance - the approach being achieved by diffusion - will the reaction occur. This is clearly not sufficient for proper understanding of reactions, as chemical reactions involve definite ratios of the constituants. In his study of *'The Mixed'*, Pierre Duhem emphasizes the opposition between this geometrical view of the reaction and that proposed by the peripathetician school of Aristotle, which envisions the product of the reaction as an entity distinct from the elements which have been involved in the process. One crucial element will decide in favor of this view. When Lavoisier writes *'Water is composed of oxygen and of an inflammable gas in proportion of 83 parts to 15'*, he illustrates by this particular example the law of definite proportions. The product of the reaction is a new entity with a rigorously defined composition. There is also clearly a limiting solubility in the association of sugar and water, but it is a function of temperature and of the presence of other elements. One should, however, moderate the victory of Aristotle over Epicurus as there exist reactions or associations controled by shapes, as in biochemistry, or weak glue as in colloids. In the various modes of reaction, the common element is the necessity of a close approach of reacting elements which is usually achieved by molecular diffusion. The relative efficiency of the processes of this close approach and of the reaction itself is measured in terms of a Damkohler number which is the ratio of transport time to reaction time. The atomist philosophers anticipated this fact through their description of the continuous agitation of molecules. This is illustrated in this text taken from the fundamental questions in physics posed by Epicurus to Herodote: *'Atoms move continuously for all eternity. They collide and, further, separate from each other. Conversely, other atoms are set into vibration as soon as they happen to be bound by interweaving or when they are surrounded by other atoms with which they bind'*. The description clearly encompasses the main ideas of the kinetic theory of gases and has also been used to describe turbulence. In addition it includes the concept of reaction resulting from the close approach of elements.

We have just been invited by Epicurus to take the dynamic view of mixing. The question of the processes taking place during mixing and, more generally, of its dynamics must be raised at this point, as, up to now, we have only had momentary views of the composite obtained in the mixing process. Indeed, mixing closely associates spatial and temporal descriptions, introduced by the philosophers dealing with atomism. Democritus also discusses the random motion of individual atoms which approach, separate and react. This kinetic theoretical view is developed in Lucretius' 'De Natura Rerum' and brings in the notion of chance in the famous image of the dust particles moving permanently as seen in a ray of light with the unpredictable pattern of the individual particle. There is necessity in the global behaviour - once again homogenized as measured by the thermodynamic temperature or pressure. These philosophers make use of a similar atomistic description in order to describe the unpredictable turbulent motions such as those that could be observed by Lucretius in the flow of the Arno river or in the fumes above mount Etna. The description makes much use of the multiple turbulent structures (turba means crowd) undergoing irregular motions as those of the atomic grains mentioned before. The clinamen describes small variations of a deterministic regular flow, which we call laminar, and which is amplified by turbulence. It is clearly the ancestor of the sensitivity to initial conditions, and the amplification of the initial differential deviation speaks to us as a Lyapounouv exponent. The concept of bifurcation embedded in this description appears throughout the philosophical studies of free choice. In the medieval scholastics, the philosophers Abelard and Buridan showed how autonomy and individual freedom results from the repeated and necessary free choice. The mule of Buridan who is equally hungry and thirsty is at an equal distance from buckets filled with water and grain. Will it starve to death? Clearly no if the drive is strong enough. In the scientific view of broken symmetry, like that found in the buckling of a compressed bar above a certain threshold pressing force, the initial symmetry will be lost and - again as a result of the sensitivity to initial conditions - the bar will buckle in one of the equally possible deformed shapes. The mule will also choose if it is in sufficient need and is likely to survive! In doing so, it will express its free will.

1.2. Historical journey

An understanding of the multiplicity of turbulent scales and their influence on the turbulent drag was already present in this quotation of Saint Venant in 1851: 'If Newton's hypothesis as reproduced by Navier and Poisson, which consists in taking internal friction proportional to the velocity of sliding filaments, can be applied to different points of the same fluid section, all known facts lead us to infer that the coefficient of proportionality must increase with the transverse dimension of the section; this can be explained, up to a certain point, by noting that the filaments do not move parallel to each other and that ruptures, eddies and other forms of complicated or oblique movements must greatly influence the value of friction and that they form and develop more in large sections!' The notion of scale similarity is introduced in the text as well as the need for the introduction of variable drag coefficient which will lead T. Von Karman to define many years later turbulence as the 'science of variable constants' /Boussinesq 1877/. In his 'Théorie des eaux courantes', from which Saint Venant's quotation has been extracted, Bousssinesq defines an eddy viscosity τ from the ratio of wall stress ϵ to the average velocity gradient across the wall: $\tau = \epsilon U / y$.

The study of mixing of a passive scalar became, due to Osborne Reynolds, intimately related to that of turbulence and also (with some precautions, particularly when the Prandtl number is large compared with unity) as a marker to the flow. Reynolds expresses his interest in mixing by writing : 'The general idea of mix-

ing is so familiar to us that the vast generalization to which these ideas afford the key remain unnoticed'. His famous experiment of the onset of turbulence in a tube makes use of a narrow filament of injected dye which visualizes flow lines and whose spreading is used as an indicator of turbulence. In an evening talk at the Royal Society with the title : *'The study of fluid motion by means of coloured bands'*, Reynolds generalizes the use of filaments of dyes made in the Poiseuille flow experiment. His work anticipated the many turbulent and chaotic flow visualizations presented during this meeting. Reynolds introduced a statistical treatment of turbulence by separating from the mean flow U_o a fluctuating component U_{fl}. In the classical case of a two-dimensional flow parallel to a plane perpendicular to y this separation is: $U(y) = U_0(y) + U_{fl}$ with $U_{fl} = \{u, v, 0\}$ and the mean of U_{fl} being zero. The exchange of momentum discussed by Saint Venant is made explicit here as the convective flux of the fluctuating momentum ρu is carried convectively with the velocity component v. Thus, the so called Reynold stress τ which gives the added friction caused by turbulent eddies is given by $\tau = \rho uv$. The expression can be extended to the case of the transport of a passive scalar $C(y)$ whose fluctuating component $c(y)$ is convectively transported, thus giving rise to an additional turbulent diffusivity.

The next main contributor to the field of turbulent mixing is G. I. Taylor who, from his first work on the subject in 1915 and for over half a century, retained an interest in many manifestations of mixing in turbulent, laminar and unstable flows. He made full use of the time he spent as a sailor since his youth and all along his life from the initial observations he made on the boat "The Scotia" which was sent to Labrador after the sinking of the "Titanic" to evaluate the risk of encountering icebergs. In this expedition he used kites, boat chimney smoke and fog to study the phenomenon of dispersion. He later used the marine current in the Irish sea as a natural hydrodynamic channel. He also developed an interest in aeronautics as a pilot during the First World War. But he is best known for his elegant model experiments emphasizing the interplay between flow geometry and convective as well as diffusive mixing. Starting from his *'Diffusion by continuous movements'* in 1921, he was a major contributor to the statistical concepts to turbulence. Around 1920, simultaneously but independently from L. Prandtl, he introduced the mixing length which makes use of the Reynolds stress tensor. In a turbulent two-dimensional gradient flow, the model assumes that the eddies convect the fluctuation of velocity on a scale l, the *mixing length*. Thus, the fluctuating component u can be related to the mean velocity profile $U_0(y)$ by $u = l\frac{dU_0}{dy}$. The fluctuating component v should also be of the order of u as can be envisioned from the picture of an eddy of radius l. Thus, the Reynolds stress is: $\tau = \rho uv = \rho l^2 (\frac{dU_0}{dy})^2$. This gives a microscopic meaning to the transport coefficient ϵ introduced by Boussinesq as: $\epsilon = \rho l^2 \frac{dU_0}{dy} \propto \rho \lambda v$. In this last form, the analogy with the kinetic theory which is at the origin of the statistical treatment of turbulence becomes explicit, λ plays the role of the mean free path of the kinetic theory model and v that of the average thermal velocity. Rather than thinking of transport of momentum by eddies as in the work of Reynolds and later Prandtl, Taylor considered that of the vorticity ω, eliminating pressure gradient terms from the evolution equation for ω. The analysis in terms of a kinetic theory of vortices was pursued one step further by Von Karman. Using the similarity hypothesis and neglecting the effect of viscosity, he obtained in particular the logarithmic velocity profile of a shear flow.

A parallel can be drawn between G. I. Taylor and his contemporary L. F. Richardson, who was also a very fine observer (he also published an article on theory of turbulence from the observation of chimney smoke!) and whose intuitions had considerable impact on the further development of the turbulence and mixing. In particular his experimental study of *'Atmospheric diffusion shown on a distance neighbor graph'* in 1926 makes use of the observations of pairs of markers injected in turbulent flows starting from an initial separation l of a few centimeters to several kilometers. He found that the rate of separation increases with l, and concluded from a log plot of the results a power law increase, now called the Richardson diffusion: $\frac{dl^2}{dt} \propto D(l)l^{4/3}$. The diffusivity starts from a value equal to twice the molecular diffusivity at small separations as expected from independent walks of two tracers. It increases with separation as more and more eddy of sizes smaller than l contribute to separation, in the spirit of the similarity hypothesis.

2. Fully-developed turbulence

2.1. Philosophical journey

When we study turbulent flows we are overwhelmed by their apparent complexity, which leads us to describe them as "disordered" or "random". But, as already stated four centuries ago by Spinoza, it is important to understand that the notion of "order" is subjective and that we call "disordered" a system whose dynamics appears to us too complicated to be described in details: *'Because those who do not understand the nature of things, but only imagine them, affirm nothing concerning things, and take the imagination for the intellect, they firmly believe, in their ignorance of things and of their own nature, that there is an order in things. For when things are so disposed that, when they are presented to us through the senses, we can easily imagine them, and so can easily remember them, we say that they are well-ordered; but if the opposite is true, we say that they are badly ordered, or confused. And since those things we can easily imagine are especially pleasing to us, men prefer order to confusion, as if order were anything in Nature more than a relation to our imagination'* /Spinoza/.

James Clerk Maxwell, in the article on *'Diffusion'* that he wrote for the 5th edition of the Encyclopedia Britannica, published in 1877, also emphasized the fact that the terms "order", "disorder" and "dissipation" are subjective notions: *'Confusion, like the correlative term order, is not a property of material things in themselves, but only in relation to the mind which perceives them. The notion of dissipated energy could not occur to a being who could trace the motion of every molecule and size it at the right moment. It is only to a being in the intermediate stage, who can hold of some forms of energy while others elude his grasp, that energy appears to be passing inevitably from the available to the dissipated state'* /Maxwell 1877/. This is why in the theory of turbulence we should be aware that the observer plays a role, and that we should take into account his scale of observation and the question he poses concerning the turbulent flow. For instance, we would not model turbulence in the same way if we evaluate the drag coefficient of an aircraft in order to compute the power of its engines, or if we study trailing vortices in order to estimate the safe distance between landing airplanes. When calculating the drag coefficient it is sufficient to consider only low-order averaged quantities (such as two-point correlation) based on frequent events which correspond to the center of the probability distribution function (PDF) of velocity or vorticity, while to find the safe distance between airplanes we must consider rare events and departure from the low-order averages, corresponding to the tails of the same PDFs.

The aim of the theory of turbulence is to define quantities whose evolution we can predict from quantities we are going to model without tracking their detailed dynamics. This programme was already clearly stated 67 years ago by Richardson in a paper entitled *'Diffusion regarded as a compensation for smoothing'* where he wrote that: *'By an arbitrary choice we try to divide motions into two classes: (a) those which we treat in detail, (b) those which we smooth away by some process of averaging. Unfortunately these two classes are not always mutually exclusive. [...] Diffusion is a compensation for neglect of detail. [...] The form of the law of diffusion depends entirely upon the arbitrarily chosen method of averaging, which is always implied when diffusion or viscosity are mentioned. This calls attention to the desirability of making much more explicit statements about smoothing operations than has hitherto been the custom'* /Richardson and Gaunt 1930/. Here Richardson emphasizes a key question, still requiring discussion in turbulence theory today, which is the definition of the averaged quantities we want to measure and predict in order to describe turbulent flows. The best discussion we know of defining appropriate averages in turbulence is found in a paper written in 1956 by Kampé de Fériet entitled *'The notion of average in turbulence theory'* /Kampé de Fériet.

The flow of a fluid is called turbulent when it exhibits an unsteady, chaotic and random behaviour; in this regime the flow is sensitive to initial conditions and therefore unpredictable. Such flows mix transported quantities, like momentum, heat or concentration of passive scalar, much more efficiently than laminar flows where only molecular diffusion is involved. This leads to enhanced diffusion, which is called "turbulent diffusion". These properties are displayed by both two- and three-dimensional turbulent flows, although some of their dynamical properties are different. In two dimensions, vorticity is a scalar advected by the fluid and enstrophy (L^2-norm of vorticity) cannot grow. In three dimensions, vorticity is a vector and enstrophy can then be generated by vortex stretching.

We should be aware that the definition of turbulence we have given here depends on our views on this phenomenon, and that the notion of turbulence has evolved in time and is still changing. Turbulence has always been a problem faced by engineers working in hydraulics and in aerodynamics, but since the beginning of the 20th century, it has also caught interest of theoretical physicists and mathematicians. This has lead to new questions such as: is there a universal statistical equilibrium state typical for turbulence which would play a role similar to Maxwell's distribution in the kinetic theory of gases? Are the solutions of Navier-Stokes equations unique and smooth for all times or do they develop singularities in finite time? We should be aware that our definition of turbulence depends on the conceptual tools we have at our disposal to describe them. If one relies on the theory of dynamical systems, one will see turbulent flows as a collection of vortices having chaotic dynamics in space and time which is characterized by a strange attractor in phase-space. If one relies on stochastic tools, one will emphasize ther randomness of turbulent flow, which can be characterized by the Probability Distribution Function (PDF) of a large ensemble of different realizations of the same flow.

Our definition of turbulence also depends on the technical tools we have at hand. For instance, the development of hot wire anemometry in the 50s has allowed experimentalists to obtain pointwise measurements of many flow realizations and therefore to calculate reliable statistics, such as energy spectra and PDFs, but this technique does not give meaningful information (for instance the instantaneous spatial distribution of velocity, pressure or temperature field) about an individual flow realization. It is the generalization of computers, beginning in the 80s, for both laboratory (data acquisition and processing) and numerical experiments (direct numerical simulation), which changed our views on turbulent flows by allowing multi-point (generally in the form of grid point sampling) measurements and by giving access to the detailed spatio-temporal structure of the different fields of interest (velocity, vorticity, pressure, temperature, concentration, ...). The change of viewpoint allowed by the computer has been advocated by experimentalists such as Günther Ahlers who testifies that: *'I believe that the most important experimental development of the 1970's was the advent of the computer in the laboratory'* [...]. Data acquisition and processing *'did not only provide us a new tool but they also gave us completely new perspectives on what types of experiments to do'* /Aubin 1997/.

In the historical review that follows, which deliberately adopts a personal and therefore limited point of view, we will focus on fully-developed turbulent flows and restrict ourselves to incompressible Newtonian fluids. We will distinguish three different perspectives to look at turbulent flows: the statistical kinetic approach (from Boussinesq to Taylor), the statistical probabilistic approach (from Kolmogorov to intermittency models) and the deterministic approach (from Helmholtz to dynamical system methods). The first approach, inspired by the kinetic theory of gases (developed by Maxwell, Boltzmann and Einstein), separates flows into mean and fluctuating motions assuming their temporal or spatial scales to be sufficiently separated (scale separation hypothesis), which leads to turbulent viscosity (Boussinesq and Reynolds) or mixing length (Prandtl) models. The second approach relies on the theory of stochastic processes (developed by Wiener, Khinchin and Kolmogorov), which involves random functions and probability measures and predicts the scaling law of the energy spectrum. The third approach focuses on the vorticity field of individual flow realizations and tries to understand the formation and interaction of coherent vortices in order to identify a low order dynamical system presenting the same chaotic behaviour as the complete flow.

2.2. Historical journey

2.2.1. The statistical kinetic approach

In order to try to master the complexity of turbulent flows, the first statistical approach was to decompose the velocity field into its mean value and fluctuations, by analogy with the kinetic theory of gases which distinguishes the mean motion from the fluctuating (thermal) motion of molecules. This statistical kinetic approach was proposed by Saint-Venant and Boussinesq supposing the existence of "fluid molecules" /Boussinesq 1877/, then by Reynolds in 1894 /Reynolds 1894/ and by Lorentz in 1896 /Lorentz 1896/. After decomposing the velocity field into a mean contribution plus fluctuations, one rewrites the Navier-Stokes equation to predict the evolution of the mean velocity as a function of fluctuations. This procedure yields the Reynolds equation. However, one encounters difficulties due to the nonlinear term $(\vec{V} \cdot \vec{\nabla} V)$ of the Navier-Stokes equation: the second-order moment of the velocity fluctuations, called the Reynolds stress tensor, depends on the third-order moment, which depends on the fourth-order moment, and so on ad infinitum. At each order one considers there are more unknowns than equations and one faces a closure problem. In order to close the hierarchy of Reynolds equations the usual strategy is to add another equation, or system of equations, chosen from some *a priori* phenomenological hypotheses. In particular, one must suppose that there exists a scale separation, namely that fluctuating motions are sufficiently decoupled from the mean motions to guarantee that the average of the product (coming from the nonlinear term of Navier-Stokes equation) is equal to the product of averages (5th Reynolds postulate /Monin and Yaglom 1965/).

To close the hierarchy of Reynolds equations, in 1925 Prandtl introduced a scale, called the "mixing length", characteristic of the velocity fluctuations (introduced in paragraph 1.2). Following the hypothesis proposed by Boussinesq /Boussinesq 1877/, and by analogy with molecular diffusion which regularizes velocity gradients for scales smaller than the molecular mean free path, Prandtl supposed that there exists turbulent diffusion, which smoothes the velocity fields at scales smaller than the mixing length; he then rewrote the Reynolds stress tensor as a turbulent diffusion term. As soon as one considers large Reynolds number flows, the mixing length hypothesis fails because the analogy with the kinetic theory of gases does not work. Molecular motions can be modelled by a diffusion equation (linear Laplacian operator $\nabla^2 \vec{V}$ applied to a mean velocity field with viscosity as transport coefficient) because they are decoupled from the large scale motions. But this is not possible for fully-developed turbulent motions because the nonlinear term $(\vec{V} \cdot \vec{\nabla} \vec{V})$ of the Navier-Stokes equation, which dominates the linear dissipation term $(\nabla^2 \vec{V})$ in this case, involves all scales, and thus there is no scale separation to decouple large scale motions from small-scale motions. This is a major obstacle in any attempt to model turbulence using moment equations and therefore the closure problem is still open. An important direction of research is to find a new representation of turbulent flows, in which there may exist such a separation, not based on scales, but on decoupling motions out of statistical equilibrium from well thermalized motions, which can then be modelled. Such a separation may be possible with a nonlinear closure procedure, based on wavelet representation and using conditional averages which depend on the local behaviour of each flow realization /Farge et al. 1992/, /Farge et al. 1997/.

Turbulence modelling should rely on a careful statistical analysis of turbulent flows, observed in the laboratory or in numerical experiments. Traditionally one studies the long time velocity correlation, supposing the turbulent flow has reached a statistically steady state, so the time covariance no longer evolves. The use of the covariance and cross-correlation functions was first proposed by Einstein, in a paper written in French and published in 1914 in Switzerland, to study the statistics of long time series of fluctuating quantities /Einstein 1914/, /Yaglom 1986/. In 1935 Taylor /Taylor 1935/ formulated, in addition to statistical stationarity, the hypothesis of statistical isotropy (and consequently homogeneity), namely he supposed that averages are invariant under both translation and rotation of the fields. Taylor, using the statistical tools introduced by Wiener (himself inspired by the earlier work of Taylor on Brownian motion /Taylor 1921/), proposed to study isotropic turbulence by measuring its energy spectrum, which is the modulus of the Fourier transform of the two-point correlation of the velocity increments /Taylor 1935/. Taylor also supposed that the spatial distribution of velocity changes slowly as it is carried past the point at which the frequency spectrum (the Fourier transform of the time correlation) is measured, which in this case gives the same energy spectrum for both time and space correlations. This is known as "Taylor's hypothesis" and is commonly used in most laboratory experiments to interpret time correlations as space correlations. These new statistical tools introduced by Taylor to analyze turbulent flows have brought about a change in turbulence research and are still in use today. Although Taylor in his work on turbulent diffusion /Taylor 1921/ was the first to introduce stochastic tools in turbulence, he always adopted a dynamical point of view. This explains why he never showed a strong interest for Kolmogorov's theory. In his famous paper, entitled *'The Spectrum of Turbulence'* /Taylor 1938/, Taylor expressed his intuition that the energy dissipation is not densely distributed in space, namely that turbulent flows are intermittent. He stated the hypothesis that this intermittency is related to the spottiness of the spatial distribution of vorticity: *'The fact that small quantities of very high frequency disturbances appear, and increase as the speed increases, seems to confirm the view frequently put forward by the author (himself) that the dissipation of energy is due chiefly to the formation of very small regions where the vorticity is very high'* /Taylor 1938/. Understanding intermittency is still a very important open problem, perhaps the essential one, to solve before we can build a satisfactory theory of turbulence.

2.2.2. *The statistical probabilistic approach*

In order to overcome the difficulty with closure, due to the fact that there is no scale separation to decouple large scale motions from small scale motions in turbulent flows, another statistical approach has been put forward. We replace the observation of individual flow realizations by the measure of the correlation between the velocities in a large number of flow realizations. We construct ensemble averages, assuming that we know the probability law of the process governing turbulent flows, and thus all its moments. This probabilist approach was initiated by Gebelein in 1935 and developed by many scientists, often independently from one another, among them Kampé de Fériet /Kampé de Fériet 1939/, Millionshchikov /Millionshchikov 1939/, Kolmogorov /Kolmogorov 1941/, Obukhov /Obukhov 1941/, Onsager /Onsager 1949/, Heisenberg and Von Weizsäcker /Heisenberg 1948/.

Since Gibbs, such a probabilistic approach has become standard in statistical physics, but the difficulty in applying it in turbulence arises from the fact that turbulent flows are open thermodynamical systems due to the injection of energy by external forces and its dissipation by viscous frictional forces. To resolve this difficulty Kolmogorov supposed the existence of an energy cascade, based on the hypotheses that external forces, which inject energy into the flow, act only on the large scales, while frictional forces, which dissipate energy, act only on the small scales. In the limit of very large Reynolds numbers, Kolmogorov supposed that there exists an intermediate range of wavenumbers, called the inertial range, for which energy is conserved and only transferred from low to high wavenumber modes, at a constant rate ϵ. As a consequence of the cascade hypothesis, Kolmogorov assumed that the skewness of the velocity increment probability distribution is non zero and does not vary with the scale, which implies that turbulent flows are non Gaussian and non intermittent. All these hypotheses led him to the prediction that the modulus of the energy spectrum scales according to a power-law $\epsilon^{2/3} k^{-5/3}$, k being the modulus of the wavenumber.

In fact, the hypothesis of a turbulent cascade and the resulting scaling of the energy spectrum can only be true for ensemble averages. If one considers each flow realization, the enstrophy (L^2-norm of vorticity) and energy are produced very locally in physical space, at locations where there are boundaries or internal shear layers, and therefore very non-locally in wavenumber space (this from the definition of the Fourier transform and the resulting Heisenberg's uncertainty principle). Likewise, the spatial support of dissipation is highly spotty for a given flow realization and therefore also non-local in wavenumber space. These observations are in contradiction with the hypothesis of a low-wavenumber injection and a high-wavenumber dissipation of energy necessary to maintain an inertial range. Such observations have already been made more than 400 years ago by Leonardo da Vinci when he wrote: *'Where turbulence of water is generated, where turbulence of water maintains for long, where turbulence of water comes to rest'* /Frisch 1995/. Obviously da Vinci was describing one turbulent flow realization in physical space and not in wavenumber space, one being dual of the other. His remark has therefore nothing to do with the notion of turbulent cascade which is a concept formulated for ensemble averages and which says nothing about individual realization. Vinci was observing the local production of vortices in the boundary layers and internal shear layers, their advection by the global velocity field and their dissipation resulting from their mutual nonlinear interactions. The possible confusion between physical space observations and the cascade concept has been discussed 23 years ago by Kraichnan in a paper entitled *'On Kolmogorov's inertial-range theories'* /Kraichnan 1974/, where he remarked: *'The terms "scale of motion" or "eddy of size l" appear repeatedly in the treatment of the inertial range. One gets an impression of little, randomly distributed whirls in the fluid, with the fission of the whirls into smaller ones, after the fashion of Richardson's poem. This picture seems to be drastically in conflict with what can be inferred about the qualitative structures of high Reynolds numbers turbulence from laboratory visualization techniques and from plausible application of the Kelvin's circulation theorem'*. Incidently we should notice that Richardson's poem described the formation of smaller and smaller structures at the interface of clouds and not in the bulk of turbulent flows /Richardson 1922/. Unfortunately the picture Richardson has proposed has been misinterpreted since then as the dynamical process responsible for the energy cascade although the resulting breaking of whirls into smaller ones does not seem to be mechanically possible in the bulk of an incompressible turbulent flow.

Due to observational evidence of small-scale intermittency introduced by Townsend in 1951 /Townsend 1951/ and following a criticism of Landau who pointed out that the dissipation rate ϵ should fluctuate, Kolmogorov proposed a new theory /Kolmogorov 1962/ which added an intermittency correction μ to the energy spectrum scaling, such that $E(k) \propto k^{-5/3-\mu}$. Kolmogorov's 1962 paper opened a debate, which is still very lively today. Twenty three years ago Kraichnan wrote /Kraichnan 1974/: *'The 1941 theory is by no means logically disqualified merely because the dissipation rate fluctuates. On the contrary, we find that at the level of crude dimensional analysis and eddy-mitosis picture the 1941 theory is as sound a candidate as the 1962 theory. This does not imply that we espouse the 1941 theory. On the contrary, the theory is made implausible by the basic physics of vortex stretching. The point is that this question cannot be decided a priori; some kind of non-trivial use must be made of the Navier-Stokes equation'*. Kraichnan, although he is a master of the statistical approach, claims that one needs to understand first the generic dynamics of the Navier-Stokes equation before being able to construct a statistical theory taking into account intermittency: *'If the Kolmogorov law $E(k) \propto k^{-5/3-\mu}$ is asymptotically valid, it is argued that the value μ depends on the details of the non-linear interaction embodied in the Navier-Stokes equations and cannot be deduced from overall symmetries, invariances and dimensionality'* /Kraichnan 1974/.

Pauli liked to say that there exist true theories, false theories and theories which are neither true nor false, namely, following Popper's terminology, theories which are unfalsifiable. In order to fit observations unfalsifiable theories prefer to add new parameters rather than to change their hypotheses. Consequently these theories can be very accurate and optimal for describing existing data, but they are not able to predict new facts and they lack predictability power. A typical example is the geocentric theory of planetary motions which was able to describe the motions of all known planets with a reduced number of epicycles. When new observations became available new epicycles were added in order to preserve the corpus of the theory. The theory was very useful in practice but was unable to predict the existence of an unknown planet as Leverrier did for Neptune. Unfalsifiable theories are poor from an epistemological point of view but they produce a large number of publications because each new refinement step in adding a new parameter leads to new models and new experimental checks. This may be what Kraichnan had in mind when he wrote: *'Once the 1941 theory is abandoned, a Pandora's box of possibilities is open. The 1962 theory of Kolmogorov seems arbitrary, from an a priori viewpoint [...]. We make the point that even in the general framework of some kind of self-similar cascade, and of intermittency which increases with the number of cascade steps, the 1962 theory is only one of many possibilities'*. One can also add a new class to Pauli's picture: "too true" theories, namely theories one cannot falsify even in situations where their hypotheses are not valid, which is illustrated by Kraichnan's comment on Kolmogorov 41's theory: *'Kolmogorov's 1941 theory has achieved an embarrassment of success. The $-5/3$ spectrum has been found not only where it reasonably could be expected, but also at Reynolds numbers too small for a distinct inertial range to exist as in boundary layers and shear flows where there are substantial departures from isotropy, and such strong effects from the mean shearing motion that the stepwise cascade appealed to by Kolmogorov is dubious'* /Kraichnan 1974/. The robustness of the energy spectrum is not entirely surprising because it is the Fourier transform of the two-point correlation of the velocity increments and therefore it is unsensitive to rare events, such as coherent vortices which are produced in shear layers and in boundary layers. But this robustness of Kolmogorov's theory is lost as soon as one considers higher-order statistics, as we will discuss in the following paragraph.

2.2.3. The deterministic approach

In parallel to the statistical probabilistic approach based on ensemble averages, there has also be the tendency to analyze each flow realization separately. Both tendencies, probabilistic and deterministic, were well represented by the participants to the 1st International Conference on Turbulence held in Marseille in 1961 /CNRS 1961/. The probabilistic approach was advocated by Kolmogorov, Obukhov, Millionshchikov, Yaglom, Batchelor, Favre, Kampé de Fériet, Kraichnan and Lumley, while the deterministic approach was defended by Liepmann, Roshko, Laufer, Von

Karman, Taylor, Saffman and Moffatt. Actually, Townsend had been the first (in 1951) to suggest that there exist organized structures, called coherent structures, which play an essential role in the transport properties of turbulent flows and which may be responsible for their chaotic and intermittent behaviour. In 1955 Theodorsen in a paper on *'The Structures of Turbulence'* /Theodorsen 1955/ formulated the hypothesis that coherent structures (in the form of bended vortex tubes), such as horseshoes or hairpins vortices, are responsible for the eddy motions of 3D turbulent flows. The presence of such coherent vortices in fully-developed turbulent flows has been observed in both laboratory and numerical experiments. For instance, many numerical experiments have shown that, when two-dimensional or three-dimensional turbulent flows are initialized with an homogeneous random distribution, vorticity tends to concentrate in a finite number of coherent vortices, which are formed during the flow evolution and seem to have their own dynamics and interaction laws quasi-independent of the background flow where they evolve. Although there is not yet a universal definition of coherent vortices (we prefer the term "vortices" rather than "structures" which is too general) on which everyone agrees, we will define them as "condensates" of the vorticity field where rotation dominates strain. They survive on time-scales much longer than the eddy turnover time characteristic of velocity fluctuations, although they can be rapidly destabilized and destroyed by strong strain. They exhibit phase-correlation over a spatial and/or temporal range significantly larger than the smallest scales of the flow. They emerge out of random initial conditions, or are created in boundary layers by instabilities. They have their own dynamics, such as same-sign vortex merging (also called vortex pairing), opposite-sign vortex binding and emission of unidimensional vorticity filaments when they are strongly strained. In two dimensions, due to the orthogonality between vorticity and velocity gradient vectors, there is no vortex stretching, while in three dimensions coherent vortices, such as vortex tubes (often called "filaments" although "tubes" may be more adequate), can be stretched by velocity gradients. This leads to either vorticity production during vortex stretching or vorticity dissipation during vortex breakdown.

For two-dimensional flows it is possible to characterize coherent vortices from a topological point of view in studying the stress (velocity gradient) tensor. Its symmetric part corresponds to the irrotational strain while its antisymmetric part corresponds to the rotation undergone by a fluid element. The eigenvalues of the stress tensor allow for the flow to be separated into different regions where the Lagrangian dynamics is different. In two dimensions there are two types of regions which are either elliptic or hyperbolic /Weiss 1992/. Imaginary eigenvalues correspond to elliptic regions where rotation dominates strain and for which fluid trajectories remain close, which characterizes coherent vortices. Real eigenvalues correspond to hyperbolic regions where strain dominates rotation and for which fluid trajectories separate exponentially, which characterizes the hyperbolic stagnation points of the background flow. In three-dimensional flows where vortex stretching plays a key role, there are more types of topological regions. Unfortunately the classical theory of turbulence is myopic to the presence of coherent vortices because they are advected by the flow in a homogeneous and isotropic random fashion. They are also highly unstable and their temporal and spatial support may be very small for three-dimensional turbulence. Consequently their presence only affects the high-order velocity structure functions (namely the high-order statistical moments of the velocity increments), which have been measured only recently. These measurements contradict Kolmogorov's theory which predicts a linear dependence (with slope 1/3) between the scaling exponent of the velocity structure functions and their order. In addition it has been found that there are actually two distinct nonlinear dependences for odd and for even order structure functions /Van der Water 1993/.

One should not forget that cascade is only a hypothesis of Kolmogorov's theory and that the observed scaling of the energy spectrum of turbulent flows can be explained without this hypothesis. An alternative viewpoint, that we will call "dynamical", is based on the assumption that the nonlinear dynamics of turbulent flows tends to form quasi-singularities in physical space, such as point vortices for 2D flows or vortex filaments for 3D flows. This dynamical view was already advocated by Kraichnan in a paper writtten in 1974 where he explained: *'The stretching mechanism has led a number of authors to conjecture that the small-scale structure should consist typically of extensive thin sheets or ribbons of vorticity, drawn out by the stirring action of their own shear field (Townsend 1951, Batchelor 1953, Kraichnan 1959, Corrsin 1962, Saffman 1968, Tennekes 1968). In this picture, the randomness lies in the distribution of thickness and extension of the thin sheets and ribbons, and in the way they are folded and tangled through the fluid. A typical small-scale structure is thought to be small in one or two dimensions only, not in the third'* /Kraichnan 1974/. This last point means that the energy spectrum of only one realization averaged over angles already exhibits a broad-band energy distribution similar to the one observed in the inertial range for ensemble averages. If each flow realization exhibits a power-law spectrum, this is *a fortiori* true for an ensemble of realizations.

This dynamical interpretation of the energy spectrum has been then developed by Saffman /Saffman 1971/ for two-dimensional turbulence and by Lundgren /Lundgren/ for three-dimensional turbulence. Saffman supposes that the nonlinear dynamics of two-dimensional turbulent flows tend to form vorticity discontinuities along lines, that we will call one-dimensional vorticity filaments, and lead to a k^{-4} energy spectrum. Lundgren supposes that the nonlinear dynamics of three-dimensional turbulent flows tend to form vortex tubes, that we will call two-dimensional vorticity filaments, which roll up into spirals in such a way that the energy spectrum scales as $k^{-5/3}$. These ideas were later developed by Gilbert /Gilbert 1988/ and Moffatt /Moffatt 1993/, who suggested that the power-law scaling of two-dimensional turbulent flows can be explained by the rolling up of vorticity filaments into spirals which, by accumulation of singularity at the center of the spiral, lead to a k^{-3} scaling of the energy spectrum. In 1991 Farge and Holschneider /Farge and Holschneider 1991/ proposed, instead of one-dimensional vorticity filaments, the formation of two-dimensional cusp-like axisymmetric coherent vortices, for which vorticity remains bounded in the core due to viscous effects. This introduces the core radius as a new length scale and the larger the Reynolds number, the smaller the core radius will be. Farge and Holschneider conjectured that these cusp-like vortices are formed from initially random vorticity field by an inviscid instability, similar to Kelvin-Helmholtz instability, which accretes vorticity onto the strongest singularities of the initial random distribution (which correspond to the tails of the vorticity PDF). The same mechanism may explain the formation of vortices at the wall. They have shown that the cusp-like quasi-singularities remain stable under Navier-Stokes evolution /Farge, Holschneider and Philipovitch 1992/ and that the strain they impose on the background flow organizes it, and inhibits further instability that otherwise

could develop in their vicinity /Kevlahan and Farge 1997/. Actual singularities may
only exist for very large or infinite Reynolds numbers, but this is still an important open question. In 1982 Cafarelli, Kohn and Nirenberg /Cafarelli 1982/ proved
that, if finite time singularities exist for three-dimensional Navier-Stokes equations,
they will be on a space-time support of measure one. Therefore, if singularities or
quasi-singularities exist, they can only be rare events (rare because if their time
support is one their space support is then zero), which can hardly be detected using
standard statistical methods or two-point correlations, since low-order statistics are
insensitive to rare events.

3. Which are the essential questions for turbulence and mixing?

After more than a century of turbulence research/Reynolds 1883/, no single convincing theoretical explanation has given rise to a consensus among physicists. In
fact, there exists a large number of *ad hoc* models, called "phenomenological", that
are widely used by fluid mechanists to compute industrial applications where turbulence plays a role, but many parameters of those turbulence models cannot be
derived from first principles and must be determined by performing experiments in
wind tunnels or water tanks. In fact it is still not known whether fully-developed
turbulence has the universal behaviour (independent of initial and boundary conditions) assumed for it in the limit of infinitely large Reynolds numbers and infinitely
small scales. Our understanding of turbulence is impaired by the fact that we do
not yet know which are the "right questions" to ask. We have not yet identified the
"right objects", namely the structures and elementary interactions from which it
would be possible to construct a satisfying statistical mechanics (or kinetic theory)
of fully-developed turbulence. Ignorance of the elementary physical mechanisms at
work in turbulent flows arises in part from the fact that we ignore coherent vortices
because we use two-point correlation functions, we think in terms of Fourier modes
that are delocalized and that we consider L^2-norm, such as energy and enstrophy,
instead of higher-order norms.

Our present lack of understanding of the dynamics of coherent vortices arises from
several reasons:

1. We focus on the velocity field and not on the vorticity field.

2. We study the flow evolution in an Eulerian frame of reference, instead of a
   Lagrangian frame attached to each fluid particule. It is more appropriate to
   follow the evolution of vorticity or of circulation, because vortex tubes are
   advected by the flow and even conserved in absence of dissipation (Helmholtz
   laws and Kelvin's circulation theorem).

3. Classically we perform point measurements and two-point correlations which
   are insensitive to rare events such as coherent vortices.

4. The infinite time limit which is usual in statistical physics is not interesting
   for many applications of turbulence where we are interested in studying the
   transient evolution and need methods, for instance, to accelerate mixing. In
   particular, the infinite time limit is irrelevant to many meteorological situations where one cannot guarantee the statistical stationarity and the time
   decorrelation (Markov's process hypothesis) of the external forcing.

5. "Real life" problems are bounded in space, time and scale. Therefore we should focus more on statistically unsteady and inhomogeneous turbulent flows, although theoreticians and mathematicians seldom consider spatial boundaries, or initial conditions, or scale cutoffs, because they destroy the homogeneity, stationarity and self-similarity they prefer to assume.

6. We consider only the modulus of the energy spectrum and not its phase, therefore we completely lose track of spatial coherence.

7. When we develop numerical codes (combinations of a Navier-Stokes solver and a turbulence model) to compute fully-turbulent flows we address neither the non-linear problem *per se* nor the fact that one does not have a statistical equilibrium. On the contrary, one supposes either a linear behaviour of the subgrid scale motions (in the case of Direct Numerical Simulation, DNS) or the existence of a scale separation, which supposes a statistical equilibrium for the subgrid scale motions (in the case of Large Eddy Simulation, LES). This last point was already noticed 24 years ago by Kraichnan when he wrote: *'Our basic point is that the inertial-range cascade represents strong statistical disequilibrium. This carries two implications. First, that analogies with equilibrium and near-equilibrium phenomena are unjustified. Second, that the structure of the inertial range depends on the actual magnitude of the coefficients coupling the degrees of freedom and not just on their overall symmetry and invariance properties. This is because cascade is a transport process and the coefficient magnitudes affect the rate of transport'* /Kraichnan 1974/.

An open question to address is: what is the importance of coherent vortices in mixing and turbulence? Do they play an essential role which should be taken into account in our models or can we neglect them? We should get rid of the present misconception which relates coherent vortices to the small wavenumbers (large eddies) of turbulent flows. This false view comes from the fact that one tries to recover some dynamical picture from averaged quantities which have already lost track of the spatial and time coherence which characterizes coherent vortices. In particular the energy spectrum in the inertial range is dominated by the background and not by the coherent structure, because their spatial and temporal support is too small to have a sufficient weight in the integral when one computes second-order structure function (see the energy spectra in figure g). This is not true for high-order structure functions and actually the presence of coherent vortices may explain the departure from Kolmogorov's prediction one observes for high-order structure functions. On the contrary, when one considers the probability distribution function (PDF) of vorticity, one finds that its non-Gaussian shape comes from the coherent vortices, which are responsible for its heavy tails (see the PDF in figure f). This is due to the fact that, although coherent vortices are quite rare in space and time, they are present in any realization of a turbulent flow. The formation of coherent vortices is probably a deep consequence of the incompressible Navier-Stokes dynamics one should try to clarify. To answer the previous question concerning the role of coherent vortices in turbulent flows, we need a clear definition, still lacking of what is a coherent vortex and we need an appropriate method to extract them. For two-dimensional flows it has been proposed /Farge et al. 1992/, /Farge et al. 1997/ to use the wavelet representation and nonlinear filtering to extract coherent vortices from the incoherent background flow (see figures a, b, c), showing that large wavelet coefficients correspond to the coherent vortices while small wavelet coefficients correspond to the background flow. The coherent vortices thus extracted have the same coherence (characterized by a functional relation between vorticity and streamfunction) as the overall vorticity field, while the background is incoherent (see the coherence function on figure d). Both regions, coherent as well as incoherent, are multiscale, and therefore cannot be separated by Fourier filtering, although they exhibit different scaling (see figure g). This programme has been extended to three-dimensional turbulent flows. It has shown /Farge et al.1990/ that coherent vortices present in mixing layer and channel flows are responsible for the flow intermittency, and that the wavelet representation can be used to detect the quasi-singularities which develop in such inhomogeneous flows. The wavelet representation has also been used to compute two-dimensional Navier-Stokes equations following only the dynamics of coherent vortices in remapping the basis functions at each time step in order to optimally follow their motion in space and scale /Schneider, Kevlahan and Farge 1997/.

The statistical probabilistic approach may be too abstract and its links with experimental observations are difficult to ascertain in most cases. In order to compare its predictions with laboratory or numerical experiments, one has to guarantee that ensemble averages converge to time or space averages, and therefore satisfy the ergodic hypothesis. One supposes that there is only one attractor which satisfies Sinai-Bowen-Ruelle conditions (for almost all initial conditions the time average exists and is the same) and that the observed turbulent flow has visited all possible phase-space configurations compatible with this attractor. Therefore, due to our limited understanding of turbulence, it has become more and more important to perform many well controlled experiments to try to get better insight and propose new models to describe the behaviour of high Reynolds number turbulent flows. There are two kinds of experimental approach, each having its own limitations. First, laboratory experiments, where it is easy to measure one- or two- (or even more) point correlations and accumulate long time statistics, but where one cannot measure in many locations and at the same time the instantaneous spatial distribution of velocity and vorticity. Secondly, numerical experiments where it is easy to measure the time evolution and spatial distribution of velocity and vorticity fields for one given flow realization, but where it is out of computational reach to compute large ensemble averages. The two approaches are in fact complementary, laboratory experiments allow us to perform statistical analysis, while numerical experiments allow us to perform dynamical analysis. Unfortunately, these two approaches are also dual, and it is therefore difficult to compare them. For instance, the statistical analysis deals with averages, describes turbulent flows in terms of fluctuations (departure from the mean field) expressed in the statistical notion of turbulent eddies, and uses random functions and probability measures, while the dynamical analysis considers each flow realization *per se*, describes the flow in terms of isolated coherent vortices, and deals with non-random functions or distributions. A lot of confusion in our understanding of turbulence is due to the fact that we try to retrieve dynamical insight out of statistical averages and statistical information out of only one flow realization. Moreover the statistical analysis relies on the Fourier representation, while the dynamical analysis relies on the spatial representation. We must be aware that we cannot reconcile these two representations, unless we use basis functions which are localized in both physical space and wavenumber space, such as wavelets or wavelet packets /Farge 1992/.

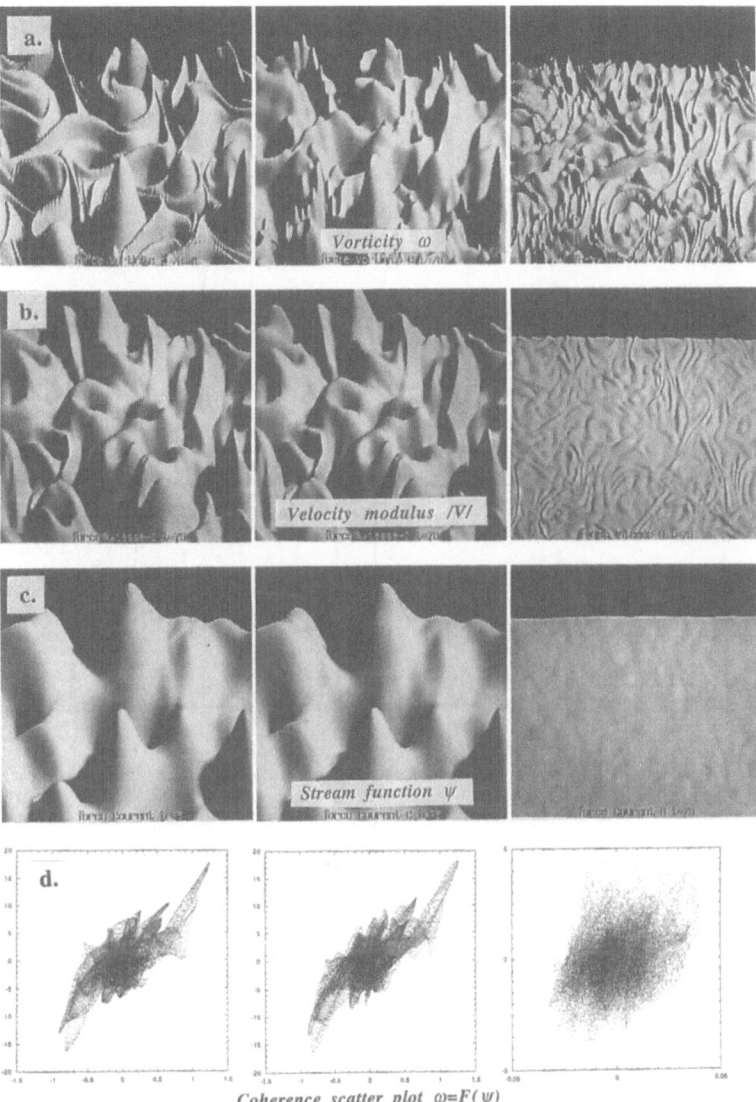

Coherence scatter plot $\omega = F(\psi)$

Fig. 1. — Wavelet compression of vorticity. (a) vorticity. (b) modulus of velocity. (c) stream function. (d) coherence scatter plot. (e) Cut of vorticity. (f) PDFs of velocity and vorticity. (f) Energy spectrum. The solid lines correspond to the total vorticity ω, the dashed lines to the coherent part $\omega_>$, and the dotted lines to the incoherent part $\omega_<$. We observe that only 0.7% of the total number of wavelet coefficients are sufficient to represent all coherent structures, while the remaining 99.3%correspond to the incoherent background flow, which is much weaker and homogeneous. The coherent vorticity $\omega_>$ contains 94.3% of the total enstrophy. Moreover, the velocity associated with the coherent structures is quasi-identical to the total velocity and contains 99.2% of the total energy. As for the coherent stream function, $\psi_>$ is perfectly identical to the total stream function ψ.

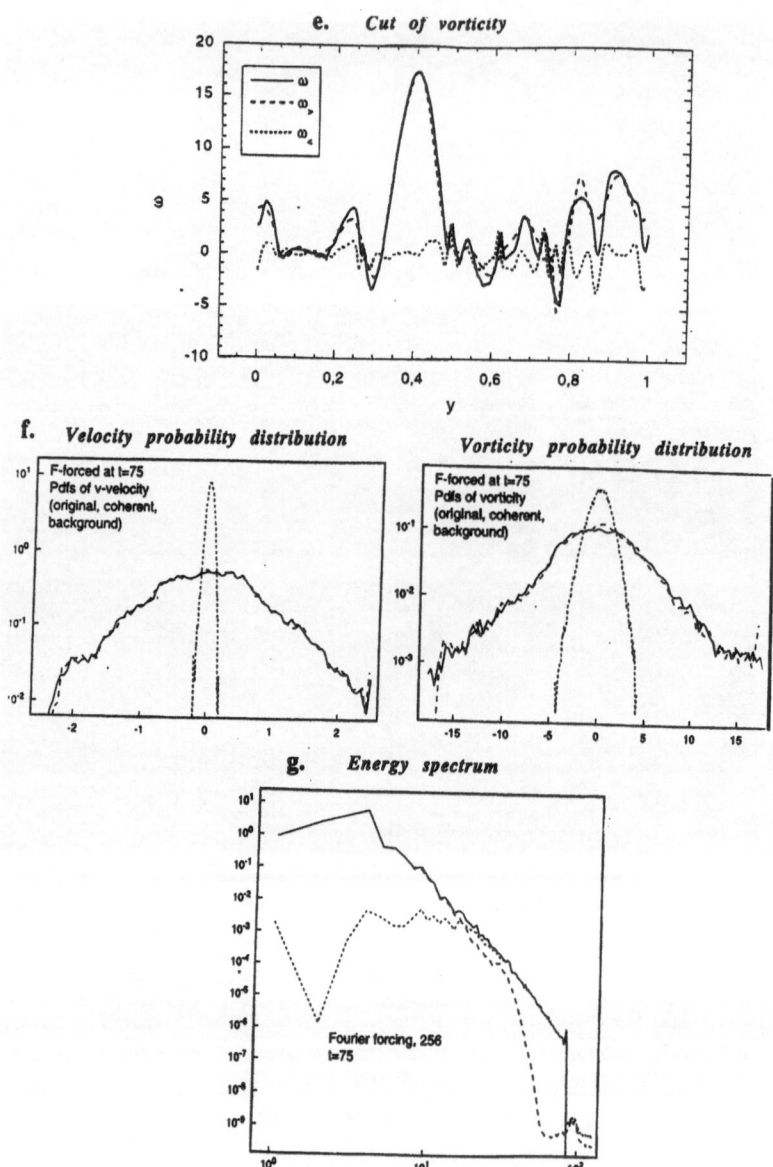

Fig. 1. — (Continued) The fact that the scatter plot of the background, $F_<$ such that $\omega_< = F_<(\psi_<)$, is isotropic proves that our method has extracted all coherent structures. The PDFs of velocity and vorticity show that only 0.7% of the wavelet coefficients are sufficient to capture the non-Gaussian one-point statistical distributions of vorticity and velocity, while the remaining 99.3% correspond to Gaussian distributions. The energy spectrum, on the contrary, is dominated at small scales by the incoherent background flow and therefore is insensitive to coherent structures because they are too rare to affect the energy spectrum (which is the Fourier transform of the two-point correlation function).

Kolmogorov's statistical theory is the simplest possible universal theory (simple in the sense of "Occam's razor" or Aristotle's logical simplicity principle). It is very well verified for second-order moments (or two-point correlation), but it fails to correctly predict higher-order moments. We believe that coherent vortices may explain this discrepancy and are essential for the understanding of turbulence. Therefore we need to find another theoretical setting, which should actually take into account the existence of coherent vortices, as characteristic features of turbulent flows. We would like to construct a statistical mechanics of turbulent flows based on the dynamics of coherent vortices, but we still do not know what should be the appropriate invariant measure for this. In the limit of infinite Reynolds number, Kolmogorov's prediction for the two-point correlation (L^2-norm) will always be verified, because the contribution of coherent vortices tends to disappear from the integral measure of their spatial support tending to zero in this limit. However, his prediction will not prove true as one measures higher-order moments (L^p-norm with p the order), because in this case the weight of coherent vortices in the integral will become significant. In this picture dissipation results from the strong nonlinear interactions between coherent vortices. The larger the Reynolds number, the more and more local in physical space, and therefore non-local in wavenumbers, dissipation will be. The Kolmogorov's dissipative scale is an averaged quantity and we have conjectured /Farge et al. 1990/ that its variance in space is large and depends on the flow intermittency. In this picture universality seems to be lost because the density of coherent vortices depends on the initial conditions and on the forcing. But there may be a universal way of describing coherent vortices in terms of internal degrees of freedom having a quantified (discrete) amount of enstrophy, while the background incoherent flow can be seen as a thermal bath which only affects the coupling between coherent vortices. The prediction of Kolmogorov's theory may be verified only for the incoherent background flow, which is homogeneous, Gaussian and well-mixed.

Today we can again try to construct a kinetic theory of turbulence, at least in the case of two-dimensional turbulence where high-resolution direct numerical simulations of the Navier-Stokes equation at large Reynolds numbers (still inaccessible for three-dimensional simulations) have given us insight into the dynamically important features of turbulent flows. These simulations, and also high resolution visualizations of laboratory experiments (such as those performed by Dimotakis and presented in this book), reveal to us the shape and elementary interactions of coherent vortices. But in order to develop a statistical kinetic theory of turbulent flows we need to identify a gap between mean and fluctuating motions, namely a sufficient decoupling between the active and passive components of motion. We know this decoupling does nor exist when we do the separation in Fourier basis, but we have shown that we have more chances to find it in a space-scale representation such as wavelet bases /Farge et al. 1992/, /Farge et al. 1997/. Moreover we think that the theory of fully-developed turbulence is in a pre-scientific phase, because we do not yet have an equation, nor a set of equations, that could be used to compute efficiently turbulent flows from first principles. The Navier-Stokes equation, which is the fundamental equation of fluid mechanics, is appropriate to study laminar flows and transition to turbulence. But it is not the right one to compute fully-developed turbulent flows because its computational complexity becomes intractable when the Reynolds number becomes too large. For example, the number of degrees of freedom necessary to compute a turbulent flow by direct numerical simulation (DNS) is proportional to $Re^{9/4}$, Re being the Reynolds number and therefore, for aerodynamics applications where typically $Re = 10^7$, it requires the solution of a linear system of 10^{16} equations. However, for large Reynolds number flows it should be possible to define averaged quantities, like in statistical mechanics, and find the corresponding transport equations to compute the evolution of these new quantities which would be the appropriate observables to describe turbulence. Like the Navier-Stokes equation that can be derived from Boltzmann's equation by considering appropriate limits (Knudsen and Mach numbers tending to zero), appropriate averaging procedures to define new coarse-grained variables (velocity and pressure) and associated transport coefficients (viscosity and density), the turbulence equations may be derived as a step further in this hierarchy of embedded approximations. Unfortunately the appropriate parameters are easier to define when we go from Boltzmann equation to Navier-Stokes equation than from the Navier-Stokes to turbulence equation. In the first case only a linear averaging procedure, namely a coarse-graining based on a known statistical equilibrium distribution of velocities, is needed. In the second case we have to find an appropriate nonlinear procedure, namely some conditional averaging yet to be found. For this we have to identify the dynamically active structures constitutive of fully-developed turbulent flows, to describe their elementary interactions and to characterize their dynamics. We should then find a nonlinear procedure to extract those elementary structures out of turbulent flows. But we may not be able to apply our statistical methods to them because they are out of statistical equilibrium and their statistics are not stationary. On the contrary, the remaining background flow is sufficiently mixed to guarantee ergodicity, stationarity and homogeneity required by Kolmogorov's theory, which can then be used to model the incoherent background flow. In conclusion, we think that the future of turbulence research will be a combination of both deterministic and statistical approaches. The deterministic approach will be necessary to compute the evolution of the low-dimensional dynamical system corresponding to the elementary structures out of statistical equilibrium, namely the coherent vortices. The statistical approach will be needed to model the incoherent background flow using an equivalent stochastic process having the same statistics.

Hans Liepman, former Von Karman professor at Caltech, likes to point out that in turbulence research we are like the drunk man who has lost his keys in a dark alley but who finds it easier to search for them under a street light. For as long as we have been studying turbulence we know that it has to do with vortex production in wall shear layers and vortex interactions. But, because we do not have a good theoretical grasp of their structure and dynamics when there are many vortices, we prefer to use the statistical formalism which, assuming we accept several hypotheses (ergodicity, low-wavenumber forcing and high-wavenumber dissipation, stationarity, homogeneity and isotropy of the statistics), gives us technical tools to predict the average spectral distribution of energy in the inertial range for an ensemble of flow realizations. But this is not the key we are looking for, because it does not enable us to grasp the elementary dynamics and compute the evolution of a given turbulent flow realization from first principles, neither to understand the near-wall dynamics. In the proceedings of the 1st International Conference on Turbulence, held in 1961 in Marseille, Hans Liepmann was already making the following remark: *'It is clear that the essence of turbulent motion is vortex interactions. In the particular case of homogeneous isotropic turbulence this fact is largely masked, since the vorticity fluctuations appear as simple derivatives of the velocity fluctuations. In general this is not the case, and a Fourier representation is probably not the ultimate answer. The proposed detailed models of an eddy structure represent, I believe, a groping for an eventual representation of a stochastic rotational field, but none of the models proposed so far has proven useful except in the description of a single process'* /Liepmann 1962/. Twenty-five years later Hans Liepmann gave us the following guidelines: *'Kolmogorov's theory has been counter-productive. It is OK for light or sound scattering by turbulent flows, but it is not useful for the main lines of turbulence. [...] In turbulence you have long range forces, and it is difficult to extrapolate from Boltzmann's gas, which has short range forces. Therefore I am uneasy about Reynolds equations. [...] As long as we are not able to predict the drag on a sphere or the pressure drop in a pipe from continuous, incompressible and Newtonian assumptions without any other complications (namely from first principles), we will not have made it!'* /Liepmann 1997/.

Acknowledgments

The ideas presented in the first part are developed in a book for the general public written by Etienne Guyon and Jean-Pierre Hulin, "Granites et fumées: un peu d'ordre dans le mélange", published by Odile Jacob (November 1997). Etienne Guyon acknowledges discussions on mixing with his philosopher colleagues at ENS, Philippe Hoffman and Frédéric Worms. Marie Farge is grateful to many friends and colleagues, in particular Konrad Bajer, Claude Bardos and Nicholas Kevlahan, for helping her to improve the second and third parts of this paper. She apologizes for not citing all of them because the list would be too long.

References

/Aubin 1997/ D. Aubin, 'A Cultural History of Catastrophes and Chaos: around the Institut des Hautes Etudes Scientifiques from 1958 through the 80s', *PhD thesis, Princeton University, 1997*

/Batchelor 1969/ G. Batchelor, 'Computation of the Energy Spectrum in Homogeneous Two-dimensional Turbulence', *Phys. Fluid, suppl. II, 12, pp. 233-239, 1969*

/Batchelor 1953/ G. Batchelor, 'Theory of homogeneous turbulence', *Cambridge University Press, 1953*

/Boussinesq 1877/ J. Boussinesq, 'Essai sur la théorie des eaux courantes', *Mémoires de l'Acad. Sci. Paris, Tome 23, 1, pp. 30-46, 1877*

/Cafarelli 1982/ L. Cafarelli, R Kohn and L. Nirenberg, 'Partial regularity of suitable weak solutions of Navier-Stokes equations', *Comm. in Pure and Applied Math., 35, pp. 771-831, 1982*

/Einstein 1914/ A. Einstein, 'Méthode pour la détermination de valeurs statistiques d'observations concernant des grandeurs soumises à des fluctuations irrégulières', *Archive des Sciences Physiques et Naturelle, vol. 37, pp. 254-255, 1914*

/Farge et al. 1990/ M. Farge, Y. Guezennec, C. M. Ho and C. Meneveau, 'Continuous wavelet analysis of coherent structures, *CTR Summer Program, NASA-Ames and Stanford University, pp. 331-348, 1990*

/Farge and Holschneider 1991/ M. Farge and M. Holschneider, 'Interpretation of two-dimensional turbulence spectrum in terms of singularity in the vortex core', *Europhys. Lett., 15, 7, pp. 737-743, 1991*

/Farge 1992/ M. Farge, 'Wavelet transforms and their applications to turbulence', *Ann. Rev. Fluid Mech., 24, pp. 395-457, 1992*

/Farge, Holschneider and Philipovitch 1992/, M. Farge, M. Holschneider and T. Philipovitch, 'Formation et stabilité des structures cohérentes quasi singulières en turbulence bidimensionnelle', *C. R. Acad. Sci. Paris, 315, Série II, pp. 1585-1592, 1992*

/Farge et al. 1992/ M. Farge, E. Goirand, Y. Meyer, F. Pascal and M. V. Wickerhauser, 'Improved predictability of two-dimensional turbulent flows using wavelet packet compression', *Fluid Dyn. Res., 10, pp. 229-250, 1992*

/Farge et al. 1997/ M. Farge, K. Schneider and N. Kevlahan, 'Coherent structure eduction in wavelet-forced two-dimensional turbulent flows', *Dynamics of Slender Vortices, ed. E. Krause, Kluwer*

/Frisch 1995/ U. Frisch, 'Turbulence, the legacy of A. N. Kolmogorov', *Cambridge University Press, p. 112, 1995*

/Gilbert 1988/ A. D. Gilbert, 'Spiral structures and spectra in two-dimensional turbulence, *J. Fluid Mech., 193, pp. 475-497*

/Heisenberg 1948/ W. Heisenberg and C. F. von Weizsäcker, 'Zur statistischen Theorie des Turbulenz', *Zeit. Für Phys., 124, pp. 628-657, 1948*

/Kampé de Fériet 1939/ J. Kampé de Fériet, 'Les fonctions aléatoires stationnaires et la théorie statistique de la turbulence homogène', *Ann. Soc. Sci. Bruxelles, 59, pp. 145-159, 1939*

/Kampé de Fériet 1956/ J. Kampé de Fériet, 'La notion de moyenne en théorie de la turbulence', *Seminario Matematico e Fisico di Milano, vol. XXVII, pp. 168-207, 1956*

/Kevlahan 1997/ N. Kevlahan and M. Farge, 'Vorticity filaments in two-dimensional turbulence: creation, stability and effect', *J. Fluid Mech., 346, pp. 49-76, 1997*

/Kolmogorov 1941/ A.N. Kolmogorov, 'Local structure of turbulence in an incompressible fluid at very high Reynolds numbers', *Doklady AN SSSR, 30, 4 pp. 299-303, 1941*

/Kolmogorov 1962/ A. N. Kolmogorov, 'A refinement of previous hypotheses concerning the local structure of turbulence in viscous incompressible fluid at high Reynolds number', *J. Fluid Mech, 13, pp. 82-85, 1962*

/Kraichnan 1967/ 'Inertial-range Ranges in Two-dimensional Turbulence', *Physics of Fluids, 10, pp. 1417-1423, 1967*

/Kraichnan 1974/ 'On Kolmogorov's inertial-range theories', *JFM 62, pp. 305-330, 1974*

/Liepmann 1962/ H. W. Liepmann, 'Free turbulent flows', *Mécanique de la turbulence, ed. Favre, Masson, pp. 211-227, 1962*

/Liepmann 1997/ *International Workshop on 'Dynamical Systems and Statistical Mechanics Methods for Coherent Structures in Turbulent Flows', University of California at Santa Barbara, 12-13th February, 1997*

/Lorentz 1896/ H. A. Lorentz, 'Ein allgemeiner Satz, die Bewegung einer reibenden Flüssigkeit betreffend, nebst einigen Anwendungen desselben', 'A general theorem on the motion of a viscous flow near a wall',*Abhandlungen über theoretischen Physik, Leipzig, 1, pp. 43-71, 1907*

/Lundgren 1982/ T. S. Lundgren, 'Strained spiral vortex model for turbulent fine structure', *Phys. Fluids, 25, pp. 2193-2203*

/Maxwell 1877/ J. C. Maxwell, 'Diffusion', *Encyclopedia Britanica, 9th edition, 1877*

/McWilliams 1984/ J. C. McWilliams, 'The emergence of isolated coherent vortices in turbulent flows', *J. Fluid Mech., 146, pp. 21-43, 1984*

/Mezic 1996/, I. Mezic, 'Lévy stable distributions for velocity and velocity dofference in systems of vortex elements', *Phys. Fluids, vol. 8, 5, pp. 1169-1180, 1996*

/Millionshchikov 1939/ M.D. Millionshchikov, *'Decay of homogeneous isotropic turbulence in viscous incompressible fluids', Doklady Akad. Nauk. SSSR, 22, 5, pp. 236-240, 1939*

/Moffatt 1993/ H. K. Moffatt, 'Spiral structures in turbulent flow', *Wavelets, Fractals and Fourier Transforms, ed. Farge, Hunt and Vassilicos, Clarendon, pp. 317-324, 1993*

/Monin and Yaglom 1965/ Monin and Yaglom, Statistical Fluid Mechanics, *The MIT Press, 1965/*

/Obukhov 1941/ A. M. Obukhov, 'Energy distribution in the spectrum of a turbulent flow, *Izvestiya AN SSSR, 4-5, pp. 453-466, 1941*

/Onsager 1945/ L. Onsager, 'The distribution of energy in turbulence, *Phys. Rev., 68, p. 286, 1945*

/Onsager 1949/ L. Onsager, 'The distribution of energy in turbulence', *Phys. Rev., 68, pp. 286, 1945, 'Statistical hydrodynamics', Suppl. Nuovo Cimento, suppl. vol. 6, pp. 279-287, 1949*

/Reynolds 1883/ O. Reynolds, 'An experimental investigation of the circumstances which determine whether the motion of water shall be direct or sinuous, and the law of resistance in parallel channels', *Phil. Trans. Roy. Soc., p. 51, 1883*

/Reynolds 1894/ O. Reynolds, 'On the dynamical theory of incompressible viscous fluids and the determination of the criterion', *Phil. Trans. Roy. Soc. London, vol. 186, pp. 123-164, 1894*

/Richardson 1922/ L. F. Richardson, 'Weather prediction by numerical process',

Cambridge University Press, 1922

/Richardson and Gaunt 1930/ L. F. Richarson and A. Gaunt, 'Diffusion regarded as a compensation for smoothing, *Memoirs of the Royal Meteorological Society, vol. 3, 30, pp. 171-175, 1930*

/Saffman 1971/ P. G. Saffman, 'A note on the spectrum and decay of random two-dimensional vorticity distribution',*Stud. Appl. Math., vol. 50, pp. 377-383, 1971*

/Schneider, Kevlahan and Farge 1997/ K. Schneider, N. Kevlahan and M. Farge, 'Comparison of and adaptive wavelet method and nonlinear filtered pseudospectral methods for two-dimensional turbulence', *Theoretic. Comput. Fluid Dynamics, 9, pp. 191-206, 1997*

/Spinoza/ B. Spinoza, *'Ethics, 1, Appendix', Penguin Classics, pp. 29-30, 1996*

/Taylor 1921/, G. I. Taylor, 'Diffusion by continuous movements', *Proc. London Math. Soc., 20, pp. 196-211, 1921*

/Taylor 1935/ G. I. Taylor, *'Statistical Theory of Turbulence', Proc. Roy. Soc. London A, vol. 151, pp. 421-478, 1935*

/Taylor 1938/ G. I. Taylor, 'The Spectrum of Turbulence, *Proc. R. Soc. London A, 164, pp. 476-490, 1938*

/Theodorsen 1955/ Theodorsen, 'The Structure of Turbulence', *50 Jahre Grenzschichtforsung, 1955*

/Townsend 1951/ A. A. Townsend, 'On the fine-scale structure of turbulence', *Proc. Royal Soc., A 208, pp. 534-542, 1951*

/Van den Water 1993/ W. Van den Water, 'Experimental study of scaling in fully developed turbulence', *Turbulence in spatially extended systems, ed. Benzi et al., pp. 189-213, 1993*

/Weiss and Mc Williams 1991/ J. B. Weiss and J. C. McWilliams, 'Nonergodicity of point vortices', *Phys. Fluids, A3, 835, 1991*

/Wiener 1933/ *'The Fourier Integral', 1933*

/Yaglom A. M., 'Einstein's Work on Methods for Processing Fluctuating Series of Observations and the Role of these Methods in Meteorology', *Izvestia, Atmospheric and Oceanic Physics, vol. 22, 1, pp. 78-82, 1986*

Mixing: Turbulence and Chaos – An Introduction

E.J. HINCH

Department of Applied Mathematics and Theoretical Physics
University of Cambridge, Silver Street, Cambridge, UK

---

1. Examples of mixing

1.1. Easy or not?

Mixing is essential for chemical reactions, in studying pollution and other processes in the atmosphere and ocean, and in very many industries. Our everyday experience of dissolving sugar in a coffee cup, which takes less than half a minute of stirring, would mislead us into thinking that mixing is easy, not worth the attention of a summer school.

First, however, we should note that the stirring is necessary: molecular diffusion would take over a day to diffuse sugar the centimeter length scale of the cup. To diffuse salt the $3\,\mathrm{km}$ depth of the oceans, molecular diffusion would take $10^8\,\mathrm{yr}$, dangerously near the age of the Earth and much longer than the $10^4\,\mathrm{yr}$ needed with the stirring available.

Second, another diner-table experiment of trying to mix oil and vinegar reminds us of problems with immiscible fluids. Many foods, particularly the modern 'lite' products, are emulsions of fat in water. Some of the new specialists synthetic fibres are blends of incompatible polymers. For such immiscible fluids, the maximum permitted size of drops a (e.g. $1\,\mu$m to ensure no separation of a colloidal under gravity) sets how vigorous the stirring must be through a minimum strain-rate $E \geq \chi/a\mu$, where χ is the interfacial tension and μ the viscosity of the continuous phase.

Third, vertical variations of density in a stratified fluid can significantly change mixing. Stirred at one level, a blob of fluid can only rise a certain height into less dense fluid before its kinetic energy is exchanged for potential energy. Thus stratification can inhibit mixing. On the other hand, if the density changes through changes in salt concentration as well as temperature, there is a 'double-diffusive' instability which causes strong mixing. This is thought to be important where the hot salty Mediterranean Sea flows out over the cold fresh Atlantic waters, and similarly for

Mixing: Chaos and Turbulence, edited by Chaté *et al.*
Kluwer Academic / Plenum Publishers, New York 1999.

Norwegian fjords. There is also the possibility of internal gravity waves propagating in a stratified fluid. Where these waves break, they cause highly localised mixing, as aircraft discover in bumpy patches of 'clear air turbulence' miles away from any cloud or other obvious provocation.

Fourth, compared with the ease of stirring sugar in coffee, stirring very viscous liquids is less effective and requires much more energy. Practical examples include glass manufacture, molten polymers, heavy crude oils, industrial processing of food and domestic cooking, and minerals moving in the convection in the Earth's mantle. There are also problems of mixing powders in the pharmaceutical and food industries, and granular materials in civil engineering.

Fifth as we shall see later, mixing by stirring is considerable more difficult for substances with low values of molecular diffusivity. It is appropriate to compare the diffusivity of a substance (temperature, chemical species, Brownian particle) with the diffusivity of momentum, i.e. the kinematic viscosity ν. In water, momentum diffuses at $\nu = 10^{-6}$, temperature at $D_\theta = 10^{-7}$, salt at $D_{NaCl} = 10^{-9}$, and a $1\,\mu m$ colloidal particle (e.g. cream globules in milk) $D_{1\mu m} = 10^{-13}$. The SI units here are $m^2\,s^{-1}$. In air, momentum diffuses at $\nu = 10^{-5}$, temperature at $D_\theta = 10^{-5}$, oxygen at $D_{O_2} = 10^{-5}$, and a $1\,\mu m$ colloidal particle (e.g. smoke) $D_{1\mu m} = 10^{-11}$. We note that substances can diffuse many orders of magnitude more slowly than momentum. Thus smoke in a smoke-ring gives a false indication of the size of the vortex core.

1.2. Passive *vs* active mixing

The majority of recent studies of mixing, reflected in the contributions in this book, have been concerned with passive scalars which have no back-effect on the flow which is mixing them. This is appropriate in many applications. There are however circumstances in which the substance being mixed significantly changes the flow field which is mixing it.

In the oceans and atmosphere, the fluid density depends on temperature, and on salt in the oceans and on humidity in the atmosphere. A change in the density structure can lead to convection and a dramatically different flow. In the dynamics of the large scale motion of the atmosphere, there is a quantity called 'potential vorticity' which is advected and so stirred by the flow, and yet this variable determines the flow itself.

The phenomena of avalanches down mountains and of turbidity currents in the oceans involve flow resuspending and mixing particles to form a dense fluid which flows down hill, creating the mixing itself.

Chemical reactions can release heat which generates pressures sufficient to drive a flow. Thus explosions can occur in coal mines and in flour mills in which the flow mixes coal or flour dust from the floor up into the air where it burns explosively, creating the flow itself. A patch of turbulence can deform a slowly moving laminar flame front (typical velocity $1\,m\,s^{-1}$) so that the rate of release of energy per unit project area is sufficient to drive a compression wave with an amplitude which raises the temperature above the ignition temperature, triggering a rapid explosion wave (typical velocity $3\,km\,s^{-1}$).

Other examples where the substance being mixed actively changes the mixing flow are:– polymeric liquids resisting stretching by the flow, immiscible drops setting the small scale in the flow, cohesion of powders, and magnetic fields in recent models of the Earth's dynamo.

1.3.   Eulerian *vs* Lagrangian chaos

The nature of stirring divides into two classes depending on whether the Reynolds number of the flow is large or small. The Reynolds number is defined as the ratio of inertial forces to viscous forces in the flow, $Re = UL/\nu$, where U is the velocity scale, L is the length scale and ν is the kinematic viscosity.

At high Reynolds number, typically $Re > 10^3$, flows become turbulent. The small value of friction means that the cost of stirring is low, and also that the flow continues sometime after stirring has stopped. Turbulence is characterised by having a wide range of length scales in the flow, and this ensures very efficient mixing.

At low Reynolds number, typically $R < 1$, friction is large, and so the cost of stirring is high. The flow stops immediately the mechanical agitation is switched off. There are only large scales in the flow, on the same length scale of the stirring device. Mixing can be very inefficient. The worst extremes are avoided if the fluid pathlines are chaotic. This so called 'Lagrangian chaos' or 'chaotic advection' can occur in regular smooth flows. To emphasise this difference, turbulence is sometimes called 'Eulerian chaos'.

1.4.   The need to mix

Good mixing is at the heart of many industrial processes, and the quality of the final product can depend entirely on the success of mixing. To make glass, tons of sand are mixed with tons of lime and some other mineral, first as dry powders of 1 mm size, and then in the molten phase in very viscous thermal convection. Any heterogeneity which survives this convection will produce optical imperfections and produce mechanical defects which disrupt the spinning of fibre-glass. The energy cost of continuing the thermal convection a little longer is measured in tens of MWs.

Mixing can be used to reduce concentrations of dangerous pollutants in the atmosphere, in estuaries and in ground water. Thus one can ask how high must a chimney be to avoid complaints from nearby housing, or where can one dump untreated sewage, or how can water quality in an aquifer be restored.

Mixing is essential for all chemical reactions. As well as the obvious chemical industry, one must remember fermentation and combustion are chemical reactions, if rather complex.

The surface temperature of the of the sea depends on the effectiveness of vertical mixing after a large storm or other large disturbances. Anomalies in the surface temperature, which persist for days, can lead to significant modifications of weather patterns, in so called 'blocking events'.

Finally, the existence of the ozone hole, localised to the winter pole of the stratosphere, depends on mixing processes in the upper atmosphere.

1.5.   Engineering

Many recent publications on mixing seem to me to have lost sight of why the subject is being studied. An understanding of how mixing takes place must produce a simple usable quantification of the process. Of course the quantities of interest will vary a little from application to application, but three themes are always present.

First, one needs to know how efficient the mixing is. In some applications, it may be required that the mixing is total, producing a uniform concentration throughout the container, for example in the pharmaceutical industry. In other applications, it may be satisfactory to produce a uniform mixing within only part of the container, e.g. to ensure a guaranteed quantity of chemical production. In pollution control or to avert an explosion, it may be necessary just to reduce the maximum concentration below a critical level. A simple measure of the degree of mixing is to compare the r.m.s. fluctuations of concentration with the mean level, $\overline{c'^2}/\overline{c}^2$.

Second, it is important to know how long it takes to achieve a specified degree of mixing. This time for mixing has to be compared with the time for chemical reactions. Similarly, continuously operated stirred-tank reactors must be designed so that the residence time exceeds the mixing time, even exceeding it several fold over when a high degree of mixing is required. (One needs here to consider the probability distribution of residence times.) A crude overall measure of the time of mixing is provided by the rate of decrease of the r.m.s. fluctuations in concentration, $\overline{c'^2}\left[d\left(\overline{c'^2}\right)/dt\right]^{-1}$.

Finally, but most important, one needs to count the cost of the mixing operation. The energy consumption in mixing at high Reynolds numbers is usually not a significant part of the total manufacturing costs. Mixing at low Reynolds numbers is however expensive, and careful design is necessary to avoid ineffectual dissipative flow. Cost may have to be reckoned per unit volume of mixture or per unit time.

1.6. Further examples

There is no general or universal theory of mixing, because applications are limitless and each brings its own peculiar twist.

Gas fluidised beds suffer from bubbling, in which packets of gas rise without contacting the catalytic particles. These undesirable but unavoidable bubbles do however play a helpful role in stirring the particles, which enables one slowly to rejuvenate them as they visit a small region.

Yield fluids, such as pastes and many semi-solid foods, have the unfortunate behaviour of flowing only in a finite region nearby the paddles. Thus it is necessary to move a kitchen whisk around a cooking bowl to ensure that all the material comes near to the mixer blades.

While most of this book is devoted to mixing within a volume, one must not forget the classical studies and results for heat and mass transfer across boundaries, including in particular Levich's $Pe^{-1/3}$ boundary layer when the diffusivity is smaller than the kinematic viscosity. One interesting geophysical problem is the heat and humidity transfer at the sea surface, which is greatly enhanced by the presence of bubbles (more than doubling the surface area of the oceans). A novel design for a heat exchanger, patented by M.R. Mackley, uses eddies shed in pulsed flow though baffles in a pipe to scour and replenish the boundary fluid – a good application of Lagrangian chaos.

A topic explored in depth at the earlier Summer School on 'Disorder and Mixing' is dispersion by the random flow in a porous medium. This mixing can usually be characterised by a diffusivity $D \sim \bar{u}a$, where $\bar{u}$ is the volume flow rate per unit surface area and a is the grain size of the medium. There is a similar phenomenon of Taylor dispersion of a passive scalar in laminar pipe flow with a diffusivity $D = D_{mol} + 0.01U^2a^2/D_{mol}$, where D_{mol} is the molecular diffusivity, U is the average velocity and a is the radius of the pipe. This Taylor dispersion can be applied to mixing fresh air in the lungs.

Finally in a laminar shear flow of a suspension, particles move across streamlines due to random collisions. This motion is described by a diffusivity $D = 0.3\gamma a^2$ if $0.1 < \phi < 0.5$, where γ is the shear-rate, a is the size of the particles, and ϕ is the volume fraction of particles.

2.   The issues

2.1.   Molecular diffusion

Consider the molecular diffusion of a passive scalar. Let $c(x, t)$ be the concentration of the substance and D its molecular diffusivity. The appropriate diffusion equation in one dimension is

$$c_t = Dc_{xx}.$$

Let the initial conditions be $c = 0$ in $x < 0$ and $c = 1$ in $x > 0$. The well known solution of the diffusion problem is

$$c(x, t) = \tfrac{1}{2} + \tfrac{1}{2}\mathrm{erf}\left(\frac{x}{\sqrt{Dt}}\right),$$

with $\mathrm{erf}(z)$ the standard error function.

From this solution, or from dimensional analysis of the governing equation, we identify the distance over which substance has diffused

$$\delta = \sqrt{Dt}.$$

The concentration has a value $c = 0.5 \pm 0.3$ at $x = \pm\delta(t)$ and a value $c = 0.5 \pm 0.45$ at $x = \pm 3\delta(t)$.

For the diffusion of heat in air, this diffusion distance is $0.14\,\mathrm{mm}$ after $1\,\mathrm{ms}$, $4\,\mathrm{mm}$ in $1\,\mathrm{s}$, $0.3\,\mathrm{m}$ in $1\,\mathrm{h}$. For the diffusion of salt in water, $\delta = 1\,\mu\mathrm{m}$ in $1\,\mathrm{ms}$, $30\,\mu\mathrm{m}$ in $1\,\mathrm{s}$, $2\,\mathrm{mm}$ in $1\,\mathrm{h}$, and $0.2\,\mathrm{m}$ in $1\,\mathrm{day}$. From these examples, we learn that molecular diffusion is very fast at early times, as $1\,\mu\mathrm{m}$ is more than 1000 molecules in size, but very slow as time increases. This slowness is the heart of the mixing problem.

2.2.   Value of the diffusivity D

For dilute gases, simple kinetic theory gives a prediction for the value of the molecular diffusivity

$$D = 0.3\bar{u}h.$$

Here $\bar{u}$ is the mean thermal velocity of the molecule, which can be found from the Boltzmann temperature kT and the mass of the molecule m, $\tfrac{1}{2}m\bar{u}^2 = \tfrac{1}{2}kT$. The mean-free-path h is related to the radius of the molecule a and the density of the gas ρ, $4\pi a^2 h\rho = m$ for a pure gas.

For colloidal particles in Brownian motion, the Stokes-Einstein formula for diffusivity of particle is

$$D = \frac{kT}{6\pi\mu a},$$

where μ is the dynamic viscosity of the suspending fluid and a is the radius of the assumed spherical particles. There are corrections to the Stokes-Einstein formula if the volume fraction of the particles is not small.

For turbulent motion, one can define an eddy diffusivity by analogy with kinetic theory

$$D = u'L,$$

where u' is the r.m.s. value of the velocity fluctuations in the fluid and L is the length scale of eddies with this fluctuating velocity.

2.3. Dispersion of a random walk

Consider a random walk of uncorrelated steps of size Δx, each taking a time Δt. After a lapse time of t, the number of steps taken is $t/\Delta t$. The variance in the displacement from the starting position will be the sum of the variances of the uncorrelated steps, i.e. $\Delta x^2 \times t/\Delta t$. Note that the variance increases linearly in time (in this normal diffusion process). The coefficient of the linear growth is defined to be the diffusivity of the random walk

$$D = \frac{d}{dt}\left(\tfrac{1}{2}\overline{x^2(t)}\right).$$

The over-bar denotes an average over several experiments. A constant value for the D is seen only after several correlation times Δt.

Taylor's calculation of the eddy diffusivity in turbulence can now be presented. Proceeding formally,

$$\frac{d}{dt}\left(\frac{1}{2}\overline{x^2(t)}\right) \;=\; \overline{x(t)\dot{x}(t)} \;=\; \int_0^t \overline{\dot{x}(t')\dot{x}(t)}\,dt'$$

The diffusivity attains its constant value only after several correlation times. (When the integral fails to converge, due to slowly decaying correlations, the diffusion becomes 'anomalous' with $\overline{x^2(t)} \sim kt^\mu$ with $\mu \neq 1$.) Thus

$$D = \overline{u'^2}T,$$

with $\overline{u'^2}$ the mean square of the velocity fluctuations and T the integral-correlation time

$$T = \int_0^\infty \overline{\dot{x}(t)\dot{x}(0)}\,dt/\overline{u'^2}.$$

Note that the velocity correlation needs to be computed as seen by a particle moving with the fluid, and hence T is known as the Lagrangian integral-correlation time. The length scale L, introduced at the end of the previous subsection as the size of the eddies, can now be defined as $L = u'T$, i.e. the distance one would move at the r.m.s. velocity u' during the correlation time T.

Taylor's calculation can also be used to derive the Stokes-Einstein diffusivity for a colloidal particle. The thermal velocity fluctuations of the particle are given by $\frac{1}{2}m\overline{u'}^2 = \frac{1}{2}kT$. The velocity remains correlated while friction forces dissipate inertia according to

$$m\ddot{x} + 6\pi\mu a\dot{x} = 0,$$

i.e.

$$\overline{\dot{x}(t)\dot{x}(0)} = \frac{kT}{m}\exp(-6\pi\mu at/m).$$

Evaluating the integral, one recovers the Stokes-Einstein formula given in the last section.

A further application of Taylor's calculation gives the diffusivity of Taylor dispersion described in §1.6. In pipe flow, the variation of the velocity between the walls and the centre of the pipe gives $u' = \bar{u}$. The velocity of a passive scalar remains correlated until it has had time to diffuse molecularly across the cross-section of the pipe, i.e. $T = a^2/D_{mol}$. Hence we estimate $D = O(\bar{u}^2 a^2/D_{mol})$. The numerical coefficient of about 10^{-2} comes from a more careful analysis.

2.4.   The problem with molecular diffusion

The problem with purely molecular diffusion is that it is far too slow, as seen in §2.1, the volume of fluid mixed at a molecular level is

$$V = A\sqrt{Dt},$$

where A is the interfacial area between the species being mixed. With the diffusion distance $\sqrt{Dt}$ increasing slowly in time, it is clear that the only way to improve mixing is to increase the interfacial area. Large factors such as 10^6 have to be contemplated.

One useful measure of the effect of stirring is the so called 'striation thickness'

$$V/A,$$

which can be thought of as the thickness of, or between, sheets of unmixed fluid. The aim is to reduce this thickness to distances over which molecular diffusion can act in the time available, i.e. to $\sqrt{Dt}$.

2.5.   Benefit and limitation of shear

Consider a dyed region of fluid, initially inside a sphere of radius a, deforming in a simple shear flow $\mathbf{u} = (\gamma y, 0, 0)$. After time t, the sphere becomes an ellipsoid

$$(x - \gamma ty)^2 + y^2 + z^2 \le a^2,$$

with semi-diameters $a\left(\sqrt{1 + \gamma^2 t^2/4} \pm \gamma t/2\right)$ and a. The surface area of the ellipsoid is a complicated elliptic function, which for large shears $\gamma t > 3$ becomes $2\pi a^2 \gamma t$ (within 10%). Thus shear increases the interfacial area between the dyed fluid inside and the clear fluid outside the initial sphere, which is beneficial to mixing. Another way to look at this is to note that the thickness of the dyed fluid, i.e. the striation thickness, is reduced, at large shears $\gamma t > 3$ becoming $2a/\gamma t$.

There are two limitations to the beneficial effect on mixing of a simple shear flow. First, the growth in interfacial area (or equivalently the decrease in the striation thickness) is only linear in time. This to achieve an increase by a factor 10^6, a time of $10^6/\gamma$ must be used. This is unreasonably long.

Second, the simple shear $\mathbf{u} = (\gamma y, 0, 0)$ is a local idealisation, and does not continue indefinitely. After many shear times, one must consider the wider global flow and this can limit mixing. For example consider a sphere of dyed fluid of radius a released in a Couette device at a radial position r_0 and axial position z_0. After a large number $O(r_0/a)$ of turns the dye will be completely spread around the torus $(r - r_0)^2 + (z - z_0)^2 < a^2$ by the shearing flow. To mix to other radial or axial positions, however, only molecular diffusion can be used.

2.6.   Stretching of lines, areas and volumes by flow

We consider the evolution of small lines, areas and volumes in a flow $\mathbf{u}(\mathbf{x}, t)$. It is the difference in velocity across these elements which causes change. Hence we need to consider the velocity gradient $\nabla \mathbf{u}$. This is split into its symmetric and antisymmetric parts, the strain-rate $\mathbf{E}$ and vorticity $\boldsymbol{\Omega}$,

$$\nabla \mathbf{u} = \mathbf{E} + \boldsymbol{\Omega}, \quad \text{with} \quad \mathbf{E}^T = \mathbf{E} \quad \text{and} \quad \boldsymbol{\Omega}^T = -\boldsymbol{\Omega}.$$

Now a short line of fluid from $\mathbf{x}$ to $\mathbf{x}+\delta\mathbf{l}$ will move in δt to the line from $\mathbf{x} + \mathbf{u}(\mathbf{x})\delta t$ to $\mathbf{x}+\delta\mathbf{l} + \mathbf{u}(\mathbf{x}+\delta\mathbf{l})\delta t$. Thus the line element $\delta\mathbf{l}$ has become $\delta\mathbf{l} + \delta\mathbf{l} \cdot \nabla\mathbf{u}\delta t$, i.e.

$$\dot{\delta\mathbf{l}} = \delta\mathbf{l} \cdot \nabla\mathbf{u}.$$

The component of the change in the direction of $\delta\mathbf{l}$ is the rate of stretching of the line, and this is entirely due to the straining motion $\mathbf{E}$. The component of the change perpendicular to $\delta\mathbf{l}$ represents a rotation of the line and comes from both the straining $\mathbf{E}$ and the vorticity $\boldsymbol{\Omega}$. To examine further the stretching, we write $\delta\mathbf{l}$ as a length δl and a direction $\mathbf{n}$ of a unit vector, $\delta\mathbf{l} = \mathbf{n}\delta l$. One then obtains the fractional change of length

$$\frac{d}{dt} \ln \delta l = \frac{\dot{\delta l}}{\delta l} = \mathbf{n} \cdot \mathbf{E} \cdot \mathbf{n}.$$

It should be noted the $\mathbf{n}$ rotates in the flow, and so the rate of stretching of the line may not be constant even in a steady flow.

Volumes of fluid remain constant in time in incompressible flows. This fact can be used to find how area elements $\delta\mathbf{A}$ change in flow. If $\delta\mathbf{l}$ is a line element, the volume produced by displacing area $\delta\mathbf{A}$ through the distance $\delta\mathbf{l}$, i.e. $\delta\mathbf{A} \cdot \delta\mathbf{l}$, remains constant in the flow. Thus

$$0 = \frac{d}{dt} (\delta\mathbf{A} \cdot \delta\mathbf{l}) = \dot{\delta\mathbf{A}} \cdot \delta\mathbf{l} + \delta\mathbf{A} \cdot \dot{\delta\mathbf{l}}.$$

Using the earlier result for $\dot{\delta\mathbf{l}}$, we obtain

$$\dot{\delta\mathbf{A}} = -\nabla\mathbf{u} \cdot \delta\mathbf{A}.$$

To examine the important stretching of the area, we write $\delta\mathbf{A} = \mathbf{n}\delta A$ with $\mathbf{n}$ the unit normal perpendicular to the surface. The fractional change of the area is then

$$\frac{d}{dt}\ln\delta A = \frac{\dot{\delta A}}{\delta A} = -\mathbf{n}\cdot\mathbf{E}\cdot\mathbf{n}.$$

We note that if at one instant a line element $\delta\mathbf{l}$ is perpendicular to an area $\delta\mathbf{A}$, then the fractional increase in the length of the line would be equal to the fraction *decrease* in the size of the area. This is however at one instant: the line and area will rotate differently and will not remain perpendicular.

In two dimensions, the stretching of lines and areas are equivalent. In three dimensions, the fractional rate of stretching lines tends on average to be greater then the fractional rate of stretching areas, because, while one principal direction in the area will be stretched as fast as the line, the orthogonal direction will not stretch as fast and may even contract.

2.7. Efficiency of stretching

Because molecular diffusion alone is slow to mix, a stirring flow is required to increase greatly the interfacial area. An appropriate measure of the rate of increase in area is the Liapounov exponent, defined for a small area moving with the flow

$$\sigma_A = \lim_{t\to\infty}\frac{1}{t}\ln\frac{\delta A(t)}{\delta A(0)}.$$

In principle this exponent depends on the initial position and the initial orientation of the small area. There is also an assumption that the initial area $\delta A(0)$ is sufficiently small for the area to remain small compared with the scale of the flow at later times. With a non-zero Liapounov exponent, the area increases exponentially in time $\delta A(t) \sim \delta A(0)\exp(\sigma_A t)$. Roughly, the exponent is the long-time average of the component of the strain-rate $\sigma_A = -\langle\mathbf{n}\cdot\mathbf{E}\cdot\mathbf{n}\rangle$, where $\mathbf{n}$ rotates with the flow and $\mathbf{E}$ varies as the area moves through the flow.

In steady simple shear, area increases linearly in time, and so $\sigma_A = 0$. In steady axisymmetric extension, however the area rotates to an orientation which benefits fully from the stretching, $\sigma_A = E$.

One measure of the *efficiency* of a stirring apparatus is to compare the average Liapounov exponent for the increase in area with the average strain-rate

$$\langle\sigma_A\rangle\,/\,\langle E\rangle.$$

The exponent would be averaged over different initial positions and orientations, while the strain-rate would be averaged over the flow. In very viscous liquids, it is undesirable to have any straining motion producing viscous dissipation while not also producing stretching. Thus a Couette device is a very inefficient mixer (zero efficiency). Well designed stirrers can achieve 30% efficiencies.

3. Turbulent mixing

3.1. Examples

To keep this brief introduction firmly anchored in reality, I would like to start by referring to the excellent collection of pictures, *An Album of Fluid Motion*, collected and published by M. Van Dyke (1982), hereafter referred to as MVD. We will look at four practical examples of turbulent mixing flows – the mixing layer, a jet, flow behind a grid, and pipe flow.

Pictures MVD 176 and 177 (by Konrad 1976 and by Rebollo 1976) show a turbulent mixing layer where two separate uniform streams join. At high Reynolds numbers $O(10^6)$, one sees a mixing zone between the two streams, with a width growing linearly with distance downstream from the splitter plate, the point where the two stream first join. Spanning the width of the layer are large scale eddies, similar in general form to the nonlinear Kelvin-Helmholtz instability. These large scale eddies change little with Reynolds number. Superimposed on the large eddies are small scale eddies, which become finer as the Reynolds number increases. Picture MVD 177 shows the length scale of the small eddies $\frac{1}{50}$th that of the width of the mixing layer.

Picture MVD 166 (by Dimotakis, Lye & Papantoniou 1981) shows a jet of dyed water immerging into a bath of water at a Reynolds number of 2300. The width of the jet of mixed water grows linearly with distance downstream from the nozzle. Superimposed on large scale eddies which span the width of the jet are smaller scale eddies of several scales, the finer ones having a length scale about $\frac{1}{50}$th the width of the jet.

Pictures MVD 152 and 153 (by Corke & Nagib) show the turbulence behind a grid, which can be dragged through a mixture to enhance the rate of mixing. At a Reynolds number of 1500, one sees near to the grid large eddies on the scale of the grid. A little behind, the flow becomes dominated by much smaller eddies. The length scale of these random fine eddies appears to increase at some distance downstream of the grid.

Van Dyke's book does not give any pictures of turbulent pipe flow, because of practical difficulties in making the images. Equivalent however are the pictures MVD 157, 158, 162 and 163 (by Corke, Guezennec & Nagib; Falco 1977; Falco; Head & Bandyapadhyay 1981) of a turbulent boundary layer on a flat plate. Here one sees fine scale eddies, possibly originating at the wall. There are no apparent large scale eddies, except perhaps for some large scale organisation of regions with and without small scales in figure MVD 157. This intermittent nature of some turbulence can be important. In pipe flow one can detect coordinated activity across the width of the pipe, i.e. large scale eddies spanning the diameter.

3.2. What is turbulence and why?

There is no generally accepted definition of turbulence and no explanation of why it exists. It is clear from the pictures referred to above that turbulent flows are random (or chaotic) in space and possess a wide range of length scales superimposed on one another. The flows are equally random in time, with a wide range of time scales superimposed. Thus the complex flow structures cannot be thought of as simply propagating. Instead each eddy seems first to evolve and then to decay during the time for fluid to go once around the eddy, the eddy turnover time. Some of the large structures may persist several turnover times. The wide range of length and time scales, which characterise turbulence and which distinguish it from the 'Lagrangian turbulence' or 'chaotic advection' in section 4, contribute to the efficiency of mixing by turbulent flows.

Why are virtually all high-Reynolds-number flows turbulent is a very good and simple question with no widely accepted, generally applicable answer. Certainly one can observe that the advection term $\mathbf{u} \cdot \nabla \mathbf{u}$, which dominates the smoothing viscous term at high Reynolds numbers, has the tendency to enhance spatial gradients, i.e. produce small scales. It is often speculated that most/all incompressible Euler flows (totally ignoring the viscous term) will produce a shock with a finite velocity discontinuity (or more plausibly a discontinuity in velocity gradient) in a finite time, although no convincing example has yet been produced.

At large Reynolds numbers, sharp shear layers are seen on the edge of the large scale eddies. These shear layers are vigorously unstable to small scale disturbances. There is the Kelvin-Helmholtz instability for fairly flat shear layers, see pictures MVD 81 and 55 (by Pierce 1961 and Werlé 1980), and for curved shear layers the Taylor-Görtler and Taylor-Couette instabilities, see pictures MVD 144 and 128 (by Bippes 1972 and Koschmieder 1979). The length scale of these fast growing instabilities is that of the thickness of the shear layer, while the velocity difference across the layer gives the velocity scale. One can envisage the eddies of these instabilities suffer themselves from secondary finer instabilities, and so on until viscosity can no longer be ignored. The instability mechanism at the largest scales of the flow may be particular to the special nature of the apparatus which produces the large scale flow. The mechanisms at smaller scales may be more universal, appearing in most flows.

3.3. Macro and micro scalings

The scales of the large eddies are set by the geometry and speed of the stirring mechanism, while the cut-off scales of the small eddies are determined by the action of viscosity. We discuss the details of the different large eddies is sections 3.6–3.9 for the different mixing flows introduced in section 3.1. Here we concentrate on the small scales for a flow with large eddies of given velocity, length and time scales, U_L, l_L, $T_L = l_L/U_L$. There is an assumption due to Kolmogorov, which is probably only partially true, that the small eddies are universal for given U_L, l_L and viscosity ν.

The important Kolmogorov microscale for the smallest eddies is based on a further assumption that the smallest eddies depend only on the rate at which energy is put into the large eddies, i.e. on one particular combination of U_L and l_L. An argument can be made that friction only acts on the smallest scale, that the sole rôle of the smallest scale is to dissipate all the energy, that the rate of dissipation must equal the supply of energy, that energy is supplied only at the large scale. The rate of dissipation is measured per unit mass, ϵ, and can be related to the macroscales by assuming that a significant fraction of the kinetic energy in the large eddies is dissipated in the turnover time of the large eddies, i.e. per unit time

$$\rho\epsilon = \rho U_L^2/T_L, \quad \text{i.e.} \quad \epsilon = U_L^3/l_L.$$

The dimensions of this dissipation per unit mass ϵ are $L^2 T^{-3}$, while the dimensions of the kinematic viscosity ν are $L^2 T^{-1}$. Hence by dimensional analysis, we obtain the velocity, length, time and strain-rate scalings of the Kolmogorov microscale

$$U_K = (\nu\epsilon)^{1/4}, \quad l_K = (\nu^3/\epsilon)^{1/4}, \quad T_K = (\nu/\epsilon)^{1/2} \quad \text{and} \quad \dot{E}_K = (\epsilon/\nu)^{1/2}.$$

Introducing the Reynolds number of the large scale eddies $Re = U_L l_L / \nu$, we have

$$U_K = U_L Re^{-1/4}, \quad l_K = l_L Re^{-3/4}, \quad T_K = T_L Re^{-1/2}, \quad \text{and} \quad E_K = E_L Re^{1/2}.$$

A common notation for the Kolmogorov microscales are η for l_K, τ for T_K and υ for U_K.

In practice the Kolmogorov micro length scale can be quite small. It is however bigger than the mean-free-path by a factor $Re^{1/4}/Ma$ where Ma is the Mach number. Here we have used the expression for the mean-free-path $h = \nu/\overline{u}$, and set the mean thermal speed equal to the speed of sound.

The Kolmogorov microscale for the smallest eddies depends on the velocity and length scales of the large eddies only in the combination $\epsilon = U_L^3/l_L$. A second microscale, the Taylor microscale uses a different combination to yield a slightly larger scale. The Taylor microscale can be thought of as the boundary layer thickness on the edge of a large eddy, i.e. $l_T = \sqrt{\nu t}$ with $t = T_L = l_L/U_L$ the turnover time of the large eddy. Hence

$$l_T = (\nu l_L/U_L)^{1/2} = l_L Re^{-1/2}$$

using the Reynolds number of the large eddies. A common notation for the Taylor microscale is λ for l_T.

So far we have considered the largest and the smallest eddies, and these two extremes are sufficient to discuss the process of turbulent mixing. An additional hypothesis of Kolmogorov is that eddies exist at all scales, and that these intermediate scales are too large to depend on viscosity and too small to depend on the details of the large eddies. Thus it is assumed that they only depend on the energy being transferred from the large to the small scales, ϵ, i.e. the energy to be dissipated 'cascades through each intermediate scale'. Dimensional analysis then gives that the velocity scale of intermediate eddies of length scale l is $(\epsilon l)^{1/3}$ and the time scale is $(l^2/\epsilon)^{1/3}$. Much more can and will be said about the spectrum of eddies, although it is not clear to me that independent dynamical entities exist at all scales.

3.4. Mixing with the Taylor eddy diffusivity

We now turn to mixing by turbulent flows. Now the turnover time of the smallest eddies T_K is shorter than the turnover time of the large eddies T_L by the factor $Re^{-1/2}$. Hence mixing takes place faster and more efficiently on small scales than on large scales. (As will be explained in the next section, the proceeding remark is a little misleading when the molecular diffusivity D is very small.) The process limiting the rate of mixing is therefore mixing the fluid from its initial separated state to become homogeneous on the scale of the large eddies.

Large scale mixing is described by the the Taylor eddy diffusivity $D = U_L l_L$, see section 2.3. For a container of height H, the time for eddy diffusion is then

$$t_{\text{macro}} = H^2/D = t_L H^2/l_L^2.$$

This is many times, H^2/l_L^2 times, the turnover time of the large eddies when the apparatus is larger than the large eddies, as might occur with small paddles or a small mixing grid. If the source of the turbulence is small, one also has to worry whether the turbulence intensity, and hence the eddy diffusivity, is uniform throughout the container.

3.5.   The problem of low molecular diffusivity D

As discussed at the end of section 1.1, the value of the molecular diffusivity D can be very much smaller that the diffusivity of momentum ν, by 6 orders of magnitude in the case of the diffusion of colloidal particles. When the diffusivity is small, turbulent mixing will homogenise the fluid down to the length scale of the large eddies l_L in a time t_{macro}, as described in the previous section. Because the small eddies turnover much faster, the fluid will be homogenised down to the scale of the small eddies l_K after a negligible time delay. Mixing down to this scale is however considerably short of mixing down to the molecular scale, or the particle scale in the case of colloidal particles.

The time for purely molecular diffusion to homogenise the fluid within the smallest eddy is $l_K^2/D = t_K \nu/D$, i.e. the small eddy turnover time multiplied by the Schmidt number $Sc = \nu/D$. Fortunately there is a faster mechanism for mixing within the smallest eddy. The rate of straining within the smallest eddy is e_K and this straining increases the interfacial area between unmixed regions exponentially like $e^{\langle \sigma_A \rangle t}$, where $\langle \sigma_A \rangle \approx 0.3 e_K$ is the Liapounov exponent of section 2.7 for area increase. The 'striation thickness' between unmixed regions within a small eddy therefore decreases like $l_K e^{-\langle \sigma_A \rangle t}$, and will equal the molecular diffusion distance $\sqrt{Dt}$ at a time

$$t_{\text{micro}} = 3 t_K \ln(\nu/D)^{1/2}.$$

By this time the interfacial area will have increased by a factor $(\nu/D)^{1/2}$, i.e. the striation thickness will have reduced from the size of the smallest eddies to $\eta_\theta = l_K (D/\nu)^{1/2}$.

Comparing the time t_{macro} to homogenise the container down to the length scale of the large eddies with the above time t_{micro} to homogenise down to the molecular diffusion scale within the smallest eddies, we see that the low value of the molecular diffusivity is a problem limiting the mixing process if

$$Re^{-1/2} \ln(\nu/D)^{1/2} > H^2/l_L^2.$$

3.6.   The turbulent mixing layer

To complete this rapid survey of turbulent mixing, it remains to give estimates for the velocity and length scales of the large eddies for the four typical turbulent flows described in section 3.1.

For the mixing layer between two streams of fluid travelling parallel at speeds U and $U + \Delta U$, the typical velocity of the large eddies is about a quarter of the slip velocity, i.e.

$$U_L = \tfrac{1}{4}\Delta U.$$

Picture MVD 177 shows the mixing zone increasing linearly with distance downstream, keeping within a wedge of angle about 20°. Thus the length scale of the large eddies is about

$$l_L = \tfrac{1}{17}x,$$

where x is the distance downstream of the splitter plate.

3.7. The turbulent jet

We consider just the case of a three-dimensional jet with volume flux Q issuing from a circular pipe. Thus at the nozzle the jet velocity is $U_0 = Q/\pi a^2$.

Picture MVD 166 shows that the turbulent jet spreads within a cone of about $20°$ in a fashion similar to the mixing layer. The large scale eddies span much of the width of the jet and so again have a length scale

$$l_L = \tfrac{1}{15}x,$$

where x is the distance downstream of the nozzle.

The typical velocity in the large eddies is about half the mean velocity of the jet on its axis. The latter decreases downstream as the initial momentum flux $\rho U_0^2 a^2$ is spread over an increasing volume of fluid within the $20°$ cone. At high Reynolds numbers, the momentum flux must be the same at each downstream section of a steady jet, and so the velocity of the mean flow must decrease inversely proportionally with distance downstream. Thus the velocity of the large eddies is

$$U_L = 13U_0 a/x.$$

The entrainment of fresh fluid into the turbulent jet as it spreads within the $20°$ cone is of course a useful mixing which is employed to dilute many times over concentrated dye. The concentration within the spreading jet reduces as $c_0 a/x$, in order that the flux is the same at each downstream section.

Thus one can ask where on the ground downstream of a chimney is the concentration of pollution greatest, assuming that the flow from the chimney can be modelled as a turbulent jet (appropriate only under certain conditions of wind and stratification). The ground concentration will be highest where the spreading cone first hits the ground, i.e. at a distance downstream approximately five times the height h of the chimney. Further the ground concentration is roughly $c_0 a/15h$, so one can decide how high the chimney must be to avoid a dangerous level.

3.8. Turbulent wake behind a cylinder

Cylinders occur in mixers most often in the form of a grid which is dragged through the fluid. Here we consider just an isolated cylinder, although the scalings can be equally applied to the turbulence behind a grid.

While one might at first think that the spreading of the momentum deficit in the wake behind a cylinder ought to be very similar to the spreading of the momentum flux in a jet, there are in fact considerable differences: the velocity scale of the large eddies in the wake increases rather than decreases downstream, and the length scale of the large eddies increases less rapidly in the wake compared with the jet. The difference in the two flows is clearly apparent in the frame in which the fluid at infinity is at rest. The jet is stationary on average, with spreading resulting from the flow downstream within the jet. The wake on the other hand is transient, with fluid being displaced only a short distance as the cylinder passes. The spreading of the wake is due to lateral diffusion in time rather than displacement downstream.

We find the length l_L and velocity U_L scales of the large eddies by combining two arguments. First, a geometrical argument that the velocity fluctuations result from deflecting the free-stream flow U_0 through an angle x/l_L, where x is the distance downstream from the cylinder, i.e. $U_L = U_0 x/l_L$. Second, the deficit in the momentum flux in the wake $\int \rho U_0 U_L \, dy$ per unit length of the cylinder must equal the drag on the cylinder $\frac{1}{2}\rho U_0^2 a$, i.e. $\rho U_0 U_L l_L = \rho U_0^2 a$. Solving we find

$$U_L = 1.6 U_0 \sqrt{\frac{a}{x}} \quad \text{and} \quad l_L = 0.25 a \sqrt{\frac{x}{a}}.$$

A consequence of these estimates is that the Reynolds number based on the large eddy scales $U_L l_L/\nu = 0.4 U_0 a/\nu$ does not change downstream in the wake, a so called self-similar property of this flow. A second consequence is that the eddy diffusivity is also constant $D = U_l l_L = U_0 a$, and so the sideways diffusion $l_y = \sqrt{Dt}$ after the time $t = x/U_0$ to flow a distance x downstream is $l_y = \sqrt{ax}$, i.e. the size of the large eddies.

3.9. Turbulent pipe flow

Turbulence in a pipe, and in other confined geometries such as the turbulent boundary layer on a rigid plate, is considerably different from the turbulence in the open geometries of the previous three sections. The origin of the turbulence in the pipe is the friction of the wall, i.e. the no-slip boundary condition exerted by viscosity. Thus energy is feed into the turbulence at the small scale, at the scale at which viscosity is important, rather than at the large scale, as in the previous three geometries. One consequence of this difference in the driving of the turbulence is that the scale of the velocity fluctuations is the same for eddies of all scales. Traditionally this magnitude of velocity fluctuations is given the symbol u_*, and is called the 'friction velocity'.

The friction velocity is related to the mean pressure gradient dp/dz in the pipe by the following argument. Consider a pipe of length L and radius a. The difference in the pressure force acting on the two ends, $\pi a^2 L dp/dz$, must balance the tangential friction stress τ exerted on the wall $2\pi a L \tau$. This friction stress is expresses as a Reynolds stress $\tau = \rho u_*^2$ using the friction velocity. Hence

$$\rho u_*^2 = a \frac{dp}{dz}.$$

The length scale of the smallest eddies l_* depends on whether the pipe has smooth walls or rough. If it is hydrodynamically smooth, as in careful laboratory experiments, then the size of the smallest eddies is such that viscous and inertial forces are comparable for the velocity fluctuations

$$l_* = \nu/u_*.$$

If the pipe has a roughness with an amplitude greater than this value, which is common in practical applications, then l_* is replaced by the height of the roughness elements.

The size of the largest eddies depends on the radial position in the pipe. In MVD 157 & 163 one can see that it is roughly half the distance from the wall, i.e.

$$l_L = 0.4(a - r).$$

Finally it is necessary to relate the frictional losses represented by u_* to the volume flow through the pipe, i.e. to the mean flow $\bar{u}$. Now the mean tangential stress, which is exerted by molecular diffusion and eddy transport of momentum, varies linearly across the pipe from its wall value τ (by using the argument for u_* applied to a cylinder of arbitrary radius within the pipe). The eddy diffusivity of momentum by the large eddies is given by section 2.3 as $0.4u_*(a-r)$. Thus

$$\rho(\nu + 0.4u_*(a-r))\frac{d\bar{u}}{dr} = \rho u_*^2 \frac{r}{a}.$$

Integrating, and adjusting the constants to fit experimental observations, we obtain the 'log law of the wall'

$$\bar{u} = u_*\left(2.5\ln\frac{a}{l_*} + 6\right).$$

An application of the above scalings is to estimate how far downstream dye must travel before is becomes well mixed in a turbulent pipe. Let the pipe be hydrodynamically smooth, so $l_* = \nu/u_*$, and take the Reynolds number to be $Re = \bar{u}a/\nu = 6000$. Solving the last equation in the previous paragraph, we find $u_* = \frac{1}{20}\bar{u}$. If the molecular diffusivity of the dye is not too small, so that we need only consider mixing on the large scales, then the time to diffuse across the radius a with eddy diffusivity u_*a is a/u_*. During this time the fluid has moved downstream at the mean velocity $\bar{u}$ a distance $20a$.

4. Lagrangian mixing

4.1. The problem

Mixing in viscous liquids at low Reynolds numbers can be difficult. Specially designed mixers are required to compensate for the lack of the small length scales in the flow that are generated internally at large Reynolds numbers. The large value of friction means that energy consumption is high, and this places a premium of the efficiency of mixing. Industrial applications include blending polymers, homogenising glass and processing food.

Mixing at low Reynolds numbers becomes difficult when the molecular diffusivity of the substance to be mixed is so small that the Péclet number UL/D is large. As seen in section 2.2, the diffusivity of colloidal particles is inversely proportional to the viscosity, and so it is not uncommon in viscous liquids for the Reynolds number to be small while the Péclet number is large.

When the Péclet number is large, material is mainly advected by the flow and it diffuses very slowly. It is therefore essential that the flow field has the property of efficiently dispersing material throughout the volume. This requires that two blobs of fluid which start nearby one another should separate rapidly. The appropriate measure of the rate of this process is the Liapounov exponent $\langle \sigma_A \rangle$ introduced in section 2.7.

4.2. Chaotic pathlines

The 'big idea' of Lagrangian chaos is the observation that velocity fields $\mathbf{u}(\mathbf{x}, t)$ which are smooth, regular and generally boring when viewed in the traditional Eulerian fixed-laboratory frame can produce fluid pathlines $x(x_0, t)$ [the solution of $\dot{x} = u(x, t)$ in $t \geq 0$ with $x = x_0$ at $t = 0$] which uniformally fill a volume in a ergodic way. These chaotic fluid pathlines disperse materially efficiently, i.e. give good mixing with nonzero Liapounov exponents.

Steady two-dimensional flows cannot have chaotic fluid pathlines: in the language of dynamical systems they are called 'integrable'. Because in a steady flow the pathlines are streamlines, and because in two dimensions streamlines cannot cross (except at points where the velocity vanishes so that fluid blobs will never reach them), almost all fluid pathlines are closed. Thus dye released in one region will remain limited to the streamlines which pass through the initial dyed region. Mixing is therefore inadequate.

Chaotic fluid pathlines can however occur in time-periodic two-dimensional flows, in steady three-dimensional flows, and in more general flows. The special case of time-periodic two-dimensional flows has been much studied due to the availability of powerful methods of analysis.

A very good introduction to the subject of mixing by Lagrangian chaos is given by the textbook by J.M. Ottino *The kinematics of mixing: stretching, chaos and transport*, hereafter denoted by JMO. I will refer to several of the excellent collection of pictures in that book.

4.3. Changing the direction of shear

We saw in section 2.5 that a steady simple shear $u = (\gamma y, 0, 0)$ will stretch the surface area of a dyed region which starts as a sphere so that it grows linearly in time like $2\pi a^2 \gamma t$. (This linear growth is attained after a total shear of about $\gamma t = 3$.) The linear growth leads to a Liapounov exponent which vanishes, and this reflect the poor mixing of steady simple shear.

Now consider applying the above simple shear for a duration Δt and then switching to a second simple shear $u = (0, \gamma' x, 0)$ in a different direction. The region of dyed fluid will remain ellipsoidal, and after a duration $\Delta t'$ of the second shear it will have a surface area of approximately $2\pi a^2 \gamma \Delta t \gamma' \Delta t'$, supposing that the total shears $\gamma \Delta t$ and $\gamma' \Delta t'$ are greater than 3. If this cycle is repeated several times, then after a time t the surface areas will be

$$2\pi a^2 \left(\gamma \Delta t \gamma' \Delta t' \right)^{t/(\Delta t + \Delta t')} .$$

Thus we have exponential growth in time with a Liapounov exponent

$$\sigma_A = \gamma \frac{\ln \gamma \Delta t}{\gamma \Delta t}$$

in the optimal case of $\Delta t' = \Delta t$ and $\gamma' = \gamma$. At $\gamma \Delta t = 3$ this is 0.37γ. Thus we see that time-periodic two-dimensional flows might lead to useful mixing.

4.4. The blinking pair of vortices

The two infinite simple shear flows are far removed from any practical device. A very interesting step towards a realisable flow is the blinking pair of line vortices suggested by Aref (1984). Consider some dyed fluid which starts near one line vortex, and let that vortex turn while the other remains at rest. The dye will be sheared by the near vortex, and after sufficient time will be spread around the circle through its initial position. Now stop the first vortex and switch on the second. The shear flow from the second vortex will be perpendicular to the shear from the first at most locations. The dye will therefore be stretched efficiently for the first few strains. One then switches back to the first vortex and continues in a cyclic manner, hence the name 'blinking pair of vortices'.

While the dye remains near the first vortex (whether this occurs depends on the durations of the 'blinks'), one can see the dye stretched out by the far vortex and then wrapped round by the near vortex. This *stretching-and-folding* is an essential ingredient of good mixing in finite devices, the stretching of interfacial area being the good mixing and the folding being necessary to fit the exponentially stretching material into a finite space. This process of stretching-and-folding is well known by bakers kneading bread by hand.

Figure 7.3.8 on page JMO 184 shows how an initial line of points spread nearly uniformly within a large region after 20 iterations of the blinking pair of vortices. While the mixing is good within this region, which encloses the initial line and circles of similar dimensions around the two vortices, it is striking that there is no mixing out of the region. Figure 7.3.11 on page JMO 188 (by Khakhar, Rising and Ottino 1986) shows with twice the frequency of blinking that most of the material starting near one vortex remains near that vortex for 25 iterations. There is thus poor communications between domains of the two vortices under certain circumstances. These two figures do illustrate that while Lagrangian chaos can produce nearly uniform mixing within part of the flow it does also suffer problems of not achieving good global mixing.

4.5. Alternating eccentric cylinders

A practical realization of mixing with cyclic switching between two shears in different directions is a fluid in the annular gap between two eccentrically mounted cylinders which rotate alternately around their fixed axes. Figure 7.4.5 on page JMO 153g shows the positions visited at the end of each of 1000 iterations of 8 particles using a cycle in which the outer cylinder is turned through a half turn and then the inner cylinder is turned backwards through a whole turn. Again one notes good uniform mixing in a large area, but roughly an equal area which is not visited and is unmixed.

4.6. Alternating helical mixers

Another practical device for efficient mixing based on the idea of alternating the direction of the flow is Kenics' static mixer, figure 8.2.2 on page JMO 223. This device consists of a circular pipe with a sequence of central dividers which turn helically along the pipe, in one direction for one turn, to be followed down the pipe by the next turning in the opposite direction and starting perpendicular to the preceding divider. The flow is thus split in two, twisted and then recombined a number of times. An interface will thus grow exponentially downstream after undergoing several cycles. Idealised studies of the device (without the helices), described in §8.2 of JMO, find thorough mixing in some parts of the flow, together with poorly mixed regions.

4.7.   One origin of chaos

This section becomes quite complicated and can be safely skipped by those interesting in mixing.

While steady two-dimensional flows cannot have chaotic fluid pathlines, time-periodic two-dimensional flows can. We consider in this section a two-dimensional flow with a small amplitude periodic variation about a steady base flow.

Almost all of the streamlines in the base flow are closed. The speed along each closed streamline will have a non-zero minimum. As long as the periodic perturbation flow is small compared with this minimum, the fluid pathlines will remain near to the streamlines of the base flow, and hence rather uninteresting.

The interesting streamlines of the base flow are the very few which go to or leave the stagnation points, 'hyperbolic points' in the language of dynamical systems. At a stagnation point the velocity of the base flow vanishes, and so small perturbations can exert a controlling influence on pathlines passing near it.

A special feature of incompressible steady two-dimensional flows is that a streamline that leaves one stagnation must go precisely to another, or re-approach itself. If this were not the case, fluid would enter a region never to exit, so violating conservation of fluid volume.

Now as the flow approaches a stagnation point it divides into two streams, each stream passing on a different side of the stagnation point. The streamlines joining stagnation points thus separate the flow into non-communicating vortices, and for this reason are called 'separatrices'.

We now consider the effect of small time-periodic perturbations on the separatrices of the base flow. The effect of the perturbation can be represented by plotting the sequence of positions of a fluid blob at the end of each period, the so called 'Poincaré section' for the flow. Consider a collection of sequences which start from a short line perpendicular to the separatrix at some position well away from the stagnation point. Some of these sequences will be deflected to one side of the stagnation point and some to the other. There must however be one starting point along the short line whose sequence is deflected neither side and instead slowly approaches a fixed point in the Poincaré section very near to the stagnation point of the base flow. By considering short lines at different places along the separatrix, one can construct a new separation line in the Poincaré section. Fluid particles do not remain on this line during the periodic perturbations, but return to it at the end of each period. Again the new separation line separates fluid (at the start of each period) which will divide and go off around vortices differently.

An important difference between a time-periodic incompressible two-dimensional flow and a steady one is that a separation line leaving one fixed point in the Poincaré section does not have to go to another fixed point, or to re-approach itself. The conservation of fluid area only requires balancing areas go each side of the second fixed point during the period. Thus the separation line leaving one fixed point must cross the different separation line which approaches the second fixed point. If it does cross once, then we can conclude that there must be an infinite number of crossings as one goes along the line to each fixed point, because the sequence of points starting from the first crossing must lie by definition on both lines as one goes along the lines to their fixed end points..

Thus in any small neighbourhood of one fixed point in the Poincaré section, the separation line leaving the fixed point is crossed an infinite number of times by the separation line which approaches the following fixed point. This means that lying on one side of the separation line leaving the first fixed point there is an infinite sequence of points exiting the following fixed point on alternating sides. This means that approaching the first fixed point one can find a point closer to the entering separation line which exits the following fixed point on either side.

This argument can be immediately applied to the fixed point following the following one. Hence one can select any arbitrary route past connected fixed points, diverting either side at each, by moving slightly nearer to the initial separating line. This random route is chaos, at least in the neighbourhood of the separatrices of the original steady two-dimensional flow.

4.8.   Observations (critical)

Lagrangian chaos is a young subject but has already made a major impact on the way we think about mixing. Lagrangian chaos leads to the possibility of efficient mixing at low Reynolds numbers, and leads to a better understanding of turbulent mixing when the diffusivity is small ($D \ll \nu$).

Despite many papers published recently on Lagrangian chaos, there is little quantitative understanding. For example, there is no theory for the value of the Liapounov exponents, and we cannot predict how much of the flow is poorly mixed and where it will be.

Some of the fascinating details of islands of chaos around islands around islands may be rapidly obliterated by the addition of a little molecular diffusion. In fact the transfer across 'barriers' between adjacent well-mixed regions, whether by molecular diffusion or by non-periodic fluctuations, is a very important issue in several applications and yet something we have hardly started to understand. Perhaps time-periodic two-dimensional flows have been over-studied and now three-dimensional and non-periodic flows deserve greater attention.

Above all, the engineering questions must not be lost sight of. How much mixing is there after just 5 cycles, not the theoretician's 1000? Open flow devices with a continuous throughput are essential in industry. What is their mixing efficiency, what is the residence time of the fluid, and of course what is the cost?

5.   References

M. Van Dyke *An Album of Fluid Motion* Parabolic Press (1982).
J.M. Ottino *The kinematics of mixing: stretching, chaos, and transport* C.U.P. (1989).

PART II

GEOMETRY OF MIXING AND ENGINEERING APPLICATIONS

Part I

THEORY OF MIXING AND ITS APPLICATIONS

Turbulence, Fractals, and Mixing

PAUL E. DIMOTAKIS[1] AND HARIS J. CATRAKIS[2]

[1] *Graduate Aeronautical Laboratories, California Institute of Technology*
Pasadena, CA 91125, USA
[2] *Mechanical and Aerospace Engineering, University of California, Irvine, USA*

Proposals and experimental evidence, from both numerical simulations and laboratory experiments, regarding the behavior of level sets in turbulent flows are reviewed. Isoscalar surfaces in turbulent flows, at least in liquid-phase turbulent jets, where extensive experiments have been undertaken, appear to have a geometry that is more complex than (constant-D) fractal. Their description requires an extension of the original, scale-invariant, fractal framework that can be cast in terms of a variable (scale-dependent) coverage dimension, $D_d(\lambda)$. The extension to a scale-dependent framework allows level-set coverage statistics to be related to other quantities of interest. In addition to the pdf of point-spacings (in 1-D), it can be related to the scale-dependent surface-to-volume (perimeter-to-area in 2-D) ratio, as well as the distribution of distances to the level set. The application of this framework to the study of turbulent-jet mixing indicates that isoscalar geometric measures are both threshold and Reynolds-number dependent. As regards mixing, the analysis facilitated by the new tools, as well as by other criteria, indicates enhanced mixing with increasing Reynolds number, at least for the range of Reynolds numbers investigated. This results in a progressively less-complex level-set geometry, at least in liquid-phase turbulent jets, with increasing Reynolds number. In liquid-phase turbulent jets, the spacings in one-dimensional records, as well as the size distribution of individual "islands" and "lakes" in two-dimensional isoscalar slices, are found in accord with lognormal statistics in the inner-scale range. The coverage dimension, $D_d(\lambda)$, derived from such sets is also in accord with lognormal statistics, in the inner-scale range. Preliminary three-dimensional (2-D space + time) isoscalar-surface data provide further evidence of a complex level-set geometrical structure in scalar fields generated by turbulence, at least in the case of turbulent jets.

Mixing: Chaos and Turbulence, edited by Chaté *et al.*
Kluwer Academic / Plenum Publishers, New York 1999.

1. Introduction

Turbulent flow, or turbulence, is found to have two important and interrelated properties. It is *chaotic* and it can *transport, stir, and mix* its constituents with great effectiveness.

By chaotic, we mean that it is characterized by irregular temporal and spatial dynamics that are unstably related to its initial and boundary conditions. The Random House Dictionary of the English Language (1971),[94] for example, offers as a definition of *turbulent flow*,

> "The flow of a fluid past an object such that the velocity at any fixed point in the fluid varies irregularly."

We may remove the requirement that an object is necessary for turbulent flow, as noted in a discussion on turbulence (cited in Hinze 1975)[52] by von Karman (1937),[119] who was quoting a 1927 lecture by G. I. Taylor:

> "Turbulence is an irregular motion which in general makes its appearance in fluids, gaseous or liquid, when they flow past solid surfaces or even when neighboring streams of the same fluid flow past or over each other."

If viscous effects are small enough, it is sufficient to impose shear on the flow, by any means, for hydrodynamic instabilities, unsteady flow, and turbulence to be sustained. While the zero-relative-velocity boundary condition on solid boundaries will suffice to generate the shear and vorticity in the flow to trigger turbulence, there are other means to do so, without recourse to solid bodies. We may also remove the restriction from the definition above that the shear zone must be sustained between "neighboring streams of the same fluid". A shear layer between two different fluids, *e.g.*, helium and nitrogen, for example, will also be turbulent.

Irregular motion, however, is an indisputable characteristic of turbulence. As we sit here on ancient Mediterranean shores, Heracleitos comes to mind, on the banks of a small river near the ancient Greek town of Ephesos, in Asia Minor, tossing little sticks on the turbulent river water, two and a half millennia ago, watching them float irregularly downstream, and remarking,([1])

> "Twice into the same river you could not enter."

Despite initial conditions defined a long time in the past (relative to the time it takes the water to reach our observation station from the spring) and steady boundary conditions (the banks are not changing), the unsteady, turbulent (flow in the) river is not the same twice. The caveat of Heracleitos summarizes a quandary of the "turbulence problem." If the goal of science and the study of natural phenomena is the ability to predict their evolution, in what sense does one "predict" turbulence, for a given set of initial and boundary conditions?

For a particular turbulent-flow configuration (initial/boundary conditions), we may list the following possible scenaria, ranging from the more to the less pessimistic:

I. The turbulent flow need not be uniquely related to its initial and boundary conditions, in any sense.

([1]) Δίς εις τὸν αυτὸν ποταμὸν ουκ αν εμβαίης.

II. The turbulent flow is a unique function of the boundary and initial conditions only in a statistical (mean) sense.

III. There exists a time, long enough for the effects of the initial conditions to have been forgotten, after which the resulting turbulent flow, in a mean sense, is determined by the boundary conditions alone.

IV. For flows in a domain of extent L that is large compared with Λ, the local outer scale of the turbulent-flow region, not only (temporally) initial conditions (after a sufficiently long time), but also inflow (spatially "initial") conditions have a negligible effect.

Scenario II, for example, would accommodate a particular turbulent flow that develops statistics that are well defined, but a function of both the boundary and initial conditions. In such a flow, different initial conditions, for the same boundary conditions, would evolve into a different flow configuration, as $t \to \infty$.([2]) Finally, we must accept that these possibilities need not be universal; what applies to one turbulent flow geometry need not apply to another.

Batchelor (1953)[4] introduces his discussion of homogeneous turbulence with the observation of its irreproducible character, but includes a statement of belief:

> "It is a well-known fact that under suitable conditions, which normally amount to a requirement that the kinematic viscosity ν be sufficiently small, some of these motions are such that the velocity at any given time and position in the fluid is not found to be the same when it is measured several times under seemingly identical conditions. In these motions, the velocity takes random values which are not determined by the ostensible, or controllable, or 'macroscopic', data of the flow, although we believe that the *average* properties of the motion are uniquely determined by the data."

The requirement that the (kinematic) viscosity, $\nu = \mu/\rho$, *i.e.*, viscous effects, must be small should, of course, be interpreted in a relative sense, *i.e.*, in terms of a criterion associated with the dimensionless Reynolds number,

$$Re \equiv \frac{U\Lambda}{\nu} , \tag{1}$$

with U and Λ the characteristic velocity and length scale of the flow, respectively. The same fluid may be turbulent, or not, depending on the flow parameters. A "small viscosity" is assessed in comparison of (the fluid property) ν and the product (flow property) $U\Lambda$. That the Reynolds number should be large, must be accepted as a *necessary* requirement for turbulence to be sustainable.

Batchelor, who states that the irregular-flow behavior is "... not determined by the ... data of the flow", here inclusively interpreted as the complete initial and boundary conditions, assumes Scenario IV. To quote Frisch (1995):[44]

> "Batchelor's (1953) *Homogeneous Turbulence* was a landmark in the field, written after considerable experimental and theoretical work had been done in

---

([2]) As P. Tabeling noted in the discussion, one may wish to consider a fifth scenario in which universal behavior may be anticipated in the limit of very high Reynolds numbers, irrespective of initial/boundary conditions.

Cambridge in the years after World War II. This monograph contributed to bringing the idea of a 'universal equilibrium' theory of turbulence (one aspect of Kolmogorov's contribution) to a wide scientific audience."

Hinze (1975),[52] in his book on turbulence, is also concerned about the implied extension of the caveat and offers the following:

"Turbulent fluid motion is an irregular condition of flow in which the various quantities show a random variation with time and space coordinates, so that statistically distinct average values can be discerned."

Scenario I must be excluded altogether from the proper study of turbulent flow, in his opinion, with the discussion in the body of the book itself predicated on Scenario IV. Nevertheless, solutions of the deterministic Navier-Stokes equations of fluid motion, that we typically accept as containing all the dynamics of turbulence,([3]) admit the possibility of non-unique solutions. Depending on the particular flow configuration, the sensitivity of the solutions to initial and boundary conditions can be such as to potentially render even average values of the resulting flow properties a non-unique function of those conditions. Batchelor's belief that the "*average* properties of the motion are uniquely determined by the data," must be accepted as an assumption. It has not been shown that chaotic solutions of the Navier-Stokes equations are ergodic, *i.e.*, that they describe stationary stochastic processes with well-defined statistics. We note the caveat(s) and proceed.

The second characteristic of turbulence is that it *can transport, stir, and mix* its constituents with great effectiveness. Such constituents can be its momentum (density), in which case efficient stirring and mixing of low-momentum fluid originating near no-slip-condition boundaries on a solid surface, for example, produces the many manifestations of drag; flow scalars, such as temperature, leading to heat transport; or chemical reactants, in which case turbulence can produce high chemical reaction rates and combustion heat release.

These two characteristics are the consequence of the dynamics of unsteady vortical structures; the hallmark of turbulence. Indeed, the two notions were intertwined in the minds of the ancient Greeks, who coined the roots of the word. *Turbulence* (situation where eddies abound), derives from the ancient Greek word, $\tau\acute{v}\rho\beta\eta$ (eddy, swirl), whence *turban* (swirled headdress), *turbine, turbid* (as in unclear water, in recognition of turbulent flow's ability to pick up dirt and muddy it), *turpitude* (lack of moral clarity), and other unspeakables.([4]) At the same time, the Greek verb, $\tau\upsilon\rho\beta\acute{a}\zeta\epsilon\iota\nu$, means: to stir things up, make a mess of, mix.([5])

([3]) Feynman, always fascinated by vortices and turbulence, states this belief: "We have no reason to think that there are any terms missing from these equations." (Feynman 1964,[41] Ch. 41, p. 11). Frisch also comments on this, stating that the N-S equation "probably contains all of turbulence." (Frisch 1995, p. 1).[44]

([4]) Also related is *tornado, turn*, etc., from the Greek $\tau\acute{o}\rho\nu\sigma\varsigma$ for lathe (wood-/metal-turning tool), or spinning wheel (as for cotton). Euripides: "$T\rho\sigma\chi\acute{o}\varsigma\ \tau\sigma\rho\nu\sigma\upsilon\ \gamma\rho\alpha\phi\acute{o}\mu\epsilon\nu\sigma\varsigma$," or, "Wheel, written by the lathe." Aeschylos: "$B\acute{o}\mu\beta\upsilon\kappa\alpha\varsigma\ \tau\acute{o}\rho\nu\sigma\upsilon\ \kappa\acute{a}\mu\alpha\tau\sigma\upsilon$," or, "Cotton, the spinning wheel's labor." Apropos turbulence and vorticity, the root also connotes twisting/bending. The word $\tau\sigma\rho\nu\epsilon\acute{\iota}\alpha$ (as in structural lumber) was used for the twisted (bent) main structural ribs of a ship.

([5]) Aristophanes: "$T\sigma\upsilon\ \pi\upsilon\lambda\grave{o}\upsilon\ \tau\upsilon\rho\beta\acute{a}\zeta\epsilon\iota\varsigma\ \beta\alpha\delta\acute{\iota}\zeta\omega\upsilon$," or, "You are dis*turb*ing the clay/dirt

At least in incompressible turbulence, mixing of a scalar field, $c(\vec{x}, t)$, *e.g.*, the jet-fluid concentration field in turbulent-jet flow, which tends to drive scalar values to their (local) spatial average, can be thought of as the result of a process involving "more or less distinct stages," as described by Eckart (1948).[38] The scalar to be mixed may be regarded as first *injected* through a bounding surface (as in a jet), or *entrained* at the largest spatial scales of the flow, as in a shear layer (Brown & Roshko 1974).[13] It is then *stirred* as it is convected by the velocity field associated with unsteady, fusing and fissioning eddies. Finally, it is *mixed* by molecular diffusion, or Brownian motion, in the case of a cloud of particles.

We will attempt a discussion of these phenomena from the vantage point of the properties of isoscalar surfaces, or (scalar) isosurfaces, *i.e.*, the instantaneous three-dimensional surface on which the value of the scalar field of interest is a constant, *i.e.*,

$$c(\vec{x}, t) = c_0 . \tag{2}$$

The irregularity in turbulent-flow motions results in scalar level-set surfaces whose geometry is exceedingly complex and suggests that an approach based on fractal ideas and analysis may be appropriate, as was suggested by Mandelbrot (1975a).[71]

Scalar transport and mixing phenomena in unsteady flow, as well as the many manifestations of chaotic advection, are interesting and worthy of study in their own right (*e.g.*, Aref 1984;[2] Ottino 1989 and 1990;[84][85] and Rom-Kedar *et al.* 1990).[98] We will restrict the present discussion to transport and mixing in turbulent flows, however, at Reynolds numbers that are sufficiently high for the turbulence to be in a fully-developed state. An estimate for the minimum Reynolds number required to achieve this state will be discussed below.

1.1. Fully-developed turbulence

As Batchelor noted, turbulence requires that viscous effects should be small. In practice, one finds that as one increases the Reynolds number of the flow, from very low values, a transition occurs from laminar to unsteady flow and, eventually, turbulence. In a free shear layer, for example, this transition will take place at a value of the local Reynolds number, based on outer flow variables,

$$Re_\Lambda(x) = \frac{U(x)\,\Lambda(x)}{\nu} , \tag{3}$$

where, for this flow, $U(x) = \Delta U \equiv U_1 - U_2$ is the difference between the two freestream velocities and $\Lambda(x) = \delta(x)$ is the local width of the shear layer, that is close to $Re_\delta = 40$, or so (Betchov & Szewczyk 1963).[8] This is not too different from the minimum value required for vortex shedding from a smooth, circular cylinder (Roshko 1954),[99] that is triggered by unsteadiness in the separation bubble behind the cylinder, *i.e.*, again, some distance from the closest solid boundary.

In wall-bounded flows, transition from steady to unsteady flow occurs at higher values of the Reynolds number and is, generally, hysteretic. For such flows, a more

by walking," or, ever sarcastic, "You are making a mess by (simply) walking." In everyday speech, *turbulent* is derogatory. "Turbulent times" are times characterized by great disturbances. The Latin *turba* derives from the closely-related, ancient Greek $\tau\acute{v}\rho\beta\alpha$, *i.e.*, *pêle-mêle* (pell-mell), helter-skelter, messy/unruly (crowd).

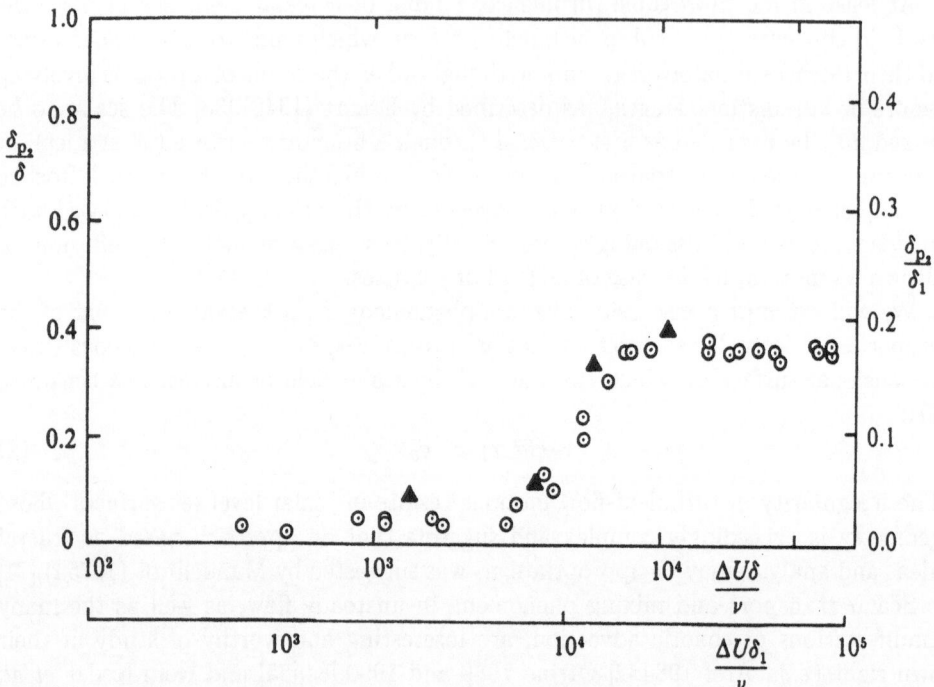

Fig. 1. — Reynolds number dependence of estimated chemical product volume fraction, in liquid-phase shear layers (Koochesfahani & Dimotakis 1986). [61]

appropriate criterion for a critical Reynolds number can be based on values below which turbulence cannot be sustained. In circular-pipe flow, for example, this critical value of the Reynolds number, based on the average mass-flow velocity and pipe diameter, is $Re_d \simeq 2 \times 10^3$, or a Reynolds number based on pipe radius of, $Re_R \simeq 1 \times 10^3$ (the latter might be regarded as the more appropriate as it is based on the length scale across which the shear is sustained). For a flat-plate boundary layer, the corresponding value, based on the free-stream velocity, U_∞, and local boundary-layer thickness, δ, is given by (Shen 1954),[103] $Re_\delta \simeq 1.2 \times 10^3$.

At Reynolds numbers beyond (but close to) these critical values, even though the flow may be unsteady, the effects of viscosity, even though insufficient to maintain steady flow, are, nevertheless, still significant and the resulting turbulent flow cannot be described as fully-developed. In all these flows, wall-bounded or not, we typically find a second transition, as the local Reynolds number is further increased, and it is beyond this second transition that the flow may be regarded as *bona fide* turbulent.

In the case of free-shear-layer flows, such a transition was documented by Konrad (1976), [60] who found a clear manifestation, as measured by the amount of fluid in the turbulent region mixed to compositions intermediate between those in the two free streams. This *mixing transition* occurred at a shear-layer Reynolds number of $Re_\delta \simeq 10^4$ and was identified as associated with a large increase in the three-dimensional interfacial area between the entrained freestream fluids (Bernal *et al.* 1979).[7]

The shear layer studied by Konrad was a gas-phase shear layer in the original

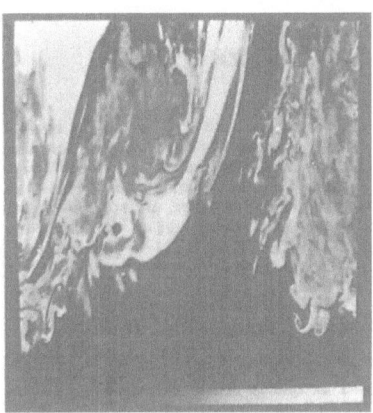

Fig. 2. — Laser-induced fluorescence space-time images in a liquid-phase shear layer. Left: $Re_\delta \simeq 1.75 \times 10^3$. Right: $Re_\delta \simeq 23 \times 10^3$. (Koochesfahani & Dimotakis 1986,[61] Figs. 7 and 9, respectively).

Brown-Roshko apparatus.[13] A similar mixing transition was identified in liquid-phase shear layers (Breidenthal 1981,[12] Koochesfahani & Dimotakis 1986),[61] in the vicinity of the same Reynolds number (see Fig. 1). This was also related to a large increase in interfacial surface area and mixing in direct flow-visualization slices of the flow (see Fig. 2). As the Schmidt number, *i.e.*,

$$Sc \equiv \frac{\nu}{\mathcal{D}} , \qquad (4)$$

where ν is the kinematic viscosity of the turbulent fluid and $\mathcal{D}$ is the species diffusivity, for the gas-phase shear-layer flows was of order unity, whereas for the liquid-phase flows of order 10^3, these results indicated that this transition was not a Schmidt-number-dependent phenomenon.

A transition in mixing behavior can also be observed in turbulent jets, as shown in Figs. 3 and 4 (reproduced from Dimotakis *et al.* 1983, Figs. 5 and 9, respectively).[37] Figure 3, below, depicts the jet-fluid concentration scalar field ($0 \lesssim z/d_j \lesssim 35$), in the plane of symmetry of the jet, recorded using laser-induced fluorescence of the jet-plenum dye, for a jet Reynolds number of $Re \simeq 2500$. Figure 4 is similar, also depicting the far field of a turbulent jet, albeit with a larger field of view ($0 \lesssim z/d_j \lesssim 200$), at a Reynolds number of $Re \simeq 10^4$. As can be seen, there is a qualitative difference in the scalar field behavior, as depicted in Fig. 3, at $Re < 10^4$, and Fig. 4, at $Re \approx 10^4$, respectively. The former indicates a substantial probability of finding unmixed reservoir fluid all the way to the axis of the jet, whereas mixing is substantially more effective at $Re \simeq 10^4$ with the jet able to homogenize the entrained fluid sufficiently quickly for that probability to be negligible. In the latter, $Re \simeq 10^4$ jet, one can also see regions of nearly homogeneous composition, gradually *decreasing* in jet-fluid concentration from downstream to upstream, separated by a progression of relatively sharp boundaries, indicating entrainment from the upstream boundary of large-scale vortical structures (Cf. Dahm & Dimotakis 1987).[31] This

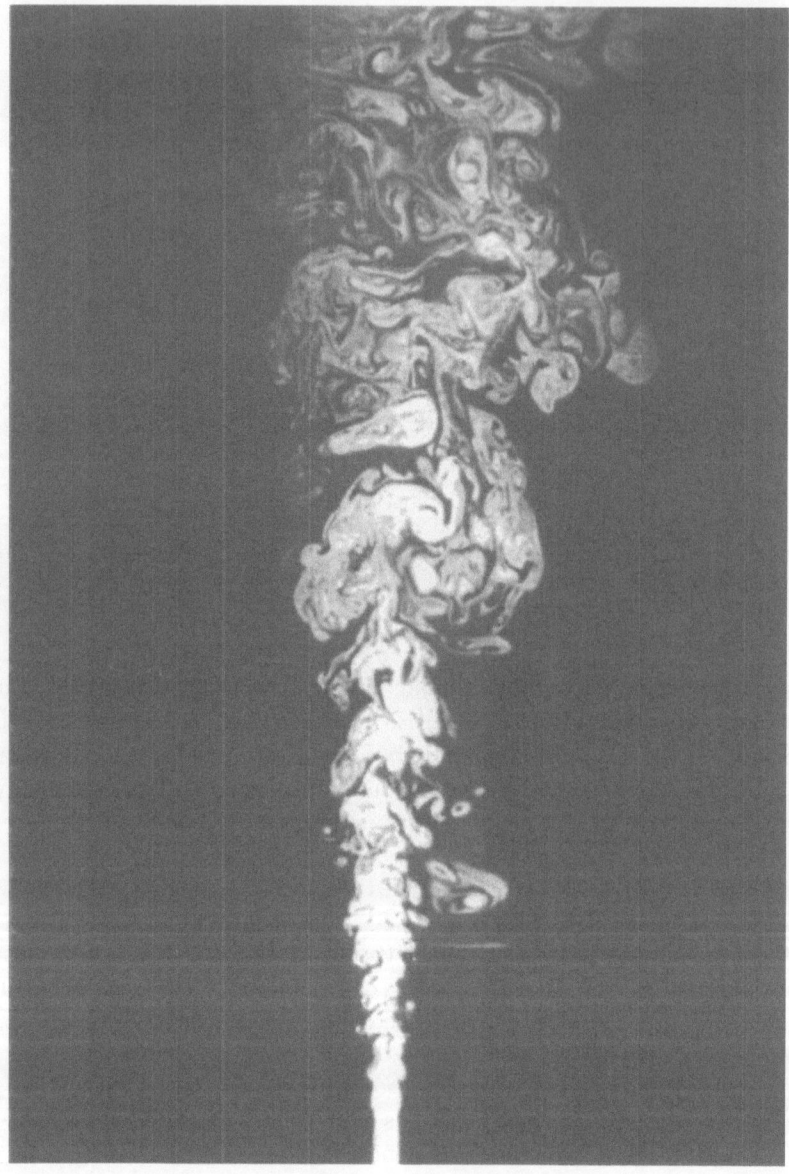

Fig. 3. — Liquid-phase turbulent jet ($0 \lesssim z/d_{\mathrm{j}} \lesssim 35$) at $Re \simeq 2.5 \times 10^3$.

transitional behavior is also confirmed by chemically-reacting experiments in which the dependence of flame length, *i.e.*, the distance required to mix and react the discharged jet fluid with the entrained reservoir fluid (Gilbrech 1991,[46] Gilbrech & Dimotakis 1992).[47]

Such transitions, albeit not always as conspicuous, have been reported to occur at similar values of the local critical Reynolds number, *i.e.*, at $Re \approx 10^4$. In flows

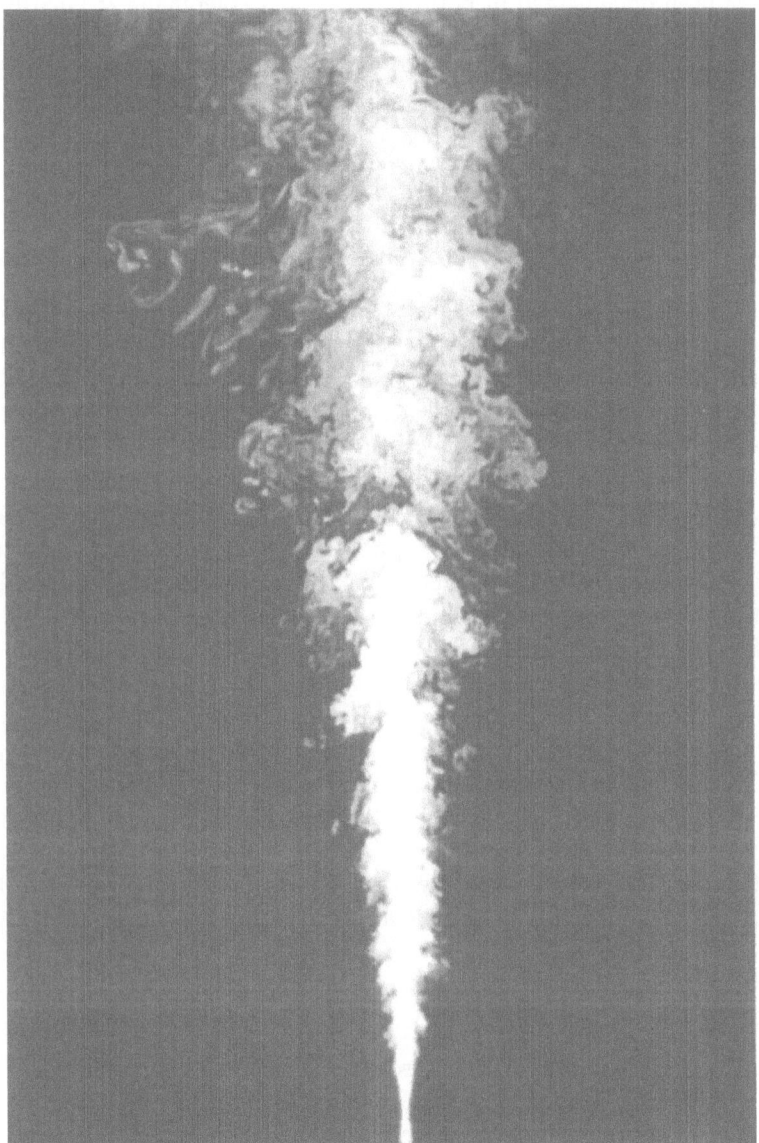

Fig. 4. — Liquid-phase turbulent jet $(0 \lesssim z/d_{\mathrm{j}} \lesssim 200)$ at $Re \simeq 10^4$.

characterized in terms of the Taylor microscale Reynolds number,

$$Re_{\mathrm{T}} \equiv \frac{u' \lambda_{\mathrm{T}}}{\nu} \, , \qquad (5)$$

where u' is the streamwise root-mean-square (rms) velocity fluctuation level and λ_{T} is the Taylor microscale, a transition is reported to occur at $Re_{\mathrm{T}} \approx 70$. In particular, in addition to shear layers and jets, as noted above, transitions to a fully-developed

turbulent state, in which flow properties are not as strongly Reynolds-number dependent, have been documented in both the near field (Liepmann & Gharib 1992)[67]
and far field of turbulent jets (as discussed above), in lifted jet flames (Hammer
1993);[50] in pipe flow (Wygnanski & Champagne 1973);[122] in flat-plate boundary
layers (Collins *et al.* 1978),[24] as manifested in the behavior of the Coles' turbulent
boundary layer wake parameter, Π (Coles 1956);[23] in the base-pressure coefficient
in circular cylinder wakes (Roshko 1992);[100] in the behavior of the scaled kinetic
energy-dissipation rate,

$$\alpha(Re_{\mathrm{T}}) = \frac{\varepsilon\,\Lambda}{u'^3}\;,\tag{6}$$

in the flow behind a grid (Sreenivasan 1984),[104] and in numerical simulations of homogeneous, cube-periodic turbulent flow (Jimenez *et al.* 1992);[53] in the transition
from "soft" to "hard" turbulence in thermal convection (Hezlot *et al.* 1978);[51] as
well as the Reynolds-number-dependence of the torque required to sustain a constant
rotation rate in Couette-Taylor flow (Lathrop *et al.* 1992a,[65] 1992b).[66]

   This list of (second) transitions, which is probably not exhaustive, indicates that
there exists a property of turbulence that results in a more-or-less distinct transition from an unsteady-flow state, at relatively low Reynolds numbers, to a state
in which viscous forces are less important. That this transition occurs in such a
wide variety of flows suggests that it is not a consequence of the large-scale behavior
of the flow, which varies substantially from one flow to the next. It is more likely
the consequence of inner-scale dynamics, that may be regarded as less sensitive to
outer-flow dynamics. This hypothesis is also supported by the observation that all
these transitions occur at, roughly, the same Reynolds number. An argument in
support of this hypothesis, based on the significance and relative magnitude of the
various spatial scales of turbulence, was discussed by Dimotakis (1993)[35] and is
summarized below for completeness.

   The dynamics of turbulence may be regarded as contained between the flow outer
scale, δ, and an inner scale in the vicinity of the kinetic-energy-dissipation (Kolmogorov 1941) scale,[57]

$$\lambda_{\mathrm{K}} \equiv \left(\frac{\nu^3}{\varepsilon}\right)^{1/4}\;.\tag{7}$$

At high Reynolds numbers, the two scales will be well separated, *i.e.*, the ratio
$\lambda_{\mathrm{K}}/\delta \sim Re^{-3/4}$ will be large. In that case, we expect that intermediate-scale behavior may be approximated as (more-or-less) uncoupled and independent of the
outer-scale dynamics, as well as inviscid, for spatial scales, λ, small compared to δ,
but large compared to λ_{K}, *i.e.*, for

$$\lambda_{\mathrm{K}} \ll \lambda \ll \delta\;.\tag{8}$$

   To refine the upper bound of this range, outer-scale-independent dynamics requires
a scale of interest, λ, that is small compared to one that can be generated directly
from the outer-scales of the flow. Such a scale would be the laminar boundary-layer
thickness, λ_{L}, that can be generated by a single δ-size sweep across the turbulent
region, for example. The size of this scale may be estimated by the 99%-thickness
of a Blasius (1908)[9] boundary layer, for example, that has been developing for a
length δ, or (see Fig. 5),

$$\frac{\lambda_{\mathrm{L}}}{\delta} \simeq 5.0\,Re^{-1/2}\;.\tag{9}$$

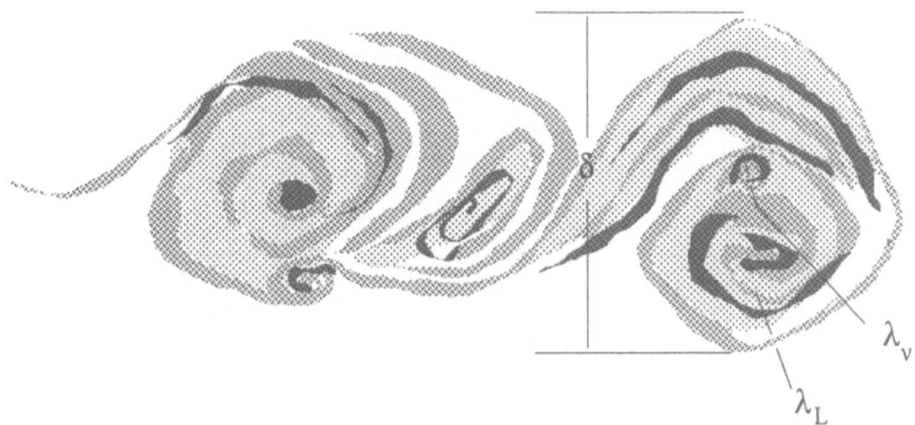

Fig. 5. — Schematic of the outer scale, δ; the Liepmann boundary-layer scale, λ_L; and the viscous scale, λ_ν, in a sheared turbulent region.

As noted by Liepmann, in private conversation many years ago, this carries the same Reynolds-number dependence and is closely-related to the Taylor scale, λ_T. We will accept that outer-scale independence, *i.e.*, no viscous coupling to outer-scale dynamics, imposes the much stricter spatial-scale-separation requirement, at large Re, of $\lambda \ll \lambda_L$.

To refine the lower bound in (8), inviscid dynamics dictate that the scale of interest, λ, must be large compared to some viscous scale, λ_ν. This can be related to the Kolmogorov scale, λ_K, and estimated in terms of the wavenumber, k_ν, where the energy spectrum has been found to deviate from a near-$\frac{5}{7}3$ power-law behavior, or $k_\nu \lambda_K \simeq 1/8$ (Cf. Chapman 1979,[21] Saddoughi & Veeravalli 1994).[102] To the extent that this may be regarded as a universal dimensionless factor, this yields a value that is a multiple of the (defined) Kolmogorov scale (Cf. (7)), *i.e.*,

$$\lambda_\nu \simeq \frac{2\pi}{k_\nu} \simeq 50\,\lambda_K \ . \tag{10}$$

At least on the centerline of a turbulent jet, for which experimental estimates of the kinetic-energy dissipation, ε, are available (Friehe *et al.* 1971),[43] the Kolmogorov scale, λ_K, can be related to the outer-scale (local jet diameter), $\delta(x)$, to yield (Cf. (7) and discussion in Dimotakis 1993),[35]

$$\frac{\lambda_K}{\delta} \simeq 0.95\,Re^{-3/4} \ , \tag{11}$$

and, therefore, if the viscous scale is estimated as the wavelength associated with k_ν, we have, at least for turbulent jets,

$$\frac{\lambda_\nu}{\delta} \simeq 50\,Re^{-3/4} \ . \tag{12}$$

We will accept that inviscid dynamics impose the stricter inner-scale requirement of $\lambda_\nu \ll \lambda$.

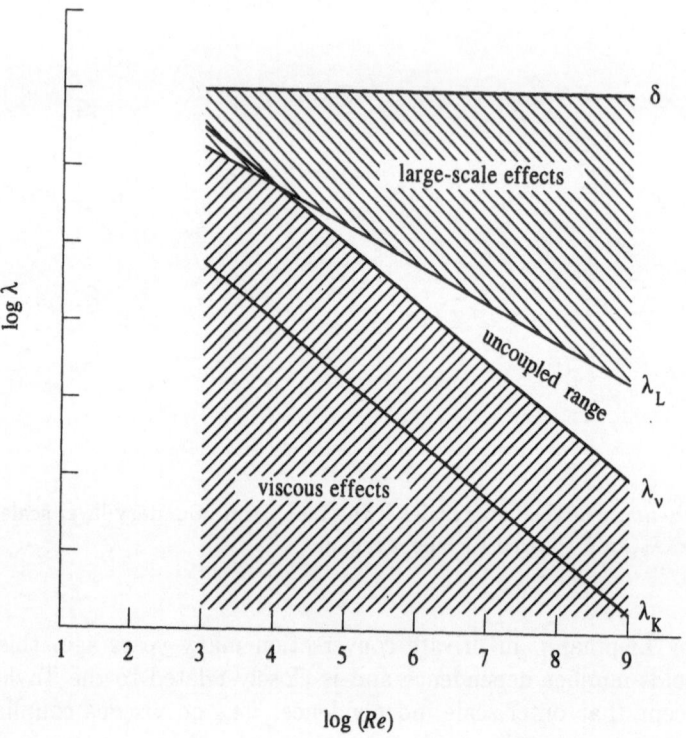

Fig. 6. — Reynolds number dependence of spatial scales for a turbulent jet.

Combining the refined inner- and outer-scale bounds (Eqs. (9) and (12)), we then have,

$$\frac{\lambda_{min}}{\delta} \simeq \frac{\lambda_\nu}{\delta} \approx 50\,Re^{-3/4} \ll \frac{\lambda}{\delta} \ll \frac{\lambda_{max}}{\delta} \simeq \frac{\lambda_L}{\delta} \approx 5.0\,Re^{-1/2}\,, \qquad (13)$$

as estimated for a turbulent jet. This condition is satisfied if,

$$\frac{\lambda_L}{\lambda_\nu} \simeq \frac{\lambda_{max}}{\lambda_{min}} \simeq 0.1\,Re^{1/4} > 1\,, \qquad (14)$$

or, if,

$$Re \gtrsim 10^4\,. \qquad (15)$$

These relations are summarized in Fig. 6.

While the numerical estimate for the energy dissipation rate, ε, was made based on empirical data for a turbulent jet, the other constants were not flow-specific. As a consequence, we may expect that the two factors of 50 and 5.0, in the λ_ν/δ and λ_L/δ scaling, respectively, should not depend strongly on the flow geometry. As indicated by the list above of flows where such a transition has been reported, it appears to be a flow-independent value.

We may conclude that a minimum Reynolds number, approximately equal to 10^4, is required for the outer scales of the flow to be regarded as inviscid and for an intermediate range of scales to exist in which the dynamics may be approximated as inviscid and outer-scale independent. It is in such an intermediate range of scales that one may expect the self-similar dynamics hypothesized by Richardson (1926),[95] Kolmogorov (1941),[57] and others to be applicable and, perhaps, produce level-sets with the scale-invariant geometric similarity described by Mandelbrot as the defining characteristic of fractals (Mandelbrot 1989),[75] as we'll discuss below. We will revisit this similarity property in the context of our discussion of experimental observations of scalar level-set geometrical properties.

1.2. Scalar transport and mixing

The convection-diffusion equation that describes the transport and mixing of a scalar field, $c(\vec{x}, t)$, is given by,

$$\frac{\partial c}{\partial t} + \vec{u} \cdot \frac{\partial c}{\partial \vec{x}} = \mathcal{D} \frac{\partial}{\partial \vec{x}} \cdot \frac{\partial c}{\partial \vec{x}} , \qquad (16)$$

where $\mathcal{D}$ is the species diffusivity ((4)), here assumed constant, or,

$$\hat{\mathcal{L}}_{\mathcal{D}}(\vec{u})\, c(\vec{x}, t) \equiv \left\{ \frac{\partial}{\partial t} + \left(\vec{u} - \mathcal{D} \frac{\partial}{\partial \vec{x}} \right) \cdot \frac{\partial}{\partial \vec{x}} \right\} c(\vec{x}, t) = 0 . \qquad (17)$$

Temperature, $T(\vec{x}, t)$, is also a scalar field with a transport described by the same equation, but with its own diffusivity, $\mathcal{D}_T$, i.e.,

$$\hat{\mathcal{L}}_{\mathcal{D}_T}(\vec{u})\, T(\vec{x}, t) \equiv \left\{ \frac{\partial}{\partial t} + \left(\vec{u} - \mathcal{D}_T \frac{\partial}{\partial \vec{x}} \right) \cdot \frac{\partial}{\partial \vec{x}} \right\} T(\vec{x}, t) = 0 . \qquad (18)$$

The convection-diffusion equation is linear, e.g., for a scalar species whose concentration field is given by $c(\vec{x}, t)$, we have $\hat{\mathcal{L}}_{\mathcal{D}}(\vec{u}) \neq \mathrm{fn}[c(\vec{x}, t)]$ and, similarly, for the temperature field. Given the evolution of the complete velocity field, $\vec{u}(\vec{x}, t)$, for a particular flow realization, and a set of initial and boundary conditions on the scalar field, it is instructive to consider the calculation, potentially after the fact, in which we compute the realization-specific evolution of a scalar field, $c(\vec{x}, t)$, as it would have been transported and diffused in the spatio-temporal environment defined by $\vec{u}(\vec{x}, t)$. Even in the case of a passive scalar, i.e., if the scalar field does not influence the velocity field and could be computed after it, however, this would still be a task of some complexity.

In fully-developed turbulence, constant-c (iso)surfaces are characterized by a very large growth rate in area, as well as a rapid increase in geometric complexity. Furthermore, in the case of a convecting-diffusing field characterized by a *high Schmidt number* ((4)), i.e., $Sc \gg 1$, we have to contend with an additional difficulty. The Schmidt number is close to unity for gas-phase species diffusion and of the order of 10^3 for liquid-phase diffusion of moderate molecular-mass species. The smallest spatial scale of the scalar field, say, $\lambda_{\mathcal{D}}$, will be smaller than the smallest (viscous) scale, λ_ν, that characterizes the velocity (vorticity) field. In the case of heat diffusion (convection-diffusion of temperature), the quantity corresponding to the Schmidt number is the *Prandtl* number,

$$Pr \equiv \frac{\nu}{\mathcal{D}_T} , \qquad (19)$$

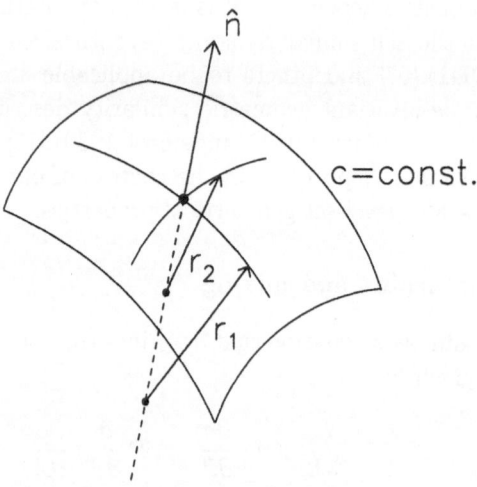

Fig. 7. — Isoscalar surface patch.

with $\mathcal{D}_T$ the diffusivity of temperature ((18)), is close to unity for heat diffusion in air, and close to 7 for water.

In the case of convection-diffusion of a high-Schmidt-number species in turbulence, one may expect (Batchelor 1959),[5]

$$\lambda_{\mathcal{D}} \approx \lambda_\nu / Sc^{1/2} \,, \tag{20}$$

where λ_ν is the viscous scale ((10)), and equivalently for temperature. By way of example, in the case of the diffusion of moderately large organic molecules in water, such as disodium fluorescein, which is often used in laser-induced fluorescence studies, $Sc \simeq 1900$ (Ware $et\ al.$ 1983) and $\lambda_{\mathcal{D}} \approx \lambda_\nu/44$. Smaller spatial scales would also render a numerical simulation temporally stiffer. The smallest Eulerian time scales associated with the scalar field would be smaller than the smallest Eulerian time scales of the convecting velocity field, by (very nearly) the same factor.

The convection-diffusion equation can be written with respect to a local coordinate system that is fixed on the scalar isosurface by noting that the only component of the local velocity that enters is the one selected by the direction of the scalar gradient. In particular, if,

$$\zeta(\vec{\mathbf{x}},\mathbf{t}) \equiv \frac{\partial}{\partial \vec{\mathbf{x}}} \mathbf{c}(\vec{\mathbf{x}},\mathbf{t}) = \hat{\vec{\mathbf{n}}}\,\frac{\partial \mathbf{c}}{\partial \mathbf{n}} = \hat{\vec{\mathbf{n}}}\,\zeta \,, \tag{21}$$

where $\hat{\vec{n}}$ is a local unit vector normal to the isosurface (in the direction of the scalar gradient), $\zeta(\vec{\mathbf{x}},\mathbf{t})$, and n is the arc length along the space-curve tangent to it, then,

$$\vec{u} \cdot \frac{\partial c}{\partial \vec{x}} = u_n \frac{\partial c}{\partial n} \,. \tag{22}$$

The scalar transport equation can then be written as, (Cf. (16)), *i.e.*,

$$\frac{\partial c}{\partial t} + \vec{u}_c \cdot \frac{\partial c}{\partial \vec{x}} = 0 , \tag{23}$$

where

$$\vec{u}_c = \vec{u} + \vec{u}_D , \tag{24}$$

and where $\vec{u}_D$ is the diffusive contribution to the isosurface transport velocity, given by,

$$\vec{u}_D = -\mathcal{D} \frac{\nabla^2 c}{|\partial c/\partial n|} \hat{\vec{n}} . \tag{25}$$

as noted by Gibson (1968).[45] More recently, Pope (1988)[86] reported on isosurface propagation, relating $\vec{u}_D$ to the local isosurface geometry (local curvature tensor), and Pope *et al.* (1989),[87] and Girimaji & Pope (1992)[48] pursuing its relation to scalar transport using velocity fields derived from direct numerical simulation of isotropic turbulence. The analysis of isosurface propagation can be simplified further, however, as illustrated below.

The isoscalar transport velocity, $\vec{u}_c$ involves only projections along the local scalar gradient. This allows the scalar transport equation to be written as,

$$\frac{\partial c}{\partial t} + u_c \frac{\partial c}{\partial n} = 0 , \tag{26}$$

with

$$u_c = u_n + u_D , \tag{27}$$

and

$$u_D = -\mathcal{D} \frac{\nabla^2 c}{\partial c/\partial n} . \tag{28}$$

The preceding indicates that isoscalar surfaces are conserved on space curves along the local scalar gradient, in a frame moving along these paths given by,

$$n(t) = n(0) + \int_0^t u_c(t') \, dt' , \tag{29}$$

with a transport velocity, u_c, that is the sum of a convective and diffusive contribution, with the latter dependent on the local isosurface geometry. In particular, writing the Laplacian in terms of the local isosurface coordinates, we have,

$$\nabla^2 c = \frac{\partial}{\partial \vec{x}} \cdot \left(\hat{\vec{n}} \frac{\partial c}{\partial n} \right) = \left(\frac{\partial}{\partial \vec{x}} \cdot \hat{\vec{n}} \right) \frac{\partial c}{\partial n} + \frac{\partial^2 c}{\partial n^2} , \tag{30}$$

where,

$$\frac{\partial}{\partial \vec{x}} \cdot \hat{\vec{n}} = \kappa = \frac{1}{r_1} + \frac{1}{r_2} . \tag{31}$$

with κ the (invariant) local mean curvature.([6]) In this expression, r_1 and r_2 are the local principal radii of curvature of the scalar isosurface (or the radii of curvature

---

([6]) This result ((31)) was communicated to one of us (PD) by G. B. Whitham, some years ago (pvte. comm.). J. M. Ottino (1989, p. 36) offers a similar result, expressing it in terms of the in-surface divergence only.

along any orthogonal plane-pair cut). We then have,

$$\nabla^2 c = \kappa \frac{\partial c}{\partial n} + \frac{\partial^2 c}{\partial n^2} . \tag{32}$$

A sign choice for r_1 and r_2, as they enter in the expression for the mean curvature, κ, is implicit in this expression, depending on whether the curvature in the corresponding direction is convex in the direction $+\hat{n}$, of the scalar gradient, or in the direction $-\hat{n}$, i.e., opposite the scalar gradient. By way of example, for the geometry depicted in Fig. 7, both curvatures are convex in the $+\hat{n}$ direction and both signs are plus. For a saddle point, the two signs would be opposite and if the two radii of curvature are equal (in magnitude) the two terms would cancel and κ would be zero. These expressions allow us to express u_D ((28)) in terms of the diffusivity and the local scalar-field geometry, i.e. (in the nondegenerate case),

$$u_D = -D \left(\kappa + \frac{1}{\zeta} \frac{\partial \zeta}{\partial n} \right) , \tag{33}$$

with $\zeta = \partial c / \partial n$ ((21)). Combining these relations leads to a scalar-transport equation that can be written as (Cf. (26)),

$$\frac{\partial c}{\partial t} + \left[u_n - D \left(\kappa + \frac{1}{\zeta} \frac{\partial \zeta}{\partial n} \right) \right] \frac{\partial c}{\partial n} = 0 . \tag{34}$$

This formulation must be amended at points where the local scalar gradient (instantaneously) vanishes (Gibson 1968).[45] Such points are singular, in this sense,([7]) with the scalar transport equation reducing to a local simple diffusion equation, i.e.,

$$\frac{\partial c}{\partial t} = D \frac{\partial}{\partial \vec{x}} \cdot \frac{\partial c}{\partial \vec{x}} , \qquad \text{wherever} \qquad \frac{\partial c}{\partial \vec{x}} = 0 . \tag{35}$$

It is interesting to scale, u_D, the diffusive contribution to the isosurface transport velocity. Accepting the diffusion scale, λ_D ((20)), as the appropriate scaling length along the scalar gradient arc-length, n, we have ((33)),

$$\frac{u_D}{u'} = -\frac{D}{u' \lambda_D} \left(\lambda_D \kappa + \frac{\lambda_D}{\zeta} \frac{\partial \zeta}{\partial n} \right) . \tag{36}$$

The prefactor is the reciprocal of a Peclet number based on the flow diffusion-scale. This can be estimated as,

$$\frac{u' \lambda_D}{D} \sim \left(\frac{u'}{U} \right) Re^{1/4} Sc^{1/2} . \tag{37}$$

For fully-developed turbulent flow, the ratio, u'/U, of the rms velocity to the outer velocity scale, can be taken as Re-independent, to an excellent approximation. Consequently, this contribution can be seen to be only weakly dependent on the Reynolds number and is lower in liquid-phase scalar transport (high Sc), for example, than in gas-phase scalar transport.

([7]) We should note, however, that they are likely to be (very) rare in a three-dimensional scalar field.

We see that local isosurface curvature, alone, contributes to isosurface transport, even in the absence of a local convecting field and a vanishing second derivative of the scalar field in the direction normal to the isosurface, *i.e.*, wherever $\partial^2 c/\partial n^2 = \partial \zeta/\partial n = 0$. It can also lead to local counter-gradient scalar transport. Conversely, in regions where isosurface curvature can be ignored, *i.e.*, if r_1 and r_2 are large, or, more precisely, if,

$$|\kappa| = \left| \frac{1}{r_1} + \frac{1}{r_2} \right| \ll \left| \frac{1}{\zeta} \frac{\partial \zeta}{\partial n} \right| , \qquad (38)$$

a locally-flat approximation can be employed. This simplifies the local convection-diffusion equation to,

$$\frac{\partial c}{\partial t} + u_n \frac{\partial c}{\partial n} \simeq \mathcal{D} \frac{\partial^2 c}{\partial n^2} . \qquad (39)$$

If diffusion occurs in a thin layer straddling, say, the $n = 0$ point along the scalar-gradient path and a local Taylor-expansion in employed, we may write,

$$u_n = u_0 + n \left(\frac{\partial u_n}{\partial n} \right)_{n=0} + O(n^2) \simeq u_0 - \sigma n , \qquad (40)$$

where $\sigma = -(\partial u_n/\partial n)_{n=0}$ is the local strain rate on the $n = 0$ (iso)surface, provides an adequate approximation for the local velocity component along the scalar gradient, the convection-diffusion equation can be simplified further to yield, locally, in a frame moving with u_0,

$$\frac{\partial c}{\partial t} - \sigma n \frac{\partial c}{\partial n} \simeq \mathcal{D} \frac{\partial^2 c}{\partial n^2} . \qquad (41)$$

This equation possesses solutions in closed form (*e.g.*, Carrier *et al.* 1975;[15] Karagozia & Marble 1986)[55] These were used in a model that estimated chemical-product formation in turbulent shear-layers, in the limit of high Damköhler numbers, *i.e.*, diffusion-limited reactions, in high Reynolds number flows, in which high local strain rates may be expected, with the preponderant fraction of molecular diffusion also expected to occur in thin diffusion layers (Dimotakis 1987).[33]

As an aside, it would appear that only the velocity component in the direction of the scalar gradient, *i.e.*, $u_n = \hat{n} \cdot \vec{u}$, enters in the scalar-field transport ((16)). This is not correct, however, as can be seen by writing the equation for the transport of the scalar gradient, *i.e.*,

$$\frac{\partial \zeta}{\partial t} + \vec{u} \cdot \frac{\partial \zeta}{\partial \vec{x}} = \omega \times \zeta - \left(\zeta \cdot \frac{\partial}{\partial \vec{x}} \right) \vec{u} + \mathcal{D} \frac{\partial}{\partial \vec{x}} \cdot \frac{\partial \zeta}{\partial \vec{x}} , \qquad (42)$$

where $\omega = \omega(\vec{x}, \mathbf{t})$ is the local vorticity field. The convected scalar gradient is rotated by the local vorticity; is amplified by a favorable local strain rate, with a term akin to the stretching term in the vorticity-transport equation; and is diffused. See also equivalent formulation in the article by P. Haynes, in this series, who used tensor notation to express the transport equation.

The preceding suggests that the local transport and mixing of a scalar field, $c(\vec{x}, t)$, can be described in terms of the evolution of the c-isosurfaces and, in turn, their local geometrical properties. While such a description also requires knowledge of the convecting velocity field, the geometrical description of the isosurfaces and, in particular, such properties as the local surface-to-volume ratio distribution, distribution of spatial scales that measure the distance to the isosurface, local-curvature distribution, etc., provide essential information, from this point of view.

The geometry of isosurfaces also arises, in a natural way, in the context of mixing in chemically-reacting systems and combustion. In the case of non-premixed, hydrocarbon diffusion flames in the high Damköhler number (diffusion-limited) regime, for example, high-activation-energy chemical kinetics confine the instantaneous burning to the unsteady, three-dimensional, stoichiometric isosurface (Burke & Schumann 1928). [14] While premixed combustion is more complicated, it can be cast in a similar language, with a burning surface that can be described as a constant-property surface. This surface, however, in addition to being propagated by the effects of the local convecting velocity field and diffusion, as in the case of a passive scalar discussed above, advances perpendicular to itself into regions of the flow occupied by unburnt reactant mixture with its own (laminar-flame) Lagrangian speed, as augmented and modified by the local, unsteady-flow, composition, and radiation environment.

2. Fractals

Characterizations of the geometry of isocontours and isosurfaces, *i.e.*, constant-property (level) sets, in turbulence and other phenomena in which complex structures are encountered, require an extension of the notions of conventional geometry. If regarded as possessing a complexity that extends over the whole range of scales, *e.g.*, the two Koch islands in two dimensions (Cf. Figs. 8,9), such objects are not *rectifiable*, *i.e.*, they are not differentiable and do not possess a finite arc length (*e.g.*, Mandelbrot 1977, p. 27).[73] They may, however, be analyzed using the language of *fractals*, an extension to conventional geometry proposed by Mandelbrot to apply to a host of natural phenomena (Mandelbrot 1967,[70] 1975a,[71] 1975b,[72] 1977,[73] and 1982).[74]

In this extension, fractional (non-integer) dimensions can be used to describe the geometrical object. The word *fractal* was coined by Mandelbrot (1975a,[71] 1975b)[72] from the Latin *fractus*, meaning "fragmented," and denotes objects that are (Mandelbrot 1989),[75] "... rough and fragmented to the same degree at all scales."

An isoelevation contour in a topographic map, or the special case of the sea-level contour, *i.e.*, a coastline, illustrates the point, as Mandelbrot noted by the question (Mandelbrot 1967, 1977)[70][73] "How long is the coastline of Britain?" expanding on a discussion by L. F. Richardson (published posthumously in 1961)[96] on the length of various coastlines. A moment's reflection will reveal that the length of a cartographer's coastline will depend on the resolution (scale) at which the coastline is rendered.

Following Richardson, one way the question may be addressed is by fixing the resolution to which the curve (or surface) is represented and then measuring the arc length (or surface area) with a commensurate yardstick, or step size, keeping track of the coastline length as a function of the step size. In particular, one can lay a ruler, open a pair of dividers, stretch a string, etc., of length, say, λ, from a point on the curve and mark the point of its intersection with the curve at the other end. That point becomes the beginning of the next step, etc. A measure of the arc length,

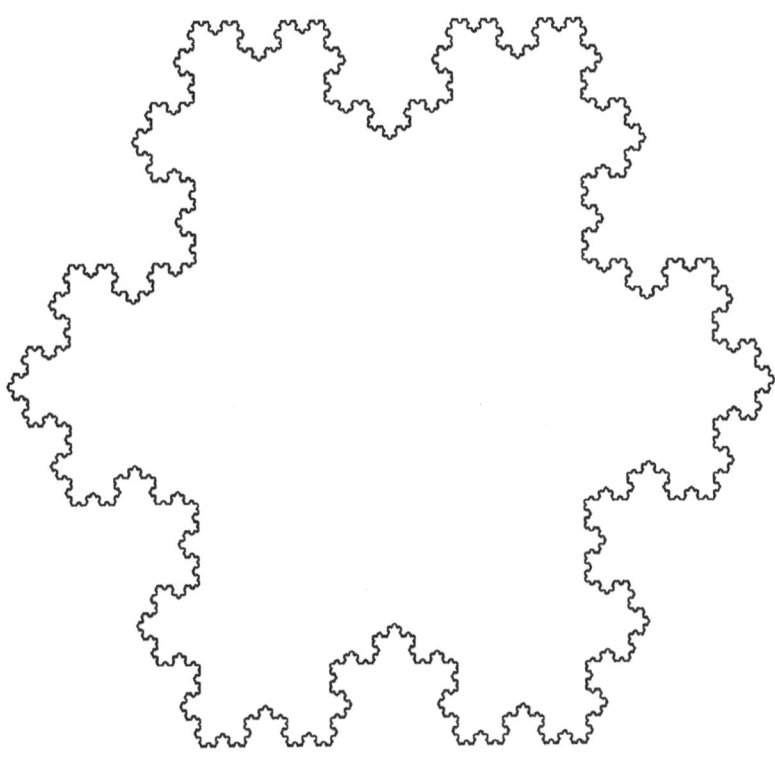

Fig. 8. — Contour of triadic Koch island with $D = 1.26$.

$L(\lambda)$, at this resolution, is then given by the product of the length of the string times the number of steps, $N(\lambda)$, required to traverse the coastline *i.e.*,

$$L(\lambda) = \lambda N(\lambda) \tag{43}$$

If the coastline may be regarded as a fractal object that is "rough and fragmented to the same degree at all scales," then the fractional increase in the number of steps required, per unit fractional increase in step size, will be a constant, *i.e.*,

$$\frac{dN/N}{d\lambda/\lambda} = -D , \tag{44}$$

and the number of λ-size steps required will follow a power-law relation,[8]

$$N(\lambda) \propto \lambda^{-D} \tag{45}$$

The (negative) exponent D is identified as the *fractal dimension* of the level set and can be seen to be the (negative) logarithmic derivative of the number of stretched-

[8] See also discussion by G. Zaslavsky, elsewhere in this series, where it is shown that such behavior occurs on phase-space island boundaries, where geometric-scale self-similarity is encountered.

Fig. 9. — A more complex Koch island with $D = 1.61$ (Mandelbrot 1977,[73] Plate 51).

string steps with respect to λ, *i.e.* ((44)),

$$D = -\frac{\mathrm{d}\log N}{\mathrm{d}\log \lambda} . \tag{46}$$

The corresponding measure of the length of the coastline can then be expressed as a function of the resolution (divider-point spacing, string-length, etc.) λ,

$$L(\lambda) = \lambda N(\lambda) \propto \lambda^{1-D} . \tag{47}$$

If the curve is a straight line, then $N(\lambda) \propto 1/\lambda$, $D = 1$, and $L(\lambda) \neq \mathrm{fn}(\lambda)$, as one would like. For a complex coastline, however, we will have the benefit of stepping across "bays" and "peninsulas", with increasing step size λ, and will require a number of steps that decreases faster than $1/\lambda$, as λ increases. If the relation between $N(\lambda)$ and λ is a power law, then the level set may be regarded as a fractal, with a fractal dimension as indicated in (46).

The western coastline of Britain was associated with an exponent for the coastline-vs-resolution (Cf. (47)) of $1 - D \simeq 0.25$, in L. F. Richardson's (1961) [96] data, corresponding to a fractal dimension of $D \simeq 1.25$. See Fig. 10, where Richardson's data on various coastlines are reproduced (from Mandelbrot 1977,[73] Plate 32). A value of $D \simeq 1.5$ was assigned by Feder (1988)[40] for Norway's fjords (Cf. Fig. 11).

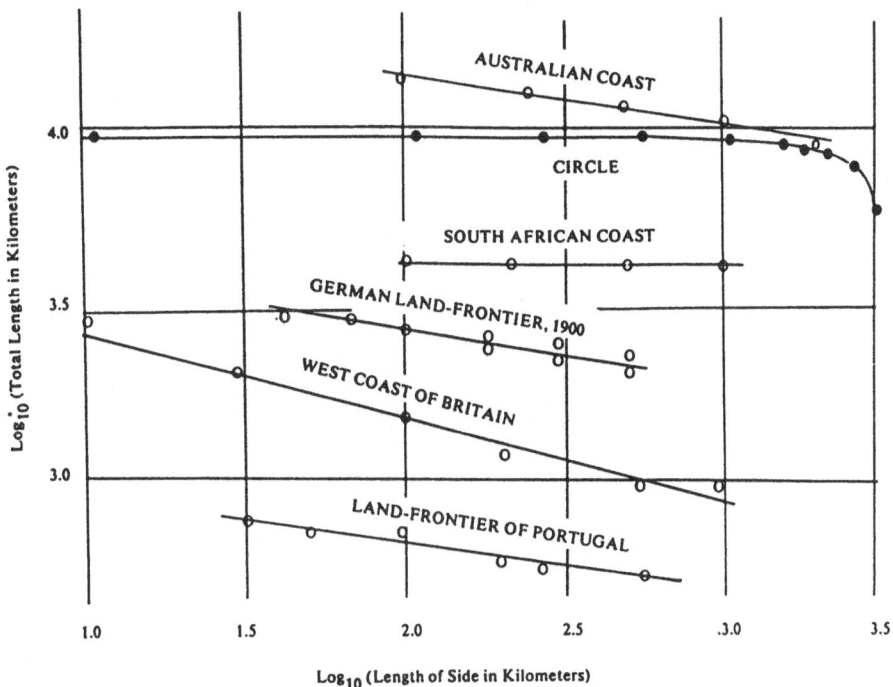

Fig. 10. — Richardson (1961) coastline data[96] (Mandelbrot 1977,[73] Plate 32).

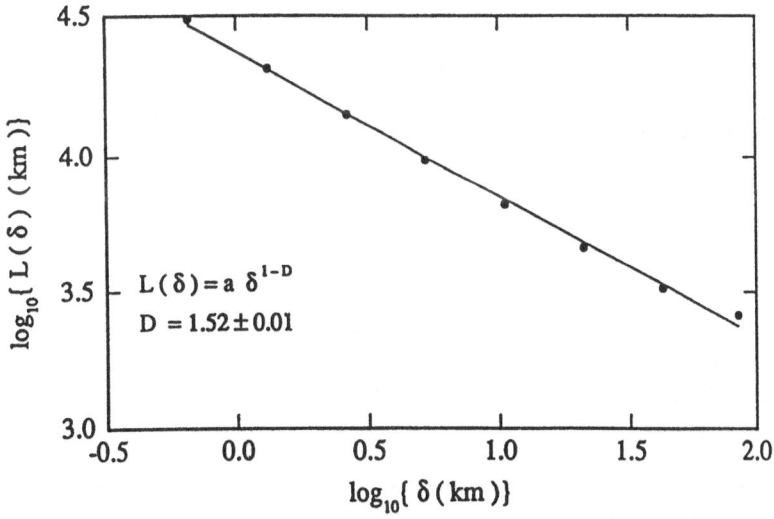

Fig. 11. — Coastline data for Norway's fjords (Feder 1988,[40] Fig. 2.2).

Generally speaking, the dimension of a fractal may be regarded as providing a measure of its geometrical complexity. A complex level set will offer the possibility

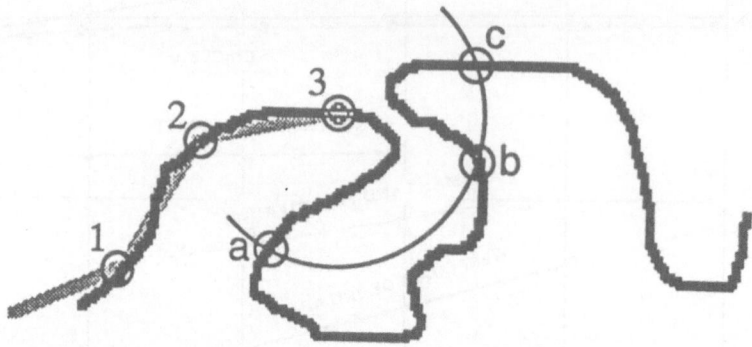

Fig. 12. — Possible λ-string endpoints along a two-dimensional level set (Russ 1994,[101] Fig. 2).

of many jumps across "bays" and "peninsulas" with a count $N(\lambda)$ that decreases rapidly as λ increases, *i.e.*, will have a large D. The higher fractal dimension of the Koch island in Fig. 9, than that of the triadic Koch Island depicted in Fig. 8, as well as the higher D assigned to the coastline in Norway's fjords, than that for the West of Britain, indicates an overall geometrical complexity for the former that is higher than the latter. Such a complexity measure would be important in our description of isosurfaces in turbulence. In the case of combustion on a stoichiometric surface, for example, a higher fractal dimension, D, could be expected to be generally associated with a higher burning rate per unit volume.

We pause, however, to note several difficulties with the "stretched-string" algorithm. In particular,

a. multiple end-point choices may exist for each step; the number of λ-steps, $N(\lambda)$, for a given level set, is not unique. See Fig. 12.

b. As λ increases, the stretched-string algorithm can result in a (negative) logarithmic derivative that exceeds 2, the dimension of the space the curve is embedded in and. for a curve embedded in a two-dimensional space, we would like the property, $D \leq 2$.

c. It is not clear how one completes the count as one goes around a closed curve; λ-steps are not likely to return to the starting point on a closed curve and for curves that exit the domain, an *ad hoc* decision must be made about how to count the exiting steps.

d. It is not clear how one deals with "islands" and "lakes" whose span is less than λ.

e. It is not clear how one extends the scheme to objects that live in three, or more, dimensions.

Some of these are noted by Mandelbrot (1977),[73] who discusses several variations, but suggests that the basic conclusions are robust and transcend these difficulties, with the fractal dimension, D, rather insensitive to these choices. This has not been the experience, however, in the application of these proposals to turbulence in the last decade-and-a-half, or so, at least at Caltech (Dimotakis 1996).[36]

Fortunately, these difficulties can be overcome by an algorithm based on the (coverage) count, $N_d(\lambda)$, of successive sub-partitions of the embedding space into λ-size boxes — segments, tiles, parallelepipeds, etc. — in d-dimensions that are required to cover the level set, as a function of their size, λ (Catrakis & Dimotakis 1996a).[18] For this purpose, the embedding space can be defined as the interior of the level-set bounding box. For a level set in two-dimensional space, for example, the bounding box can be defined as the smallest rectangle (or square, if need be) that contains the set (Tricot 1995).[114] If such a two-dimensional box has an extent λ_x and λ_y along the x and y directions, respectively, then the bounding box size, δ_b, can be defined as the square root of its area,

$$\delta_b \equiv A^{1/2} = \sqrt{\lambda_x \lambda_y} \, , \tag{48}$$

the geometric mean of the two sides. For a square bounding box, δ_b is the length of its edge. In d-dimensions, the size of the bounding box can be defined as the dth root of its d-dimensional volume, i.e.,

$$\delta_b \equiv V^{1/d} \, . \tag{49}$$

This can be illustrated on some isoscalar data derived from a slice in the far field of a turbulent jet (Catrakis 1996),[17] along a plane perpendicular to the jet axis (Cf. Fig. 13), at a jet Reynolds number of $Re = 4.5 \times 10^3$. The bounding box for an isoscalar (constant jet-fluid concentration) contour is shown as the shaded region in Fig. 14. This shaded region may also be considered as the largest tile required to cover the level set, corresponding to a level-set coverage count of $N_2(\lambda) = 1$, for $\lambda = \delta_b$.

We now cut each direction of the bounding box in half, partitioning it into four equal tiles (Fig. 15), corresponding to $\lambda = \delta_b/2$. The jet-fluid concentration level set contours visit all four regions covered by the $(\delta_b/2)$-size tiles and, therefore, $N_2(\lambda) = 4$, for $\lambda = \delta_b/2$. If we were to seek a power-law relation, $N_2(\lambda) \propto \lambda^{-D}$ ((45)), between the two-dimensional coverage count, $N_2(\lambda)$, and the tile size λ, we will have $D = 2$, at least in this large-tile-size range, since cutting the largest λ-tiles in half quadruples the two-dimensional coverage count, $N_2(\lambda)$. The resulting $D = 2$ value for these large tiles can be traced to the fact that all the boxes (tiles) required to cover the two-dimensional embedding space are also required to cover the level set that lives in it. That these fractal-dimension values are derived by counting the number of contiguous, non-overlapping tiles (boxes) required to cover the set (that also cover the embedding space exactly once), ensures that they are bounded by the dimension of the embedding space, i.e.,

$$D_d \leq d \, . \tag{50}$$

Further into the sequence, Fig. 16 depicts the same bounding box partitioned into a total of $N_{2,\text{tot}} = 2^{10} = 1024$ tiles of size $\lambda = \delta_b/2^5$. Counting the number of tiles

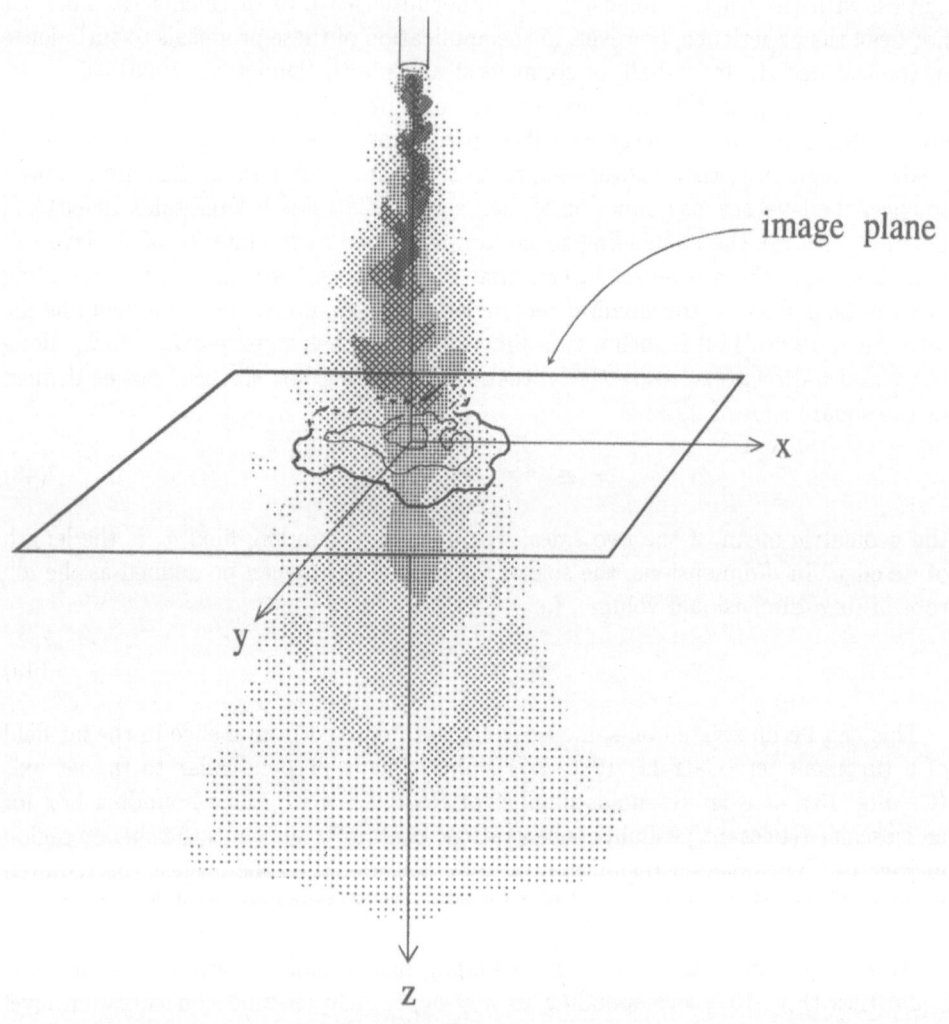

Fig. 13. — Turbulent-jet far-field geometry in relation to imaged scalar-field (jet-fluid concentration) plane.

required to cover the jet-fluid concentration level set, we now find $N_2(\lambda) = 580$. These are still rather large tiles, however, and we may prefer to go further into the binary embedding-space subpartition sequence, to smaller λ's yet, to estimate a fractal dimension from the coverage count at those scales.

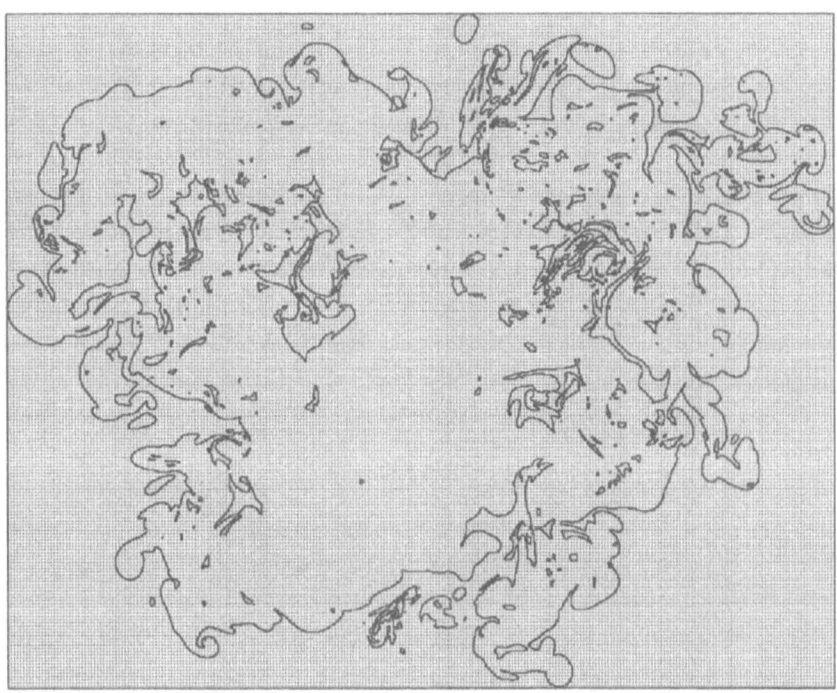

Fig. 14. — Jet-fluid isoconcentration (level) set in the far field ($z/d_j = 275$) of a $Re = 4.5 \times 10^4$ turbulent jet (Cf. Fig. 13). Shaded region indicates the level-set bounding box, of area $A = \lambda_x \lambda_y = \delta_b^2$; for $\lambda = \delta_b$, $N_2(\lambda) = 1$.

For a size $\lambda = \delta_b/2^7$, for which we have a total of $N_{2,\text{tot}}(\lambda) = 2^{14} = 16,384$ tiles, we find that $N_2(\lambda) = 4363$ are required to cover the level set. These are the ones indicated in Fig. 17. The next binary subpartition of the bounding box, in tiles of size $\lambda = \delta_b/2^8$, yields a total of $N_{2,\text{tot}}(\lambda) = 2^{16} = 65,536$ tiles and a coverage count of $N_2(\lambda) = 11,238$, as indicated in Fig. 18. If a power-law describes the relation of the coverage count, $N_2(\lambda)$, to the tile size, λ, at this intermediate range of scales (tile sizes), the fractal dimension would be given by ((46)),

$$D_2 \equiv -\frac{d \log N_2(\lambda)}{d \log \lambda} \simeq -\frac{\log N_2(\lambda_2) - \log N_2(\lambda_1)}{\log \lambda_2 - \log \lambda_1} = \frac{\log(11238/4363)}{\log(2)} = 1.36,$$
(51)

where the logarithmic derivative has been approximated here as a (first-order) finite-difference ratio. This is the slope of the straight line in the $\log N$ vs $\log \lambda$ plot and, for a D that is constant and if the statistical fluctuations in the corresponding coverage counts are negligible, this estimate would be, of course, exact.

We open a parenthesis here to note that in cases where the level set may be regarded as extending to infinity in all directions, possessing, for example, a spatially-homogeneous stochastic geometry, coverage statistics may be derived by partitioning (tiling, in two-dimensions) the interior of a δ-size box that covers a selected/measured

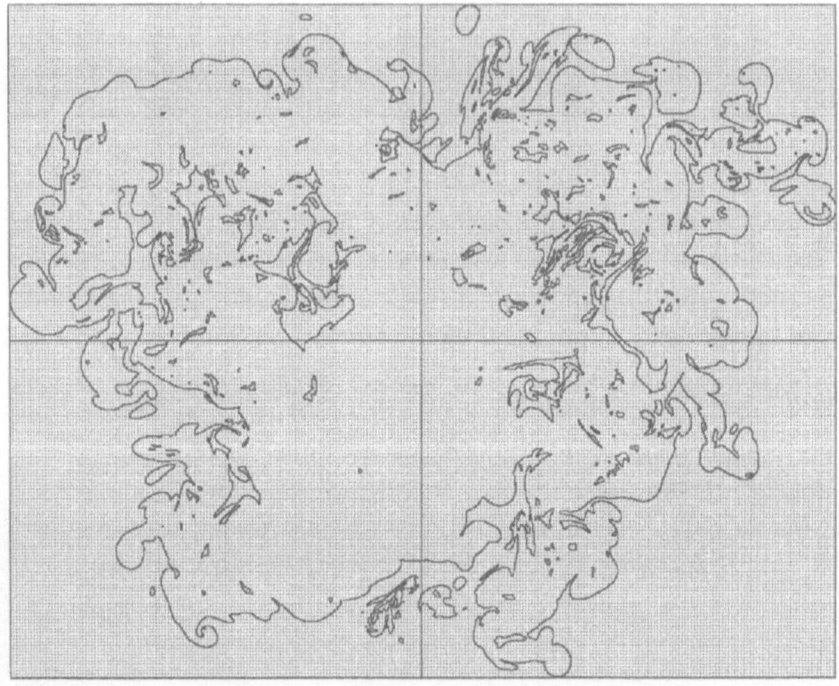

Fig. 15. — Legend as in Fig. 14. Bounding box partitioned into four tiles, all of which are required to cover the level set, i.e., for $\lambda = \delta_b/2$, $N_2(\lambda) = 4$.

portion of the level set. In that case, the level set dimension may, but need not, approach the embedding dimension (Cf. (50)), as $\lambda \to \delta$.

   Isoscalar surfaces, or contours, generated by turbulence are describable by a scalar-transport equation, as we've discussed ((16)). In such a process, the effects of diffusivity, however small, will ensure differentiability of the level set and a finite arc length (rectifiable curves), in two dimensions, or a finite surface area, for isosurfaces in three dimensions. Strictly speaking, it is clear, in other words, that level sets generated by turbulence cannot be "rough and fragmented to the same degree at all scales." For such level sets in two dimensions, as in our example, we expect that $D_2 \to 1$ as $\lambda \to 0$, i.e., the level-set dimension should attain a value equal to the topological dimension value of a smooth curve, as guaranteed by scalar diffusion, and, in general (Cf. Dimotakis 1991,[34] Catrakis & Dimotakis 1996a,[18] and (50) and related discussion),

$$\lim_{\lambda \to 0} D_d = d_t , \qquad (52)$$

where d_t is the topological dimension of the level set. This is also a property that one would also like a dimension to satisfy. For scalar data derived from turbulent flows, this limiting value for D may be expected to be approached for small λ, or, more precisely, for $\lambda \lesssim \max\{\lambda_\mathcal{D}, \lambda_{res}\}$, where $\lambda_\mathcal{D}$ is the scalar species diffusion scale ((20)) and λ_{res} is the resolution scale of the measurement (e.g., the pixel size, λ_p, of

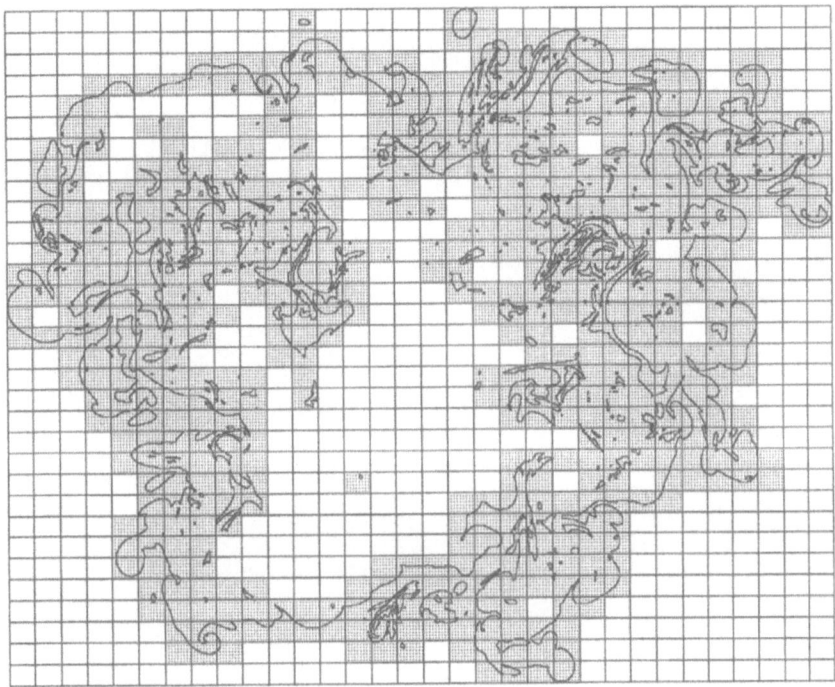

Fig. 16. — Bounding box partitioned into a total of $N_{2,\text{tot}}(\lambda) = 1024$ tiles of size $\lambda = \delta_b/32$. Of these, $N_2(\lambda) = 580$ are required to cover the level set.

a digital image, or the spatial extent of the measuring volume).

We should open a second parenthesis here to note that the limiting value indicated by (52) need not also be a lower bound. This may be illustrated by considering a level set comprised of a sparse cluster of small "islands" of, roughly, the same size, λ_0, such that $\lambda_0 \ll \delta_b$, sprinkled in the interior of a two-dimensional bounding box of size δ_b and separated by distances large compared to λ_0 (Fig. 19). To successive subdivisions of the bounding box, for a range of scales (tile sizes) $\lambda_0 \ll \lambda \ll \delta_b$, this sparse island archipelago will look like a collection of points, with a level-set dimension, D_2, that may be smaller than unity. For $\lambda < \lambda_0$, however, the islands will be covered by tiles, as is a coastline, yielding a level-set dimension that may be expected to be in the range, $1 < D_2 < 2$. Such multiscale sets can arise in the distribution of matter in clusters of galaxies, for example, as well as other natural phenomena.

To return, Mandelbrot's proposal that level sets generated by turbulence possess a coverage describable by a power-law relation must be interpreted as implying that there exists an intermediate range of scales of (tile) size λ, such that,

$$\lambda_D \lesssim \lambda_{\min} \ll \lambda \ll \lambda_{\max} \lesssim \delta_b , \qquad (53)$$

in which the fractal dimension, D, assumes an intermediate value (e.g., (51)) that

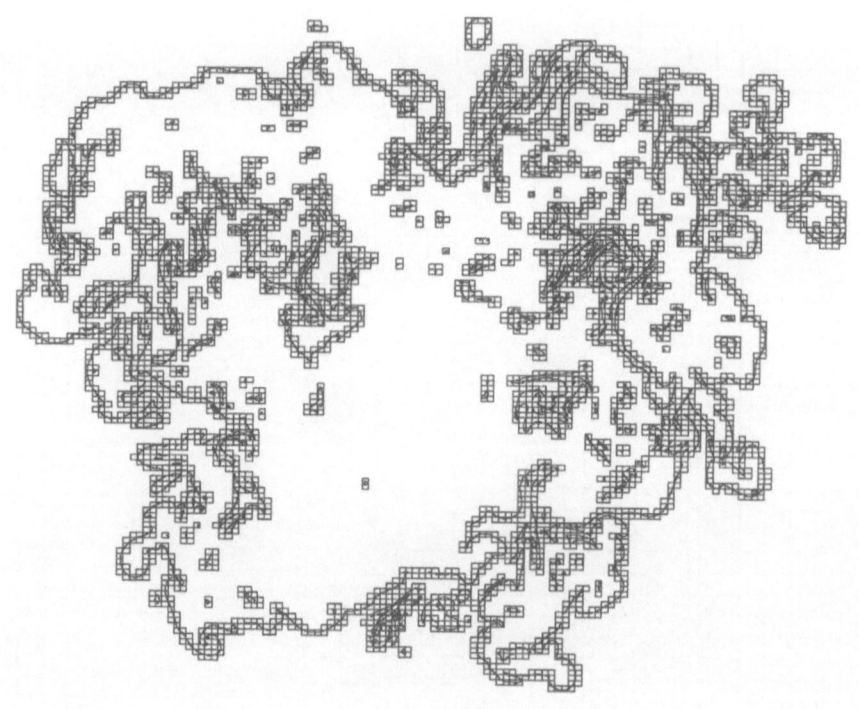

Fig. 17. — Bounding box partitioned into a total of $N_{2,\text{tot}} = 2^{14} = 16{,}384$ tiles of size $\lambda = \delta_b/2^7$, with a coverage count, $N_2(\lambda) = 4363$. Only tiles that cover are indicated.

may be accepted as a constant over this λ-range. To quote Falconer (1990, p. xxi),[39] the underlying point of view is that, for naturally-generated complex geometrical objects, "... over certain ranges of scale they appear very much like fractals and at such scales may usefully be regarded as such." The required self-similarity in behavior, for fractal scaling, at least over some restricted range of scales, as indicated in (53), for example, was suggested by Vicsek (1992)[116] as, "a property we can assume for all objects arising as a result of any physical process." In what follows, we will discuss these proposals in the context of turbulence, in particular, as well as the consequences as regards their capacity to describe natural phenomena, in general.

2.1. An overview of reports on fractals for turbulence

Welander, in 1955,[120] discussed (what would now be termed) fractal properties of model objects like Koch islands, which he called "snow-flake curves". In a discussion focused on scalar transport in two-dimensional, liquid-phase, unsteady flow, he analyzed the representation of interfacial arc length, which he recognized could, in principle, increase to very high values. Welander suggested that methods of analysis, reminiscent of today's fractal-geometry tools, could be applied to the description of "snow-flake" curves, as well as scalar interfaces advected in turbulence. Figure 20

Fig. 18. — Bounding box partitioned into a total of $N_{2,\text{tot}}(\lambda) = 2^{16} = 65{,}536$ tiles of size $\lambda = \delta_{\text{b}}/2^8$, with a coverage count, $N_2(\lambda) = 11238$. Only tiles that cover are indicated.

is reproduced from his paper, and depicts the deformation of a small (originally-square) dyed fluid region on the surface of a fluid, that he agitated and then set into rotation to maintain quasi-two-dimensional flow.

There probably exist other reports, as well, of discussions with similar proposals to apply fractal ideas. There can be no question, however, that it is Mandelbrot who must be credited for bringing the subject of fractals to attention, proposing that they can be expected to find application and that they constitute the natural formalism for the description of the geometrical properties of level sets generated in many natural phenomena. In the case of turbulence (Mandelbrot 1975a),[71] he noted,

"turbulent shapes ... almost cry out for proper geometrical description,"

and, in his book on fractals (Mandelbrot 1977, p. 146),[73]

"... we have already noted that sophisticated stochastic geometry is not used in any science, and includes the specific random surfaces and lines of turbulence. It is a pity. I believe that more imaginative stochastic geometry would be helpful in describing the truly disordered aspects of turbulence. Stochastic fractal geometry should be of particular interest."

Fig. 19. — Sparse archipelago of Poisson-distributed, randomly-oriented, uniform-size (triadic) Koch islands.

For isoscalar surfaces in three-dimensional homogeneous turbulence, he argued for a fractal dimension,

$$D_3 = 3 - 1/3 = 8/3 , \qquad (54)$$

if turbulence could be described as possessing "Kolmogorov-Gauss" scaling, a value he obtained by analyzing random, Gaussian fields with Kolmogorov variance (Mandelbrot 1975a, [71] 1975b,[72] and 1977),[73] and a fractal dimension,

$$D_3 = 3 - 1/2 = 2.5 , \qquad (55)$$

for "Gauss-Burgers turbulence" (Mandelbrot 1977).[73]

Mandelbrot also suggested that intermittency in turbulence is a direct consequence of the fractal nature of the geometry of regions in the flow where kinetic energy dissipation occurs, offering an alternative interpretation to the phenomenon than the latter theories proposed by Kolmogorov (1962)[59] and, subsequently, others. See discussion in Monin & Yaglom (1975)[82] as well as the more recent review by Frisch (1975,[44] Ch. 8) of intermittency theories based on cascade, fractal, and multifractal models.

On this topic, comparing the contributions of this formalism to turbulence, vs (low-order) dynamical systems, Frisch (1995,[44] p. 205) notes,

Fig. 20. — Observed deformation of a small, dyed square fluid region on the surface of an agitated fluid (Welander 1955,[120] Fig. 3).

"Paradoxically, the multifractal character of fully developed turbulence is still a controversial matter, whereas attractors of chaotic dissipative dynamical systems are known, and sometimes proven rigorously, to be multifractal."

Intermittency and multifractal models deal with the *amplitude* distribution and support of the field of interest and are beyond the scope of the present discussion, which is focused on the stochastic geometrical properties of level sets of turbulence-generated scalar fields.

Following Mandelbrot's proposals, experimental measurements investigating fractal behavior of level sets in turbulence were first reported by Sreenivasan & Meneveau (1986),[107] for a variety of flows including shear layers, jets, wakes, and boundary layers. Sreenivasan & Meneveau accepted that level sets in turbulence, "can be expected to be fractal-like only in an intermediate range of scales ... bounded [from above] by scales comparable with or smaller than the large scale of the flow ... and [from below] by the Kolmogorov scale."[9] They reported a value of $D_1 = 0.32$,

---

[9] In the case of scalar fields, the lower bound would be associated with the (Batch-

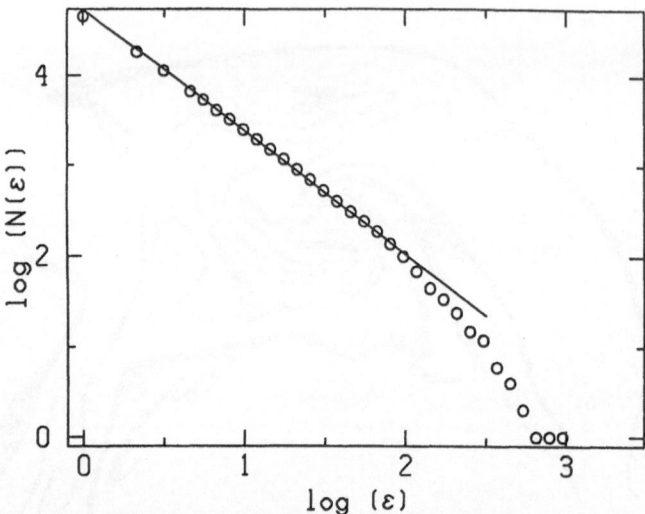

Fig. 21. — "Log-log plot of the number of boxes $N(\epsilon)$ containing the boundary [in a longitudinal section] vs the size of the box ϵ; from the straight line portion of the graph the fractal dimension is found to be 1.36" (square brackets ours). From Prasad & Sreenivasan (1989,[88] Fig. 10).

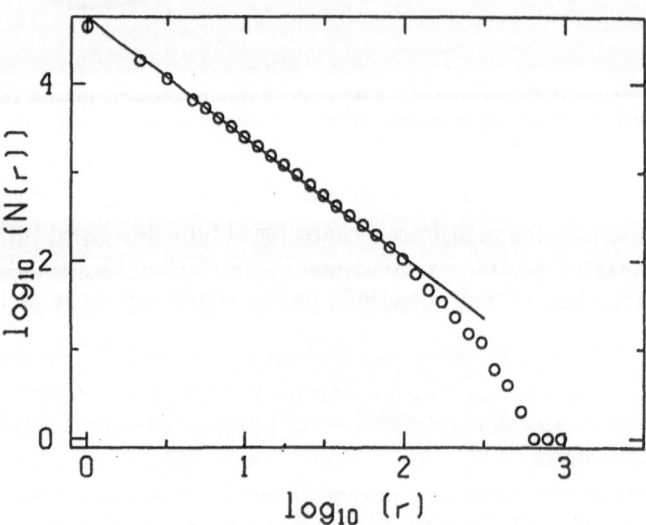

Fig. 22. — "The log-log plot of the number of boxes $N(r)$ containing the boundary in an orthogonal section ($z/d_j = 21$) [in our notation] versus the size r of the box. As before, the slope of the straight region in the plot gives the fractal dimension of the boundary. The value of the dimension is the same as that calculated from longitudinal sections" (square brackets ours). From Prasad & Sreenivasan (1990,[89] Fig. 12).

for one-dimensional (temporal) cuts of isoscalar data, and $D_2 = 1.33$, for two-dimensional measurements in a turbulent jet. Sreenivasan & Meneveau also accepted Mandelbrot's suggestion that, for isotropic level sets, a two-dimensional slice of a three-dimensional isosurface whose fractal dimension is given by D_3 would have a fractal dimension given by,

$$D_2 = D_3 - 1 , \tag{56}$$

and suggested that, therefore, three-dimensional isoscalar surfaces should be characterized by a fractal dimension, $D_3 = D_2 + 1 = 2.33$. In contrast to these findings, however, Sreenivasan & Meneveau pointed out, in the same paper, that long records (of the order of 500 large scales) of 1-D signals of the streamwise velocity in turbulent boundary layers do not appear fractal but, rather, exhibit statistics resembling those of random (Poisson) points on a line. On the whole, however, Sreenivasan & Meneveau concluded that "several aspects of turbulence can be described roughly by fractals."

Sreenivasan $et\ al.$ (1989)[108] reported a fractal dimension of $D_2 = 1.36$ for isoscalar jet data, with $D_2 = 1.35 \pm 0.05$ as a mean value for various turbulent shear flows, and offered arguments for a value of $D_3 = 7/3$ based on Reynolds-number similarity. Prasad & Sreenivasan (1989,[88] 1990)[89] analyzed two-dimensional cuts of turbulent jets along the mean flow direction ($i.e.$, including the jet axis), as well as transverse to the jet axis (as in Fig. 13, above). Figure 21 reproduces their coverage data for a $Re = 4000$ jet, identified as corresponding to their streamwise cut ($8 \leq z/d_j \leq 24$, in our notation), from which the,

> "... value of D ... turns out to be 1.36. Orthogonal sections also yielded 1.36 within the experimental error, showing the independence of the result on the orientation of the intersecting plane."

The latter can be ascertained from their data, reproduced in Fig. 22 from their 1990 report,[89] where they are identified as an example of coverage data derived from a transverse cut at $z/d_j = 21$ (in our notation), at the same Reynolds number. As can be seen by a comparison of the data in Figs. 21 and 22, the behavior is, indeed, very similar. Prasad & Sreenivasan (1990b)[90] also analyzed three-dimensional data of the isoscalar surfaces in turbulent jets, reporting a dimension of $D_3 = 2.35 \pm 0.04$.

A lack of (constant-D) fractal scaling was later noted by Sreenivasan (1991),[105] however, for isoscalar surfaces in the interior of the jet. Sreenivasan pointed out that, for level sets of concentration in turbulent jets at intermediate thresholds, "the log-log [coverage] plots are somewhat rounded," concluding that level sets "close to the mean concentration level ... are not good candidates for fractallike description" (Sreenivasan 1991,[105] p. 553). Sreenivasan (1991,[105] Fig. 6b) includes an illustration of this lack of fractal scaling, in the same paper.

A theoretical estimate for a dimension of $D_3 = 2.5$ was obtained by Constantin (1989,[25] 1990),[26] in accord with the Mandelbrot Gauss-Burgers turbulence model ((55)). An analysis of level-set geometry, based on the equations of motion, was put forth by Constantin $et\ al.$ (1991),[30] based on Constantin's 1989 work.[25] Surface area estimates of a scalar isosurface portion in a fluid element advected with the flow were obtained using the co-area rule of geometric-measure theory. A

---

elor 1959)[5] diffusion scale. For water (high-Schmidt-number fluid), for example, this is substantially smaller than the (Kolmogorov) viscous scale (Cf. (20)).

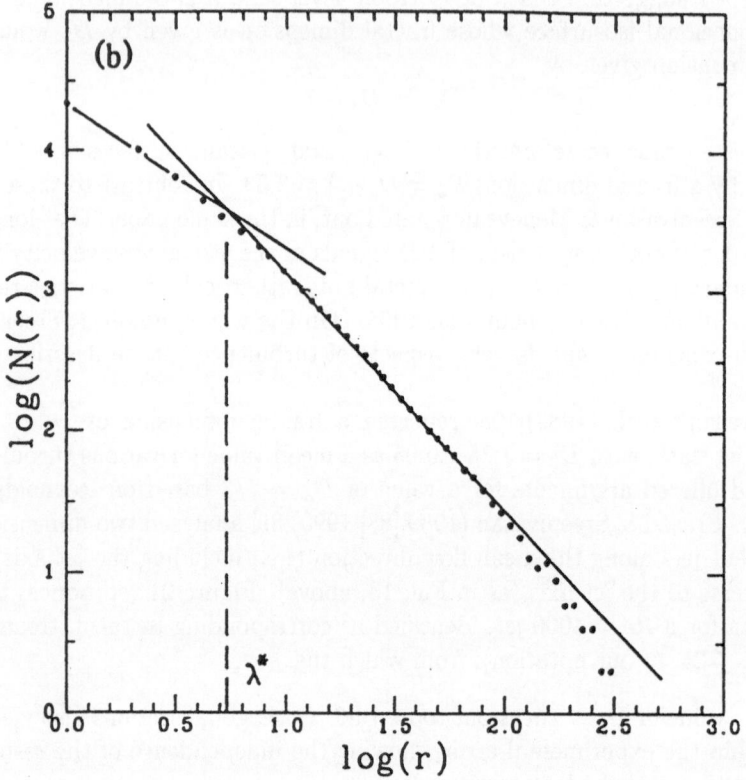

Fig. 23. — Theory of Constantin *et al.* (1991)[30] applied to experimental results for level sets of concentration in a turbulent jet.

comparison with the experimental data is reproduced in Fig. 23.([10]Constantin *et al.* (1991)[30] suggested a fractal dimension of $D_3 = 8/3$, for isoscalar surfaces in the jet interior, in accord with Mandelbrot's Kolmogorov-Gauss turbulence model ((54)), and $D_3 = 7/3$, for isoscalar surfaces near the jet boundary.

Procaccia *et al.* (1992),[93] pointed out, however, that "the theory [of Constantin *et al.* (1991)][30] cannot exclude the possibility that the scaling exponent, D, depends on [the scale] r" (the scale λ, in our notation — inserts in square brackets ours). Procaccia *et al.* (1992) analyzed isosurfaces of vorticity in three-dimensional homogeneous turbulence, using the direct-numerical-simulation data of Vincent & Meneguzzi (1991),[118] and concluded that, "... it is impossible to state with confidence that the [fractal] behavior [of the vorticity isosurfaces] is clear-cut."

Lane-Serff (1993)[64] reported a threshold-dependent fractal dimension, for iso-

---

([10]) It would appear that the data found in agreement with the theory are the same as the ones that yielded the coverage data in Sreenivasan (1991, Fig. 6b), that led to the conclusion of a "rounded" coverage plot, *i.e.*, a lack of "fractallike description" (recall discussion above).

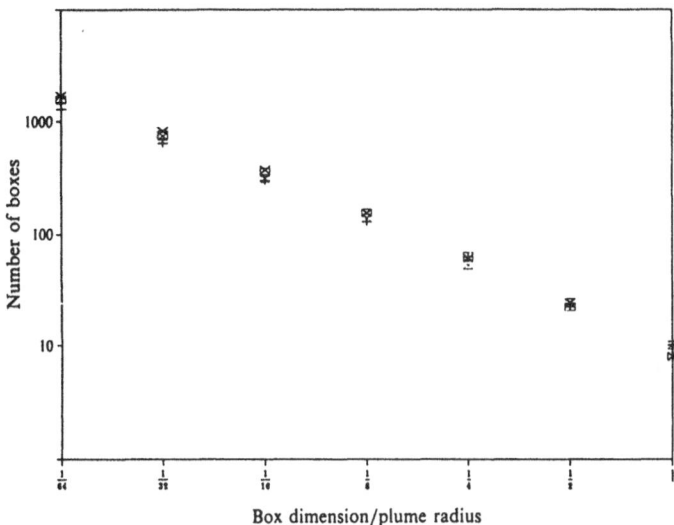

Fig. 24. — Coverage counts vs scale, normalized by local liquid-phase jet (plume) radius. Different symbols denote different image realizations. Systematic curvature in log-log plot noted by author (Lane-Serff 1993, Fig. 7).

scalar contours estimated from images in liquid-phase jet and plume flows. A minimum value of $D_2 = 1.23$, with respect to the threshold choice, was estimated by fitting a least-squares straight line to the coverage data. Figure 7 of the Lane-Serff paper is reproduced here as Fig. 24. As the author noted, "The data are close to a straight line but there is a distinct curve." *i.e.*, curvature, in the log-log coverage plots. This was attributed, however, to "... the small range between integral and Kolmogorov scales at the Reynolds numbers of [the] experiments." Praskovsky *et al.* (1993)[91] analyzed high-Reynolds-number turbulence level-set data of velocity, noting the absence of any wide range of scales characterized by a constant fractal dimension, but reported an average value of $D_1 \approx 0.4$.

The 1991 analysis by Constantin *et al.* [30] was revisited and expanded by Constantin (1994a,[27] 1994b)[28] and Constantin & Procaccia (1994),[29] with isosurface-area bounds obtained from the scalar transport equation, with Kolmogorov scaling for the advecting velocity field. In these analyses, as with the earlier discussion, it was assumed that level sets of the scalar fields were characterized by a constant fractal dimension. The 1994 update of this work identified the estimate of $D_3 = 8/3$ as a sharp estimate (rather than an upper bound), in the limit of $Sc \to \infty$.

Flohr & Olivari (1994)[42] analyzed isoscalar surfaces in gas-phase turbulent jets and reported, "constant scaling behavior over a wide range [of scales]," with a threshold-dependent fractal dimension exhibiting a maximum value.([11])For the outer isoscalar surfaces, they suggested that $D_2 = 1.30 \pm 0.05$. Their Fig. 7 is reproduced here as Fig. 25. Sreenivasan (1994)[106] suggested that $D_3 = 2.35 \pm 0.05$ for outer

---

([11]) Recall that Lane-Serff (1993), however, reported data with a *minimum* value of the fractal dimension, with respect to scalar threshold.

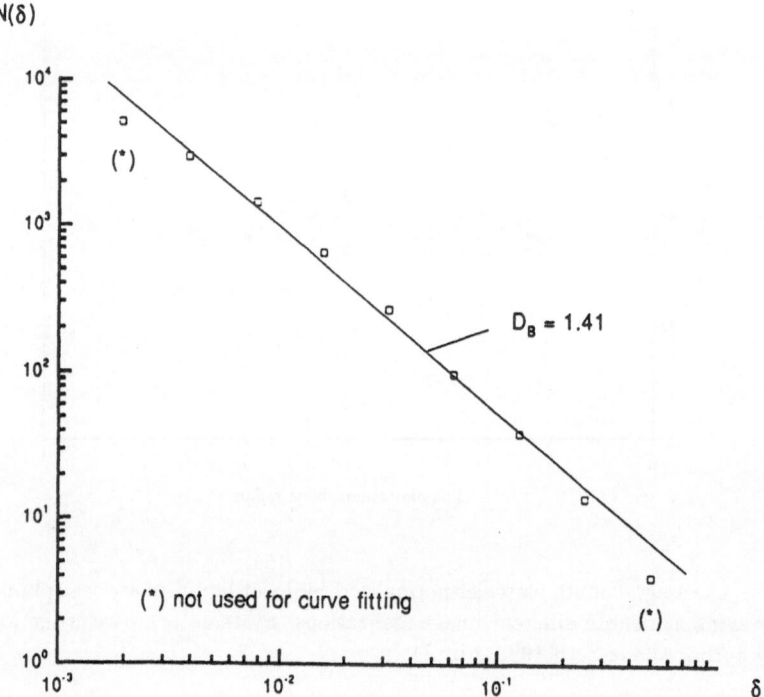

Fig. 25. — Isoscalar (smoke concentration) box-counting statistics derived from a $Re \simeq$ 2000 jet. Planar image normal to the jet axis at $z/d_j \simeq 15$ (Flohr & Olivari 1994, Fig. 7).

isoscalar surfaces in turbulent jets, with fractal scaling, "over much of the interval between the integral scale and the Kolmogorov scale," and a fractal dimension of $D_3 = 2.67 \pm 0.05$ for the inner isoscalar surfaces, in a scaling range stated to be "smaller" than for outer isosurfaces. He characterized the degree of confidence of these conclusions as, "fairly certain." More recently, Baldyga & Bourne (1995)[3] offered fractal and multifractal models of turbulent mixing, assuming constant-D behavior.

An analysis of the (temporal) evolution of a line element in grid-generated turbulence was reported by Villermaux & Gagne (1994).[117] Their measurements of the wrinkling of a smoke filament, shed from a thin wire downstream of a grid in a wind tunnel, were interpreted in terms of a fitted power-law coverage dimension that increased (linearly) with the distance downstream from a thin wire. Invoking Kolmogorov scaling, they proposed that the slope of the fractal dimension vs (Lagrangian) time was proportional to $Re^{1/2}$, reporting good agreement with their data.

The reports mentioned above offered theoretical and experimental evidence in support of constant-D descriptions of level sets arising in turbulence. Over the same period, several others, however, arrived at the opposite conclusion, also based on both theoretical and experimental work. Namely, that a constant-D coverage description is not applicable to level-sets in turbulent flows and in other contexts. These latter reports will be briefly discussed below.

Takayasu (1982),[111] argued on theoretical grounds that the dynamics of turbulent flow necessarily vary with scale. He suggested that descriptions of turbulent diffusion, in particular, may be expected to require coverage dimensions that are functions of scale and not constant, $i.e.$, in terms of the coverage, $N_d(\lambda)$,

$$D_d(\lambda) = - \frac{\mathrm{d}\log N_d(\lambda)}{\mathrm{d}\log \lambda} \tag{57}$$

of fluid-particle Lagrangian trajectories. Takayasu invoked Reynolds-number similarity to support his conjecture.([12])In his study of the geometry of the trajectory (as opposed to level set) of a one-dimensional random-walk particle as a function of step-size, λ, he employed a scale-dependent $D(\lambda)$, in our notation, and conjectured that such variable dimensions should be applicable to turbulent diffusion and other natural phenomena. He argued that, in random walks with a fixed step size, particle trajectories appear (inertially) correlated, when viewed at scales larger or smaller than the step size. G. I. Taylor (1921)[113] had also considered a modified random walk as a model of turbulent diffusion, in which he allowed for particles with inertia, $i.e.$, a random walk with correlated steps. See discussion by McComb (1991),[78] of G. I. Taylor's proposal. Takayasu used a real-space renormalization argument to derive an expression for the successive (spatial) coverage of a one-dimensional, random walk with finite step size. He initially used the term, "differential fractal dimension," later employing the term, "scale-dependent fractal dimension" (Takayasu 1992).[112] In 1993, Borgas[10] offered Lagrangian-statistic arguments for a scale-dependent $D(\lambda)$ for particle trajectories in turbulent flow and that, in particular, a power-law (constant-D) region should not be expected.

Regarding turbulence-generated level sets, Miller & Dimotakis (1991)[81] reported on high signal-to-noise-ratio measurements of the jet-fluid concentration in the far field of liquid-phase turbulent jets, conducted at various Reynolds numbers, with level sets computed for a range of scalar thresholds. A constant D was not found, for either one-dimensional temporal data at a fixed point, one-dimensional spatial data, or two-dimensional space-time data. Instead, their coverage statistics indicated a variable, scale-dependent $D(\lambda)$, which they computed as the local slope (logarithmic derivative) of the coverage ((57)). For a wide range of scalar thresholds straddling the local mean, the coverage dimension for one-dimensional temporal data varied smoothly from 0 to 1; for one-dimensional spatial data, from near 0 to near 1; and, for two-dimensional space-time data, from near 1 to near 2.

Sreenivasan (1991),[105] in a review that followed shortly, commented on the Miller & Dimotakis experimental findings, suggesting they could be attributable to differences between temporal and spatial data.([13])Kerstein (1991)[56] also commented on the Miller & Dimotakis, offering an alternate possible explanation for variable-D behavior.

([12]) As noted above, the principle of Reynolds-number similarity had also been invoked by Sreenivasan $et\ al.$ (1989),[108] to argue for a (constant) fractal dimension of level sets.

([13]) Some previous reports of constant-D findings, based on one-dimensional data, such as those by Sreenivasan & Meneveau (1986),[107] had also relied on temporal data. In addition to temporal data, variable-D behavior was also reported by Miller & Dimotakis (1991)[81] for (purely) spatial measurements.

A general discussion on constant-D behavior was offered by Dimotakis (1991),[34] on the basis of dimensional analysis and similarity. He suggested that power-law expressions, like (45), on which fractal similarity (constant-D) is based, imply a relatively strong dependence of the level-set geometry on some characteristic scale, in contrast to accepted notions. In the case of turbulence, it was argued that such a scale was unlikely to apply to fully-developed turbulent flows.

Gluckman *et al.* (1993),[49] in their experiments in thermal turbulence, reported that temperature isosurfaces do not display constant-D scaling, while species concentration isosurfaces show a limited range of "approximately-fractal" scaling. Recent experiments to investigate the geometry of scalar level sets were reported by Catrakis & Dimotakis (1996a),[18] who obtained two-dimensional, high signal-to-noise ratio, spatial measurements of the concentration field in the far field of liquid-phase, turbulent jets, in a plane perpendicular to the jet axis. These investigations were conducted for a range of scalar thresholds and values of the Reynolds number below, near, and above the estimated mixing-transition Reynolds-number threshold of $Re \approx 10^4$ (Cf. Sec. 1.1). The geometry of the jet-fluid concentration level sets were found to be described by a scale-dependent fractal dimension, $D_2(\lambda)$, increasing smoothly, with increasing scale, from unity, at small scales, to 2 at scales comparable to the local jet diameter (bounding box size).

As noted above, the reports summarized above, and others not listed here, fall in two categories. A set that is generally supportive of the (constant-D) fractal scaling proposals for turbulence-generated level sets and a set not supportive of these proposals. The reports that may be regarded as supportive, include experimental and theoretical analyses of level-set geometry derived from scalar, velocity, and vorticity fields in turbulence, extending the original proposals that fractal geometry may be applicable to isoscalar surfaces in turbulent flows.

Nevertheless, some of these generally supportive reports also cast some doubt on the notion of a power-law relation between the coverage count, $N(\lambda)$, versus the covering-box size, λ, *i.e.*, that a constant fractal dimension, D, exists over some significant range of scales. These doubts were either explicitly stated by the authors, or can be gleaned from the actual data presented and analyzed for the purpose, as suggested in the data and analysis reproduced in Figs. 21 and 22, as well the data in Fig. 23. The reader is invited to sight along the data and the fitted straight line in Figs. 21 and 22 to ascertain the systematic curvature in the coverage data. In particular, one can see a continuously-increasing slope, for most of the scale range, in the corresponding $\log N(\lambda)$ vs $\log \lambda$ plots. Secondly, a part of the experimental evidence that was accepted as leading to this conclusion was derived from relatively low Reynolds number flows, *i.e.*, not fully-developed turbulence (Cf. discussion in Sec. 1.1), for which, as was noted by some of the authors (*e.g.*, Lane-Serff 1993),[64] the various similarity laws and, presumably, (constant-D) fractal similarity would not be expected to apply.

The reports that concluded that the fractal dimension must be treated as a scale-dependent variable, *i.e.*, that $D_d = D_d(\lambda)$, also included theoretical and experimental work. Interestingly, the need to extend the (power-law) fractal framework first appeared in a theoretical analysis of random-walk processes (Takayasu 1982),[111] on which the conjecture that it should apply to other processes, such as turbulence, was made. To date, however, it has not proven possible to argue, theoretically, whether fully-developed (high-Re) turbulence is characterized by constant-D fractal geometry.

As noted above, the theoretical analyses that have dealt with the general issue have focused on fractal-dimension bounds and estimates, assuming constant-D fractal geometry. Recall, however, the negative statement by Dimotakis (1991),[34] who argued that if co-dimension values, $d - D_d$, are in the range proposed, constant-D fractal behavior would require a strong dependence on some characteristic scale, whose existence is unlikely in fully-developed turbulence. The analysis, by Procaccia *et al.* (1992),[93] of the isovorticity surfaces derived from the Vincent & Meneguzzi (1991)[118] direct numerical simulations, was inconclusive on this question and, in any event, based on simulations at a (Taylor) Reynolds number that may not have been high enough to settle the issue either way. It was left to experiments to address the problem.

2.2. Other reports of scale-dependent geometry

Suzuki (1984)[110] reported variable-D behavior for Japanese coastlines and discussed a scale-dependent Koch curve model. Suzuki used the term "transient fractal dimension" and "transient fractals" to describe scale-dependent geometrical properties. Mark & Aronson (1984)[76] reported similar behavior in the analysis of topographic surfaces and used the term, "scale-dependent fractal dimension."

Matsuura *et al.* (1986)[77] reported variable-D behavior in measurements of random walk (Brownian motion) trajectories of particles in solution, in accord with Takayasu's (1982) one-dimensional analysis, as shown in Fig. 26.

Chilés (1988)[22] studied fractured rocks, finding a continuously-varying, "local similarity dimension," as a function of scale, in both 1-D and 2-D surveys, as shown in Fig. 27. Chilés suggested several models for such scale-dependent fractal behavior, including scale-dependent Cantor dust.

In his study of biological tissues using microscopic biometry, Rigaut (1991)[97] also reported scale-dependent fractal behavior, as shown in Fig. 28. In particular, he found, in his words, a "drifting fractal dimension" with scale, and used the term "semi-fractals" as applicable to the analysis of the alveolar geometry of lungs of prematurely-born rabbits.

In characterizing the distribution of galaxies in the universe, Castagnoli & Provenzale (1991)[16] suggested that, "... it is probably necessary to consider models whose scaling and fractal properties vary with the spatial scale." Brandt *et al.* (1991)[11] reported in their analysis of solar granulation data, "... a smooth transition of the fractal dimension from small to large granules." More recently, Laherrère (1996)[63] reported scale-dependent geometry in the distribution of town sizes and Andrle (1996)[1] found similar behavior in the geometry of coastlines of fjord, volcanic, as well as tectonic origin.

2.3. Scale-dependent fractals

Before proceeding, we note that, in the original context of Mandelbrot, "scale-dependent fractal dimension" represents a contradiction in terms. In particular, if a fractal is to be regarded as, "rough and fragmented to the same degree at all scales," as Mandelbrot defined it, it must possess a constant (scale-independent) fractal dimension, as indicated in (44). In adopting the term "scale-dependent frac-

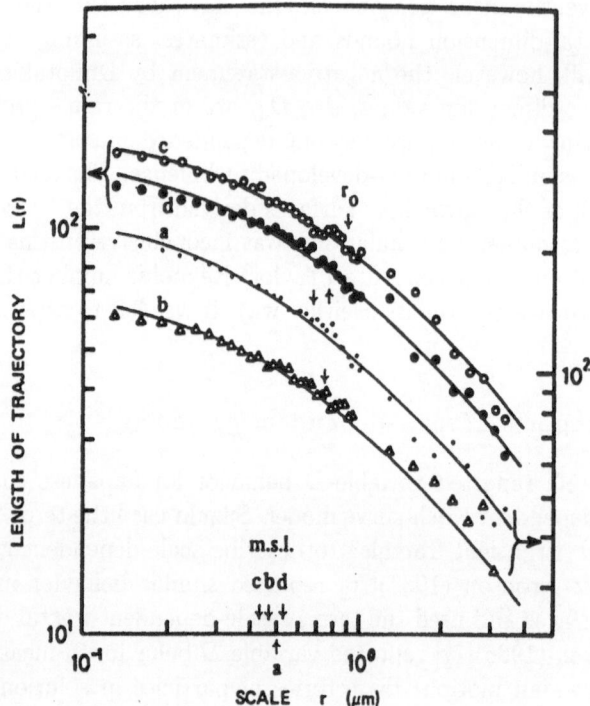

Fig. 26. — Length of random-walk (Brownian-motion) trajectory of particles in solution, as a function of scale. (a) polystyrene latex; (b) bacteriophage T4; (c) T4 particles with host *E. coli B* cells; (d) as in (c) with measurements over a shorter time; m.s.l. denotes mean segment length (step size). From Matsuura *et al.* (1986,[77] Fig. 1).

tal dimension", employed by Mark & Aronson (1984),[76] as well as by Takayasu (1992),[112] and the proposed extension, we recognize that the requirement of scale-independent properties has been relaxed. However, the proposal is to retain the term "fractal", even though scale-dependent geometry is accepted, in deference to Mandelbrot's (1975a,[71] 1975b)[72] adoption of the latin root, *fractus* which (only) means "fragmented," with no implied requirement for scale self-similarity.

The proposed extension allows a more flexible description of the geometry of level sets and has a utility that extends beyond the descriptive, as will be demonstrated in the discussions that follow. In any event, constant-D fractal behavior is not excluded and may naturally be incorporated as a special case. The proposed extended framework is referred to as "scale-dependent-fractal" (SDF) geometry and retains the notion of quantifying complex geometries in terms of fractional dimensions, while allowing for scale-dependent geometric behavior, with the original, scale-invariant, fractal geometry of level sets referred to as "power-law-fractal" (PLF) geometry (Catrakis & Dimotakis 1996a).[18]

As the chronology of the brief overview of the turbulence literature indicates, in this context, the earlier experimental work was, generally, supportive of the Man-

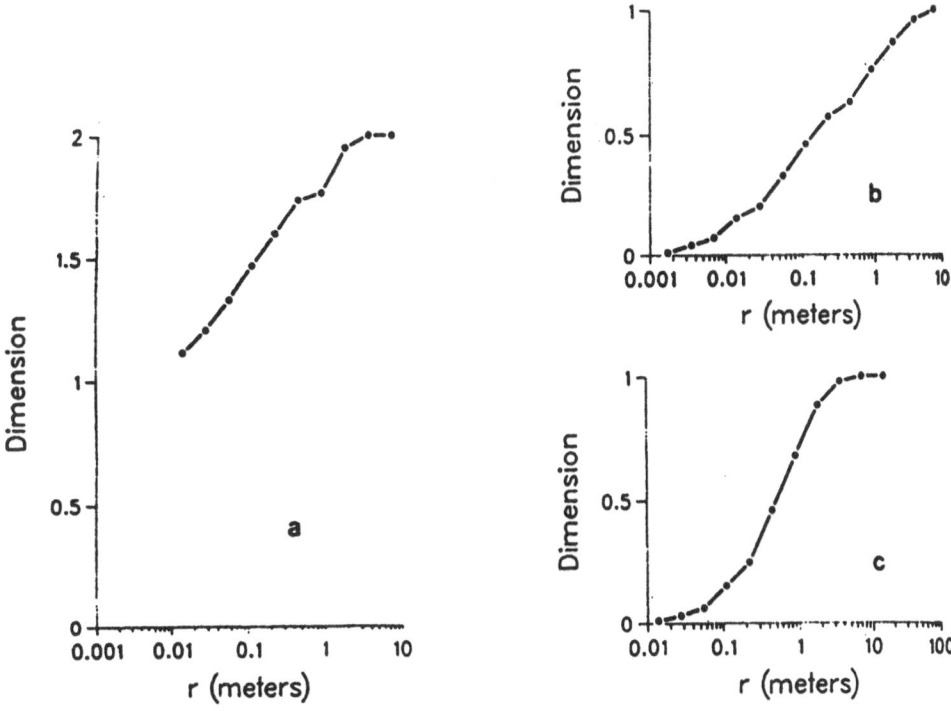

Fig. 27. — SDF dimension of the fracture network of rocks. (a) 2-D survey of a 10 m × 10m section; (b) 1-D survey through the 2-D section of (a); (c) 1-D survey of a 80 m × 2m section of a drift wall (Chilès 1988,[22] Fig. 6).

delbrot (constant-D) PLF scaling proposals, with the experimental work indicating a need for a scale-dependent fractal (SDF) description appearing later, for the most part. Coming in the wake of the first set, the latter work had the benefit of the earlier experience and, in seeking an explanation for the discrepancy, explored additional questions that could influence the results. These will be discussed below.

3. Scale-dependent geometry of isoscalars in turbulent jets

Shortly after Mandelbrot's (1977)[73] book, several investigations of fractal properties of level sets in turbulence were undertaken. Laser-induced-fluorescence images of turbulent jets (Dimotakis et al. 1983),[37] such as the one reproduced in van Dyke's book (1982,[115] p. 97), and others, were already available (Cf. Figs. 3,4. In the second edition of his book, Mandelbrot refers to Fig. 3 (1982,[74] p. 52b) as evidence corroborating fractal proposals for turbulence. Our own attempts at the time, however, to perform a fractal analysis on the lower-Reynolds-number images (e.g., Fig. 3), that had the more easily identified scalar interfaces, did not yield constant fractal dimensions. The difficulties of the "stretched-string" algorithm that had been employed (recall Fig. 12 and related discussion), in conjunction with the limited abil-

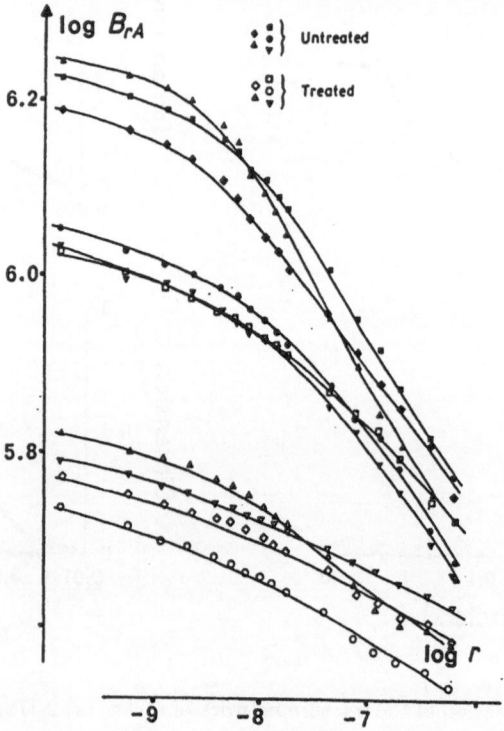

Fig. 28. — The perimeter, B_{rA}, in a reference area, A, of the outline of microscopic sections of the pulmonary alveoles of prematurely born rabbits, as a function of scale, r. Untreated (underdeveloped) as well as treated (with surfactant at birth) alveoles were analyzed (Rigaut 1991,[97] Fig. 14.1).

ity to acquire and analyze enough data for reliable statistics, discouraged us from sharing these early results, which were regarded as inconclusive. A few other groups had made similar informal attempts, some using the same laser-induced photos of the turbulent jet, who reached the same conclusion. Later efforts at Caltech, based on linear-CCD images also failed to provide conclusive results (Dimotakis 1996).[36]

The first reports by Sreenivasan and his group, and subsequently others, in support of Mandelbrot's fractal proposals for turbulent-flow-generated level sets, cited above, appeared shortly thereafter. However, in our opinion at least, some of the issues we were worried about had not been addressed.

To investigate the matter in a way that would settle some of these, an experiment was designed for the purpose: to produce data of adequate statistical confidence, have a large enough dynamic range of scales, permit a more reliable analysis of the data using coverage-based (rather than "stretched-string") algorithms, explore the effects of flow Reynolds number, investigate the dependence of the results on the

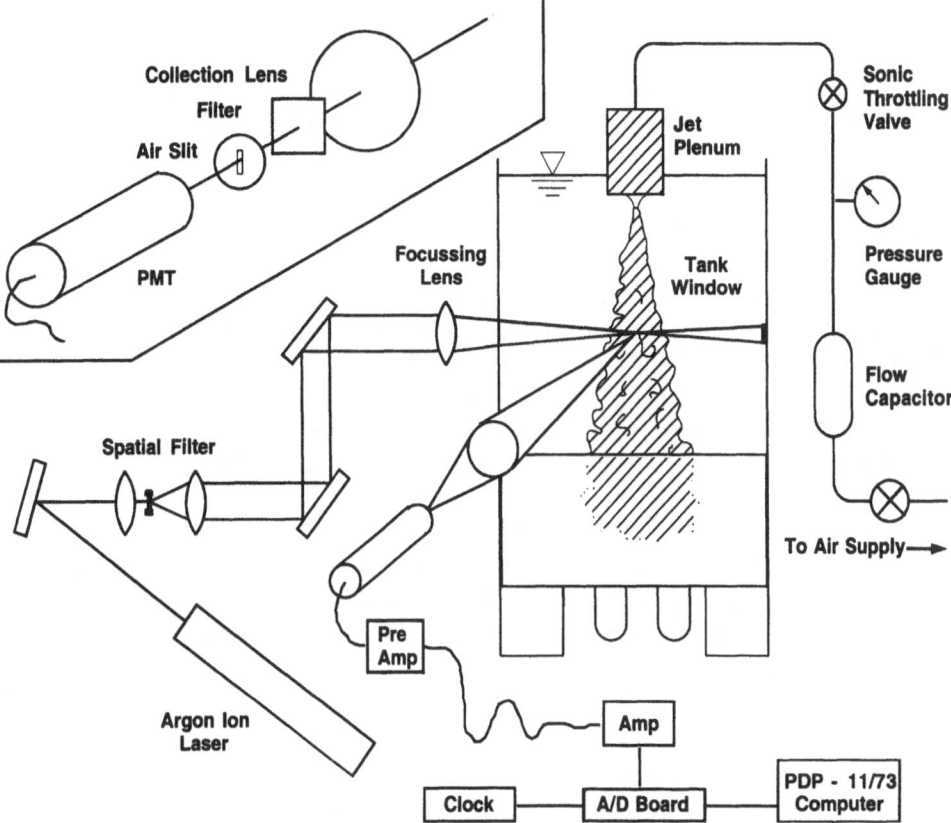

Fig. 29. — Schematic of turbulent-jet apparatus used for 1-D temporal and spatial scalar measurements (*ibid.* Fig. 1).

scalar threshold chosen to define the level set. It was also designed to address the important issue of fixed-level crossings by a signal with a finite signal-to-noise ratio. The experiments were reported by Miller & Dimotakis (1991)[81] and are discussed below.

3.1. 1-D temporal and spatial isoscalar behavior

The Miller & Dimotakis (1991)[81] experiments to investigate scalar level set behavior were carried in the facility schematically shown in Fig. 29. The discussion and figures in the brief account below are extracted, or derived, from their paper. A water tank with glass windows on all four sides, with an interior volume of about $1.1\,\mathrm{m}^3$, acted as the (discharge) reservoir that provided the (unlabeled) fluid scalar entrained by the jet. To establish the flow, air was sonically metered to drive the jet fluid at constant velocity through a $d_j = 2.54\,\mathrm{mm}$ (0.1 in) diameter nozzle at the base of the plenum, that was filled with water labeled with a fluorescent laser dye (disodium fluorescein), at a concentration less than 10^{-6} M.

For the single-point (one-dimensional) time-series measurements, an argon-ion laser beam was expanded, collimated, and aligned, crossing the jet centerline axis at right angles (radially) at the measuring station at $z/d_j = 100$. The optics were designed to avoid spurious reflections and generated a small Gaussian waist at the focus formed on the jet centerline. The laser beam power was maintained at 1.0 W to avoid heating of the dyed fluid in the very small focal volume and to prevent saturation (photo bleaching). The much lower concentrations of dye at $z/d_j = 100$ did not measurably attenuate or steer the beam. The resulting signal-to-noise ratio was 60 dB, as estimated on the basis of the scalar time series power spectrum (Cf. Fig. 30).

The single-point time-series measurements were made utilizing a low f# lens to collect light from the very short segment centered at the waist of the focused laser beam at the measuring station, through an optical low-pass filter to eliminate the laser-frequency (green) light, onto a photomultiplier tube, yielding single-point jet-fluid concentration values vs time. A slit spatial filter defined the length of the laser line segment sampled, with a width chosen such that the sampling volume was roughly cubic in shape, about 80 μm on a side.[14] The signal chain incorporated a three-pole Butterworth filter, with a cutoff frequency set slightly under 10 kHz. The data were intentionally oversampled in time, by some margin (at a sampling frequency of 20 kHz), as can be seen by the frequency where the white noise floor crossed the concentration spectrum. The same sampling frequency was used in all the runs, permitting the reliable definition of a Wiener-filter kernel, as will be explained below.

The Reynolds number was varied in these experiments, with measurements in the range, $2.9 \times 10^3 \leq Re \leq 23 \times 10^3$. Special consideration was given to the treatment of noise. Specifically, power spectra of the data were computed, along with the optimal (least mean-square error) Wiener filter in each case (Wiener 1949,[121] Press et al. 1986).[92] The data were then convolved with the Wiener kernel to obtain the optimally-estimated signal, consistent with the detection noise level in each run.

The Wiener-filtered data were thresholded and transitions, i.e., crossings at the selected scalar threshold value, were located. Figure 31 compares the level crossings, over a short time interval, of the raw (unfiltered) signal to those of the Wiener-filtered signal, at a scalar threshold equal to the local mean, $c = \bar{c}$. The choice of threshold is important and will be discussed below. As can be seen, the difference in both the number and spacing of level crossings between the two signals is very large. It should be recalled that it is the Wiener-filtered signal whose spectrum is concave downwards, as expected for a scalar time series, and is depicted as the dashed curve in Fig. 30.

The resulting level set, i.e., the record of transition locations, was then processed using a 1-D box-counting algorithm. The algorithm determines the number, or coverage count, $N_1(\lambda)$, of contiguous, constant-length segments, or "tiles", required to cover the transition locations on a record, as a function of the tile size, λ. The fractal dimension, $D_1(\lambda)$ is then computed as the logarithmic derivative ((57)) of the coverage count.

The log-log plots of $N_1(\lambda)$ and plots of the resulting $D_1(\lambda)$ for the single-point, on-axis concentration measurements in the jet, using a threshold value equal to the mean concentration, are shown in Figs. 32 and 33. The length scale, λ, was estimated by

---

([14]) A slit, rather than a pinhole, was employed so that any small beam movements in the vertical direction did not alter the size of the measurement volume (and, by extension, the intensity of the signal).

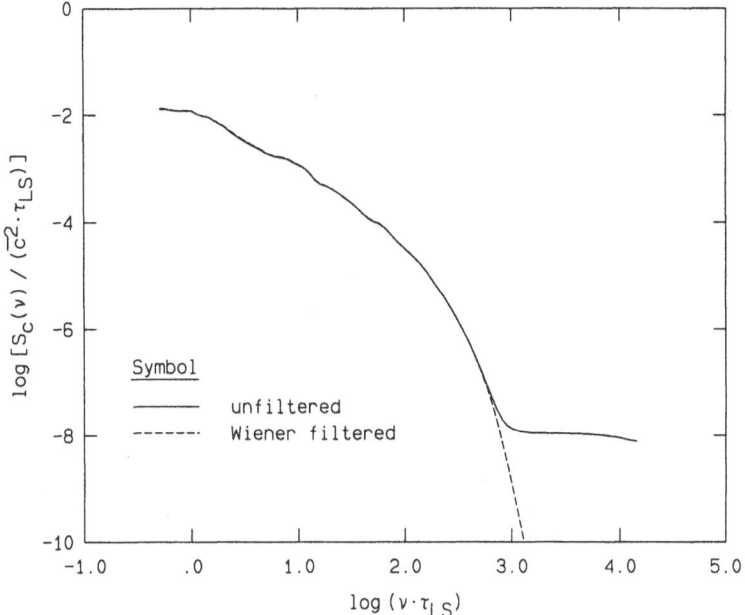

Fig. 30. — Temporal power spectra of unfiltered and Wiener-filtered data (Miller & Dimotakis 1991, Fig. 2).

multiplying the time interval with the calculated mean (centerline) velocity and has been expressed in absolute length (meters). The local jet diameter at the measuring location was estimated as equal to $\delta(x) \simeq 0.1\,\mathrm{m}$, *i.e.*, $\log_{10}(\lambda/\mathrm{m}) = -1.0$ (Cf. Fig. 33) corresponds to a scale $\lambda \approx \delta$.

If the level set possessed a constant fractal dimension, a horizontal portion of the curve would be observed in Fig. 33. As can be seen, no evidence of such a constant value is evident in the $D_1(\lambda)$ plots, other than the limiting values of 0 and 1 that can be argued for *a priori*. The observed continuous increase of $D_1(\lambda)$ occurs over a range of equivalent spatial scales from below the Kolmogorov scale, up to the outer large scales of the flow ($\log_{10} \lambda/\mathrm{m} \simeq 0.1$). The limiting value of 1 at the large tile sizes indicates that every tile of sufficient length covers (some) transitions. This is to be expected for scales on the order of the local jet diameter ($\delta(z) \simeq 11\,\mathrm{cm}$ for the data in Fig. 33, or larger, since over such a distance (or corresponding time) a crossing of the mean concentration level is almost certain. Failure to reach an asymptotic value of unity would indicate that either the data record was of insufficient length to capture the largest scales of the flow, or that the processing algorithm stopped at a tile size shorter than the largest scales. All four curves in Fig. 33 approach $D_1(\lambda) \simeq 0$, merging in the vicinity of the spatial resolution estimate, *i.e.*, $\lambda \simeq \lambda_{\mathrm{res}} \simeq 80\,\mu\mathrm{m}$, as noted above, for these measurements. The limiting behavior for $D_1(\lambda)$, for large and small λ, is consistent with what should be expected (Dimotakis 1991),[34] as noted above (recall Eqs. (50) and (52), and related discussion).

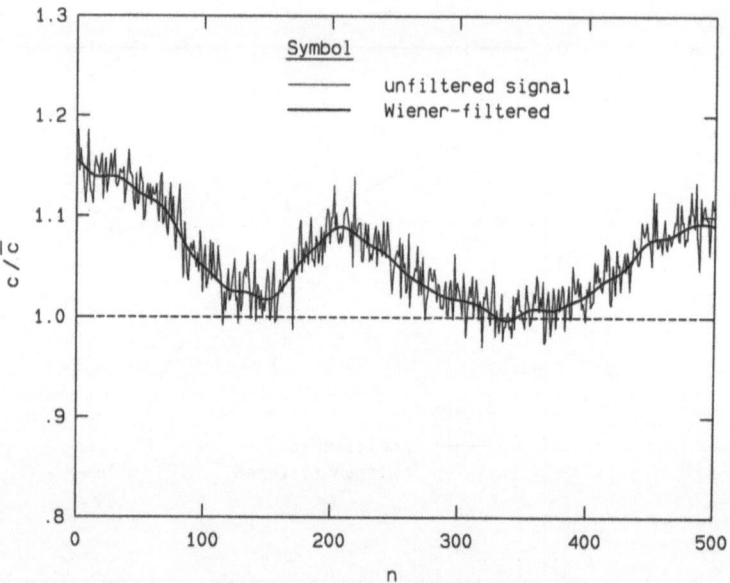

Fig. 31. — Effect of noise on threshold transitions. Data recorded for $Re = 2.9 \times 10^3$ (*ibid.* Fig. 15). See Fig. 30, above.

Measurements were also made off the jet centerline, on the rays $\eta = r/z = 0.06$ and 0.13, at a Reynolds number of 8600. The results, from level sets derived for the threshold choice, $c = \bar{c} = \mathrm{fn}(\eta)$, are shown in Fig. 34, along with the corresponding centerline curve, *i.e.*, $\eta = 0$. The similarity in the three results is noteworthy, despite the relatively large variation in the value of the threshold, $\bar{c}(\eta)$.

It had been suggested by Sreenivasan & Meneveau (1986),[107] that if long records are used in the box-counting algorithm, they may mask local power-law fractal behavior. Miller & Dimotakis (1991, Fig. 17 and related discussion) addressed this issue by processing a sequence of short records independently, and, aside from the impaired statistics, as expected from the shorter records, found a similar $D_1(\lambda)$ dependence on the scale, λ, as for the longer records.

Scalar-threshold effects were also addressed. A range of thresholds for the single-point data, both on-axis and off-axis at $\eta = 0.13$, were examined. The dependence of $D_1(\lambda)$ on the scalar-threshold is shown in Fig. 35. The continuous variation of the $D_1(\lambda)$ curves with λ can be seen to persist for a rather large range of threshold values on either side of the local mean concentration. As the threshold is either increased or decreased, two effects are observed. The sloping region shifts position, achieving the asymptotic value of 1 at progressively larger (time) scales, as would be expected for temporal data.([15]) Secondly, a bump appears at smaller values of λ. While the linear

---

([15]) The probability of encountering a transition within a certain tile size (time interval) decreases for scalar thresholds away from the mean, increasing the tile size required to

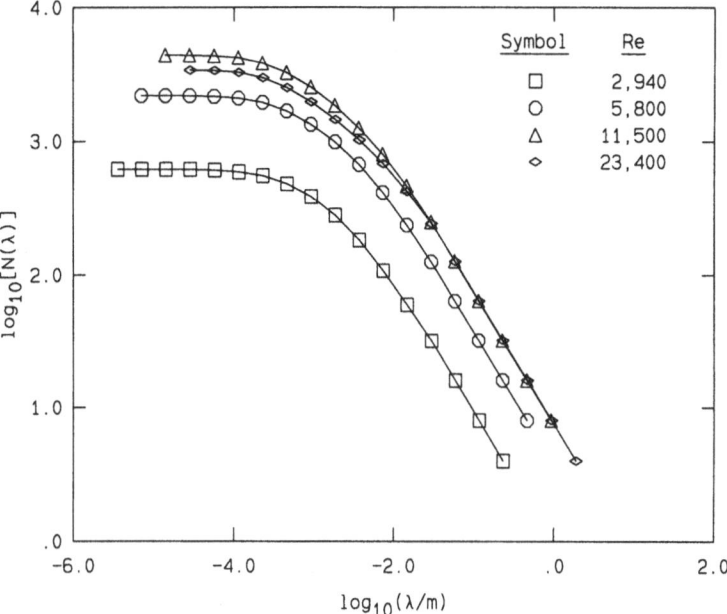

Fig. 32. — Coverage, $N_1(\lambda)$, as a function of scale for 1-D temporal scalar measurements in a turbulent jet (*ibid.* Fig. 4).

scale may not make it clear, the off-axis data exhibit a similar bump at small scales as the on-axis data, when viewed at smaller values in logarithmic coordinates. This behavior reflects the stochastic character of the concentration signal (rather than the fluid dynamics of the turbulent jet), as indicated, by processing measurements of laser light scattered from a very dilute, constant-concentration fluorescent-dye solution.

To further understand the three-dimensional plots, the nature of the concentration signal must be considered. Possible scalar values are bounded by zero and some upper limit and the measured concentration time history exhibits many maxima and minima. Near the mean, relatively few of these extrema are encountered. On the other hand, for thresholds approaching the highest or lowest values detected, many such turning points are found, and they may dominate the statistics. Picture a local minimum in concentration, with a roughly parabolic dependence of $c(t)$ in its vicinity. Imagine then a threshold level that is slowly decreased toward the minimum value. Two threshold crossings are encountered, which, as the threshold is lowered, move closer together. These eventually (almost) join before the threshold drops below the minimum (recall that the signal is discretely sampled in time). Thus, near the turning points there can be a separation of scales; one length is associated with the typical distance between extrema, and the other is a much smaller scale associated with the spacing of crossings within pairs at each extrema.

cover a threshold crossing with some certainty.

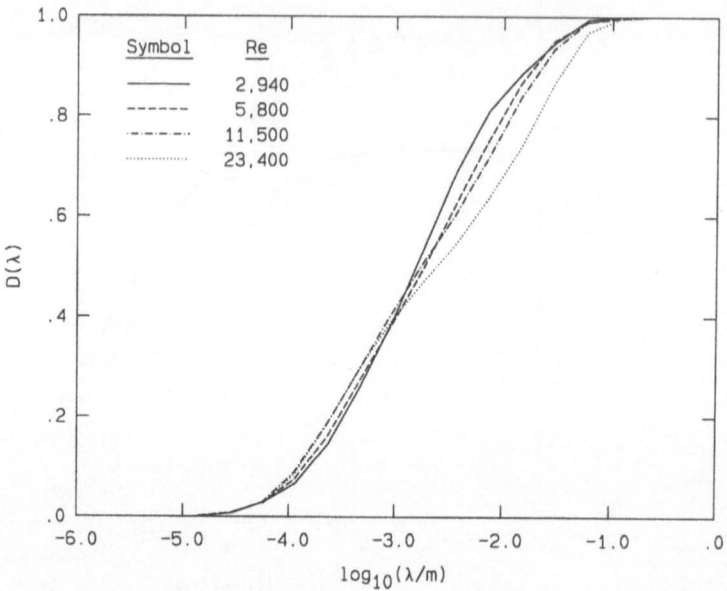

Fig. 33. — Coverage dimension, $D_1(\lambda)$, as a function of scale, derived from the coverage data in Fig. 32 (*ibid.* Fig. 5).

This scale separation manifests itself in the $D_1(\lambda)$ plots in two primary ways. One is the appearance of a bump at the smallest scales, traceable to the double crossings at each extremum. This bump is indicative of a characteristic length at that scale. The second is that the rise of $D_1(\lambda)$ to its asymptotic value of unity is shifted to larger scales as the threshold is increased, or decreased, from the local mean. This is ascribable to what could be called conservation of spacing. Pairs of crossings drop out as the threshold moves past their extrema, producing a larger length scale associated with the distance between crossings to either side of the pair. This is enhanced by crossings within a pair coming closer together as the threshold approaches their extremum, causing the spaces between adjacent pairs to correspondingly increase. Both the bump at small scales and the shift in the rise are evident in all of the three-dimensional $D_1(\lambda)$ plots.

One-dimensional spatial measurements were also recorded (with different beam-expansion optics to produce a collimated beam; Cf. Fig. 29), imaging a line element of the concentration field on a 512-pixel linear photodetector array. The line element spanned one tenth of the local jet diameter, at $z/d \simeq 300$, centered on the jet axis, at $Re \simeq 1.0 \times 10^3$, 2.0×10^3, and 3.0×10^3. The coverage dimension was computed for these data as a function of spatial scale, $D_1(\lambda)$, without invoking Taylor's hypothesis (Fig. 36). The behavior is similar to that of Fig. 33, even though, for these, smaller-dynamic-range data, the range of scales is insufficient to allow $D_1(\lambda)$ to attain the expected limiting values of 0 and 1 (Cf. Eqs. (50), (52), and related discussion). In

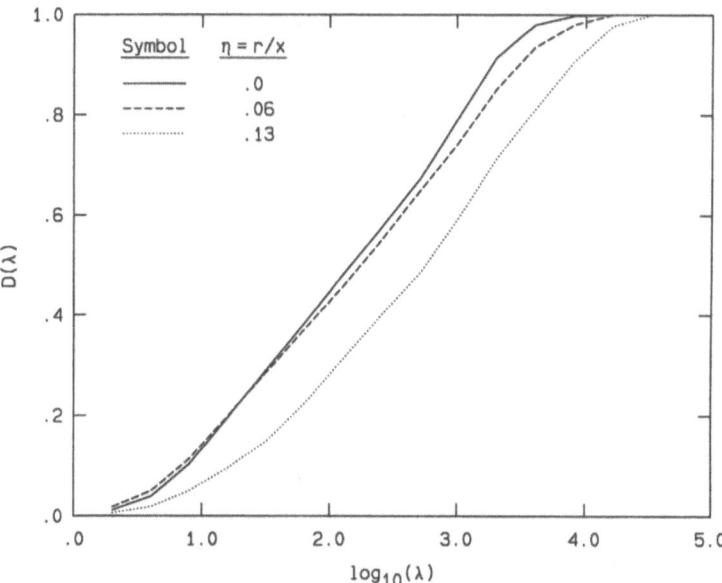

Fig. 34. — Coverage dimension, $D_1(\lambda)$, derived from off-axis, 1-D temporal scalar measurements in a turbulent jet (*ibid.* Fig. 8).

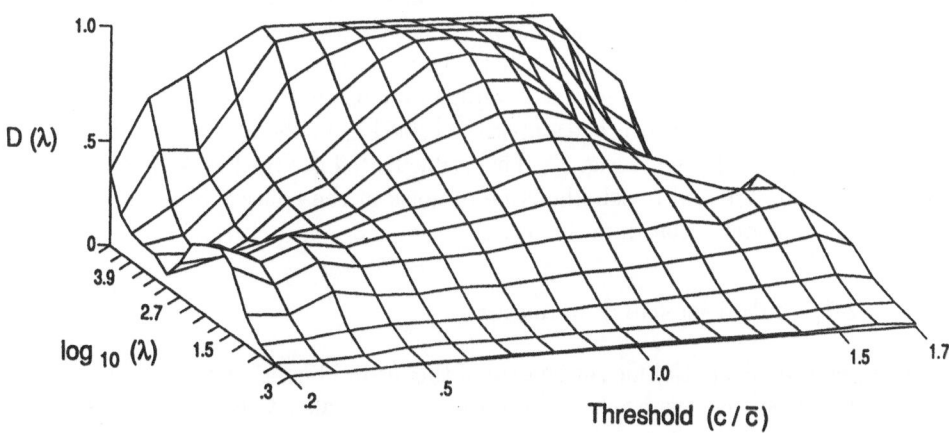

Fig. 35. — Scalar-threshold effects on $D_1(\lambda)$ (*ibid.* Fig. 10).

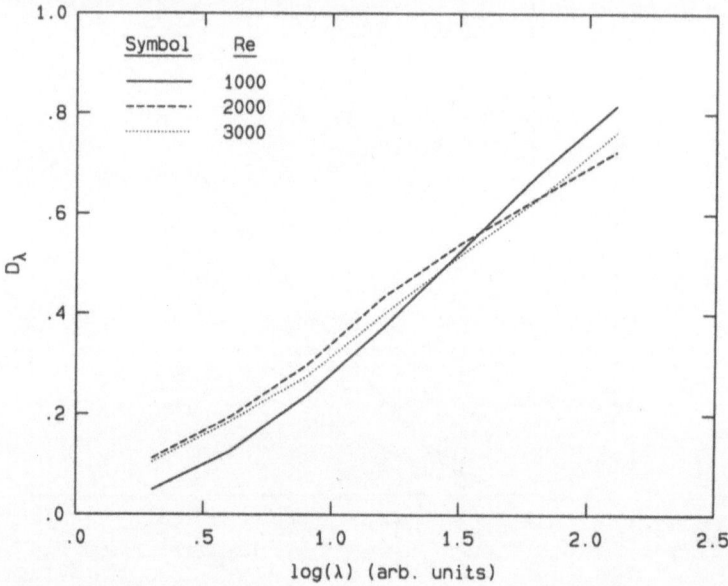

Fig. 36. — Coverage dimension, $D_1(\lambda)$, derived from 1-D spatial scalar measurements in a turbulent jet (*ibid.* Fig. 6).

particular, no constant-D scale region is observed for these spatial data either.

Before ending this part of the discussion, we should cite the investigation of the effects of noise on the coverage statistics by Miller & Dimotakis (1991). Briefly, they computed the coverage of level crossings derived from the unfiltered data (Cf. Figs. 30 and 31). The much-higher number of crossings computed from the unfiltered data dominated the coverage counts, for a surprisingly large scale range, which then reflected the behavior of the noise, rather than that of the scalar signal. In particular, a spurious region of near-constant coverage dimension was indicated by the noisy data, in contrast to the coverage statistics derived from the optimally-reconstructed scalar time trace, plotted in Fig. 32. This is illustrated in Fig. 37 (Miller & Dimotakis 1991, Fig. 16). We will revisit the issue of noise in the context of level-set analysis of two-dimensional scalar data, below.

3.2. 2-D spatial isoscalar behavior

Two-dimensional, spatial measurements of the jet-fluid concentration field in liquid-phase, turbulent jets were recently reported by Catrakis & Dimotakis (1996a).[18] The experiments, which relied on laser-induced fluorescence digital-imaging, investigated transverse (jet-axis-normal) planar cuts in the far field ($z/d_j = 275$) of turbulent jets and included a coverage analysis of isoscalar geometry.

The experiments were conducted in the same facility used for the Miller & Di-

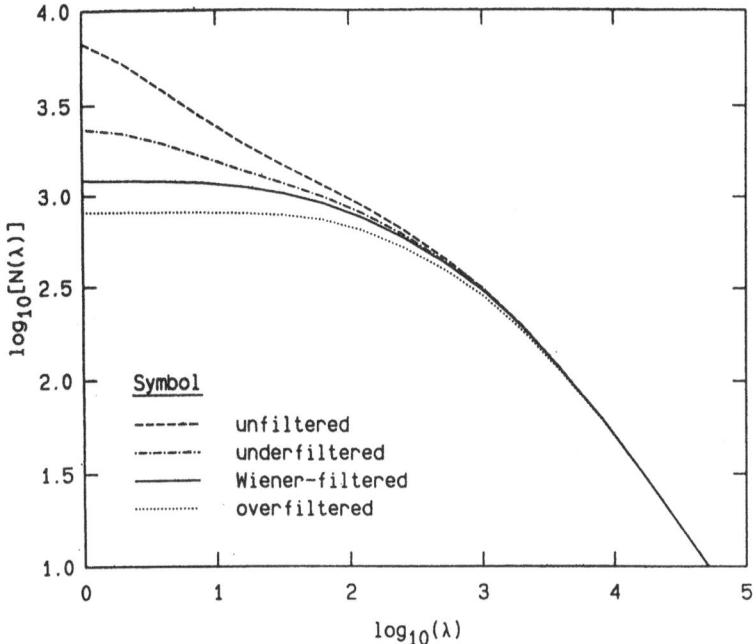

Fig. 37. — Influence of noise on 1-D coverage statistics (*ibid.* Fig. 16).

motakis (1991)[81] experiments (Cf. Fig. 29), employing a galvanometrically-driven mirror to (linearly) sweep a collimated laser beam that formed the illumination sheet (Cf. Fig. 13). The sectional images were recorded on a cryogenically-cooled, (1024×1024)-pixel, CCD camera. The laser beam waist was estimated as $w_0 \lesssim 300\,\mu\mathrm{m}$, with a Rayleigh range of $\pm 12.5\,\mathrm{cm}$ (on either side of the waist). The field of view spanned $\ell_0 \simeq 42\,\mathrm{cm}$, contained the full transverse spatial extent of the turbulent-jet fluid at the measuring station, and corresponded to a pixel resolution of $\lambda_\mathrm{p} \simeq 420\,\mu\mathrm{m}$. Consequently, the image-plane-normal resolution was smaller than the in-plane (pixel) resolution in the center, and comparable to it near the edges of the field of view.

At the downstream station in these experiments ($z/d_\mathrm{j} = 275$), it was verified that the low jet-plenum dye (disodium fluorescein) concentration, $c_0 \simeq 2.0 \times 10^{-6}\,\mathrm{M}$, produced negligible laser attenuation across the field of view. The image calibration procedure employed ensured that the measured scalar-field values were referenced, in absolute value, to the (pure) jet-plenum concentration, c_0. Spatial power spectral measurements indicated a digital-image signal-to-noise ratio in excess of 50 dB, *i.e.*, in excess of 300:1.

The jet far field was investigated for three Reynolds numbers: $Re = 4.5 \times 10^3$, 9.0×10^3, and 18×10^3. Images of individual realizations of jet slices, for each of the three Reynolds numbers, are depicted in Figs. 38, 39, 40. In these figures, grey levels have been assigned to the scalar values, starting from black for zero jet-fluid concentration (unmixed reservoir fluid), through dark grey, for low-concentration fluid, on to lighter grey shades for increasing concentration levels.

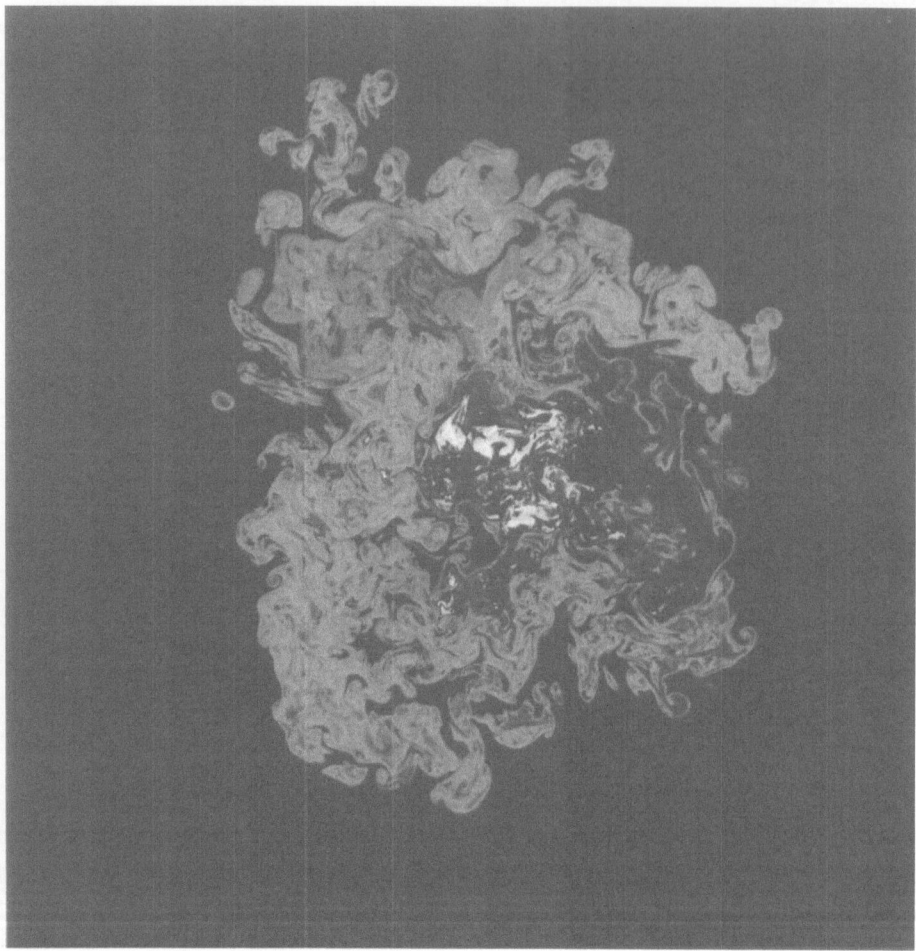

Fig. 38. — Jet-fluid concentration in the far-field ($z/d_j = 275$) of a turbulent jet at $Re \simeq 4.5 \times 10^3$.

While the data in Figs. 38, 39, 40 are individual realizations and should be used with some caution in drawing general conclusions, they are representative and help illustrate some aspects of the Re-dependence of the geometry of the scalar field. In particular, the lowest Reynolds number flow ($Re \simeq 4.5 \times 10^3$) field data (Fig. 38) indicate relatively high scalar gradients in the periphery (outer, low-speed region) of the jet. The grey shades in the image data can be seen to transition from medium grey, corresponding to moderately-high scalar concentrations, almost directly to black, corresponding to zero (negligible) scalar concentration, with very thin regions of intermediate scalar values in between. In contrast, the highest Reynolds number flow ($Re \simeq 18 \times 10^3$) image indicates a relatively large portion of the field occupied by low-scalar-concentration fluid with correspondingly lower scalar gradients (almost no direct transitions to black) in the periphery of the jet. The intermediate Reynolds number ($Re \simeq 9 \times 10^3$) flow slice also indicates lower scalar gradients in the periphery,

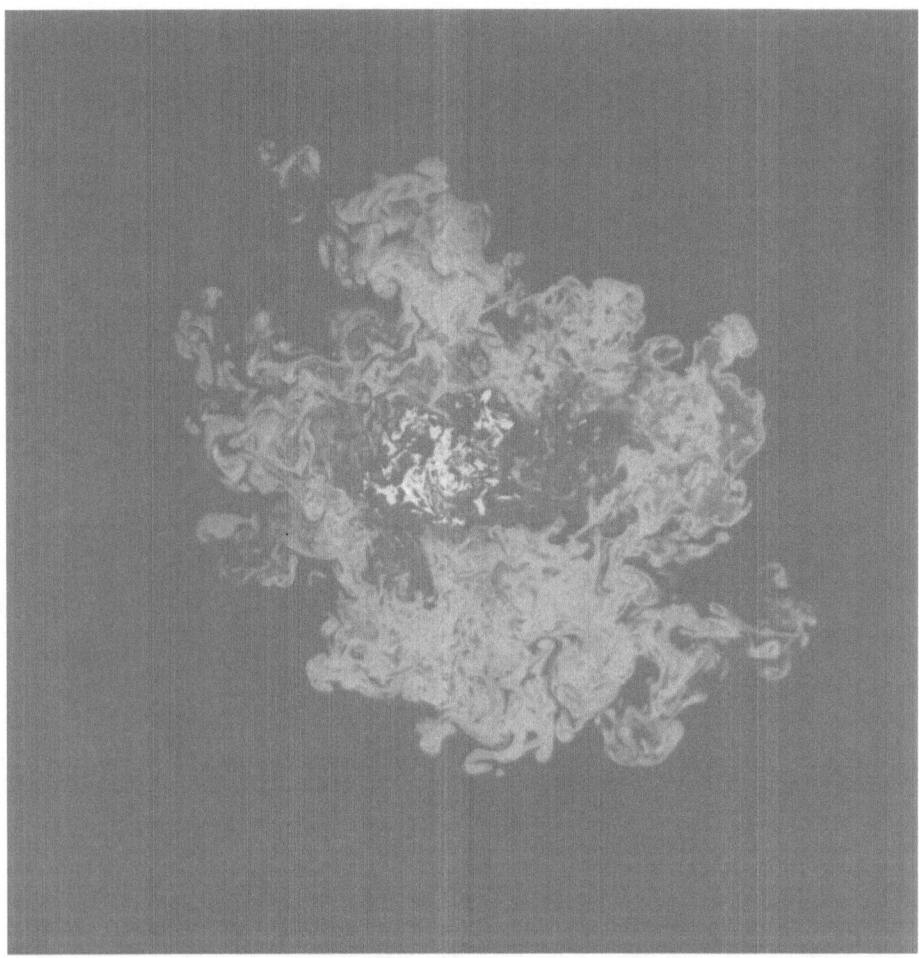

Fig. 39. — Jet-fluid concentration in the far-field ($z/d_\mathrm{j} = 275$) of a turbulent jet at $Re \simeq$ 9.0×10^3.

with a low-concentration region. However, a substantially smaller area is occupied by it than at $Re = 18 \times 10^3$.

At the intermediate Reynolds number, $Re \simeq 9.0 \times 10^3$ (Cf. Fig. 39), the scalar-diffusion (full-wavelength) scale, $\lambda_\mathcal{D}$, at the jet centerline conditions, is estimated to be approximately equal to the pixel size, λ_p, *i.e.*, underresolved by a factor of 2 (per Nyquist's criterion), but much larger than the pixel size in the outer region of the jet, where the level sets have been computed. Also, at this Reynolds number, the time for the passage of the scalar-diffusion scale is estimated to be a factor of 30 times larger than the exposure time of an individual pixel, on the jet axis, and larger yet in the outer regions of the jet. The images were acquired maintaining a constant product of the beam-scanning time, which scaled the time exposure per pixel, and the local flow velocity, over the Reynolds numbers investigated. These choices provided temporally- as well as spatially-resolved measurements of the scalar

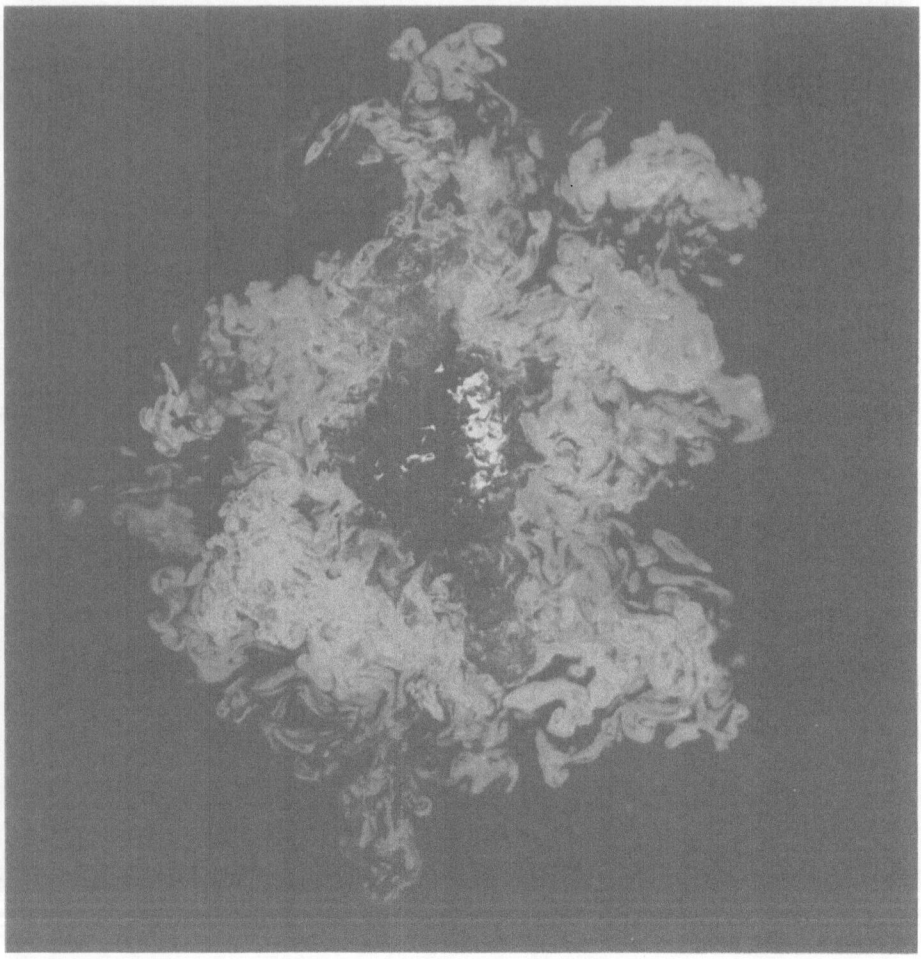

Fig. 40. — Jet-fluid concentration in the far-field ($z/d_\mathrm{j} = 275$) of a turbulent jet at $Re \simeq 18 \times 10^3$.

field, throughout the Reynolds number range, certainly in the intermediate-to-outer regions of the jet where level-set behavior was investigated.

Figures 41,42,43, compiled from the individual realization data reproduced in Figs. 45,46,47, provide another view into these differences. These are histograms of the number of pixels found in the designated scalar range, normalized by the total number of pixels in each image, with scalar concentration values normalized by $c_\mathrm{ref} \simeq c_0/220$, for all the data. Again, it should be borne in mind that these represent the statistics of the three individual realizations. As can be seen, the most probable value (mode) of jet-fluid concentration, in the lowest Reynolds number data (Fig. 41), is at $c/c_\mathrm{ref} \simeq 2$, in contrast to that for the highest Reynolds number data (Fig. 43), for which the mode is at $c/c_\mathrm{ref} \simeq 1$, reflecting the qualitative conclusion that was drawn on the basis of the grey-level representation of the scalar values in Figs. 38,39,40.

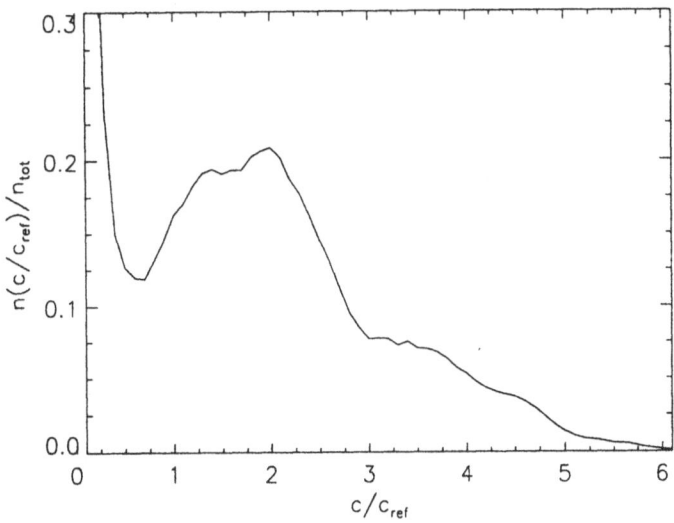

Fig. 41. — Jet-fluid concentration histogram derived from individual $Re \simeq 4.5 \times 10^3$ jet slice data (Fig. 38).

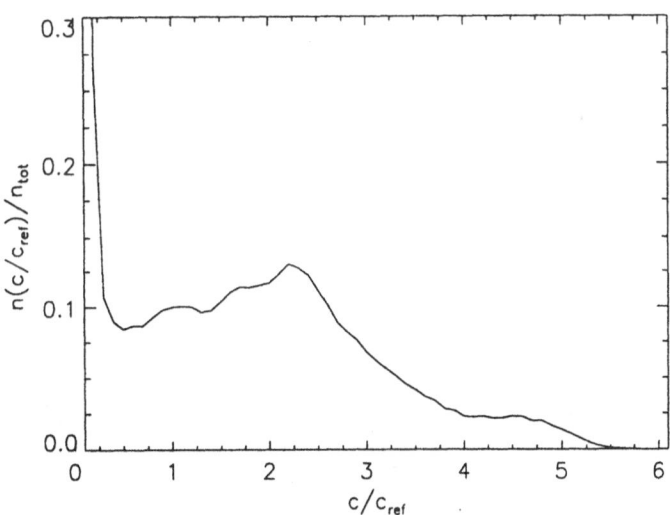

Fig. 42. — Jet-fluid concentration histogram derived from individual $Re \simeq 9 \times 10^3$ jet slice data (Fig. 39).

An estimate of the scalar field (jet-fluid concentration) probability-density functions was made and are depicted in Fig. 44. Recall that these are spatially-normalized, representing the fraction of the area in each slice given to a range of scalar values,

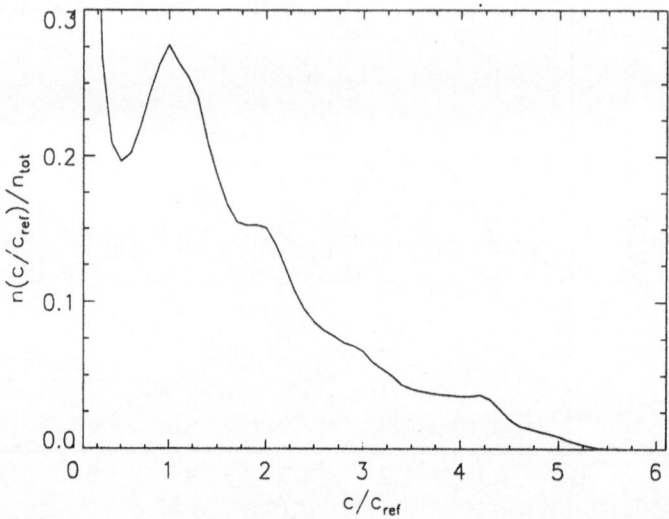

Fig. 43. — Jet-fluid concentration histogram derived from individual $Re \simeq 18 \times 10^3$ jet slice data (Fig. 40).

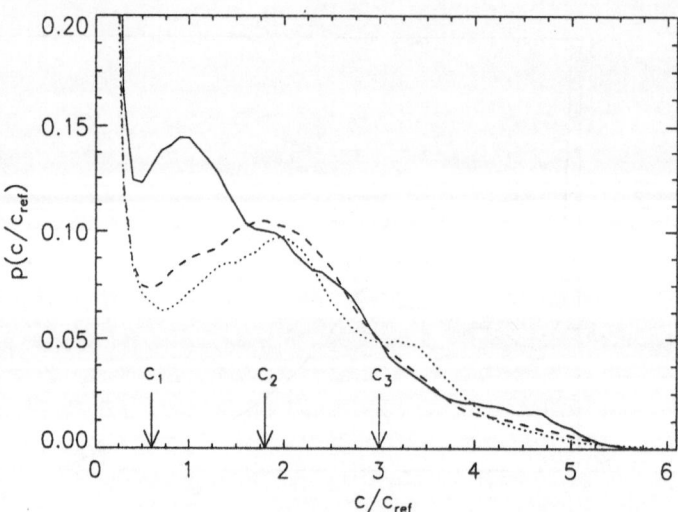

Fig. 44. — Jet-fluid concentration pdf in the far-field ($z/d_j = 275$) of a turbulent jet. $Re \simeq 4.5 \times 10^3$: dotted line; $Re \simeq 9.0 \times 10^3$: dashed line; $Re \simeq 18 \times 10^3$: solid line. Three scalar-threshold values, c_1, c_2, and c_3, are also indicated.

i.e.,

$$p(c)\,dc = \frac{|dA(c)|}{A_0}\,, \tag{58}$$

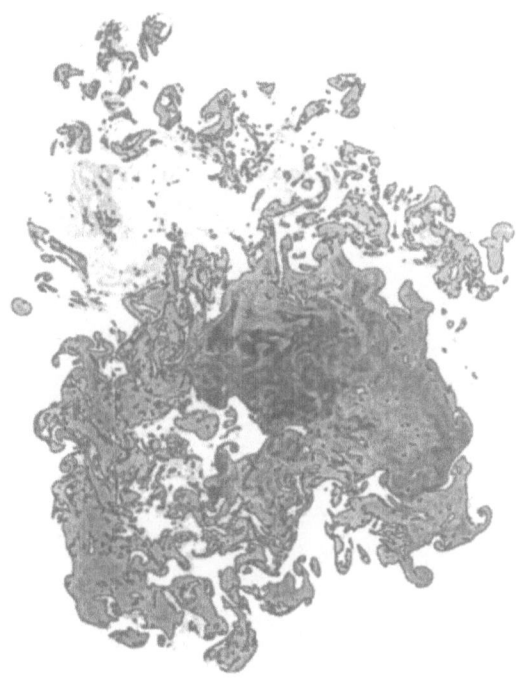

Fig. 45. — Level set at $c = c_2$ of the jet-fluid concentration superimposed on scalar-image data in the far-field ($z/d_j = 275$) of a turbulent jet. $Re \simeq 4.5 \times 10^3$.

where $dA(c)$ is the (ensemble-averaged) image slice area over which the scalar values are between c and $c + dc$ and A_0 is the total image slice area (*e.g.*, Kuznetsov & Sabel'nikov 1990, p. 27).[62] The estimated pdf's indicate that scalar level-set measures may be expected to depend on both the threshold value employed to compute the level set, as well as the flow Reynolds number. Three scalar values are marked on the pdf abscissa: c_2, in the vicinity of the local c-maximum for the two lower-Re flows, c_1, in the vicinity of the local c-minimum — corresponding to the outer isosurface values — where (constant-D) fractal behavior has been reported (Cf. Sreenivasan 1991),[105] and a high-level, c_3, chosen such that $c_2 - c_1 = c_3 - c_2$, to investigate inner isosurface behavior.

Isoscalar contours (level sets) corresponding to the $c = c_2$ intermediate threshold, for the data depicted in Figs. Fig. 38,39,40, for each of the three Reynolds numbers, are depicted in Figs. 45,46,47, respectively, superimposed on the scalar-field data.

Several features can be deduced from these spatial pdf data, that also manifest themselves in the associated isoscalar behavior.

1. The scalar pdf values are very nearly matched for $c \gtrsim c_2$, within statistical

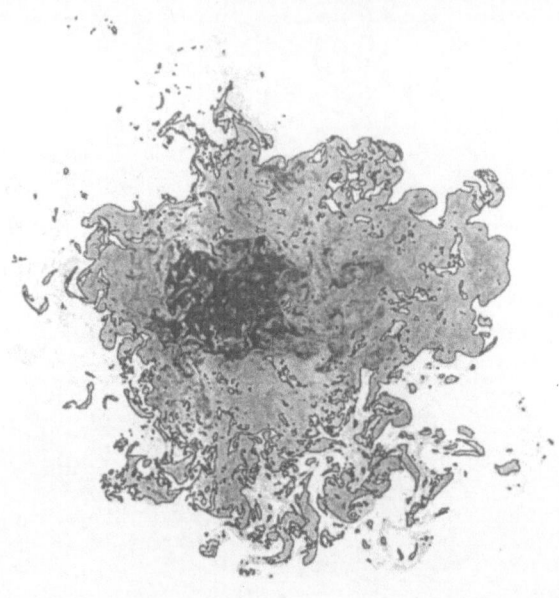

Fig. 46. — Legend as in Fig. 45: $Re \simeq 9.0 \times 10^3$.

confidence.

2. The data indicate a progressively increasing probability of encountering jet-fluid mixed to concentration levels $c < c_2$, with increasing Reynolds number.

3. As a consequence of the spatial normalization employed, the spatial extent of the jet, corresponding to low-concentration-level contours, may be expected to be Re-dependent.

4. The data indicate a qualitative change between the lower-Re scalar pdf's ($Re = 4.5 \times 10^3$, 9.0×10^3), which are characterized by a local maximum at $c \approx c_2$, and the $Re = 18 \times 10^3$ pdf, consistent with the expectation of a mixing transition at $Re \approx 10^4$ (recall discussion in Sec. 1.1).

5. The isoscalar contours appear to be more convoluted at *lower* Reynolds number, becoming less complex with increasing Reynolds number.

These observations will be discussed further below.

In what follows, the spatial scale of the coverage boxes (tiles), λ, is normalized by $\delta_b(c; Re)$, the ensemble-averaged, threshold- and Re-dependent, bounding-box

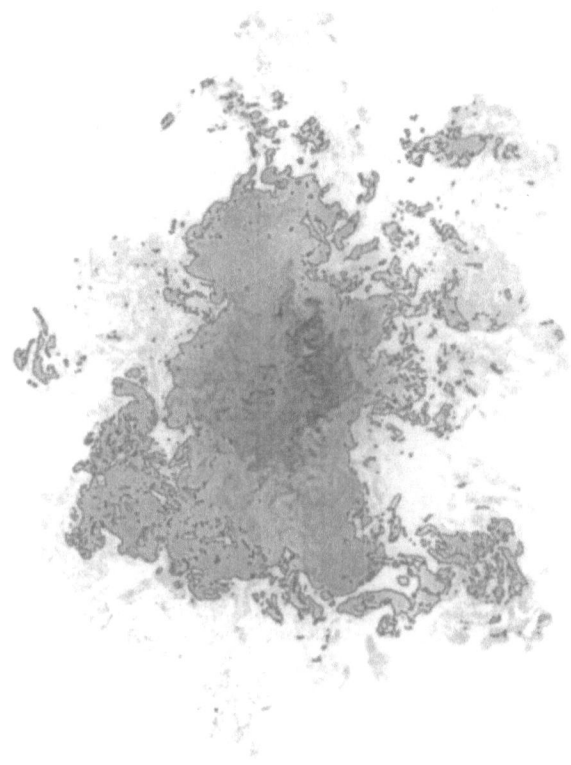

Fig. 47. — Legend as in Fig. 45: $Re \simeq 18 \times 10^3$.

size (Cf. (48)). It encloses the portion of the scalar field whose value exceeds the threshold, c (*e.g.*, Fig. 14). The continuous dependence of δ_b on the threshold, c, is plotted in Fig. 48, for each of the three Reynolds numbers investigated. It is interesting that, at low Reynolds numbers, the data indicate a weak dependence of δ_b on threshold, for low threshold values, *i.e.*, diminished mixing that would have otherwise produced intermediate scalar values in the outskirts of the turbulent jet region.

The data suggest that, at high Reynolds numbers, δ_b approaches a straight line, *i.e.*,

$$\frac{\delta_b(c; Re)}{z} \rightarrow \beta_0 - \beta_1 \frac{d_j \, c}{z \, c_0} \, , \tag{59}$$

where $\beta_0 \simeq 0.6$ and $\beta_1 \simeq 0.08$.[16] This also indicates that the increased mixing in the outer regions of the jet has increased the transverse extent, or "reach", of the

_____

[16] Equation (59) provides the correct scaling with z/d_j. It also appears in Catrakis & Dimotakis (1996a, Eq. 3.3), but as written there only applies to the particular downstream location, $z/d_j = 275$, of those experiments.

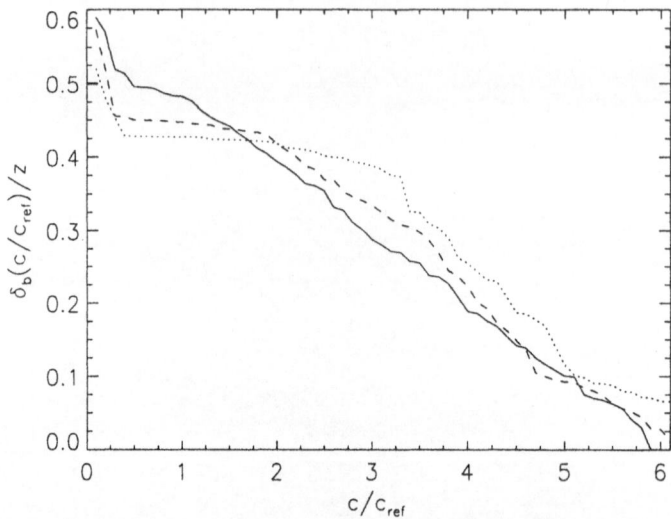

Fig. 48. — Isoscalar bounding box size, $\delta_b(c; Re)$. $Re \simeq 4.5 \times 10^3$: dotted line; $Re \simeq$ 9.0×10^3: dashed line; $Re \simeq 18 \times 10^3$: solid line.

mixed (jet-plenum) fluid away from the jet axis. This can be directly discerned by comparing the mixed fluid region, for c-levels lower than c_2, for each of the three Reynolds numbers, in Figs. 45,46,47. In particular, fluid mixed to $c < c_2$ (outside the c_2 level set) occupies an increasing area in the image with a substantial jump in going from $Re = 4.5 \times 10^3$ and 9.0×10^3, to $Re = 18 \times 10^3$, in accord with a mixing transition across this Re interval, as also noted earlier.

Figure 49 shows the ensemble-averaged, two-dimensional coverage count, $N_2(\lambda)$, of the level sets, for $Re \simeq 9.0 \times 10^3$. These are plotted for the three thresholds, c_1, c_2, and c_3, with lines of increasing solidity denoting increasing scalar threshold. The points are joined by straight-line segments and represent the coverage counts computed at the indicated λ-scales of the (binary) subpartitions of the bounding-box, in each case. Six images were ensemble averaged to estimate this statistic, permitting error bars (standard deviation of the mean count) to be estimated. They were smaller than the size of the symbols employed in the plot of Fig. 49. The coverage counts for the intermediate threshold, $c = c_2$, are seen to be larger than for $c = c_1$, or $c = c_3$, in accord with the scalar pdf behavior at this Reynolds number.

The coverage dimension, $D_2(\lambda)$, computed as a direct, second-order, (centered) finite difference of the $N_2(\lambda)$ data in Fig. 49, is shown in Fig. 50. The threshold dependence of the coverage dimension also reflects the scalar pdf behavior at this Reynolds number with $D_2(\lambda)$ largest, over most of the scale range, at the intermediate threshold, c_2. It is seen that the dimension is a function of scale, increasing monotonically and continuously with scale, from near unity, at the smallest scales, to 2, at the largest scales, $i.e.$,

$$\frac{dD_2(\lambda)}{d\lambda} > 0 , \quad \text{with,} \quad 1 < D_2(\lambda) < 2 . \tag{60}$$

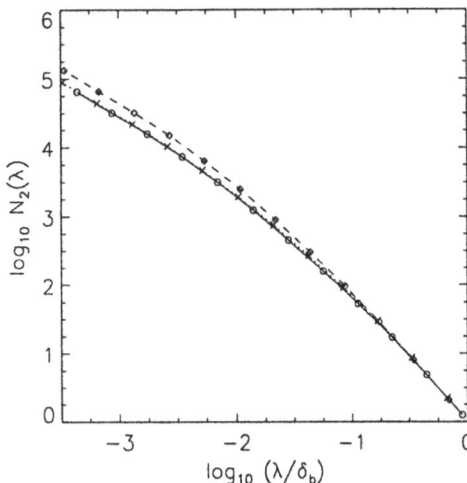

Fig. 49. — Coverage, $N_2(\lambda)$, of isoscalar surfaces at $Re \simeq 9.0 \times 10^3$. $c = c_1$: dotted line, crosses; $c = c_2$: dashed line, diamonds; $c = c_3$: solid line, circles. Recall that $\delta_b = \delta_b(c; Re)$.

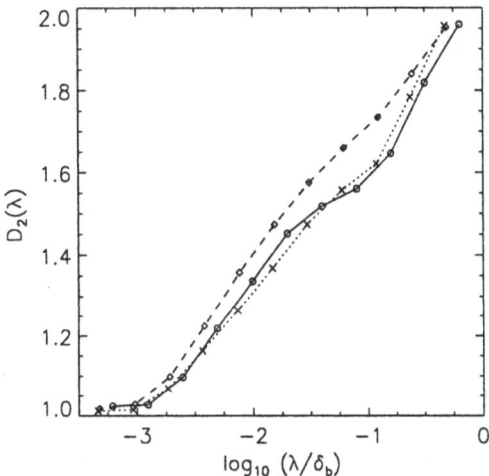

Fig. 50. — Coverage dimension, $D_2(\lambda)$, of isoscalar surfaces at $Re \simeq 9.0 \times 10^3$ (Cf. Fig. 49). Line/symbol legend as in Fig. 49.

The bounds are the topological dimension, $d_i = 1$, and the embedding dimension, $d = 2$, as expected for a nondecreasing coverage dimension with scale (recall Eqs. (50) and (52), and related discussion).

Figure 51 shows the Re-dependence of the coverage dimension, derived from the coverage data for the c_2 level sets. These data are in substantial agreement with the Miller & Dimotakis (1991) 1-D temporal on-axis (Fig. 33) and off-axis (Fig. 34)

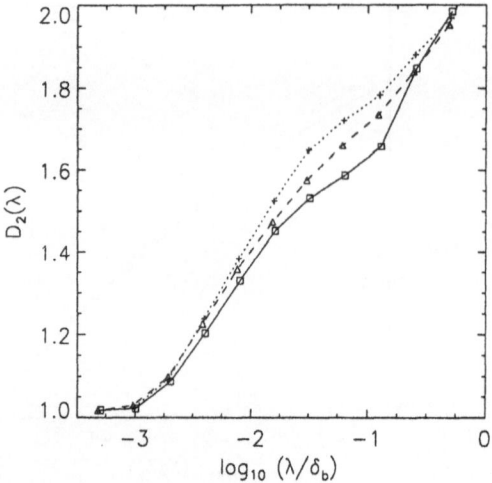

Fig. 51. — Reynolds-number dependence of coverage dimension, $D_2(\lambda)$, at the intermediate scalar threshold, c_2. $Re \simeq 4.5 \times 10^3$: dotted line, crosses; $Re \simeq 9.0 \times 10^3$: dashed line, triangles; $Re \simeq 18 \times 10^3$: solid line, squares.

coverage data as well as with their 1-D spatial (Fig. 36) data. In particular, the 2-D spatial data do not support the proposal of constant-D fractal scaling. They also indicate that the coverage dimension decreases, in the range of moderate-to-large scales, as Re increases, as also indicated by the Miller & Dimotakis (1991) coverage-dimension data (reproduced in Fig. 33, above). This is interesting in that it suggests that the complexity of the level sets, as measured here by $D_2(\lambda)$, *decreases* (Cf. Figs. 8 and 9) with increasing Reynolds number at intermediate scales.

While the quantitative results depend on scalar threshold, the qualitative behavior is relatively insensitive to this choice. As noted above (Sec. 2.1), it has been argued (Sreenivasan 1991)[105] that constant-D scaling should be applicable to jet-fluid isoscalars in the outer regions of the turbulent jet. Such scalar level sets correspond to lower thresholds (*e.g.*, $c = c_1$) one of which is depicted in Fig. 52. A compilation of the coverage dimension for the (outer) c_1 level sets, is plotted in Fig. 53. No constant-D scaling range is indicated by the data for these outer contours either. Notably, the dependence of $D_2(\lambda; c = c_1)$ on Reynolds number is relatively weak and does not reflect the jump in behavior seen in the scalar pdf, for example, between the lower and higher Reynolds numbers; coverage statistics capture different measures than scalar pdf's, for example.

It is interesting to study the Reynolds-number dependence of the (normalized) isoscalar coverage length, *i.e.*,

$$\frac{L(\lambda)}{\delta_b} \equiv \frac{\lambda}{\delta_b} N_2(\lambda) \ . \tag{61}$$

This notion is derived from the posthumous publication of Richardson's (1961) proposal, that provides a measure of the length of coastlines (recall Figs. 10 and 11 and related discussion) and brought to general attention by Mandelbrot (1967, 1977):

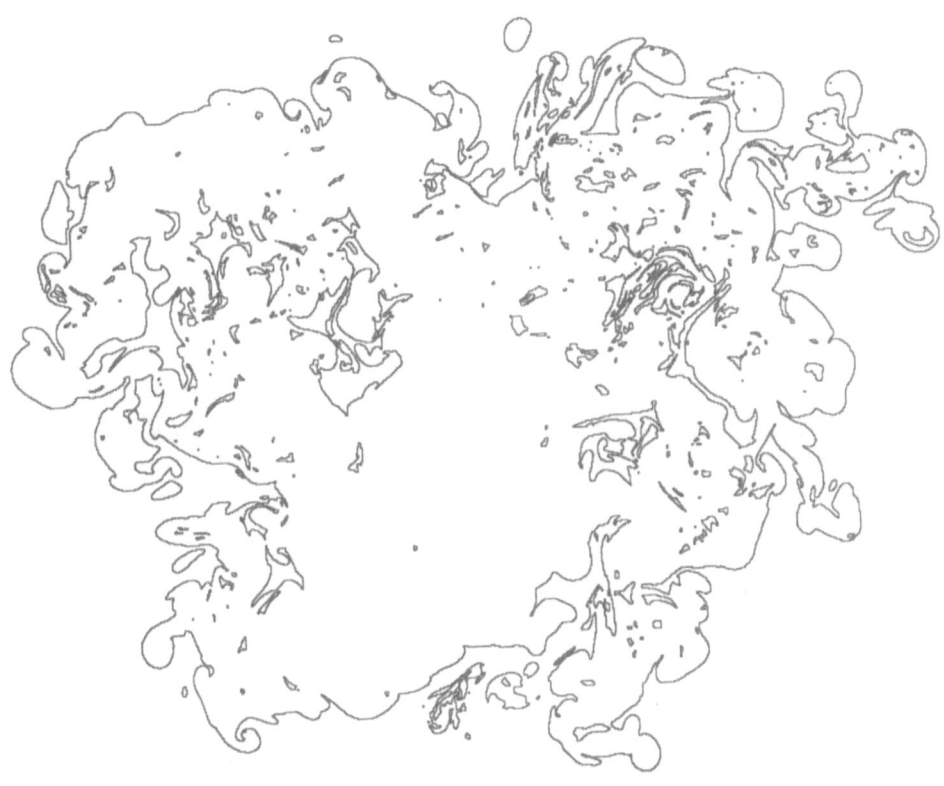

Fig. 52. — Level set for $c = c_1$, derived from $Re = 9.0 \times 10^3$ image data.

In Richardson's analysis, the count $N_2(\lambda)$ was estimated by the number, $N(\lambda)$, of λ-steps in the "stretched-string" algorithm (recall Eqs. (43) and (47) and related discussion). In the case of the turbulent-jet level-set contours, it was derived from the coverage counts depicted in Fig. 49, as was also done in the Feder (1988) analysis of the Norwegian fjords (Fig. 11). The results for the c_2 level sets are depicted in Fig. 54. Note that the limit,

$$f_2(\lambda \to 0) = \varsigma_2 = \frac{L(\lambda \to 0)}{\delta_b^2} , \qquad (62)$$

represents the (coverage) arc-length per unit area within the bounding box. At this intermediate threshold, the coverage-length data indicate a *decreasing* level-set length-to-area ratio, with increasing Reynolds number. This is an important result and also seen in the $D_2(\lambda)$ behavior. It can be attributed to increased *mixing* (and homogenization) of the scalar field, relative to *stirring*, for this turbulent-jet

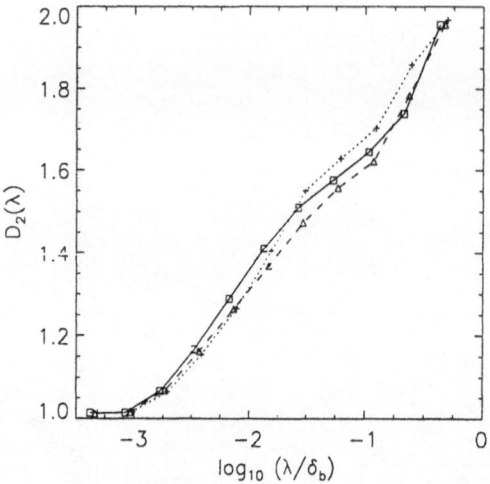

Fig. 53. — Reynolds-number dependence of coverage dimension, $D_2(\lambda)$, at the lower threshold, c_1 (outer contours). Legend as in Fig. 51.

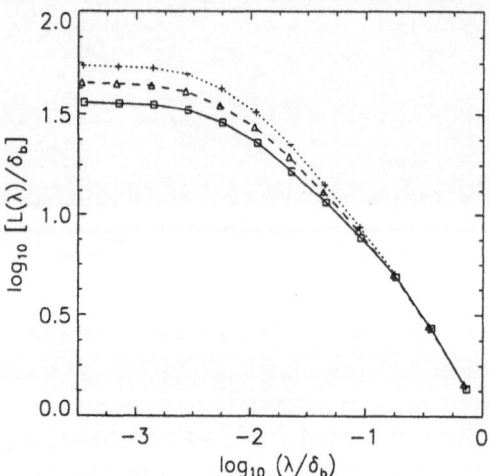

Fig. 54. — Normalized coverage length, $L(\lambda)/\delta_b = \lambda N_2(\lambda)/\delta_b$, derived from coverage data used to compute the Re-dependent $D_2(\lambda)$, at c_2 (Cf. Fig. 51).

flow, with increasing Reynolds number. This is also apparent in the streamwise slices of the scalar-field depicted in Figs. 3,4 and is in accord with previous, high-resolution point measurements, on the jet axis, that indicate a decrease in scalar fluctuations (increased mixing) with increasing Reynolds number (Miller & Dimotakis 1991, Fig. 4).[81] We'll discuss other manifestations of this behavior below. The small-scale limit of the coverage length, as normalized by the perimeter of

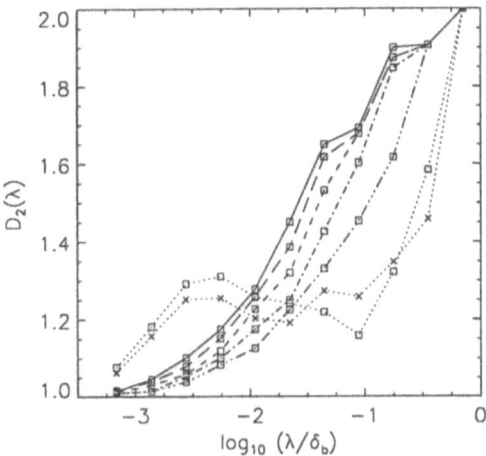

Fig. 55. — $D_2(\lambda)$, computed for level sets at low thresholds. $c/c_{ref} = c_1/c_{ref} = 0.6$ (solid line), 0.5 (long-dash line), 0.4 (short-dash line), 0.3 (dot-dash line), 0.2 (triple-dot-dash line), 0.1 (dotted line, squares), and 0.104 (dotted line, crosses).

the bounding box, yields $L(\lambda \to 0)/(4\delta_b) \approx 11$. In other words, the turbulent convection-diffusion process generates scalar level sets approximately an order of magnitude longer (by this measure) than the bounding-box perimeter.

The effects of noise on the two-dimensional level set coverage statistics were also investigated (Catrakis 1996, Appendix B). In particular, it was found that for threshold levels substantially lower than c_1 (Cf. Fig. 44), the level sets extended progressively beyond the main body of the mixed fluid, *i.e.*, in regions of low (local) signal-to-noise ratio. The c_1-level set for a $Re = 9.0 \times 10^3$ run is depicted in Fig. 52, for reference (substantially free of noise).

A progression of level sets, for very low threshold values, $c \leq c_1$ (Cf. Fig. 44), was processed to extract the corresponding coverage dimension, $D_2(\lambda; c)$. This is depicted in Fig. 55. As can be seen, as the threshold is lowered below c_1, $D_2(\lambda; c)$ evolves, passing through a near-constant coverage-dimension regime, which was encountered at $c/c_{ref} = 0.104 \approx c_1/6$. The effective coverage dimension, at this threshold, can be seen to be very nearly constant, with $D_2 \approx 1.25 \pm 0.05$, in a scale range, $-2.5 \lesssim \log_{10}(\lambda/\delta_b) \lesssim -1.0$.

Considering the interest in constant-D fractal scaling, it is illustrative to examine the level set that begets this result. It is reproduced in Fig. 56. As can be seen, the level set of the measured image field, at this very low threshold, in addition to the evident imprint of the low-level isoscalar contour, has a significant contribution stemming from the noise "dust" in the image. As was noted by Miller & Dimotakis (1991), level sets derived from low signal-to-noise data can be dominated by the characteristics of noise and yield coverage behavior more reflective of it, rather than the geometry of the (scalar) field under investigation.

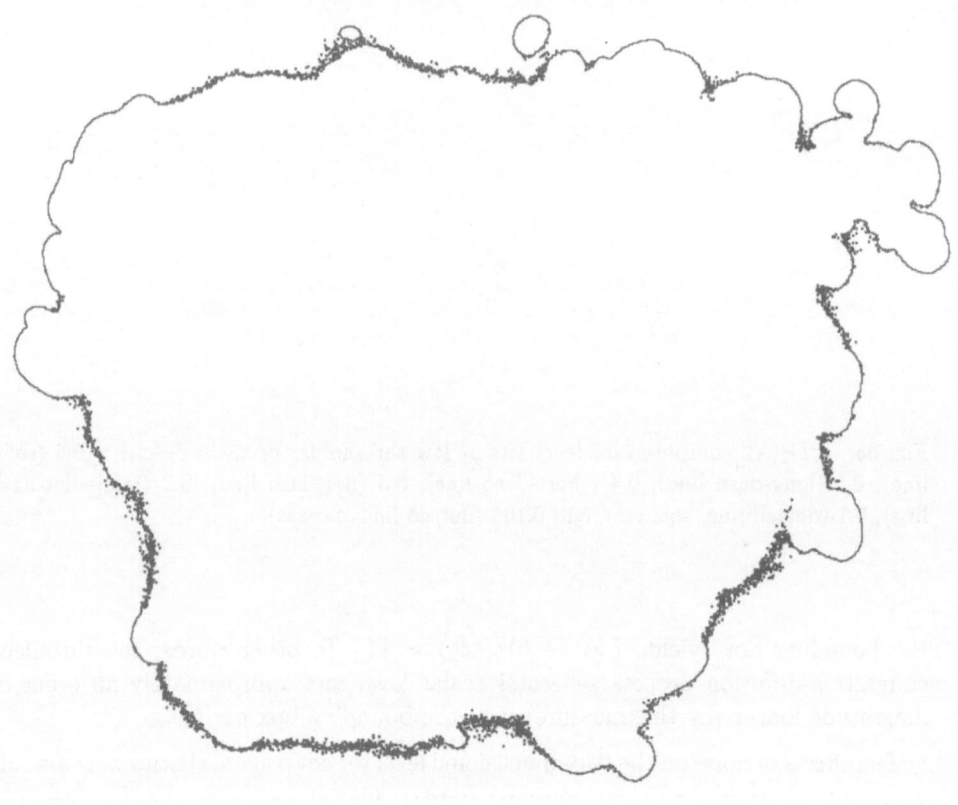

Fig. 56. — Level set for $c = 0.104\,c_{\mathrm{ref}} \approx c_1/6$, yielding near-constant-D behavior (due to noise). Isoscalar derived from same image data used in Fig. 52.

4. Consequences of scale-dependent geometry

The preceding overview of the experimental results on level-set behavior in turbulent jets, as well as those stemming from reports in both fluid-mechanics and other contexts (Secs. 2.1 and 2.2) necessitate a generalization of the framework of coverage dimensions that allows scale-dependent level-set behavior. This is a natural extension of the constant-D framework that is dictated by phenomena whose dynamics are not scale invariant. In general, scale-dependent objects may be regarded as possessing a higher level of complexity than constant-D sets. The notion was addressed by Mikhailov & Loskutov (1991),[80] for example, who discussed a three-level hierarchy of complexity, in which Level 1 can be assigned to Euclidean objects (circles, squares), Level 2 objects are (constant-D) fractals, with Level 3 reserved for more complex objects yet, with a scale-dependent structural complexity.

Revisiting the Koch island archipelago (Fig. 19) illustrates the point. At larger scales, the coverage statistics would be dominated by the Poisson-distributed statistics of positions of the Koch islands. On the other hand, in the limit of scales smaller than the Koch island extent, the fractal character of the individual islands will prevail. Coverage statistics that span the two scaling regimes will yield a coverage dimension, $D(\lambda)$, that reflects the two-scale character of this set, transitioning from the constant (triadic) Koch-island coverage dimension, at small scales, to a variable-D scale regime, at larger scales, reflecting the Poisson-distributed positions, as we'll discuss below.

For both turbulent-flow level sets, as well as other geometrical structures, such as the Koch island archipelago, the level-set geometry statistics can be captured, in a uniform manner, by the proposed scale-dependent-fractal (SDF) framework. This was presented by Catrakis & Dimotakis (1996a, Sec. 5) and is described below.

4.1. SDF framework

Scale-dependent-fractal (SDF) objects exhibit a variable coverage dimension ((57)),

$$D_d(\lambda) = -\frac{d \log N_d(\lambda)}{d \log \lambda} . \tag{63}$$

Objects exhibiting scale-invariant ((44)), constant-D, or power-law-fractal (PLF) behavior, over some range of scales, can then be regarded as a special case. The geometry of an object will be considered as SDF if,

$$\frac{dD_d(\lambda)}{d\lambda} \neq 0 , \tag{64}$$

while it will be PLF (constant-D), in a range of scales, if,

$$\frac{dD_d(\lambda)}{d\lambda} = 0 , \tag{65}$$

in that scale range.

The SDF dimension, $D_d(\lambda)$, can be expected to tend to the topological dimension, d_t, at the smallest scales, and to the embedding dimension, d, at the largest scales, *i.e.*,

$$D_d(\lambda) \longrightarrow \begin{cases} d_t, & \text{as } \lambda \to 0; \\ d, & \text{as } \lambda \to \delta, \end{cases} \tag{66}$$

where δ is the largest characteristic scale of the set (Cf. Dimotakis 1991[34] and discussion in Sec. 2, above). For data confined in a bounding box, the SDF dimension will approach the embedding dimension at the scale $\delta = \delta_b$ (the bounding-box size).[17]

The SDF geometric-coverage law that follows from (63) is given, in differential form, by

$$\frac{dN_d(\lambda)}{N_d(\lambda)} = -D_d(\lambda)\frac{d\lambda}{\lambda} . \tag{67}$$

---

[17] If $D_d(\lambda)$ is monotonic with scale, the limiting values in equation ((66)) will also be the bounding values. As noted above, non-monotonicity of the SDF dimension is possible, however, for cluster-like structures, for example. Scale-local clustering may lead to SDF dimension values that are below the topological dimension, in a range of scales.

Notably (Cf. (45)),

$$\text{if} \quad D_d(\lambda) \neq \text{const.}, \qquad \text{then} \quad N_d(\lambda) \not\propto \lambda^{-D_d(\lambda)} . \tag{68}$$

Integrating the differential coverage relation ((67)) from some reference scale, λ_1, we obtain (Cf. Takayasu 1982,[111] 1992)[112]

$$\frac{N_d(\lambda)}{N_d(\lambda_1)} = \exp\left\{ -\int_{\lambda_1}^{\lambda} D_d(\lambda') \frac{d\lambda'}{\lambda'} \right\} \tag{69}$$

and, if the set possesses a bounding scale, δ_b, we also have, since $N_d(\delta_b) = 1$,

$$N_d(\lambda) = \exp\left\{ \int_{\lambda}^{\delta_b} D_d(\lambda') \frac{d\lambda'}{\lambda'} \right\} . \tag{70}$$

In contrast to PLFs, SDF coverage is, in general, a nonlocal function of scale, with geometric structure across the whole range of scales potentially contributing to the coverage at any one scale.

A useful coverage measure is the coverage fraction, $F_d(\lambda)$, or (embedding space) volume-fill fraction, of the set at a scale λ, defined as (Cf. Dimotakis 1991),[34]

$$F_d(\lambda) \equiv \frac{N_d(\lambda)}{N_{d,\text{tot}}(\lambda)} = \left(\frac{\lambda}{\delta_b}\right)^d N_d(\lambda) , \tag{71}$$

where, for a finite domain, $N_{d,\text{tot}}(\lambda)$ is the total number of d-dimensional partitioning boxes of the δ_b-bounding box.[18] The coverage fraction can be identified as the geometric probability that a randomly placed λ-box, interior to the bounding δ_b-box, contains part of the set. The coverage fraction, as opposed to the SDF dimension, must be a nondecreasing function of scale (Cf. (64)), i.e.,

$$\frac{dF_d(\lambda)}{d\lambda} \geq 0 , \tag{72}$$

at all scales. The logarithmic derivative of $F_d(\lambda)$ follows from Eqs. (63) and (71),

$$\frac{d\log F_d(\lambda)}{d\log \lambda} = d - D_d(\lambda) , \tag{73}$$

which is the SDF codimension. The limiting behavior of the coverage fraction is (Cf. (66)),

$$F_d(\lambda) \begin{cases} \sim \lambda^{d-d_t} , & \text{as } \lambda \to 0; \\ \to 1 , & \text{as } \lambda \to \delta_b . \end{cases} \tag{74}$$

Integrating (73) from a coverage scale, λ, to the largest scale, δ_b, the SDF relation for the coverage fraction becomes (Cf. (69)),

$$F_d(\lambda) = \exp\left\{ -\int_{\lambda}^{\delta_b} \left[d - D_d(\lambda') \right] \frac{d\lambda'}{\lambda'} \right\} , \tag{75}$$

---

[18] These relations can be extended to spatially-unbounded sets, in which case the δ_b-bounding box size is selected to exceed the scale sizes of interest and $N_{d,\text{tot}}(\lambda) = (\delta_b/\lambda)^d$.

since $F_d(\delta_b) = 1$ (Cf. (74)). The extent to which a SDF set fills space, therefore, varies with scale and, potentially, depends on the geometric behavior at all scales.([19

The above SDF statistics can be connected to the distribution of (various measures of) scales of the set S. The coverage fraction, $F_d(\lambda)$, can be identified as the geometric probability that a (randomly-placed) λ-box covers part of S, as noted above ((71)). It can also be interpreted as a cumulative distribution function of a measure of spatial scales, in the following sense. For a scale increment, $\Delta\lambda$, the coverage fraction can be written as,

$$F_d(\lambda + \Delta\lambda) \equiv F_d(\lambda) + \int_{F_d(\lambda)}^{F_d(\lambda+\Delta\lambda)} dF_d(\lambda') \ . \tag{76}$$

The differential coverage fraction, in this integral, can be associated with a probability density function of a measure of scales, $f_d(\lambda)$, where

$$f_d(\lambda) \equiv \frac{dF_d(\lambda)}{d\lambda} \ . \tag{77}$$

In this expression, $f_d(\lambda)$ is the probability density function of the *largest-empty-box* (LEB) scale, λ, *i.e.*, the size of the largest box that is empty, *i.e.*, covers no part of S, as discussed below.

Consider a λ-box and a $(\Delta\lambda/2)$-wide strip, around the λ-box, as illustrated in Fig. 57. The identification of $f_d(\lambda)$, in ((77)), with the largest-empty-box scales can be established by considering the probabilities of the following three events:

$$
\begin{aligned}
\mathcal{A} &\equiv \ \{\, (\lambda + \Delta\lambda)\text{-box covers part of } S \,\} \\
\mathcal{B} &\equiv \ \{\, \lambda\text{-box covers part of } S \,\} \\
\mathcal{C} &\equiv \ \{\, (\Delta\lambda/2)\text{-wide strip, around } \lambda\text{-box covers part of } S \,\}
\end{aligned}
$$

The geometric probabilities of events $\mathcal{A}$, $\mathcal{B}$, and $\mathcal{C}$ are related as follows,

$$\mathcal{P}\mathcal{A} = \mathcal{P}\mathcal{B} \cup \mathcal{C} \equiv \mathcal{P}\mathcal{B} + \mathcal{P}\mathcal{C} \cap \overline{\mathcal{B}} \ . \tag{78}$$

Since,

$$F_d(\lambda + \Delta\lambda) = \mathcal{P}\mathcal{A} \quad \text{and} \quad F_d(\lambda) = \mathcal{P}\mathcal{B} \ , \tag{79}$$

we have (Cf. (76)),

$$\int_{F_d(\lambda)}^{F_d(\lambda+\Delta\lambda)} dF_d(\lambda') = \mathcal{P}\mathcal{C} \cap \overline{\mathcal{B}} \ , \tag{80}$$

i.e., the probability that the $(\Delta\lambda/2)$-wide strip, around a λ-box, covers part of S *and* that the λ-box is empty.

This relation allows the connection between the coverage statistics and the distribution of this (multidimensional) measure of spatial scales, λ. In this context, this scale is identified as the size of the LEB containing a randomly-located point, P, but no part of S, *i.e.*, is empty. Equivalently, the scale λ is a measure of (twice) the distance from a point P to the nearest element of S.

([19]) For PLF sets, for which $d - D_d(\lambda) = d - D_d = \text{const.}$, (75) recovers, $F_d(\lambda) \propto \lambda^{d - D_d}$.

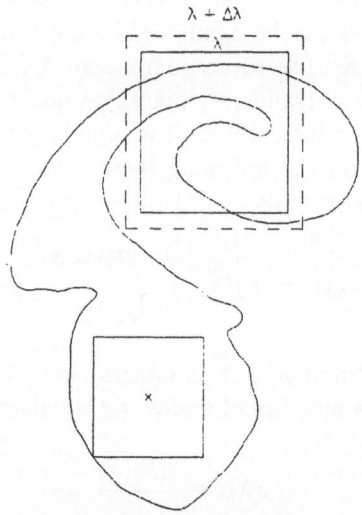

Fig. 57. — Illustration of a λ-box and a $(\lambda + \Delta\lambda)$-box, for relating geometric probabilities of coverage. An example of a LEB is also depicted.

From (77), we see that $f_d(\lambda)$ satisfies the required normalization condition over the range of spatial scales, *i.e.* (Cf. (74)),

$$\int_0^\infty f_d(\lambda)\,d\lambda \;=\; F_d(\infty) - F_d(0) = 1 \;, \tag{81}$$

Integrating (77), we have,

$$F_d(\lambda) \;=\; \int_0^\lambda f_d(\lambda')\,d\lambda' \;. \tag{82}$$

This allows a direct connection between the SDF dimension, $D_d(\lambda)$, and the distribution of LEB scales, $f_d(\lambda)$, *i.e.*,

$$D_d(\lambda) \;=\; d - \frac{\lambda\,f_d(\lambda)}{\int_0^\lambda f_d(\lambda')\,d\lambda'} \;. \tag{83}$$

This can be inverted to yield the LEB scale pdf from the SDF dimension, *i.e.* (Cf. Eqs. (75) and (77)),

$$\begin{aligned}
f_d(\lambda) &= \tfrac{d - D_d(\lambda)}{\lambda}\, F_d(\lambda) \\
&= \tfrac{d - D_d(\lambda)}{\lambda}\, \exp\left\{ -\int_\lambda^\infty [d - D_d(\lambda')]\,\tfrac{d\lambda'}{\lambda'} \right\}
\end{aligned} \tag{84}$$

Equations (83) and (84) may be regarded as a SDF transform pair in d-dimensional space.

From these relations, we can derive the small-scale behavior of $f_d(\lambda)$. In particular (Cf. Eqs. (74) and (77)),

$$f_d(\lambda) \sim \lambda^{d-d_t-1} \rightarrow \begin{cases} \varsigma_d, & \text{as } \lambda \rightarrow 0 \text{, for } d_t = d-1 \text{ ;} \\ 0, & \text{as } \lambda \rightarrow 0 \text{, for } d_t < d-1 \text{ ,} \end{cases} \qquad (85)$$

In this expression, ς_d is a constant we will discuss below. The case $d_t = d-1$ is an important special case and applies, for example, to isoscalar surfaces, as created and described by the scalar convection-diffusion equation ((16)). The case $d_t < d-1$ corresponds to points in 2-D, space-curves in 3-D. etc. The large-scale behavior of $f_d(\lambda)$ is given by (Cf. (74)),

$$f_d(\lambda) \rightarrow 0, \text{ as } \lambda \rightarrow \infty , \qquad (86)$$

unconditionally.

4.2.   One-dimensional scale distribution

Exploratory investigations of the mean spacing, l_m, of zero-crossings of 1-D velocity signals in turbulence were reported and discussed by Liepmann (1949).[68] Measurements of the corresponding spacing pdf have been approximated by Poisson statistics in turbulent boundary layers by Sreenivasan et al. (1983),[109] and Kailasnath & Sreenivasan (1993).[54]

Lognormal spacings were also reported by Sreenivasan et al. (1983),[109] for some of their data, for level crossings in 1-D in various turbulent flows, including level-crossings of 1-D velocity measurements in turbulent boundary layers. One-dimensional records of the streamwise velocity in turbulent boundary layers were found by Sreenivasan & Meneveau (1986)[107] to have level sets qualitatively resembling Poisson spacings and exhibited a (nonconstant-D) coverage behavior that the authors attributed to the random point spacings they found were described by (near-) Poisson statistics. In a subsequent discussion, however, Sreenivasan (1991)[105] reported constant fractal dimensions from such signals. In what follows, we will discuss the relation between level-set spacings and coverage statistics.

In an effort to model the coverage behavior of level sets of their 1-D measurements of concentration in turbulent jets, Miller & Dimotakis (1991)[81] noted that random, lognormally-spaced points are characterized with a SDF dimension that can provide a good (two-parameter) fit to $D_1(\lambda)$, as experimentally determined from their turbulent-jet data. Pdf's of the measured crossing spacings independently showed that the lognormal distribution is a good approximation. Figure 58, from their paper, includes the result for a Reynolds number of 5800. Level sets of 1-D scalar signals derived from plumes dispersing in the atmospheric surface layer (Yee et al. 1995)[123] were also found to be described by lognormal statistics.

The preceding discussion suggests that there may exist a relation between the spacing statistics of 1-D signals and the associated coverage dimension. It also points to an apparent contradiction stemming from reports of constant fractal dimensions derived from the analysis of data reported to be well-described by Poisson, or lognormal, statistics.

The connection between the spacing statistics in 1-D level sets and the coverage dimension can be established (Catrakis & Dimotakis 1996b).[19] Based on the relation of $F_1(\lambda)$ and the probability, $p_{1,e}(l)$, of an empty l-interval on the 1-D level-set

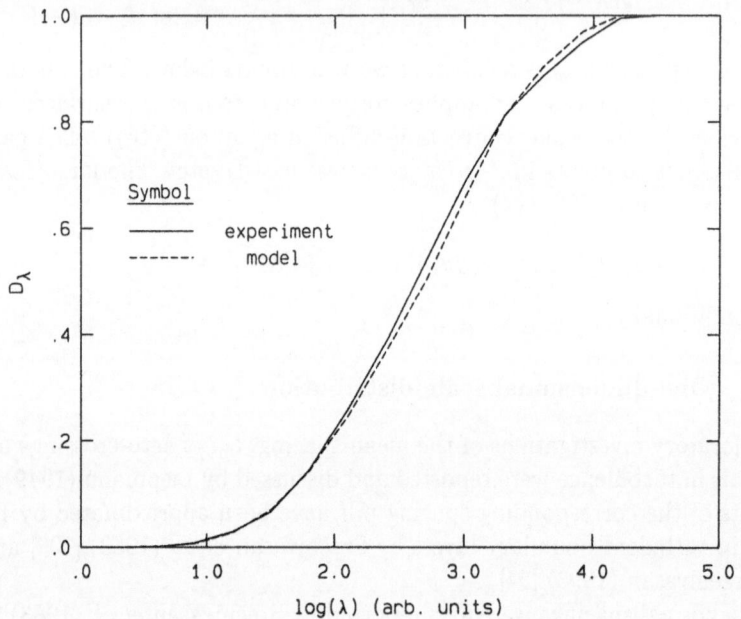

Fig. 58. — Comparison of $D_1(\lambda)$, derived from 1-D scalar measurements in turbulent jets ($Re = 5800$), with a model based on a lognormal pdf of spacing scales (Miller & Dimotakis 1991, Fig. 9).

record, the general result,

$$D_1(\lambda) = 1 - \frac{\lambda \int_\lambda^\infty p_1(l)\,\mathrm{d}l}{\int_0^\lambda \int_{\lambda'}^\infty p_1(l)\,\mathrm{d}l\,\mathrm{d}\lambda'}\,, \tag{87}$$

follows. This can be inverted (analytically) to obtain $p_1(l)$, given $D_1(\lambda)$. By way of example, for a Poisson distribution,

$$p_1(l) = \frac{e^{-l/l_m}}{l_m}\,, \quad \text{and} \quad D_1(\lambda) = 1 - \frac{\lambda/l_m}{e^{\lambda/l_m} - 1}\,. \tag{88}$$

Expressions for lognormal and (truncated) power-law spacings pdf's were also derived. Significantly, while constant-D fractal scaling generally implies power-law spacing statistics, the converse is not true. The nonlocal relation ((87)) between $D_1(\lambda)$ and $p_1(\lambda)$ may result in a scale-dependent $D_1(\lambda)$, even if $p_1(\lambda)$ is power-law for a substantial range of scales.

The spacing pdf, $p_1(l)$, can also be connected to the one-dimensional largest-empty-box (LEB) size pdf, $f_1(\lambda)$, and is given by (Cf. (77)),

$$f_1(\lambda) = \frac{1}{l_m} \int_\lambda^\infty p_1(l)\,\mathrm{d}l = \frac{\mathrm{d}F_1(\lambda)}{\mathrm{d}\lambda}\,.$$

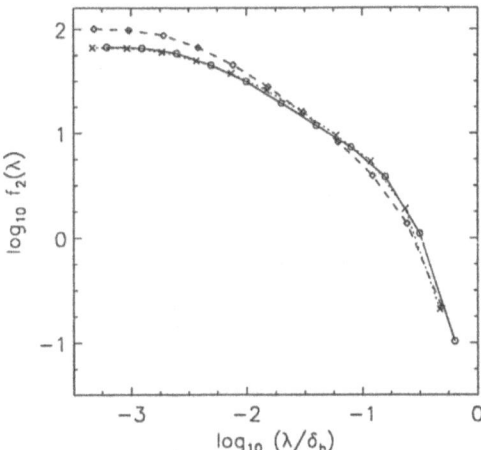

Fig. 59. — Pdf of LEB scales, $f_2(\lambda)$, for isoscalar surfaces at $Re \simeq 9.0 \times 10^3$. $c = c_1$: dotted line, crosses; $c = c_2$: dashed line, diamonds; $c = c_3$: solid line, circles. Recall (Fig. 48) that $\delta_b = \delta_b(c; Re)$.

The generalization to two (and higher) dimensions can be made in terms of the LEB size pdf, $f_d(\lambda)$, as was discussed in the previous section. The notion of "spacings", as used here, is restricted to one-dimensional data.

An account of this analysis was reported by Catrakis & Dimotakis (1996b)[19] and is also discussed elsewhere in this NATO/ASI series (Catrakis & Dimotakis 1996).[20]

4.3.   Two-dimensional scale distribution

Using the analytical results discussed in Sec. 4.1, the pdf of the LEB scale distribution, $f_2(\lambda)$, was computed from the two-dimensional scalar data, discussed in Sec. 3.2, at $Re \simeq 9.0 \times 10^3$, for the same three scalar thresholds: c_1, c_2, and c_3. This is depicted in Fig. 59, with lines of increasing solidity denoting increasing threshold, as before (recall that the intermediate threshold, c_2, corresponds to the peak of the scalar pdf). This is a normalized pdf over the range of scales. For a given threshold, $f_2(\lambda)$ is seen to be larger at smaller scales, approaching a constant value at the smallest scales,

$$\lim_{\lambda \to 0} \{f_d(\lambda)\} = \varsigma_d(c; Re) , \qquad (89)$$

as expected for level sets consisting of lines (curves) in a plane, i.e., for geometric sets with $d_t = d - 1$ (recall (85) and related discussion).

The data indicate a higher probability density of LEB scales, at small scales, for the c_2 threshold corresponding to the neighborhood of the peak of the scalar pdf. The small-scale limit, $f_2(\lambda \to 0) = \varsigma_2(c; Re)$, is a measure of the surface-to-volume ratio (perimeter-to-area ratio in 2-D) of the isosurface. The highest surface-to-volume ratio is observed at the c_2 threshold.

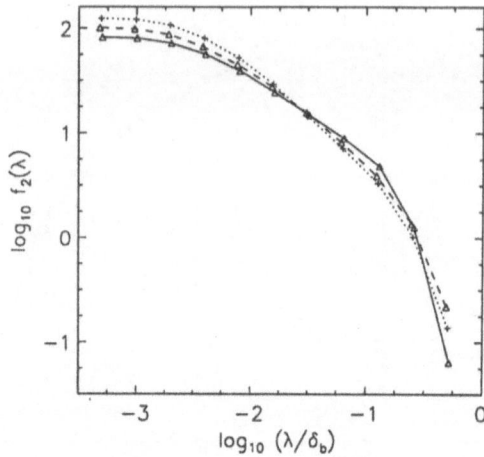

Fig. 60. — Re dependence of LEB scale pdf, $f_2(\lambda)$, at the intermediate threshold, c_2. Lines of increasing solidity denote increasing Re.

The Re dependence of the LEB scale pdf, $f_2(\lambda)$, is shown in Fig. 60. A systematic trend with increasing Re is evident. Specifically, the small-scale limit of the LEB scale pdf *decreases* with increasing Re, *i.e.*, the surface-to-volume (perimeter-to-area) ratio is decreasing with increasing Re. This, perhaps, surprising result can be seen in the SDF dimension behavior (Cf. Fig. 51) — from which it is derived ((84)) — with $D_2(\lambda)$ decreasing with increasing Re. It is also in accord with the conspicuous Re-dependence of the coverage length, $L(\lambda)$ (Cf. Fig. 54 and related discussion), which is derived directly from the coverage counts ((61)).

The scale dependence on Reynolds number is better illustrated in terms of the pdf, $\tilde{f}_2(\log \lambda)$, of $\log \lambda$. This is depicted in Fig. 61. These data indicate that small-scale regions of the flow are more likely to be visited by the isosurfaces; a lower probability of finding a largest-empty-box region of that size, as the Reynolds number is increased. Additionally, the expectation value of the LEB scale (as well as the most probable) is *increasing* with increasing Reynolds number, *i.e.*, the mean distance from a point in the interior of the bounding box to the c_2 isosurface is *increasing* with increasing Reynolds number.

These observations, taken collectively, indicate enhanced molecular *mixing*, that is responsible for (local) scalar-field homogenization, relative to *stirring* (recall discussion in Sec. 1.2), that is responsible for isoscalar surface-area generation, with increasing Reynolds number. This occurs at thresholds· corresponding to isoscalar surfaces (contours) mostly to be found in the intermediate-radius (high-shear) regions of the jet. This can be gleaned directly, in retrospect, from the level-set contours in Figs. 45,46,47. An examination of these three images confirms this (possibly counterintuitive) conclusion, *i.e.*, that the geometric complexity of the c_2 isocontour is decreasing, as the Reynolds number increases.

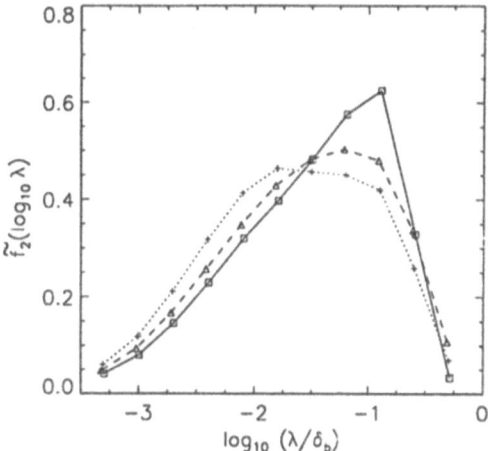

Fig. 61. — Re dependence of pdf of the logarithm of LEB scales, $\tilde{f}_2(\log \lambda)$, for $c = c_2$. Lines of increasing solidity denote increasing Re.

4.4.   Two-dimensional size distribution

Level sets of jet-fluid concentration consist of multiple, disjoint, closed, isoscalar contours of varying size and shape. These isoscalar contours may be viewed as either "islands", or "lakes", depending on whether the neighboring interior isosurfaces are at a higher, or lower, threshold, respectively (Fig. 52). The largest island in each image realization may be regarded as the "continent".

The pdf of the island/lake sizes, $p_A(A^{1/2}/\delta_b)$, with size defined as the square-root of the area enclosed by the contour, $A^{1/2}$, is depicted in Fig. 62. The lake/island data were compiled from all the 2-D slice data recorded at a Reynolds number of 9.0×10^3 and computed for level-set contours at the intermediate threshold, c_2. See Fig. 46 and Catrakis & Dimotakis (1996, Fig. 5)[20] for examples of individual realizations. Several hundred islands and lakes are found in each such image, with several thousand included in the statistics in Fig. 62. A Gaussian fit of the data (in these coordinates) is included in Fig. 62. The quality of the fit indicates that a lognormal distribution provides a good description of isoscalar island/lake sizes, at the inner scales (Catrakis & Dimotakis 1996b).[19]

This result is in accord with measurements of the (horizontal) size of clouds and radar-echo regions, for which a lognormal distribution was found in a variety of atmospheric conditions (Lopez 1977).[69] Also consistent are the previously-cited findings of lognormal distributions of level-crossing (spacing) scales derived from 1-D scalar measurements in liquid-phase turbulent jets (Miller & Dimotakis 1991)[81] and in plumes in the atmospheric surface layer (Yee et al. 1995).[123]

Arguments for lognormal statistics have been put forth by Kolmogorov (1941b,[58] 1962),[59] Oboukhov (1962),[83] Lopez (1977),[69] Bernal (1988),[6] and others, as applicable to stochastic fragmentation (or growth/amalgamation) processes, such as may be expected to occur in turbulence and mixing.

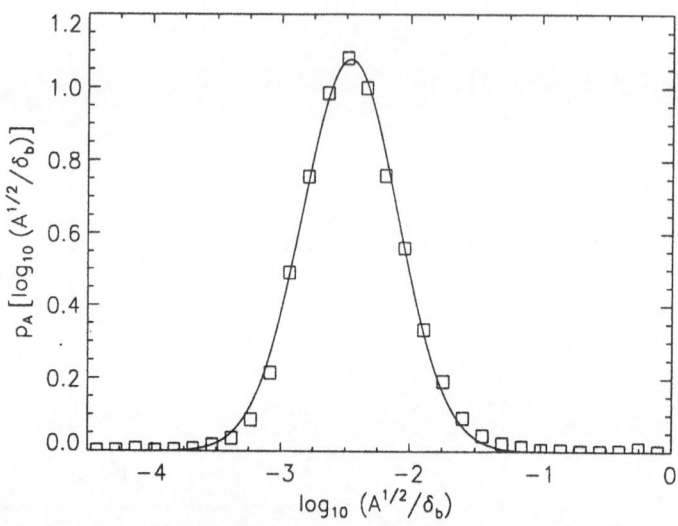

Fig. 62. — Probability density function of size, $A^{1/2}$, of isoscalar islands/lakes ($Re = 9.0 \times 10^3$). Solid curve: Gaussian fit (in these coordinates).

5. Three-dimensional scalar data

Before concluding, we should recall that turbulence — and the fields it generates — lives in three-dimensional space, $\vec{x} = (x, y, z)$, and, by being unsteady, generates scalar and other fields that are four dimensional (three space dimensions, plus time), e.g., concentration fields, $c = c(\vec{x}, t) = c(x, y, z, t)$. The advent of high framing-rate digital-image techniques, in conjunction with laser-induced fluorescence diagnostics, offer the promise of recording three- and four-dimensional, quantitative information of turbulent flows. Reports to date have documented imaging of small portions of turbulent flows at the inner scales (e.g., Prasad & Sreenivasan 1990b;[90] Dahm et al. 1991;[32] Merkel et al. 1996),[79] as well as investigations of large-scale structure by whole-field imaging (e.g., Yoda et al. 1994).[124]

Preliminary experiments to investigate the geometry of outer scalar isosurfaces from 3-D whole-field measurements are also in progress in our laboratory. This work represents a collaborative effort with J. Janesick, A. Collins, and T. Elliott of JPL; and D. Lang and D. Laidlaw of Caltech. In these experiments, three-dimensional space-time measurements (2-D space, plus time) of the jet-fluid concentration were obtained at the same far-field station ($z/d_j = 275$) of a liquid-phase turbulent jet by imaging the laser-induced fluorescence in the plane perpendicular to the jet axis and recording successive images as a function of time. Figure 63 shows a schematic of the flow geometry and associated imaging, as well as of the 3-D (fixed-z) slice space-time data, i.e., $c(x, y, t; z = 275d_j)$. The jet Reynolds number was $Re \simeq 9.0 \times 10^3$. For the data presented here, the framing rate was, approximately, 1-frame/s, with 1024^2 pixels/frame.

The successive-frame image data (30 frames total) were normalized and cali-

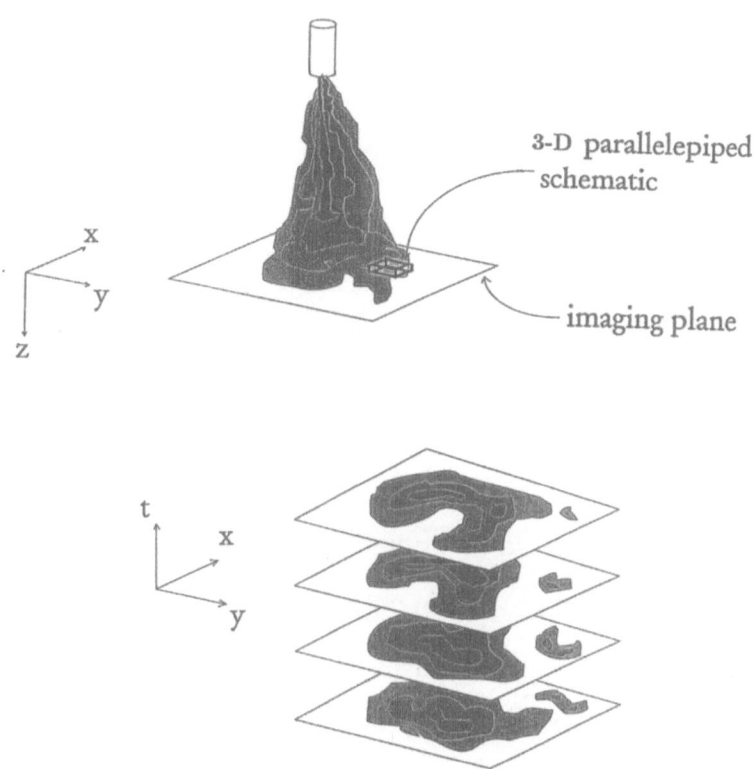

Fig. 63. — Schematic of flow geometry, associated imaging, and space-time data for 3-D (2-D in space, plus time) scalar-field measurements in a turbulent jet.

brated, pixel by pixel, using an ensemble-average of 8 background-noise images and 8 uniform-concentration images, using the same procedure as with the data in Figs. 38, 39,40. The signal-to-noise ratio of the data was also approximately 300:1, as estimated from the $\approx 50\,\mathrm{dB}$ amplitude range in the scalar power spectra. The total image-data were then comprised of $1024 \times 1024 \times 30 = 30\,\mathrm{Mpixels}$ of 12-bit data.

A $256 \times 256 \times 25$ portion of the image data was selected, near the outer boundary of the jet, to form the 3-D space-time images. In that region, the fluid velocity can be substantially lower than in the interior. Additionally, the successive, 256×256, subframe regions that were processed were selected as exhibiting a low rate of change from one temporal slice to the next. While, even for that portion, the scalar field must be regarded as temporally undersampled, in the strict sense, the resulting space-time image can be used to visualize the structure of the (outer) isosurfaces.

A 3-D rendering of an outer (low scalar threshold) isosurface portion, in this three-dimensional space-time volume, computed by D. Laidlaw of the Caltech Computer Science Department, is depicted in Fig. 64. In the computer visualization, a light source was positioned in this space-time volume and illuminated the, approximately,

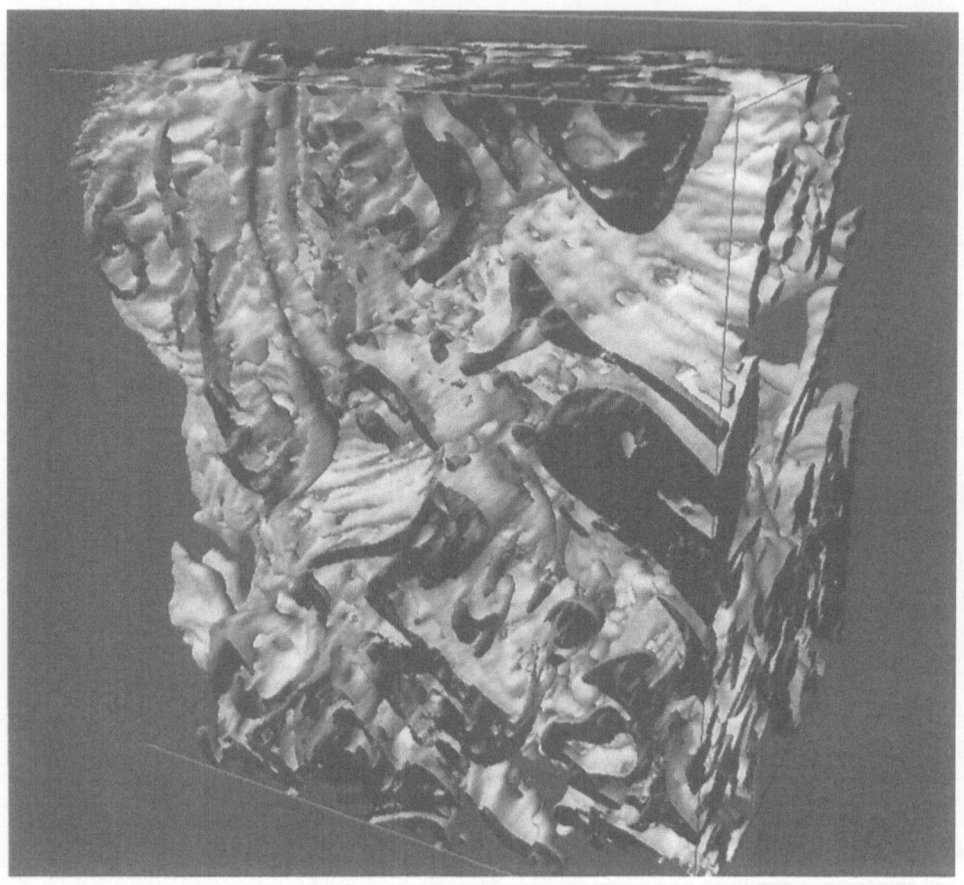

Fig. 64. — 3-D rendering of an outer (low scalar threshold) isosurface portion, derived from (2+1)-D space-time measurements of the jet-fluid concentration in the far field of a turbulent jet.

three million polygonal facets that were required to render the isosurface. Grey shades code time, with dark marking early time (rear) and light late time (front).

The selected scalar isosurface is seen to have a complex geometry, indicating the existence of both sheet-like ("lasagna") structures, with evidence of developing Kelvin-Helmholz instabilities (top right), isoscalar surfaces wrapped around tube-like ("spaghetti") structures (top left), as well as composite topologies, not readily classifiable, near the bottom. The three-dimensional geometrical structure of such isosurfaces can be analyzed by the natural extensions of the framework that was used for the 1-D and 2-D data discussed above.

6. Summary and conclusions

Proposals and experimental evidence, from both numerical simulations and laboratory experiments, regarding the behavior of level sets in turbulent flows have been discussed. A few important conclusions can be drawn from this overview.

Isoscalar surfaces in turbulent flows, at least in liquid-phase turbulent jets, where sufficient experimental evidence is available, appear to have a geometry that is more complex than (constant-D) fractal. Their description requires an extension of the original, scale-invariant, fractal framework that can be cast in terms of a variable (scale-dependent) coverage dimension, $D_d(\lambda)$. We have accepted the terms, scale-dependent-fractal (SDF) set and scale-dependent-fractal dimension, the latter originally employed by Mark & Aronson (1984),[76] to accommodate the description of sets exhibiting such behavior.

Sets exhibiting constant-D behavior have been termed "power-law-fractal" (PLF) sets and may be regarded as a special case. It is found that while constant-D scaling, generally, implies power-law, point-spacing statistics, in one-dimension, for example, the converse is not true. Power-law spacing statistics, over a significant (but finite) range of scales, may not produce constant-D fractal scaling.

The extended, scale-dependent framework allows level-set coverage statistics to be related to other quantities of interest. In addition to the pdf of point-spacings (in 1-D), it can be related to the scale-dependent surface-to-volume (perimeter-to-area in 2-D) ratio, as well as the distribution of sizes of empty-space regions, *i.e.*, in the case of level sets, the distribution of the length-scale of regions not visited by the isosurface. This latter statistic yields a measure of the distance from a random point in the domain to the set. This framework allows level-set geometry, in general, and scalar mixing, in particular, to be studied in terms of quantitative measures that are useful in many contexts.

The application of this SDF framework to the study of turbulent-jet mixing indicates that the geometric measures of the scalar (jet-fluid concentration) level-set field depend on both scalar threshold and flow Reynolds number. There is every reason to believe that this conclusion may be expected to apply to other turbulent flows as well and that no universal behavior need exist.

As regards mixing, the quantitative isoscalar analysis facilitated by the new tools indicate increased mixing with increasing Reynolds number, in liquid-phase turbulent jets, at least for the range of Reynolds numbers for which reliable data are available. This results in a progressively less-complex level-set geometry, at least in liquid-phase turbulent jets, with increasing Reynolds number. This, perhaps surprising, result is discernible in the scalar and isoscalar image data directly, in retrospect, and can be attributed to the increasingly effective homogenization of the scalar field and the attendant simpler isoscalar geometry with increasing Reynolds number.

By a number of measures, liquid-phase turbulent-jet flows exhibit a mixing transition, at a Reynolds number in the vicinity of 10^4, as has been documented to also occur in many other flows. Turbulent flows at Reynolds numbers below this value should, probably, not be regarded as fully-developed.

In liquid-phase turbulent jets, the spacings in one-dimensional records, as well as the size distribution of individual "islands" and "lakes" in two-dimensional isoscalar slices, where size is defined as the square root of the area of the individual islands and lakes, are found to be well described by lognormal statistics in the small-scale range. The scale-dependent coverage dimension derived from such sets is also in accord with lognormal statistics, in the small-scale range. Such statistics may be expected to apply to other flows, as well.

The technology that has yielded the preliminary three-dimensional data in turbulent flows, by us and others, along with data from the computational world, hold the promise of quantitative analysis of the full dimensionality of turbulence and its dynamics, across the whole range of scales, for the first time. If, as we have argued, a minimum Reynolds number is required for turbulence to attain its fully-developed character, perhaps the laboratory will be able to help us sooner than numerical simulations. Even in the laboratory, however, it is also on the computer we must rely to acquire, store, and analyze the demanding volume and rate of the new multi-dimensional experimental data.

In conclusion, recent data and a new geometric framework have provided additional information and tools that have helped probe the complex geometric signatures of turbulence. While the insight, thus far, derived from the new vantage point has been primarily descriptive, one may hope that it will also add to the experimental and theoretical means by which the dynamics of turbulence can be analyzed. It is too early to say whether these new tools will result in substantive progress in our quest to understand the behavior of turbulence, qualitatively and quantitatively. An important part of the complex natural world we live in, turbulence has intrigued observers from the earliest instances of recorded thought and has, for the most part, defied an adequate description. Perhaps because, as Heracleitos remarked many years ago,

$$H \ \phi\grave{v}\sigma\iota\varsigma \ \kappa\rho\grave{v}\pi\tau\epsilon\sigma\theta\alpha\iota \ \phi\iota\lambda\epsilon\acute{\iota},$$

or, "Nature likes to conceal itself," or, to be more precise, "... likes to conceal *herself*." The word *nature* is feminine and there can be no doubt that, in the eyes of the ancient Greeks, nature was a woman.

Acknowledgments

The work on fractals and their relation to mixing, described in this overview, has benefited from research performed at Caltech over the last 15 years, or so. The informal, unpublished, early efforts had contributions by R. Miake-Lye, D. Dowling, and G. Losi. The development of high signal-to-noise, temporal and spatial imaging, data-acquisition technology, on which this work has heavily relied, has had the benefit of many and valuable contributions by D. Lang. This tutorial has benefited from discussions with P. Miller, originally at Caltech and presently at the Lawrence Livermore National Laboratories, as well as with C. Bond, M. Cross, A. Leonard, D. Meiron, D. Pullin, and P. Saffman, at Caltech, and P. Constantin of the University of Chicago, I. Procaccia of the Weizmann Institute in Israel, and K. Sreenivasan of Yale University.

The preliminary 3-D space-time measurements discussed in Sec. 5 have not been published previously. They were recorded using the "Cassini camera", a digital-imaging system developed by D. Lang, P. Dimotakis, and P. Svitek at Caltech, with assistance and advice by J. Janesick, S. A. Collins, and T. Elliott, of Caltech's Jet Propulsion Laboratory. It is based on a 1024×1024 CCD focal-plane array developed by the above JPL team for the upcoming Cassini space mission. The camera is capable of higher framing rates and is presently being used to acquire additional and better-resolved space-time data. The resulting 3-D visualization image was computed by D. Laidlaw of Caltech.

The work at Caltech described here is part of a larger effort to investigate turbulent mixing and chemical reactions in free-shear flows, initially supported by AFOSR and GRI and, more recently, under AFOSR Grants F49620–92–J–0290 and F49620–94–1–0353. Finally, the authors would like to thank H. Chaté, J.-M. Chomaz, E. Villermaux, and the organizers of this NATO/ASI summer school for the invitation, their gracious hospitality, and financial support that made our participation possible.

References

[1] R. Andrle, Math. Geol. **28**, 275 (1996).

[2] H. Aref, J. Fluid Mech. **143**, 1 (1984).

[3] J. Baldyga and J.R. Bourne, Cehm. Eng. Sci. **50**, 381 (1995).

[4] G.K. Batchelor, *The Theory of Homogeneous Turbulence* (Cambridge U.P., Cambridge, 1953).

[5] G.K. Batchelor, J. Fluid Mech. **5**, 113 (1959).

[6] L.P. Bernal, Phys. Fluids **31**, 2533 (1988).

[7] L.P. Bernal, R.E. Breidenthal, G.L. Brown, J.H. Konrad, and A. Roshko, "On the Development of Three-Dimensional Small Scales in Turbulent Mixing Layers", in *2nd. Int. Symposium on Turb. Shear Flows*, Springer-Verlag, New York, 1979), pp. 305–313.

[8] R. Betchov and A. Szewczyk, Phys. Fluids 6, 1391 (1963).

[9] H. Blasius, Z. Math. u. Phys. **56**, 1 (1908).

[10] M.S. Borgas, Phil. Trans. Roy. Soc. Lond. A **342**, 379 (1993).

[11] P.N. Brandt, R. Greimel, E. Guenther, and W. Mattig, "Turbulence, Fractals, and the Solar Granulation", in *Applying Fractals in Astronomy*, A. Heck and J. M. Perdang eds., (Springer, Berlin, 1991), pp. 77–96.

[12] R.E. Breidenthal, J. FLuid Mech. **109**, 1 (1981).

[13] G.L. Brown and A. Roshko, J. Fluid Mech. **64**, 775 (1974).

[14] S.P. Burke and T.E.W. Schumann, Ind. Eng. Chem. **20**, 998 (1928).

[15] G.F. Carrier, F.E. Fendell, and F.E. Marble, SIAM J. Appl. Math. **28**, 463 (1975).

[16] C. Castagnoli and A. Provenzale, Astro & Astrophysics **246**, 634 (1991).

[17] H.J. Catrakis, Ph.D. thesis, California Institute of Technology, 1996.

[18] H.J. Catrakis and P.E. Dimotakis, J. Fluid Mech. **317**, 369 (1996).

[19] H.J. Catrakis and P.E. Dimotakis, Phys. Rev. Lett. **77**, 3795 (1996).

[20] H.J. Catrakis and P.E. Dimotakis, "Scale-dependent fractal geometry", this Volume.

[21] D.R. Chapman, AIAA J. **17**, 1293 (1979).

[22] J.P. Chilés, Math. Geol. **20**, 631 (1988).

[23] D. Coles, J. Fluid Mech. **1**, 191 (1956).

[24] D.J. Collins, D.E. Coles, and J.W. Hicks, "Measurements in the turbulent boundary layer at constant pressure in subsonic and supersonic flow. Part I. Mean flow," AEDC-TR-78-21 (1978).

[25] P. Constantin, "Remarks on the Navier-Stokes equations", in *New Perspectives in Turbulence*, L. Sirovich, ed., (Springer-Verlag, New York, 1991), pp. 229–261.

[26] P. Constantin, Comm. Math. Phys. **129**, 241 (1990).

[27] P. Constantin, SIAM Rev. **36**, 73 (1994).

[28] P. Constantin, "Geometric and Analytic Studies in Turbulence", in *Trends and Perspectives in Applied Mathematics*, L. Sirovich, Ed., (Springer-Verlag, New York, 1994), pp. 21–54.

[29] P. Constantin and I. Procaccia, Nonlinearity **7**, 1045 (1994).

[30] P. Constantin, I. Procaccia, and K.R. Sreenivasan, Phys. Rev. Lett. **67**, 1739 (1994).

[31] W.J.A. Dahm and P.E. Dimotakis, AIAA J. **25**, 1216 (1987).

[32] W.J.A. Dahm, K.B. Southerland, and K.A. Buch, Phys. Fluids A **3**, 1115 (1991).

[33] P.E. Dimotakis, "Turbulent shear layer mixing with fast chemical reactions", in *Turbulent Reactive Flows*, (Springer-Verlag, New York, 1989), pp. 417–485.

[34] P.E. Dimotakis, Nonlinear Sci. Today **2**, 1, 27–31 (1991).

[35] P.E. Dimotakis, "Some issues on turbulent mixing and turbulence", GALCIT Report FM93–1a (1993).

[36] P.E. Dimotakis, Engineering & Science**59**, 22 (1996).

[37] P.E. Dimotakis, R.C. Miake-Lye, and D.A. Papantoniou, Fluid Dyn. Trans. **11**, 47 (1983).

[38] C. Eckart, JMR **VII**, 265 (1948).

[39] K. Falconer, *Fractal Geometry: Mathematical Foundations and Applications*, Wiley, Chichester, UK, 1990).

[40] J. Feder, *Fractals*, (Plenum Press, New York, 1988).

[41] R.P. Feynman, R.B. Leighton, and M. Sands, *The Feynman Lectures on Physics* **2** (Addison-Wesley Publishing Co, Reading, MA, 1964).

[42] P. Flohr and D. Olivari, Physica D **76**, 278 (1994).

[43] C.A. Friehe, C.W. Van Atta, and C.H. Gibson, "Jet turbulence: Dissipation rate measurements and correlations", in AGARD report *Turbulent Shear Flows* CP-93, 18.1–7 (1971).

[44] U. Frish, *Turbulence. The Legacy of Kolmogorov*, (Cambridge U.P., Cambridge, 1995).

[45] C.H. Gibson, Phys. Fluids **11**, 2305 (1968).

[46] R.J. Gilbrech, *An Experimental Investigation of Chemically-Reacting, Gas-Phase Turbulent Jets*, Ph.D. thesis, California Institute of Technology (1991).

[47] R.J. Gilbrech and P.E. Dimotakis, "Product formation in chemically-reacting turbulent jets", AAIA 30th *Aerospace Sciences Meeting*, Paper 92-0581 (1992).

[48] S.S. Girimaji and S.B. Pope, J. Fluid Mech. **234**, 247 (1992).

[49] B.J. Gluckman, H. Willaime, and J.P. Gollub, Phys. Fluids **5**, 647 (1993).

[50] J.A. Hammer, *Lifted Turbulent Jet Flames*, Ph.D. thesis, California Institute of Technology (1993).

[51] F. Heslot, B. Castaing, and A. Libchaber, Phys. Rev. A **36**, 5870 (1987).

[52] J.O. Hinze, *Turbulence*, 2nd ed., (McGraw-Hill, 1975).

[53] J. Jimenez, A.A. Wray, P.G. Saffman, and R.S. Rogallo, "The structure of intense vorticity inhomogeneous isotropic turbulence", in *Studying Turbulence Using Numerical Simulations Databases – IV*, Center for Turbulent Research, NASA Ames & Stanford University (1992), pp. 21–45.

[54] P. Kailasnath and K.R. Sreenivasan, Phys. Fluids A **5**, 2879 (1993).

[55] A.R. Karagozian and F.E. Marble, Combust. Sci. and Technol. **45**, 65 (1986).

[56] A.R. Kerstein, Phys. Fluids A **3**, 1110 (1991).

[57] A.N. Kolmogorov, Dokl. Akad. Nauk SSSR **30**, 299 (1941). Reprinted in: Usp. Fiz. Nauk **93**, 476 (1967). Translated into English in: Sov. Phys. Usp. **10**, 734 (1968).

[58] A.N. Kolmogorov, Akad. Nauk SSSR **31**, 99 (1941). Translated, V. Levin, 1969, NASA TT F–12287.

[59] A.N. Kolmogorov, J. Fluid Mech. **13**, 82 (1962).

[60] J.H. Konrad, *An Experimental Investigation of Mixing in Two-Dimensional Turbulent Shear Flows with Applications to Diffusion-Limited Chemical Reactions*, Ph.D. thesis, California Institute of Technology (1976).

[61] M.M. Koochesfahani and P.E. Dimotakis, J. Fluid Mech. **170**, 83 (1986).

[62] V.R. Kuznetsov and V.A. Sabel'nikov, *Turbulence and Combustion*, (Hemisphere Publishing, New York, 1990).

[63] J. Laherrère, C. R. Acad. Sci. Paris **322**(IIa), 535 (1996).

[64] G.F. Lane-Serff, J. Fluid Mech. **249**, 521 (1993).

[65] D.P. Lathrop, J. Fineberg, and H.L. Swinney, Phys. Rev. Lett. **68**, 1515 (1992).

[66] D.P. Lathrop, J. Fineberg, and H.L. Swinney, Phys. Rev. A **46**, 6390 (1992).

[67] D. Liepmann and M. Gharib, J. Fluid Mech. **245**, 643 (1992).

[68] H.W. Liepmann, Helv. Phys. Acta **22**, 119 (1949).

[69] R.E. Lopez, Mon. Wea. Rev. **105**, 865 (1977).

[70] B.B. Mandelbrot, Science **155**, 636 (1967).

[71] B.B. Mandelbrot, J. Fluid Mech. **72**, 401 (1975).

[72] B.B. Mandelbrot, *Les Objets Fractals: Forme, Hasard, et Dimension*, (Flammarion, Paris, 1975).

[73] B.B. Mandelbrot, *Fractals, Form, Chance, and Dimension*, (W. H. Freeman & Co., San Fransisco, 1977).

[74] B.B. Mandelbrot, *The Fractal Geometry of Nature*, (W. H. Freeman & Co., San Fransisco, 1982).

[75] B.B. Mandelbrot, "Fractal geometry: What is it and what does it do?", in *Fractals in the Natural Sciences*, M. Fleischmann, D. J. Tildesley, and R. C. Ball, Eds., (Princeton U.P., 1989), p. 7.

[76] D.M. Mark and P.B. Aronson, Math. Geol. **16**, 671 (1984).

[77] S. Matsuura, S. Tsurumi, and N. Imai, J. Chem. Phys. **84**, 539 (1986).

[78] W.D. McComb, *The Physics of Fluid Turbulence*, (Clarendon Press, Oxford, 1991).

[79] G.J. Merkel, T. Dracos, T., and P. Rys, "Two-dimensional and three-dimensional imaging of passive scalar fields in a turbulent jet", in *Atlas of Visualization*, Y. Nakayama, and Y. Tanida. Eds., (CRC Press, 1996), pp. 67–78.

[80] A.S. Mikhailov and A.Y. Loskutov, *Foundations of Synergetics II: Complex Patterns*, (Springer-Verlag, Berlin, 1991).

[81] P.L. Miller and P.E. Dimotakis, Phys. Fluids A **3**, 168 (1991); 1156 (1991).

[82] A.S. Monin and A.M. Yaglom, *Statistical Fluid Mechanics: Mechanics of Turbulence II*, (MIT Press, Cambridge, 1975).

[83] A.M. Obukhov, J. Fluid Mech. **13**, 77 (1962).

[84] J.M. Ottino, *The Kinematics of Mixing: Stretching, Chaos, and Transport* (Cambridge University Press, 1989).

[85] J.M. Ottino, Ann. Rev. Fluid Mech. **22**, 207 (1990).

[86] S.B. Pope, Int. J. Eng. Sci. **26**, 445 (1988).

[87] S.B. Pope, P.K. Yeung, and S.S. Girimaji, Phys. Fluids A **1**, 2010 (1989).

[88] R.R. Prasad and K.R. Sreenivasan, Exp. in Fluids **7**, 259 (1989).

[89] R.R. Prasad and K.R. Sreenivasan, Phys. Fluids A **2**, 792 (1990).

[90] R.R. Prasad and K.R. Sreenivasan, J. Fluid Mech. **216**, 1 (1990).

[91] A.A. Praskovsky, J.F. Foss, S.J. Kleis, and M.Y. Karyakin, Phys. Fluids A **5**, 2038 (1993).

[92] W.H. Press, B.P. Flannery, A.A. Teukolsky, and W.T. Vetterling, *Numerical Recipes. The Art of Scientific Computing* (Cambridge University Press, 1986).

[93] I. Procaccia, A. Brandenburg, M.H. Jensen, and A. Vincent, Europhys. Lett. **19**, 183 (1992).

[94] J. Stein (Ed. in Chief), *The Random House Dictionary of the English Language*, (Random House, New York, 1971).

[95] L.F. Richardson, Proc. Roy. Soc. London A **110**, 709 (1926).

[96] L.F. Richardson, General Systems Yearbook. **6**, 139 (1961).

[97] J.-P. Rigaut, "The problem of contiguity: an appendix of statistics of deadly quarrel", in *Fractals: Non-integral Dimensions and Applications*, G. Cherbit, Ed., (Wiley, Chichester, UK, 1991), p. 151.

[98] V. Rom-Kedar, A. Leonard, and S. Wiggins, J. Fluid Mech. **214**, 347 (1990).

[99] A. Roshko, "On the development of turbulent wakes from vortex sheets", NACA Report 1191 (1954).

[100] A. Roshko, "Perspectives on bluff body aerodynamics", 2nd *Int. Coll. on Bluff Body Aerodynamics*, Melbourne, Australia (1992).

[101] J.C. Russ, *Fractal Surfaces* (Plenum, New York, 1994).

[102] S.G. Saddoughi and S.V. Veeravalli, J. Fluid Mech. **268**, 333 (1994).

[103] S.F. Shen, J. Aeronaut. Sci. **21**, 62 (1954).

[104] K.R. Sreenivasan, Phys. Fluids **27**, 1048 (1984).

[105] K.R. Sreenivasan, Ann. Rev. Fluid Mech. **23**, 539 (1991).

[106] K.R. Sreenivasan, Fractals, **2**, 253 (1994).

[107] K.R. Sreenivasan and C. Meneveau, J. Fluid Mech. **173**, 357 (1986).

[108] K.R. Sreenivasan, R.R. Prasad, C. Meneveau, and R. Ramshankar, Pure & Appl. Geoph. **131**, 43 (1989).

[109] K.R. Sreenivasan, A. Prabhu, and R. Narasimha, J. Fluid Mech. **137**, 251 (1983).

[110] M. Suzuki, Prog. Theor. Phys. **71**, 1397 (1984).

[111] H. Takayasu, J. Phys. Soc. Japan **51**, 3057 (1982).

[112] H. Takayasu, *Fractals in the Physical Sciences*, (Wiley, Chichester, UK, 1992).

[113] G.I. Taylor, Proc. London Math. Soc. **20**, 196 (1921).

[114] C. Tricot, *Curves and Fractal Dimension*, (Springer-Verlag, New York, 1995).

[115] M. van Dyke, *An Album of Fluid Motion*, (Parabolic Press, Stanford, 1982).

[116] T. Vicsek, *Fractal Growth Phenomena*, 2nd Ed., (World Scientific, Singapore, 1992).

[117] E. Villermaux and Y. Gagne, Phys. Rev. Lett. **73**, 252 (1994).

[118] A. Vincent and M. Meneguzzi, J. Fluid Mech. **225**, 1 (1991).

[119] T. von Karman, J. R. Aero. Soc. **41**, 1109 (1937).

[120] P. Welander, Tellus **7**, 141 (1955).

[121] N. Wiener, *Extrapolation, Interpolation and Smoothing of Stationary Time Series* (Wiley, New York, 1949).

[122] I. Wygnanski and F.H. Champagne, J. Fluid Mech. **59**, 281 (1973).

[123] E. Yee, R. Chan, P.R. Kosteniuk, G.M. Chandler, C.A. Biltoft, and J.F. Bowers, Bound. Layer Met. **73**, 53 (1995).

[124] M. Yoda, L. Hesselink, and M.G. Mungal, J. Fluid Mech. **279**, 313 (1994).

Scale-Dependent Fractal Geometry

HARIS J. CATRAKIS[1] AND PAUL E. DIMOTAKIS[2]

[1] Mechanical and Aerospace Engineering, University of California, Irvine, USA
[2] Graduate Aeronautical Laboratories, California Institute of Technology
Pasadena, CA 91125, USA

———————

A generalization of fractal geometry that allows for scale-dependent behavior is described. The resulting framework, termed "scale-dependent fractal" (SDF) geometry, can be used to quantify and model various phenomena whose geometric complexity may depend on scale. Relations are derived between scale distributions and coverage dimensions in the SDF framework, for one- as well as multi-dimensional sets, and applied to level sets arising in turbulence and mixing.

1. Introduction

Various phenomena exhibit complex multiscale structure over a wide range of spatial and/or temporal scales (*e.g.*, Mandelbrot 1982,[24] Cross & Hohenberg 1993).[11] To quantify the geometric behavior, several measures may be employed. Of particular importance are such measures as, for example, surface-volume measures of scalar isosurfaces, that are useful in turbulent mixing and combustion (*e.g.*, Sreenivasan 1991),[31] or in nephron-cell boundaries, employed to estimate fluid transport driven by osmotic gradients (*e.g.*, Welling *et al.* 1996),[38] etc. Analyses of the geometry of complex phenomena have primarily been put forth in terms of (constant-dimension) fractal descriptions (*e.g.*, Mandelbrot 1982,[24] Sreenivasan 1991).[31]

More general descriptions are possible, however, in terms of a scale-dependent extension of the fractal framework in which the fractal dimension may be a function of scale. The extended framework, termed scale-dependent fractal (SDF), will

Mixing: Chaos and Turbulence, edited by Chaté *et al.*
Kluwer Academic / Plenum Publishers, New York 1999.

be discussed below. The SDF framework retains the notion of quantifying complex structures in terms of fractal (noninteger) dimensions, while allowing for scale-dependent behavior. As a consequence, it can describe phenomena whose geometry may vary with scale. In particular, SDF behavior can be connected to classical descriptions of multiscale geometry that have employed distributions of scales in 1-D, e.g., zero-crossing spacings analyzed by Rice (1945) and[28] Liepmann (1949).[19] In this report, a recently-developed framework relating scale distributions to coverage dimensions, in one- as well as multi-dimensional spaces, is described along with examples arising in turbulence and mixing.

There are many reports of (constant-dimension) fractal behavior in nature. This is indicated by power-law scaling inferred from the coverage count, $N_d(\lambda)$, of the set embedded in a d-dimensional space, over a range of scales, with the exponent identified as the fractal dimension, D_d, e.g., Mandelbrot (1982),[24]

$$N_d(\lambda) \propto \lambda^{-D_d} . \tag{1}$$

Various phenomena, however, are found to exhibit scale-dependent fractal (SDF) behavior, over a range of scales, as measured by the coverage dimension, $D_d(\lambda)$, e.g., Takayasu (1982),[34]

$$D_d(\lambda) \equiv - \frac{d \log N_d(\lambda)}{d \log \lambda} . \tag{2}$$

In these expressions, the coverage, $N_d(\lambda)$, counts the number of nonoverlapping, contiguous boxes, of size λ, that partition the embedding space, needed to cover the set.

SDF behavior accommodates a dependence of the coverage dimension, $D_d(\lambda)$, on the scale, λ. The scale λ may be spatial or temporal, accordingly. Such dimensions have been obtained, for example, for random-walk trajectories (as opposed to level sets) in 1-D by Takayasu (1982),[34] for Japanese coastlines by Suzuki (1984),[33] topographic surfaces by Mark & Aronson (1984),[25] random-walk (Brownian motion) trajectories of particles in solution by Matsuura et al. (1986),[26] one- and two-dimensional surveys of the fracture network of drifts of a uranium mine in a granite massif by Chilès (1988),[9] the galaxy distribution in the universe by Castagnoli & Provenzale (1991),[5] solar granulation by Brandt et al. (1991),[4] as well as the alveolar structure of lung tissues of prematurely-born rabbits by Rigaut (1991).[30] Rigaut, in particular, studied many other biological organisms and reported that, "the [coverage] graphs were concave [curved] for all ... objects studied," including "optical microscope [studies of] nuclear contours of erythrocytes and human blood lymphocytes, ... electron micrographs [of] the cellular outline of a foetal blood erythroblast of a rat, ... the nuclear outline of a human pulmonary macrophage, ... " for example (Rigaut 1991, p. 159).[30] SDF behavior has also been reported by Laherrère (1996)[18] for the distribution of town sizes. See also Dimotakis & Catrakis (1996, Sec. 2.2) for a discussion of reports of SDF behavior.

Amplifying Richardson's (1961)[29] notions of self-similarity and scaling in his analysis of coastline and border data, Mandelbrot proposed (1967)[22] that the West Coast of Britain may be regarded as statistically self-similar, i.e., possessing a (constant) fractal dimension. A recent study of Britain's West Coast by Andrle (1996a),[2] however, reports that, "... although the resulting log-log plot of measured length against steplength appears linear, statistical tests for linearity strongly suggest that the coastline is not statistically self-similar." Andrle (1996b)[3] also reports SDF behavior for various other coastlines of fjord, volcanic, as well as tectonic origin. Andrle (1996a) suggests that, regarding reports of constant fractal dimensions, in general,

"

> ... apparently linear segments (fractal elements) often found in Richardson [coverage] plots may contain systematic curvature revealed only by more rigorous tests for non-linearity."

Different investigators have employed various terms to denote (what is here termed) SDF behavior: "differential fractal" by Takayasu (1982),[34] "transient fractal" by Suzuki (1984),[33] "superfractal" by Denley (1990),[12] "semi-fractal" by Rigaut (1991),[30] "partially scaling fractal" by Amritkar (1994),[1] "nonlinear scaling" by Tsonis & Elsner (1995),[36] "parabolic fractal" by Laherrére (1996),[18] and "characteristic scale [behavior]" by Andrle (1996),[2, 3] for example. The word "fractal" was coined by Mandelbrot (1975)[23] from the Latin *fractus*, meaning "fragmented," as he noted (Mandelbrot 1982, p. 4).[24] While conventionally used to denote power-law (fragmented) behavior, "fractal" need not be excluded from referring to scale-dependent fragmented behavior. The latter will be referred to here as "scale-dependent fractal" (SDF) behavior, a term previously employed by Mark & Aronson (1984)[25] and Takayasu (1992),[35] for example.

Some investigators proposed analytical expressions for the observed SDF behavior. Takayasu (1982)[34] derived an expression for the coverage dimension of a random walk trajectory in 1-D, given by,

$$D(\lambda) = 2 - \frac{1}{1 + \lambda/\lambda_0} , \qquad (3)$$

where λ_0 is a scale proportional to the random-walk step size. The dimension, $D(\lambda)$, increases continuously with scale from unity, at the small scales (relative to the step size), to 2, at the large scales. This expression was found to compare well with measurements of Matsuura *et al.* (1986)[26] of the (3-D) random walk (Brownian motion) of particles in solution (polystyrene latex, bacteriophage T4, as well as T4 particles with host *E. coli B* cells) observed through a microscope. In their data analysis, Matsuura *et al.* employed the expression,

$$D(\lambda) = 2 - \frac{1}{1 + (\lambda/\lambda_0)^n} , \qquad (4)$$

where n is a "sharpness parameter" of the variation of $D(\lambda)$ from unity, at the small scales (relative to the step size), to 2, at the large scales, and reported that $n \simeq 1$. Rigaut (1991)[30] proposed, empirically, an expression for the perimeter of lung tissues of prematurely-born rabbits as a function of scale, which is equivalent to a coverage dimension,

$$D(\lambda) = 2 - \frac{1 + (1 - \beta)\,(\lambda/\lambda_0)^\beta}{1 + (\lambda/\lambda_0)^\beta} , \qquad (5)$$

in a range of scales. In this expression, $D(\lambda)$ increases from unity, at small scales, to

$1 + \beta$, at larger scales. In addition to such expressions for the coverage dimension,([1]) models of SDF behavior have been proposed in terms of scale-dependent generalizations of classical fractal models, such as SDF Koch curves (*e.g.*, Suzuki 1984),[33] or SDF Cantor dust (*e.g.*, Chilès 1988).[9]

Experimental evidence necessitating the generalization of (constant-dimension) fractal geometry, to accommodate scale-dependent behavior in turbulence, was presented in Miller & Dimotakis (1991)[27] and further discussed in Dimotakis (1991).[13] In the former, using a Monte-Carlo simulation, it was also shown that the measured behavior of the scale-dependent coverage dimension of threshold-crossings of scalar (jet-fluid concentration) time series could be well-matched by the coverage of points with lognormal-spacing statistics. Additional evidence in turbulence and a development of the mathematical framework were reported in Catrakis & Dimotakis (1996a,b).[7][8] The resulting framework deals with scale-dependent geometry. Relations were also established between coverage statistics and scale distributions. Specifically, a transform pair was derived connecting the distribution of spacing scales to the coverage dimension for one- as well as multi-dimensional geometries.

2. Scale distributions and coverage dimensions

In the SDF framework, objects can exhibit a coverage dimension, $D_d(\lambda)$, that varies continuously with scale, λ, in a range of scales; cf. (2). This coverage dimension, at a scale λ, can be identified as the fractional decrease in coverage, $-\,\mathrm{d}N_d/N_d$, per unit fractional increase in scale, $\mathrm{d}\lambda/\lambda$, *i.e.*,

$$\frac{\mathrm{d}N_d(\lambda)}{N_d(\lambda)} = -D_d(\lambda)\frac{\mathrm{d}\lambda}{\lambda}\,. \tag{6}$$

SDF objects may be regarded as scale-dependent generalizations of (power-law) fractals (PLFs) and offer a description in which variable complexity is present at different scales.

SDF behavior is associated with a coverage relation given by inverting (2) (cf. Takayasu 1982,[34] Catrakis & Dimotakis 1996a),[7] *i.e.*,

$$N_d(\lambda) = \exp\left\{\int_\lambda^{\delta_\mathrm{b}} D_d(\lambda')\frac{\mathrm{d}\lambda'}{\lambda'}\right\}, \tag{7}$$

where δ_b is the largest scale of the object, *e.g.*, bounding-box size. For objects confined in a d-dimensional bounding volume V_b, $\delta_\mathrm{b} = V_\mathrm{b}^{1/d}$, with $N_d(\delta_\mathrm{b}) = 1$ by construction.

([1]) Amritkar[1] suggested "additive and multiplicative corrections" to power-law scaling, to describe objects that do not exhibit exact power laws, and wrote, for the coverage,

$$N(\lambda) \sim \lambda^{-D}\,e^{K_1/\lambda}\,[\log(1/\lambda)]^{K_2}\left(A + B\lambda^{-K_3}\right)\,,$$

stating that D is the fractal dimension, K_1 is the "exponential dimension," K_2 is the "logarithmic dimension," and K_3 is the "metadimension" (Amritkar 1994, p. 596).[1] With such corrections, however, D is no longer a scale-local measure (logarithmic derivative) of the coverage behavior.

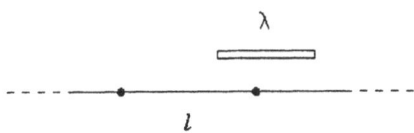

Fig. 1. — Spacing scale, l, and coverage scale (tile size), λ, for a stochastic point process (indicated by filled circles) on the real line.

In contrast to PLFs, the coverage of SDFs is a nonlocal function of scale, with geometric structure across the whole range of scales contributing to the coverage at any one scale.[2] In terms of the coverage fraction, $F_d(\lambda)$, defined as (Dimotakis 1991),[13]

$$F_d(\lambda) \equiv \frac{N_d(\lambda)}{N_{d,tot}(\lambda)} = \left(\frac{\lambda}{\delta_b}\right)^d N_d(\lambda) , \tag{8}$$

where $N_{d,tot}(\lambda)$ is the total number of λ-boxes that can fit in the bounding box (for data of finite extent), the SDF relation (7) can be written as,

$$F_d(\lambda) = \exp\left\{-\int_{\lambda}^{\delta_b}\left[d - D_d(\lambda')\right]\frac{d\lambda'}{\lambda'}\right\} , \tag{9}$$

since, at the largest scale, $F_d(\delta_b) = 1$. The coverage fraction, $F_d(\lambda)$, in the limiting case of statistics compiled over an ensemble of realizations, represents the probability that a randomly-placed d-dimensional box, of size λ, will cover the set. We will return to this property in the discussion below.

The degree to which a SDF set fills space, therefore, varies with scale (spatial resolution) and is dependent on the geometric behavior at other scales. The coverage fraction can be identified as the geometric probability that a randomly placed λ-box, interior to the bounding box, covers part of the set. This can be related to the distribution of (various measures of) scales in one- as well as multi-dimensional spaces, as described below.

For point sets in 1-D, the coverage dimension can be connected to the probability density function (pdf) of the point spacing (Catrakis & Dimotakis 1996b).[8] Consider a stochastic point process, in space or time, with a distribution of spacing scales, i.e., interval-lengths between successive events of the process, described by a probability density function (pdf), $p(l)$, where l denotes a (nonnegative) spacing scale (cf. Fig. 1). The fraction of length spanned by a l-scale will be $m(l) \propto l\,p(l)$. The probability density that a random location, with uniform measure on the real line, lies in a l-spacing is, therefore (cf. Fig. 1),

$$m(l) = \frac{l\,p(l)}{l_m} , \quad \text{where,} \quad l_m \equiv \int_0^{\infty} l\,p(l)dl \tag{10}$$

_____

[2] If $D_d(\lambda) \neq$ const., (7) does not imply a power-law-like coverage, i.e., $N_d(\lambda) \not\propto \lambda^{-D_d(\lambda)}$.

(assuming a finite mean scale, l_{m}). The geometric probability that a λ-tile, randomly located on the real line (with uniform measure), is empty, *i.e.*, covers no points, can be written as,

$$p_{\mathrm{e}}(\lambda) = \int_0^\infty p_{\mathrm{e}}(\lambda \mid l)\, m(l)\, dl \;, \tag{11}$$

where $p_{\mathrm{e}}(\lambda \mid l)$ is the conditional probability that a randomly-placed λ-tile covers no points, given that it lies in a l-spacing, and is given by (cf. Fig. 1),

$$p_{\mathrm{e}}(\lambda \mid l) = \begin{cases} 0 \;, & \text{for } \lambda > l \;; \\ 1 - \lambda/l \;, & \text{for } \lambda \le l \;. \end{cases} \tag{12}$$

The probability that a λ-tile lies in a l-spacing is given by $m(l)$; cf. (10). By a λ-tile lying in a l-spacing we mean that a chosen (reference) point of the tile, *e.g.*, the left endpoint of the tile, lies in a l-spacing, cf. Fig. 1. The probability that a λ-tile does not contain any transitions is then (cf. Eqs. (10–12)),

$$p_{\mathrm{e}}(\lambda) = \int_\lambda^\infty \left(\frac{l - \lambda}{l_{\mathrm{m}}} \right) p(l)\, dl \;. \tag{13}$$

To make the connection to coverage statistics, consider a L-record, $L \gg l_{\mathrm{m}}$, partitioned in λ-tiles, $\lambda \le L$. Let $N_1(\lambda; L)$ be the coverage count, *i.e.*, the (ensemble-averaged) number of nonoverlapping, contiguous λ-tiles necessary to cover all points (*e.g.*, threshold-level transitions) in the L-record. Consider the coverage fraction, $F_1(\lambda)$, *i.e.*, the ratio of the number of tiles that cover transitions to the total number of tiles (Dimotakis 1991, Catrakis & Dimotakis 1996a).[13] [7] We then have,

$$F_1(\lambda) = 1 - \lim_{L/\lambda \to \infty} \left\{ \frac{N_{1,\mathrm{e}}(\lambda; L)}{N_{1,\mathrm{t}}(\lambda; L)} \right\} = 1 - p_{\mathrm{e}}(\lambda) \;, \tag{14}$$

where $N_{1,\mathrm{t}}(\lambda; L) = N_1(\lambda) + N_{1,\mathrm{e}}(\lambda; L)$ is the total number of λ-tiles and $N_{1,\mathrm{e}}(\lambda; L)$ is the number of empty λ-tiles in the L-record.[3] The coverage fraction can then be expressed as (cf. (13)),

$$F_1(\lambda) = 1 - \int_\lambda^\infty \left(\frac{l - \lambda}{l_{\mathrm{m}}} \right) p(l)\, dl = \frac{1}{l_{\mathrm{m}}} \int_0^\lambda \int_{\lambda'}^\infty p(l)\, dl\, d\lambda \;, \tag{15}$$

which yields,

$$\begin{aligned} F_1(\lambda) &= \tfrac{1}{l_{\mathrm{m}}} \int_0^\lambda \int_{\lambda'}^\lambda p(l)\, dl\, d\lambda + \tfrac{1}{l_{\mathrm{m}}} \int_0^\lambda \int_\lambda^\infty p(l)\, dl\, d\lambda \\ &= \tfrac{1}{l_{\mathrm{m}}} \int_0^\lambda l\, p(l)\, dl + \tfrac{\lambda}{l_{\mathrm{m}}} \int_\lambda^\infty p(l)\, dl \;, \end{aligned} \tag{16}$$

or, in terms of the (cumulative) scale distribution function, $P(l) \equiv \int_0^l p(l')\, dl'$,

$$F_1(\lambda) = \frac{1}{l_{\mathrm{m}}} \int_0^\lambda [1 - P(\lambda')]\, d\lambda = \frac{\lambda}{l_{\mathrm{m}}} - \frac{1}{l_{\mathrm{m}}} \int_0^\lambda P(\lambda')\, d\lambda \;. \tag{17}$$

---

[3] This relation is also discussed by Vassilicos & Hunt (1991),[37] who propose, however, in place of (14) (in the present notation), $F_1(\lambda) = 1 - \int_\lambda^\infty p_{\mathrm{e}}(\lambda')\, d\lambda'$.

The limiting behavior of $F_1(\lambda)$, at small or large coverage scales, is then,

$$F_1(\lambda) \quad \begin{cases} \sim \lambda/l_m \to 0 \,, & \text{as } \lambda \to 0 \,; \\ \to 1 \,, & \text{as } \lambda \to \infty \,, \end{cases} \tag{18}$$

where the latter limit follows from the (normalization) (14). The corresponding coverage dimension,

$$D_1(\lambda) \equiv -\frac{\mathrm{d}\log N_1(\lambda)}{\mathrm{d}\log \lambda} = 1 - \frac{\mathrm{d}\log F_1(\lambda)}{\mathrm{d}\log \lambda} \,, \tag{19}$$

can be expressed in terms of the pdf of spacing scales, i.e.,

$$D_1(\lambda) = 1 - \frac{\lambda \int_\lambda^\infty p(l)\,\mathrm{d}l}{l_m - \int_\lambda^\infty (l - \lambda)\,p(l)\,\mathrm{d}l} = 1 - \frac{\lambda \int_\lambda^\infty p(l)\,\mathrm{d}l}{\int_0^\lambda \int_{\lambda'}^\infty p(l)\,\mathrm{d}l\,\mathrm{d}\lambda'} \,, \tag{20}$$

which can be viewed as a SDF transform, connecting $p(l)$ to $D_1(\lambda)$. Two limiting cases follow from (20),

$$D_1(\lambda) \to \begin{cases} 0 \,, & \text{as } \lambda \to 0 \,; \\ 1 \,, & \text{as } \lambda \to \infty \,, \end{cases} \tag{21}$$

as required for the coverage of 1-D transition sets (cf. Dimotakis 1991).[13]

Inverse relations connecting the pdf of spacing scales to the coverage dimension (and coverage fraction) can also be obtained. In particular ((15)), we have,

$$\frac{\mathrm{d}F_1(\lambda)}{\mathrm{d}\lambda} = \frac{1}{l_m} \int_\lambda^\infty p(l)\,\mathrm{d}l \,, \tag{22}$$

and, therefore,[4]

$$p(l) = -l_m \frac{\mathrm{d}^2 F_1(l)}{\mathrm{d}l^2} \,. \tag{23}$$

The mean spacing scale, l_m, can then be written as (cf. (22)),

$$\frac{1}{l_m} = \lim_{\lambda \to 0} \left\{ \frac{\mathrm{d}F_1(\lambda)}{\mathrm{d}\lambda} \right\} \,, \tag{24}$$

in agreement with the small-scale behavior of the coverage fraction (cf. (18)). Since the coverage fraction can also be written in terms of the coverage dimension as,

$$F_1(\lambda) = \exp\left\{ -\int_\lambda^\infty \left[1 - D_1(\lambda') \right] \frac{\mathrm{d}\lambda'}{\lambda'} \right\} \,, \tag{25}$$

(cf. (19)) it follows that,

$$\lim_{\lambda \to 0} \left\{ \frac{\mathrm{d}F_1(\lambda)}{\mathrm{d}\lambda} \right\} = \lim_{\lambda \to 0} \left\{ \frac{1}{\lambda} \exp\left[-\int_\lambda^\infty \left[1 - D_1(\lambda') \right] \frac{\mathrm{d}\lambda'}{\lambda'} \right] \right\} \,. \tag{26}$$

---

[4] A relation similar to (23) was obtained by Longuet-Higgins (1958)[20] for the special case of zero-crossings of stochastic Gaussian functions.

Combining Eqs. (23), (25), and (26) we obtain the (inverse) relation connecting the pdf of spacing scales to the coverage dimension, *i.e.*,

$$p(l) = \frac{l_m}{l^2}\left\{D_1(l)\left[1 - D_1(l)\right] + l\frac{dD_1(l)}{dl}\right\}$$
$$\times \exp\left\{-\int_l^\infty\left[1 - D_1(l')\right]\frac{dl'}{l'}\right\},$$

(27)

where the mean scale, l_m, is given by (cf. Eqs. (24) and (26)),

$$l_m = \lim_{l\to 0}\left\{l\,\exp\left[\int_l^\infty\left[1 - D_1(l')\right]\frac{dl'}{l'}\right]\right\}.$$

(28)

An alternative measure of geometric scale, in 1-D, that can be connected to coverage statistics and can be generalized to d-dimensional sets (see below), is the largest-empty-tile scale. This scale is the size of the largest tile, (centered) at a random location, that is empty (does not cover points of the set). The pdf of this scale, $f_1(\lambda)$, is also the probability density that a random point is a distance $\lambda/2$ away from the nearest element of the point set and is given by,

$$f_1(\lambda) = \int_0^\infty h_e(\lambda\,|\,l)\,m(l)\,dl,$$

(29)

(proportional to the fraction of intervals greater than λ). The conditional probability density, $h_e(\lambda\,|\,l)$, that a randomly-placed λ-tile is the largest-empty-tile, given that it lies in a l-interval, is given by,

$$h_e(\lambda\,|\,l) = \begin{cases} 0, & \text{for } \lambda > l; \\ 1/l, & \text{for } \lambda < l, \end{cases}$$

(30)

since $m(l)$ is the probability density that a λ-tile lies in a l-spacing (cf. (11) and related discussion). For finite $p(l)$, the pdf of largest-empty-tile scales can be related to the pdf of spacing scales and the coverage fraction, *i.e.*,[5]

$$f_1(\lambda) = \frac{1}{l_m}\int_\lambda^\infty p(l)\,dl = \frac{dF_1(\lambda)}{d\lambda}.$$

(31)

The distribution of largest-empty-tile scales can then be connected to the (1-D) coverage statistics (combining Eqs. (15), (20), and (31)), *i.e.*,

$$D_1(\lambda) = 1 - \frac{\lambda f_1(\lambda)}{\int_0^\lambda f_1(\lambda')\,d\lambda'},$$

(32)

---

[5] We must also have $h_e(\lambda = l\,|\,l) = 1/(2l)$, since, in that case, there is only one point $\lambda/2$ away from the nearest element of the set (the point at the middle of the l-spacing). Accordingly, accommodating the possibility that the spacing-scale pdf, $p(l)$, may have a delta function at $l = \lambda$, we have, in general,

$$f_1(\lambda) = \frac{1}{l_m}\lim_{\epsilon\to 0}\int_{\lambda+\epsilon}^\infty p(l)\,dl + \frac{1}{2l_m}\lim_{\epsilon\to 0}\int_{\lambda-\epsilon}^{\lambda+\epsilon} p(l)\,dl,$$

in place of the first equality in (31) (cf. (30) and related discussion).

and,

$$f_1(\lambda) = \frac{1 - D_1(\lambda)}{\lambda} \exp\left\{ -\int_\lambda^\infty [1 - D_1(\lambda')] \frac{d\lambda'}{\lambda'} \right\} , \qquad (33)$$

which offer alternative SDF transforms (cf. Eqs. (20) and (27)).

In d-dimensional space, a natural extension of the notion of (1-D) largest-empty-tile scales allows the connection between coverage statistics and the distribution of the corresponding multidimensional largest-empty-*box* scales (Catrakis & Dimotakis 1996a,b).[7][8] Consider a set comprised of points, lines, surfaces, etc., embedded in a d-dimensional space, E_d. Recall that $F_1(\lambda)$ represents the probability that a λ-box covers part of the set (cf. (8) and related discussion). The coverage fraction can also be interpreted as a cumulative distribution function of a measure of spatial scales, in the following sense. For a scale increment, $\Delta\lambda$, the coverage fraction can be written as,

$$F_d(\lambda + \Delta\lambda) \equiv F_d(\lambda) + \int_{F_d(\lambda)}^{F_d(\lambda+\Delta\lambda)} dF_d(\lambda') . \qquad (34)$$

The differential coverage fraction in this integral can be associated with a probability density function of a measure of scales, $f_d(\lambda)$, where

$$f_d(\lambda) \equiv \frac{dF_d(\lambda)}{d\lambda} . \qquad (35)$$

In this expression, $f_d(\lambda)$ is the probability density function of the *largest-empty-box* (LEB) scale, λ, *i.e.*, the size of the largest box that is empty, *i.e.*, covers no part of the set (cf. (31) and related discussion).

This allows the connection between the coverage statistics and the distribution of this (multidimensional) measure of spatial scales, λ. In this context, this scale is identified as the size of the largest-empty-box that contains a randomly-located point, P, but contains no part of the set, *i.e.*, is empty. Equivalently, the scale λ is a measure of (twice) the distance from a point P to the nearest element of the set.

From (35), we see that $f_d(\lambda)$ satisfies the required normalization condition over the range of spatial scales, *i.e.*,

$$\int_0^\infty f_d(\lambda)\, d\lambda = F_d(\infty) - F_d(0) = 1 \qquad (36)$$

(cf. Eqs. (18)a,b). Integrating (35), we have the relation for the coverage fraction,

$$F_1(\lambda) = \int_0^\lambda f_d(\lambda')\, d\lambda' . \qquad (37)$$

The coverage dimension, $D_d(\lambda)$, can be expressed, therefore, in terms of the distribution of LEB scales, $f_d(\lambda)$, *i.e.*,

$$D_d(\lambda) = d - \frac{\lambda f_d(\lambda)}{\int_0^\lambda f_d(\lambda')\, d\lambda'} . \qquad (38)$$

This can be inverted to yield the LEB scale pdf from the coverage dimension, $D_d(\lambda)$, directly, *i.e.*,

$$f_d(\lambda) = \frac{d - D_d(\lambda)}{\lambda} F_1(\lambda) = \frac{d - D_d(\lambda)}{\lambda} \exp\left\{ -\int_\lambda^\infty [d - D_d(\lambda')] \frac{d\lambda'}{\lambda'} \right\} ; \qquad (39)$$

cf. Eqs. (9) and (35), which constitutes, therefore, the SDF transform pair in d-dimensional space.

This framework applies to both homogeneous and inhomogeneous geometries. In the latter case, the function $F_1(\lambda)$ represents the probability of coverage for a λ-tile placed in the δ_b-box, without regard to its location.

The implications of SDF geometry, as illustrated in (9), are that geometric structures across a wide range of scales can contribute to the scaling behavior at any one scale, λ. This is manifested, for example, in the coverage, coverage-length, or volume-fill fraction. For these measures, the scaling becomes nonlinear (in logarithmic coordinates).

In SDF geometry, the coverage dimension, $D_d(\lambda)$, is no longer a scaling exponent. It does, however, quantify the departure of the set from the topological dimension at a given scale, and indicates the type of structure (or, more precisely, the complexity of structure type) present at any one scale. By definition, it measures the (fractional) rate of increase of the coverage with descreasing scale, $i.e.$, inreasing resolution (recall (6)).

3. Scale-dependent surface-volume measures

The scale distribution, $f_d(\lambda)$, provides a scale-dependent surface-volume measure, as discussed below. The LEB scale can be interpreted as a measure of (twice) the distance to the nearest element (point, in 1-D) of the level set. The LEB scale pdf, $f_d(\lambda)$, therefore, has units of reciprocal length and indicates the surface-to-volume (perimeter-to-area in 2-D) ratio of surfaces (contours), located a distance $\lambda/2$ (within a proportionality constant) from the level set. In particular, the small-scale limit, $f_d(\lambda \to 0)$, is a measure of the (fully-resolved) surface-to-volume ratio of the level set.

While the large-scale behavior of $f_d(\lambda)$, for a complex object in general, is given by (cf. (18)b), $f_d(\lambda \to \delta_b) \to 0$, independently of the embedding-space dimension, d. The small-scale behavior of the scale distribution, $f_d(\lambda \to 0)$, is, however, dependent on the embedding-space dimension (cf. Eqs. (18)a and (35)),

$$f_d(\lambda) \sim \lambda^{d-d_t-1} \to \begin{cases} \varsigma_d, & \text{as } \lambda \to 0, \quad \text{for } d_t = d-1 \text{ ;} \\ 0, & \text{as } \lambda \to 0, \quad \text{for } d_t < d-1 \text{ ,} \end{cases} \qquad (40)$$

where ς_d is a constant. As will be shown below, this constant is a measure of the surface-to-volume ratio of the set, for the case $d_t = d-1$. The case $d_t < d-1$ corresponds to points in 2-D, points or space-curves in 3-D, etc. The case $d_t = d-1$ is significant in the context of turbulent mixing, because it can be applied, for example, to (smooth) isoscalar surfaces (generated by advection/diffusion).

Extending Richardson's (1961)[29] idea to d-dimensional space, a useful scale-dependent measure of the size of a set of topological dimension d_t can be defined as,

$$\mathcal{S}_d(\lambda) \equiv \lambda^{d_t} N_d(\lambda) \text{ ,} \qquad (41)$$

where d_t denotes the topological dimension of the set. For level sets in 2-D, for example, $\mathcal{S}_2(\lambda)$ becomes a scale-dependent measure of the (coverage) length, denoted here as $L_2(\lambda)$, of the level sets (distinct from arc-length, because of the box character of the coverage process), with,

$$\frac{S_2(\lambda)}{\delta_b} \equiv \frac{L_2(\lambda)}{\delta_b} \equiv \frac{\lambda}{\delta_b} N_2(\lambda) , \tag{42}$$

where the coverage length is scaled by the size of the bounding-box, δ_b. In the small-scale limit, $S_d(\lambda \to 0)$ provides a measure of the surface-to-volume ratio. In the 2-D case, the small-scale limit of the coverage-length is,

$$\frac{L_2(\lambda \to 0)}{\delta_b} = \lim_{\lambda \to 0} \left[\frac{\lambda}{\delta_b} N_2(\lambda) \right] , \tag{43}$$

and provides a finite (dimensionless) measure, at the smallest scales, for 1-D ($d_t = 1$) level sets embedded in a 2-D space ($d = 2$). In particular, it is a dimensionless measure of the perimeter-to-area ratio of the level sets. For level sets in 3-D, this would be a measure of the surface area per unit volume, etc.

For level sets of 2-D smooth fields, in the small-scale limit,

$$f_2(\lambda \to 0) = \frac{L_2(\lambda \to 0)}{\delta_b^2} , \tag{44}$$

or, in d-dimensional space, for $d = d_t - 1$ (recall (40)),

$$f_d(\lambda \to 0) = \varsigma_d = \frac{S_d(\lambda \to 0)}{V_b} = \frac{S_d(\lambda \to 0)}{\delta_b^d} , \tag{45}$$

(cf. (40)) where $V_b \equiv \delta_b^d$ is the bounding-box volume. The small-scale limit of the scale distribution, $f_2(\lambda \to 0)$, therefore, is a measure of the surface-to-volume ratio. Equivalently,

$$\delta_b f_d(\lambda \to 0) = \delta_b \varsigma_d = \frac{\delta_b S_d(\lambda \to 0)}{V_b} = \frac{S_d(\lambda \to 0)}{\delta_b^{d-1}} , \tag{46}$$

provides a dimensionless surface-volume measure.

4. Some examples

Some examples of scale distributions and their relation to coverage dimensions will be discussed below. In 1-D, for example, the SDF framework can be applied to the study of Poisson, lognormal, as well as power-law statistics (Catrakis & Dimotakis 1996b).[8] For Poisson point processes,

$$p(l) \, dl = \frac{1}{l_m} \exp(-l/l_m) \, dl , \tag{47}$$

the preceding analysis yields a coverage dimension given by (cf. (20)),

$$D_1(\lambda) = 1 - \frac{\lambda/l_m}{e^{\lambda/l_m} - 1} . \tag{48}$$

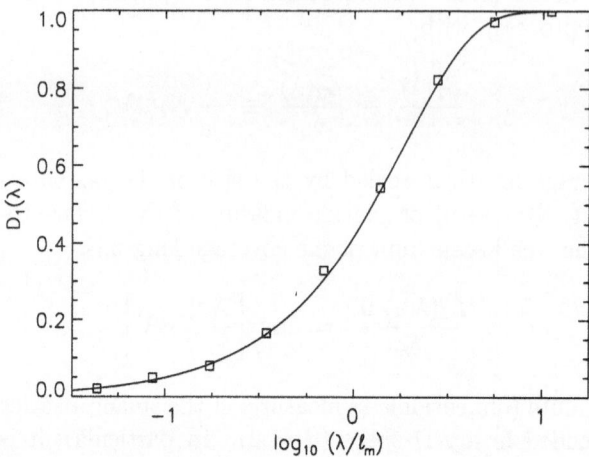

Fig. 2. — Dimension as a function of scale for Poisson point processes. Theory: solid line; simulations: squares.

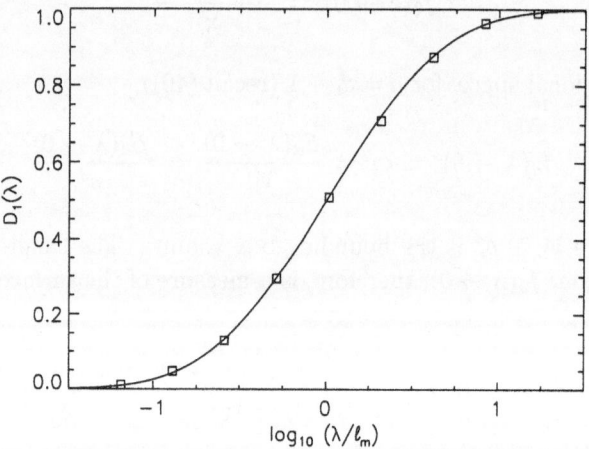

Fig. 3. — $D_1(\lambda)$ for a lognormal spacing pdf ($\sigma = 1$). Theory: solid line; simulations: squares.

Figure 2 compares $D_1(\lambda)$, from (48), to results from five Monte-Carlo simulations. For each simulation, a randomly-placed L-length record, where $L/l_m = 1000$, was successively partitioned into smaller λ-intervals and the coverage fraction, $N_1(\lambda)$, computed for each λ. The standard deviation of the ensemble-averaged Monte-Carlo estimates is smaller than the symbol size (Figs. 2–4). Measurements of the pdf of zero-crossing spacings derived from 1-D velocity signals in turbulent boundary layers have been reported to be well-approximated by Poisson statistics (Sreenivasan et al. 1983,[32] Kailasnath & Sreenivasan 1993).[15]

Lognormal statistics can be expected for fragmentation processes, in general (Kolmogorov 1941),[16] and are of interest in turbulence, with reports that apply them to spacings derived from level-crossings of 1-D signals in various turbulent flows.

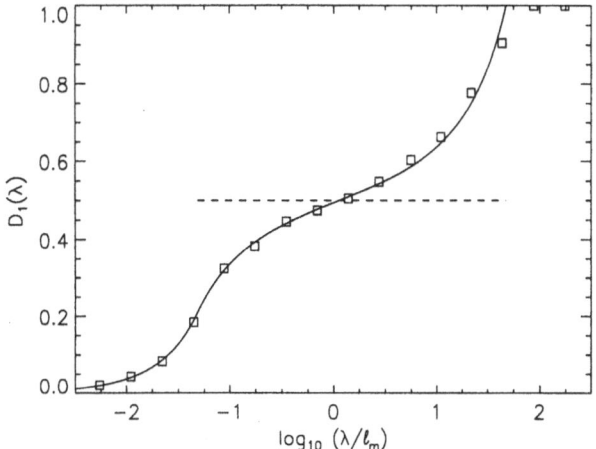

Fig. 4. — $D_1(\lambda)$ for a power-law spacing pdf ($\nu = 3/2$, $l_2/l_1 = 1000$). Theory: solid line; simulations: squares.

Sreenivasan *et al.* (1983)[32] applied them to zero-crossings of (1-D) velocity time series, measured in turbulent boundary layers; Miller & Dimotakis (1991)[27] reported that lognormal statistics provided a good description of level-crossings of jet-fluid concentration measurements in liquid-phase turbulent jets; Yee *et al.* (1995)[39] also reported lognormal statistics in turbulent plumes dispersing in the atmospheric surface layer. For a lognormal pdf,

$$p(l)\,dl \;=\; \frac{1}{\sqrt{2\pi}\,\sigma l}\,\exp\left\{-\left[\ln(l/l_{\rm m})/\sigma + \sigma/2\right]^2/2\right\}\,dl\;, \tag{49}$$

we obtain a coverage dimension (cf. (20)),

$$D_1(\lambda) = 1 - \left\{1 + \frac{l_{\rm m}}{\lambda}\left[\frac{1 + \mathrm{erf}\left[(\ln(\lambda/l_{\rm m})/\sigma - \sigma/2)/\sqrt{2}\right]}{1 - \mathrm{erf}\left[(\ln(\lambda/l_{\rm m})/\sigma + \sigma/2)/\sqrt{2}\right]}\right]\right\}^{-1}. \tag{50}$$

A comparison to results from five Monte-Carlo simulations, with $L/l_{\rm m} = 3000$, is shown in Fig. 3. Such behavior was previously noted by Miller & Dimotakis (1991),[27] who showed in a Monte-Carlo simulation that it provided a good representation of the coverage dimension computed from scalar measurements in liquid-phase turbulent jets.

Power-law statistics can also be similarly studied, *e.g.*, for a $p(l)$, with power-law behavior over a finite range of scales, $l_1 < l l_2$, *i.e.*,

$$p(l)\,dl \;=\; \begin{cases} \dfrac{a\,dl}{l_1}\,, & l < l_1\;; \\[2ex] \dfrac{a\,l_1^{\nu-1}\,dl}{l^\nu}\,, & l_1 < l < l_2\;; \\[2ex] 0\,, & l_2 < l\,, \end{cases} \tag{51}$$

for $1 < \nu < 2$, correspond to a SDF dimension given by (Catrakis 1996),

$$
D_1(\lambda) = \begin{cases} \dfrac{\lambda/l_1}{2\left(\nu - \alpha^{1-\nu}\right)/(\nu - 1) - \lambda/l_1}, & \lambda < l_1 \,; \\[2mm] \dfrac{\beta l_1/\lambda + (1 - \nu)\,(l_2/\lambda)^{\nu-1}}{2 - \nu + \beta l_1/\lambda - (l_2/\lambda)^{\nu-1}}, & l_1 < \lambda < l_2 \,; \\[2mm] 1, & l_2 < \lambda, \end{cases} \tag{52}
$$

where, $\alpha \equiv l_2/l_1$, and $\beta \equiv \nu(\nu - 1)\,\alpha^{\nu-1}/2$. This is plotted in Fig. 4, for $\nu = 3/2$ and $l_2/l_1 = 1000$. A comparison to five Monte-Carlo simulations with record lengths $L/l_m = 4000$ is also shown. In the limit of $l_2/l_1 \gg 1$ and for scales $l_1 \ll l \ll l_2$ (cf. (52)), we see that,

$$
D_1(\lambda) \to \nu - 1 \,, \tag{53}
$$

for $1 < \nu < 2$ (cf. dashed line in Fig. 4 for $\nu = 3/2$).

Conversely, if $D_1(\lambda) = D_1 = \text{const.}$ (or $F_1(\lambda) \sim \lambda^{1-D_1}$), for $\lambda_1 \ll \lambda \ll \lambda_2$,

$$
p(l) \sim l^{-D_1-1} \tag{54}
$$

(cf. (27)), i.e., the scale dependence of $D_1(\lambda)$ in Fig. 4 is a finite scale-range effect. A very large scaling range would be required to produce a near-constant coverage dimension. In other words, while a power-law coverage fraction, over a range of scales, implies a power-law $p(l)$ in the same range (cf. (54)), the converse is not true, as reflected in the non-local (integral) relation in the forward transform ((15)).

The SDF framework can also be used to compute the LEB-scale pdf for level sets derived from multidimensional measurements in turbulence, for example. Experiments were conducted to measure the jet-fluid concentration in the far field of liquid-phase turbulent jets for Reynolds numbers, $4.5 \times 10^3 \le Re \le 18 \times 10^3$, at a Schmidt number, $Sc \simeq 1.9 \times 10^3$ (Catrakis & Dimotakis 1996a).[7]

Figure 5 depicts a level set of 2-D spatial measurements of concentration at $Re \simeq 9 \times 10^3$, recorded perpendicular to the jet axis ($z/d_0=275$, where d_0 is the jet-nozzle diameter) using laser-induced fluorescence and digital-imaging techniques. This level set corresponds to the peak of the concentration pdf at this Re. The geometric complexity of such level sets is attributable, in part, to the large number (~ 700, in each realization, on average) of islands and lakes, i.e., closed contours whose immediate interior is at higher or lower concentration, respectively, at this Reynolds number.

Figure 6 shows the ensemble-averaged dimension, $D_2(\lambda)$, for level sets of concentration at $Re \simeq 9 \times 10^3$ (six images) at a threshold corresponding to the peak of the concentration pdf. The smallest (diffusion) scale of the concentration field is estimated to be $\log_{10}(\lambda_D/\delta_b) \simeq -3.0$, on the jet axis. For each level set, the δ_b-sized bounding box was identified and partitioned into contiguous λ-boxes, to compute the (coverage) fraction of the number of boxes that cover the level set, as a function of scale (Catrakis & Dimotakis 1996a).[7] The dimension increases smoothly with the scale, λ, spanning the full range of possible values for such data (error bars indicated if larger than symbol size).

Coverage statistics can now be used to compute the LEB-scale pdf (Fig. 6). While the jet is not (spatially) statistically homogeneous, $f_2(\lambda)$ retains its meaning, i.e., it is the pdf of the size of LEBs, randomly placed, interior to the δ_b-box. The data

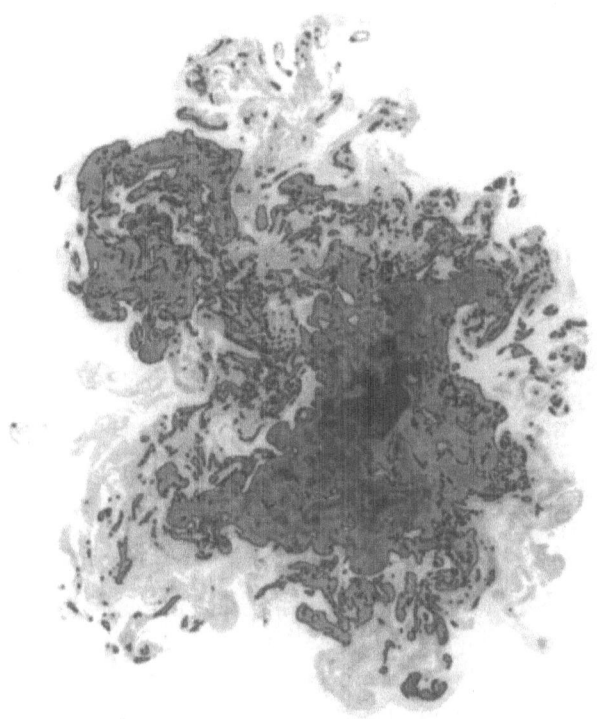

Fig. 5. — Level set of jet-fluid concentration in the far field ($z/d_0 = 275$) of a liquid-phase turbulent jet at $Re \simeq 9 \times 10^3$.

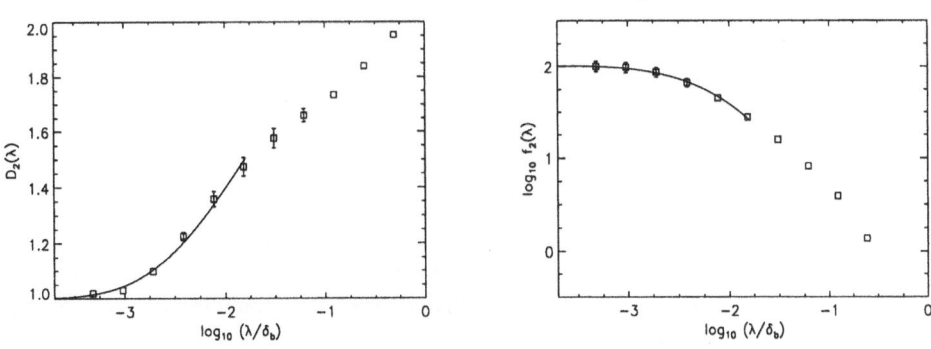

Fig. 6. — Left: Dimension, $D_2(\lambda)$, as a function of scale for level sets of concentration in a turbulent jet at $Re \simeq 9 \times 10^3$. Right: Pdf of LEB scales, $f_2(\lambda)$.

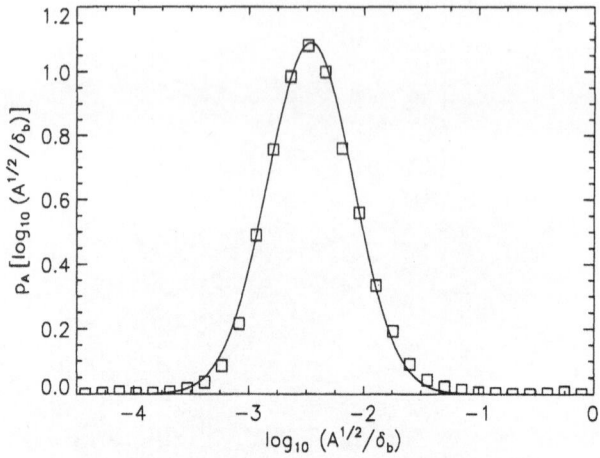

Fig. 7. — Size pdf of islands and lakes at $Re \simeq 9 \times 10^3$.

indicate that the probability density of a LEB scale increases continuously with decreasing scale, tending to a constant as $\lambda \to 0$.

At this stage, several models could be employed to fit the data in Fig. 6. Analysis of the size pdf of the islands/lakes, however, indicates lognormal statistics at inner scales (Fig. 7), where size is defined here as $A^{1/2}$, where A is the area of each island/lake. Such statistics are consistent with fragmentation and growth (fission/fusion) processes (Kolmogorov 1941, Lopez 1977).[16] [21]

This finding suggests that lognormal statistics may be used to model the level set at inner scales. A 2-D lognormal model (derived using the LEB-scale pdf computed for 1-D lognormal spacings, cf. (50)) is shown in Fig. 6 at inner scales (solid lines) for the scale pdf (and for the dimension),

$$f_2(\lambda) \ \propto \ \frac{1}{2l_{\mathrm{m}}} \, \mathrm{erfc} \left[\left\{ \ln\left(\lambda/l_{\mathrm{m}}\right)/\sigma + \sigma/2 \right\} /\sqrt{2} \right] , \tag{55}$$

with $\log_{10}(l_{\mathrm{m}}/\delta_{\mathrm{b}}) \simeq -1.5$ and $\sigma \simeq 1.2$, as fitted to the inner scales. The departure from the lognormal model, at large scales, indicates a break in behavior, suggesting that an alternate description at the outer scales of the flow is appropriate, as expected.

As noted above, the LEB scale is a measure of (twice) the distance to the nearest element of the level set. The LEB pdf, $f_2(\lambda)$, therefore, is a scale-dependent perimeter-area measure, indicating the perimeter-to-area ratio of contours located a distance $\lambda/2$ (within a proportionality constant) from the level set. The small-scale limit, $f_2(\lambda \to 0)$, is a measure of the (resolved) level-set perimeter-to-area ratio (cf. Fig. 6).

This analysis can be applied to any surface with complex geometry, with the LEB scale pdf, $f_d(\lambda)$, providing a scale-dependent surface-volume measure. In the context of turbulent combustion, in particular, the LEB scale pdf, $f_d(\lambda)$, is a scale-dependent surface-volume measure indicating the the surface-to-volume ratio of surfaces located a distance $\lambda/2$ (again, within a proportionality constant) from the instantaneous burning (level-set) surface. The small-scale limit, $f_d(\lambda \to 0)$, is, again, a d-dimensional measure of the surface-to-volume ratio of the isoscalar surface. It is the probability, $f_d(\lambda)\,d\lambda$, that is dimensionless, $i.e.$, $f_d(\lambda)$ has units of reciprocal λ (length), as does the surface-to-volume ratio. The latter can be scaled (made dimensionless), however, as indicated in (46) and related discussion.

5. Conclusions

A scale-dependent generalization of (constant-dimension) fractal geometry has been presented. The framework of scale-dependent-fractal (SDF) geometry can be used to quantify multiscale phenomena, whose behavior may be a function of scale. SDF statistics are invertible and can be used to deduce the scale distribution. Level sets of concentration, velocity, or vorticity, arising in turbulence, for example, can be analyzed and modeled with this framework. In 2-D slices of turbulent jets, for example, the LEB-scale pdf of level sets of concentration, at small scales, is consistent with a lognormal pdf of sizes of regions forming islands and lakes. Arguments for lognormal statistics have been put forth, for example, by Kolmogorov (1941, 1962)[16][17] for stochastic fragmentation (or growth) processes, as may be expected to occur in turbulence and other phenomena.

Acknowledgments

We would like to acknowledge the support of this work under AFOSR Grant F49260–94–1–0353, as well as discussions with, C. Bond, M. Cross, H. Liepmann, A. Leonard, P. Miller, D. Pullin, and P. Saffman. We are also grateful to H. Chaté, J.-M. Chomaz, E. Villermaux, and the organizers of this NATO/ASI summer school for the invitation, their gracious hospitality, and financial support that made our participation in this summer school possible.

References

[1] R. E. Amritkar, Indian J. Pure & Appl. Phys. **32**, 595 (1994).

[2] R. Andrle, Earth. Surf. **21**, 955 (1996).

[3] R. Andrle, Math. Geol. **28**, 275 (1996).

[4] P.N. Brandt, R. Greimel, E. Guenther, and W. Mattig, "Turbulence, Fractals, and the Solar Granulation," in *Applying Fractals in Astronomy*, A. Heck and J. M. Perdang, Eds., (Springer, Berlin, 1991), pp. 77–96.

[5] C. Castagnoli and A. Provenzale, A&A **246**, 634 (1991).

[6] H. J. Catrakis, *Mixing and the Geometry of Isosurfaces in Turbulent Jets*, PhD Dissertation, California Institute of Technology (1996).

[7] H. J. Catrakis and P.E. Dimotakis, J. Fluid. Mech. **317**, 369 (1996).

[8] H. J. Catrakis and P.E. Dimotakis, Phys. Rev. Lett. **77**, 3795 (1996).

[9] J. P. Chilés, Math. Geol. **20**, 631 (1988).

[10] P. Constantin, I. Procaccia, and K.R. Sreenivasan, Phys. Rev. Lett. **67**, 1739 (1991).

[11] M. C. Cross and P.C. Hohenberg, Rev. Mod. Phys. **65**, 851 (1993).

[12] D. R. Denley, J. Vac. Sci. Technol. A **8**, 603 (1990).

[13] P. E. Dimotakis, Nonlinear Sci. Today **2** 1, 27–31 (1991).

[14] P. E. Dimotakis and H.J. Catrakis, "Turbulence, fractals, and mixing", this Volume.

[15] P. Kailasnath and K.R. Sreenivasan, Phys. Fluids A **5**, 2879 (1993).

[16] A. N. Kolmogorov, Akad. Nauk SSSR **31**, 99 (1941). Translated, V. Levin, 1969, NASA TT F-12287.

[17] A. N. Kolmogorov, J. Fluid Mech. **13**, 82 (1962).

[18] J. Laherrère, C. R. Acad. Sci. Paris **322**(IIa), 535 (1996).

[19] H.W. Liepmann, Helv. Phys. Acta **22**, 119 (1949).

[20] M. S. Longuet-Higgins, Proc. Roy. Soc. London A **246**, 99 (1958).

[21] R. E. Lopez, Mon. Wea. Rev. **105**, 865 (1977).

[22] B. B. Mandelbrot, Science **155**, 636 (1967).

[23] B. B. Mandelbrot, J. Fluid Mech. **72**, 401 (1975).

[24] B. B. Mandelbrot, *The Fractal Geometry of Nature* (W. H. Freeman & Co., New York, 1982)

[25] D. M. Mark and P.B. Aronson, Math. Geol. **16**, 671 (1984).

[26] S. Matsuura, S. Tsurumi, and N. Imai, J. Chem. Phys. **84**, 539 (1986).

[27] P. L. Miller and P.E. Dimotakis, Phys. Fluids A **3**, 168 (1991).

[28] S. O. Rice, Bell System Technical J. **24**, 46 (1945).

[29] L. F. Richardson, General Systems Yearbook **6**, 139 (1961).

[30] J.-P. Rigaut, "Fractals, semi-fractals, and biometry", in *Fractals: Non-integral Dimensions and Applications*, G. Cherbit, Ed., (Wiley, Chichester, UK, 1991), pp. 151–187.

[31] K. R. Sreenivasan, Ann Rev. Fluid Mech. **23**, 539 (1991).

[32] K.R. Sreenivasan, A. Prabhu, and R. Narasimha, J. Fluid Mech. **137**, 251 (1983).

[33] M. Suzuki, Prog. Theor. Phys. **71**, 1397 (1984).

[34] H. Takayasu, J. Phys. Soc. Japan **51**, 3057 (1982).

[35] H. Takayasu, *Fractals in the Physical Sciences* (Wiley, Chichester, UK, 1992)

[36] A. A. Tsonis and J.B. Elsner J. Stat. Phys. **81**, 869 (1995).

[37] J. C. Vassilicos and J.C.R. Hunt Proc. Roy. Soc. London A **435**, 505 (1991).

[38] D. Welling, J. Urani, L. Welling, and E. Wagner, Cell Physiol. **39**, 953 (1996).

[39] E. Yee, R. Chan, P.R. Kosteniuk, G.M. Chandler, C.A. Biltoft, and J.F. Bowers, Bound. Layer Met. **73**, 53 (1995).

Comparing Extremes:
Mixing of Fluids, Mixing of Solids

JULIO M. OTTINO AND TROY SHINBROT

*Robert R. McCormick School and Engineering and Applied Science
Northwestern University, Evanston, Illinois 60208-3120, USA*

Mixing's reach is remarkably wide; examples appear in the liquid-liquid, solid-liquid, and solid-solid areas. Traditional engineering approaches tackle all of these problems on a case-by-case basis, no description being expected to cover all possible situations. Much can be gained, however, by juxtaposing extremes; *fluid-fluid mixing* and *solid-solid mixing* provide such bounds. For fluids mixing, basic understanding at a continuum level is firmly established: Navier-Stokes equations provide a first principles description valid on macroscopic scales for most problems. Somewhat paradoxically, basic understanding for solids mixing is much less developed despite the fact that a first principles description – particle dynamics – is arguably better for solids than for fluids. Shortcomings of continuum descriptions manifest themselves on macroscopic scales, particle segregation being an instance where physical meso-scale processes are not well understood. Slow mixing of powders and slow mixing of fluids can be described by maps; in powders the map is a succession of distinct avalanches; in fluids the repetition of stretching and folding. In both cases rather simple pictures can be extended to the point that non-trivial conclusions can be obtained. In mixing of fluids repeated stretching and folding leads to chaos, the disorder often being accompanied by symmetries and regularity; in powders by unmixed cores and systems that mix or do not mix depending upon the sense of rotation.

Mixing: Chaos and Turbulence, edited by Chaté *et al.*
Kluwer Academic / Plenum Publishers, New York 1999.

1. Introduction

> "Suppose I apply myself to a complicated calculation and with much difficulty
> arrive at a result. I shall have gained nothing by my trouble if this has not
> enabled me to foresee the results of other analogous calculations, and to direct
> them with certainty, avoiding the blind groping with which I had to be contented
> the first time... On the contrary, my time will not have been lost if this very
> ... groping has succeeded in revealing to me the profound analogy between the
> problem just dealt with and a much more extensive class of other problems ...
> then it will not be merely a new result that I have acquired, but a new force."

Jules Henri Poincaré
Science et Méthode

Mixing is so widespread in technology and nature that one might expect that a comprehensive theory would have been developed long ago. This, however, has not been the case. Part of the reason is the daunting spectrum of problems covered. A rough classification would divide problems into liquid-liquid, solid-liquid, solid-solid; all of these admit sub-classifications. Thus, liquid-liquid mixing can be divided into laminar (or viscous) mixing, transitional or moderate Reynolds number mixing, and turbulent or high Reynolds number mixing. Each, in turn, may involve single or multiphase media. Viscous mixing is important in the context of food processing, reactive and non-reactive polymer processing, and stabilization of hazardous wastes. Transitional mixing is of relevance in bio-reactors, and turbulent reactive mixing is critical to the understanding of NOx production and petrochemical processing. Liquid-solid mixing plays a critical role in pharmaceutical manufacturing and in the paper, rubber and cosmetic industries. Solid-solid mixing is important in catalyst preparation, blending and processing of ceramic precursors, and the pharmaceutical industry. The list can go on and on.

Traditional engineering approaches have attempted to tackle each of these problems on a one-by-one basis. This method is manifestly inefficient and many arguments may be advanced for a more encompassing approach. It is nevertheless obvious that no single description could possibly cover all nuances and contingencies.

Much can be gained by juxtaposing and contrasting extremes. Two cases belonging to fluid-fluid mixing and solid-solid mixing are selected. The *fluids mixing* case corresponds to the mixing of a fluid with itself; that is, a fluid region being marked by a passive tracer and followed in space and time. Typically the Reynolds number is taken to be small, the flow is driven by boundary motions, and the velocity field is time-periodic. The *solids mixing* case focuses on non-cohesive particles, possibly differing in size, density, shape, etc. Mixing is achieved by rotation and a fluid, typically air, is present. However, to the extent that lubrication forces become important at length scales below the typical roughness of the particles, the fluid may be safely ignored. [1] Two subcases are considered: *slow flow*, where distinct avalanches occur, and *continuous flow*, where flow is restricted to a thin continuously avalanching layer, the rest of the granular material rotating as a single solid body.

Both problems – mixing of fluids and mixing of powders – can be described by maps. In powders the map is a succession of distinctive avalanches; in fluids the repetition of stretching and folding suggested by Osborne Reynolds [2] over a century ago. In both cases rather simple pictures can be extended to the point that non-trivial conclusions can be obtained. In mixing of fluids repeated stretching and folding leads to chaos, the disorder often being accompanied by symmetries and regularity; in powders by unmixed cores and systems – blenders with rotational, but not reflectional, symmetries – that may or may not mix depending upon the sense of rotation. Continuum descriptions apply to granular materials as well, in at least one instance with far more success than might reasonably be expected.

2. Mixing of fluids

A complete description of mixing of fluids is far from trivial and probably will never be attained. The underlying physical processes are clear however: exceptions notwithstanding, the Navier-Stokes equations govern flow; diffusional processes, if present, are well described by Fick's law; interfacial conditions by the Young-Laplace equation, and so on. This is not where problems reside though. The issue is that – until recently – it was unclear how to apply the considerable mathematical apparatus of fluid mechanics to mixing. Knowing the equations is clearly not enough.

Most fluid mechanical problems are attacked from an Eulerian perspective: what happens to a fluid in coordinates fixed in a laboratory frame. Whereas some successes can be attained by this approach, it is clearly not the right framework for many problems in mixing. Illustration 1 exemplifies this point .

Illustration 1: Frames of reference [3]

Consider a particle of mass m with coordinate Y(t) under the action of a force F. It obeys

$$m\frac{d^2Y}{dt^2} = F .$$

Implicit in this formulation is the recognition that t is the independent variable. On the other hand if Y is regarded as the independent variable, now t(Y), and Newton's law becomes

$$F + m \left(\frac{dY}{dt}\right)^3 \frac{d^2t}{dY^2} = 0 .$$

The underlying physics is still the same but viewing variables in a different way has made the problem much harder. This is clearly not the right way to look at this problem.

---

3. Flows and mappings

Let us start by asserting that knowing where particles are advected by a prescribed velocity field is a solution to the mixing problem. Advection of fluid particles can be interpreted from a continuum viewpoint as a point transformation, particles at position $\mathbf{x}_0$ at time 0 are mapped to position $\mathbf{x}$ at time t.

The solution to the equation of motion of a fluid particle in the Eulerian velocity field, i.e.

$$\frac{d\mathbf{x}}{dt} = \mathbf{v}(\mathbf{x}, t) \text{ with } \nabla \cdot \mathbf{v} = 0 \tag{1}$$

gives the flow or motion,

$$\mathbf{x} = \Phi_t(\mathbf{x}_0), \quad \mathbf{x}_0 = \Phi_{t=0}(\mathbf{x}_0) \tag{2}$$

That is, a particle with position $\mathbf{x}_0$ at time $t = 0$ is mapped to the position $\mathbf{x}$ after a time t via a flow $\Phi_t(\cdot)$. (It is common to refer to a specific fluid particle as "particle $\mathbf{x}_0$", when in fact we mean the fluid particle that was initially located at $\mathbf{x}_0$; this subtle distinction is particularly important in the case of periodic points.) The basic question is the following: Under what conditions are $\mathbf{v}(\mathbf{x}, t)$ or its corresponding motion $\mathbf{x} = \Phi_t(\mathbf{x}_0)$ able to produce widespread and efficient stretching of an arbitrarily designated region of fluid throughout the space occupied by the flow?

A geometrical view helps. A somewhat classical interpretation is that streamlines confine flows. This is certainly true in 2D systems and this is the reason that steady flows are poor mixers. This however, does not even start to reveal the complexities associated with time periodic flows (i.e., $\mathbf{v}(\mathbf{x}, t) = \mathbf{v}(\mathbf{x}, t + T)$) or steady 3D flows where streamsurfaces may fail to confine particle motions.

Stretching and folding – horseshoe maps

The basic mixing map consists of stretching and folding, as shown in Fig. 1 (as we have mentioned, such a mechanism was advanced by Osborne Reynolds over a hundred years ago. [2]) Stretching and folding, and its close companion cutting and pasting, did in fact sporadically appear in the mixing literature. [4] These efforts did not, however, take root. Stretching and folding as the fingerprint of chaos formally appeared in the mathematical literature in the late 1960's through the horseshoe map of Smale[5], and the connection between fluid mixing, stretching and folding, and horseshoes – demonstrated by means of experiments – was made only a decade ago[6].

Illustration 2: Historical digression

It is now standard to visualize the internal motion of fluids by deformation of "coloured bands" as Reynolds called them in his 1894 paper. It has also become less of a heresy to apply geometrical thinking to rationalize internal motions of fluids, including the once forbidden turbulent flows – "a perfectly random motion of particles [where] no basic pattern should or could exist". [7] Similar thinking applied until recently to mixing, conceptual pictures relying mostly on eddy diffusivities and the like, something that can be traced back, most clearly, to Reynolds himself and his averaged Navier-Stokes equations.[8]

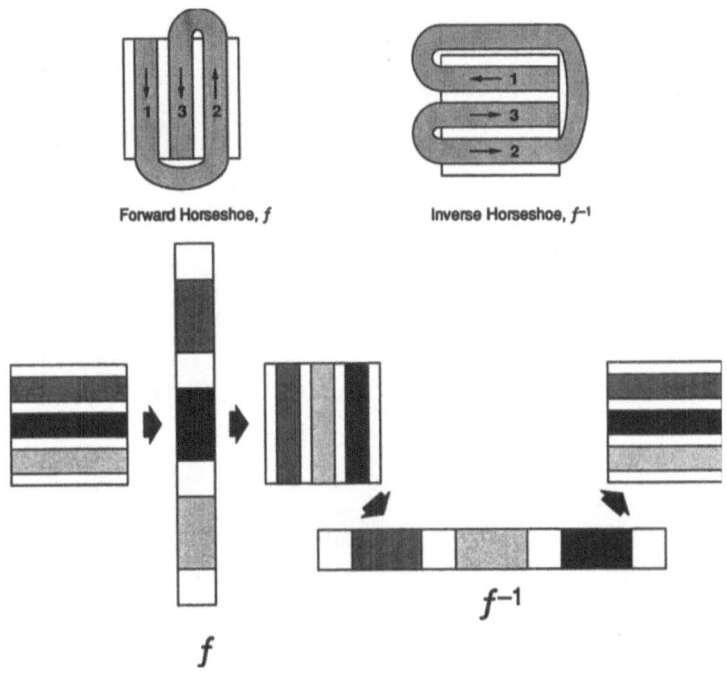

Fig. 1. — Horseshoe and its inverse

Lesson 1: Time periodicity leaves unmixed islands; symmetries may prevent mixing

---

Illustration 3: Symmetries in the egg-beater flow

Time-periodicity implies symmetries of various kinds. Consider for example the case of egg-beater flows – the composition of two shear flows acting on a square with periodic boundary conditions (Fig. 2). The first flow acts in the x-direction for a time T with a velocity $v_x(y)$; the second flow, $v_y(x)$, acts in the y-direction, also for a time T. Either velocity field, $v_x(y)$ and $v_y(x)$, may be odd or even about $x = 0$ and $y = 0$, respectively and may be combined to create "rotational" or "extensional" flow patterns in the square. Elliptic periodic points in Poincaré sections clearly display rotational and reflectional symmetries which may be predicted in the absence of a precise description of $v_x(y)$ and $v_y(x)$.[9]

Consider a time periodic flow (i.e., $\mathbf{v}(\mathbf{x}, t) = \mathbf{v}(\mathbf{x}, t+T)$) In such a case, the system

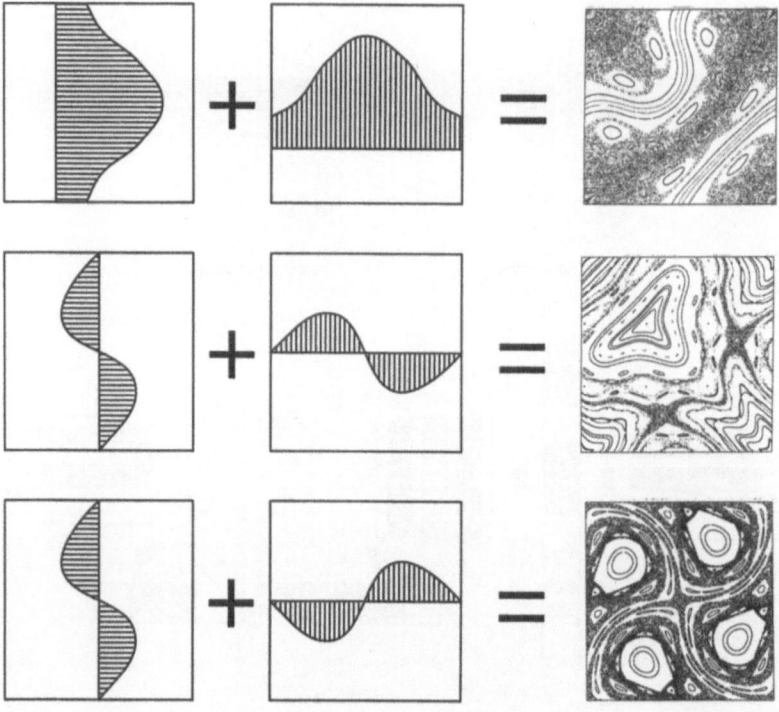

Fig. 2. — Eggbeater flows: equivalent spatially periodic flows (left) and Poincaré sections (right).

might be represented by a map

$$\mathbf{x}_{n+1} = F(\mathbf{x}_n) \quad \text{or} \quad \mathbf{x}_n = F^n(\mathbf{x}_0) \tag{3}$$

where n successive applications of the (nonlinear) point transformation F gives the position of the fluid particle initially located at $\mathbf{x}_0$. As we shall see, the same formulation applies to powder flows. A periodic point, $\mathbf{p}$, of the (nonlinear) mapping F is a point such that a particle initially located at $\mathbf{p}$ returns to $\mathbf{p}$ after n periods, that is $\mathbf{p} = F^n(\mathbf{p})$, where the periodicity n is the smallest value satisfying the equality.

Let us, for simplicity, restrict discussion to the case of 2D maps. Periodic points of 2D maps can be classified as hyperbolic, elliptic, or parabolic, according to the nature of deformation of the fluid in the neighborhood of the periodic point (and, in the case of parabolic points, the presence of a boundary). The center of a small circle surrounding an elliptic point returns to its original position while the circle undergoes a net rotation. Elliptic points are surrounded by *islands*, or *regular regions*, where there is no chaos. Typically, the lowest period points are associated with the largest islands.

These points are nicely illustrated by experiments [10] (Fig. 3). Dye structures in time-periodic flows evolve in an iterative fashion: an entire structure is mapped into

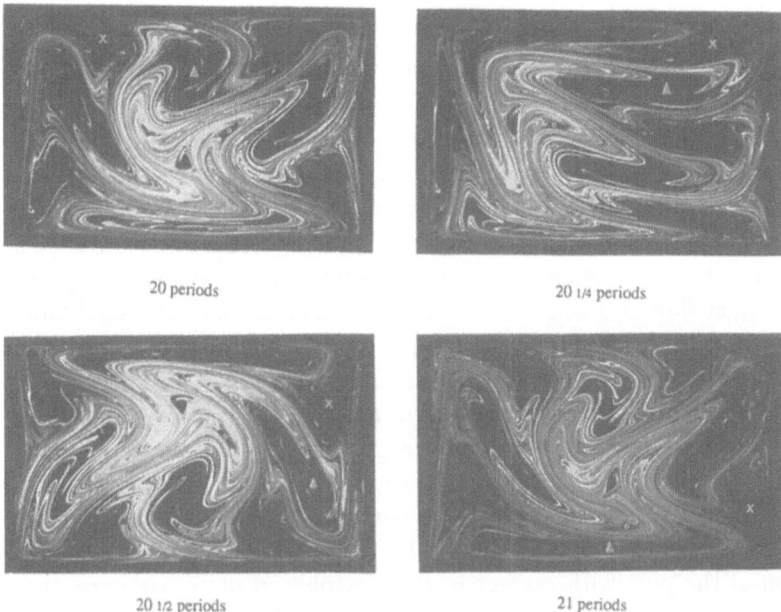

20 periods

20 1/4 periods

20 1/2 periods

21 periods

Fig. 3. — Time evolution of a passive tracer in a low Reynolds number two-dimensional time-periodic flow. The initial condition is a drop of fluorescent dye located at the center of a cavity filled with glycerine. The flow is induced by moving the top and bottom walls of the cavity while the other two walls are fixed. In this experiment the top wall moves from left to right and the bottom wall moves from right to left; both velocities are of the form $U \sin^2(2\pi t/T)$, with the same amplitude U and the same period T, but with a phase shift of $90°$. The dye stretches and folds in chaotic regions and reveals the existence of islands, which cannot be penetrated by the tracer. The motion of the islands can be followed by the markers placed in the flow (triangle, circle, and cross). The photographs are taken when the islands display vertical and horizontal symmetries. Note that the overall pattern becomes reestablished after one period, however the position of the islands is not the same.

a new structure with persistent large-scale features, and with finer and finer scale features revealed at each period of the flow. Thin striations are produced at the expense of thicker ones, and length scales (characterized by a striation thickness s) decrease exponentially in time. Islands translate, stretch, and contract periodically. Although islands undergo a net rotation, they preserve their identity, and represent the primary obstacle to efficient mixing. On the other hand, particle trajectories in chaotic regions separate exponentially fast, and material filaments are continuously stretched and folded by means of horseshoes. There may be much regularity underlying the chaos. For example, if the instantaneous streamline patterns are symmetric — as often happens in low Reynolds number flows — it is possible to show, theoretically an d experimentally, that the placement of the islands in the flow region becomes symmetric at periodic intervals of time [11]. It is apparent that more energy does not imply more mixing – islands will not disappear no matter how long the mixing is performed.

Lesson 2: Lessons from 2D can be extended to 3D flows - New designs suggested

The above findings suggest several practical questions. One is, "How may one get rid of islands?" One answer to this question is to systematically destroy symmetries. Another question is how to design mixing flows of engineering importance, such as continuous flows.

The prototypical continuous throughput flow is a duct flow. Duct flows are two-dimensional in cross section, and are augmented by a unidirectional axial flow. Fluid is mixed in the cross-section while it is simultaneously transported down the duct axis. A duct flow is defined such that the cross-sectional and axial flows are independent of both time and distance along the duct axis, and material lines stretch linearly in time, in very much the same manner as in two-dimensional steady (or regular) flows such as those presented in Fig. 3.

Chaos in duct flows can be achieved by time-modulation or by spatial changes along the duct axis. One example of a spatially-periodic duct flow, the partitioned-pipe mixer (PPM), consists of a pipe partitioned with a sequence of orthogonally placed rectangular plates (see Fig. 4). The cross-sectional motion is induced through rotation of the pipe with respect to the assembly of plates, whereas the axial flow is caused by a pressure gradient; the behavior of the system is characterized by the ratio of cross-sectional twist to axial stretching. Other duct flows of technological importance, which possess mixing mechanisms similar to the PPM, are the T-mixer [sequences of twisted pipes] and the K-mixer [an idealized version of a static mixer]. Yet another class of spatially periodic flow is suggested by cavity flows [9].

Lesson 3: Steady flows are poor mixers; a secondary baffle can improve mixing

Steady cavity flows are a poor mixers [6]. As an example, in Fig. 5(a), we show a photograph of dye mixed by steadily moving opposing walls in a 2D cavity filled with glycerine. Mixing here can, however, be significantly improved by means of time-dependent changes in geometry. In Fig. 5(b), we move the upper and lower walls *alternately* in the same apparatus and for the same total distance as before. This idea can be readily implemented in the context of duct flows by adding a secondary baffle [12] (see Fig. 5(c)). Such a concept has real applications in polymer processing; for example, single screw extruders can be imagined as a channel with a moving lid, very much as in Fig. 4.

Lesson 4: Faster mixing – more energy – does not imply more mixing

Driving a system faster, for example by moving boundaries more under a fixed protocol, does not imply better mixing: islands survive and do not go away. Yet another instance where faster action may actually lead to *worse* mixing is provided by the case of viscoelastic fluids. Niederkorn and Ottino [13] studied experimentally and computational mixing of Boger fluids – a viscoelastic fluid with a constant shear viscosity – in the flow between two concentric cylinders. In the limit of slow flow, a Boger fluid behaves as Newtonian; faster flows lead to viscoelastic effects quantified in terms of Weissenberg number (We), the ratio of the relaxation time of the fluid to a time scale of the flow; e.g. the inverse of the shear rate. Spectacular effects

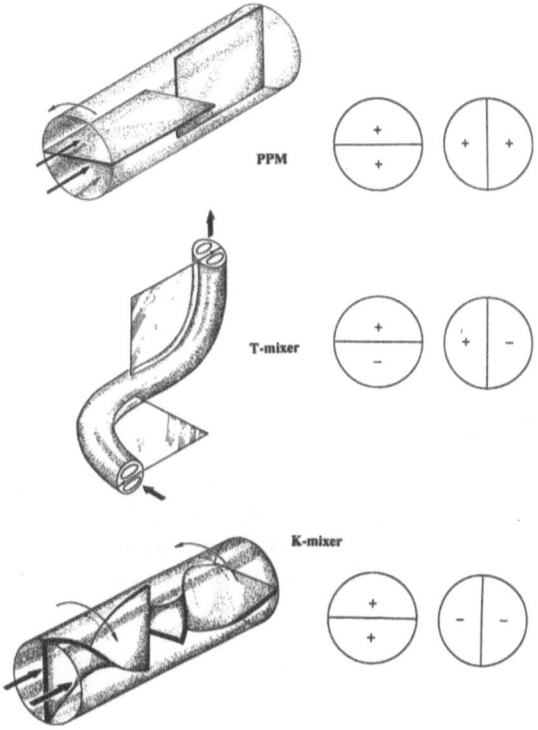

Fig. 4. — Spatially periodic duct flow: the partitioned pipe mixer

occur at moderate We; Fig. 6 shows the contrast between the Newtonian (We $\simeq$ 0) and the non-Newtonian case (We = 0.06).

4.   Mixing of solids

It may naively be argued that, as in the case of mixing of fluids, the underlying physical processes governing granular mixing are clear as well – one knows how particles interact: normal forces depend on restitution coefficients, tangential force are Coulombic. Much less is known about cooperative phenomena and it is at this length scale that granular materials part company with their fluid counterparts; averaging, the cornerstone of continuum field descriptions, may not work.

Illustration 4: Hydrostatic pressure

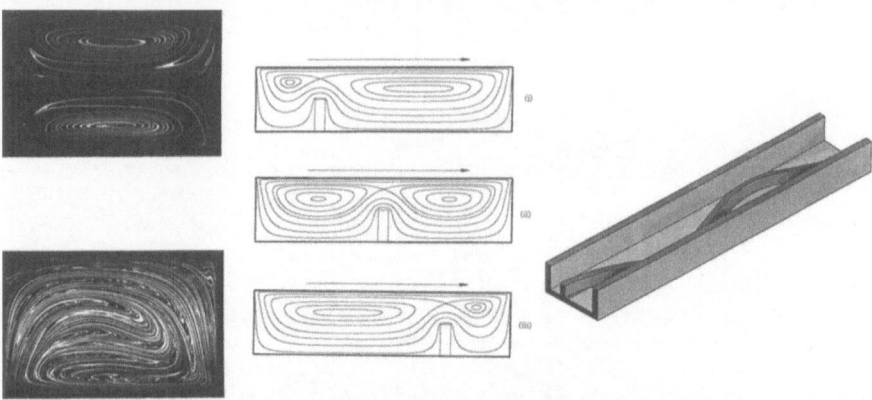

Fig. 5. — (a) Above: steady cavity flow; upper and lower walls moved continuously; below: chaotic cavity flow; upper and lower walls moved alternately. In both cases the total distances moved by the walls are identical. (b) Baffled cavity flow; (c) secondary baffle in duct flow.

As an example, consider one of the simplest possible kinematic problems: hydrostatic pressure in a vertical pipe. In the fluid case the pressure increases with depth, due simply to the increased weight of supported fluid. For the solid problem this argument does not hold. Pressure in this case increases with depth to a fixed asymptotic value, beyond which it no longer increases. The reason for this behavior underlies a characteristic difference between solids and typical fluids – Bingham plastics excepted. That is, solids can support a load by distributing stresses: although the weight of granular material above any given level does increase with depth into the column, the weight is transferred out to the walls where friction between the grains and the walls causes the walls themselves to support some of the vertical stress.

Qualitative differences in behavior between fluids and granular solids are not restricted to static problems; very often solid flows are seen to exhibit behaviors quite different from those seen in their fluid counterparts. In fluidized beds, for example, a fluid medium (typically a gas) is driven past beads with sufficient velocity to levitate them. Superficially the motion of the beads appears to be a close analog to gas kinetics problems. In fluidized beds, however, it is well known that increasing the flow speed of the fluid medium – which ought to make the problem an even closer fit to gas kinetics by increasing both mean free path and fluctuation velocities – in fact results in a 'slugging' state, in which the beads cooperatively entrain and all rise and fall in unison. No such behavior is seen in gas kinetics.

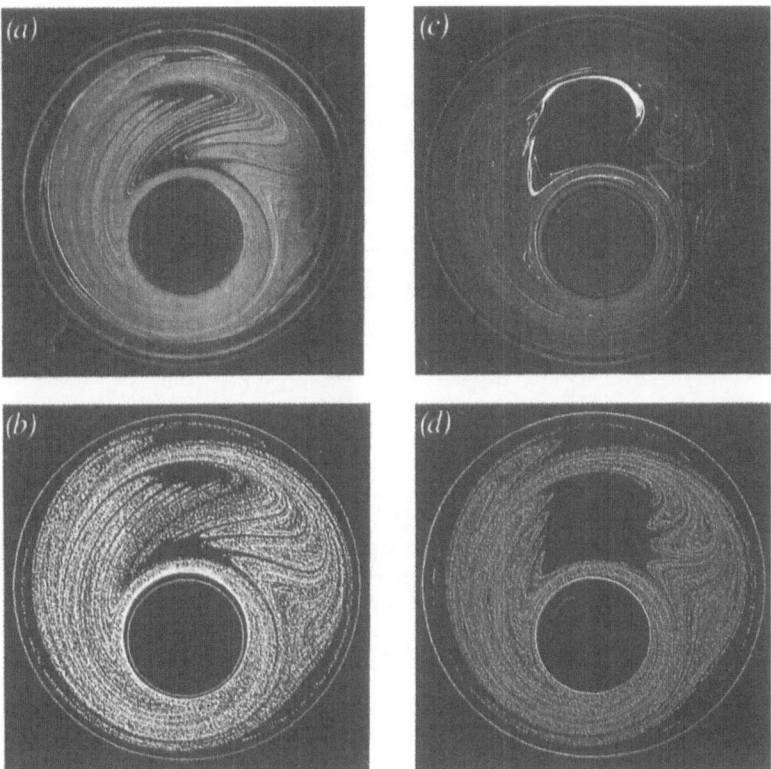

Fig. 6. — Experiments and computations on the advection of a dye blob in an eccentric cylinder apparatus[13]. (a) experiment using Newtonian fluid; (b) numerical simulation of same situation; (c) identical experiment as in (a), but using viscoelastic fluid (We≃ 0.06); (d) numerical simulation using We= 0.04.

In the case of slugging, interactions between the fluid medium and the solid beads are certainly important. Cases of non-fluidic behavior of solids flows in the complete absence of gas abound nevertheless. As a final example, consider the problem of a vertically shaken container partially filled with grains and evacuated of all gas [14]. This problem has been well studied experimentally [15, 16, 17] over the past decade, and it is uniformly observed that for shaking accelerations typically much larger than the acceleration of gravity, convective rolls – superficially resembling those seen in many fluid problems – are set up, in which grains rise in the center and fall along the walls of the container (Fig. 7). Here again, the particle motion appears superficially to be 'fluidized', in that particles are in continual motion and shear stresses are transmitted much as in a viscous fluid. Indeed, analyses along the lines of Navier-Stokes equations have been considered. The actual mean velocity field, however, is nothi ng like that seen in a fluid [18], unless extraordinary boundary conditions [19] are invoked: the flow is extremely rapid near the boundaries, and slows suddenly

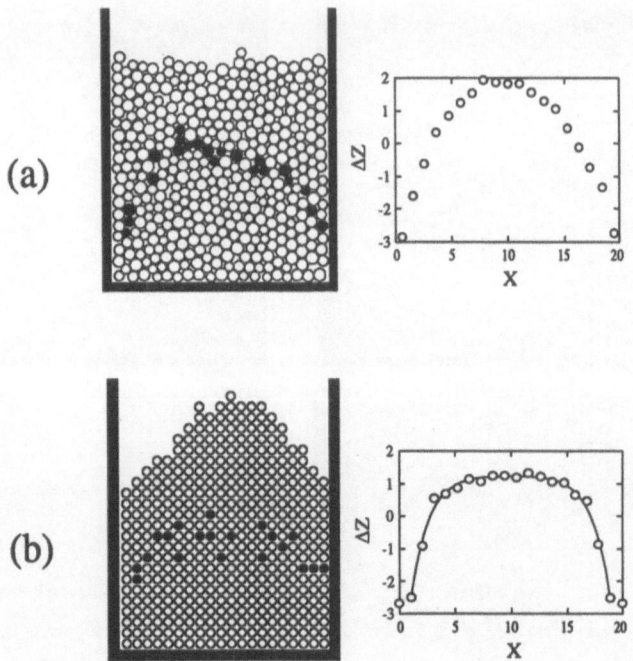

Fig. 7. — Granular convection. (a) Particle dynamics simulation showing displacement profiles caused by convective flow. On left, the black particles were nearly horizontal and placed mid-way up the stack prior to a shake; the figure shows the position following one shake. On right are average vertical displacements (ΔZ) over 300 shakes as a function of horizontal location (X), both expressed in units of particle diameters. (b) For comparison, lattice-based results for the same situation (discussed in later section) are displayed. Both the heaped surface and the detailed displacement shape for the lattice model actually agree better with experiments than for the particle dynamics simulations.

toward the center of the container. Thus far from the fluid case, where fluid sticks in a boundary layer, in the granular case the 'fluidized' grains flow *fastest* in the boundary layer.

Reasons for differences

The underlying sources of differences between granular and fluid behaviors are several and depend on context. Three sources seem to be central.

First, as we have already mentioned, static configurations of grains can support a load and distribute stresses. An example of this effect can be seen in granular 'arches', in which grains spontaneously configure themselves to fully support a load. The way that grains accomplish this is to assemble 'stress chains' along which compressive stresses are supported, and details of these chains have been studied intensively over the past several years. [20] Arches are common practical situations, and in fact industrial granular flow equipment very often includes moving impellers or tapping devices to prevent flow interruptions.

Second, and more importantly from the standpoint of mixing studies, granular flow requires voids to be present in advance of flow. One can push a fluid through an orifice by pressurizing the fluid from upstream, however for granular materials one must typically first evacuate material from the downstream. This is observed daily in salt shakers: salt flows through holes only so long as voids are produced through the mechanism of shaking. This phenomenon can be graphically confirmed by filling two syringes, one with water and the second with salt. Water is easily squirted out the first syringe, but no amount of pressure drives the salt from the second syringe. On the contrary, only by *releasing* pressure on the plunger and creating voids near the orifice (for example by shaking or by jiggling a thin wire inserted into the orifice) can the salt be made to flow. This observation dates to Reynolds, who noted that the volume occupied by sand must increase in order for it to flow. [21]

Third, and most significant from the aspect of gas kinetics comparisons, granular collisions are typically highly inelastic. This differs from the case of typical molecular collisions where inelasticity is small, and herein lies a source of many intriguing collective behaviors of solid flows.

Beyond these, ubiquitous, sources of differences between granular and fluid flows, several other effects can be important in particular problems. For example, forces during collisions can be non-central – i.e. *torques* can be transmitted. This seems to be particularly important for non-symmetrical particles, however even for perfectly spherical particles surprising differences have been revealed. For example if particles in a 2D tumbler are allowed to slide, but are prevented from rotating, angles of repose are found to exceed observed angles by as much as 60°.

Illustration 5: Billiards

The importance of non-central forces is easily revealed on the billiard table, by comparing the results of two extremely similar shots, as shown in Fig. 8. On the left we show two object balls touching, and on the second the object balls are very slightly separated. When the object balls are touching, torques can be directly transmitted to the black ball, and it acquires much of the momentum of the cue as a result. By comparison in the case on the right the collisions cue→grey and grey→black are separated, and the resultant velocity of the black ball is quite different. Similar effects play out in particle dynamics simulations, and it has been found that neglecting rotational motion of particles can have measurable effects on mean flows of granular materials.

Segregation

No summary of comparisons between fluids and granular materials would be complete without mention of segregation; mixing and segregation often come together. More agitation in solids does not imply better mixing. Granular mixtures of dissimilar (and not-too-dissimilar) materials often segregate when shaken or tumbled.

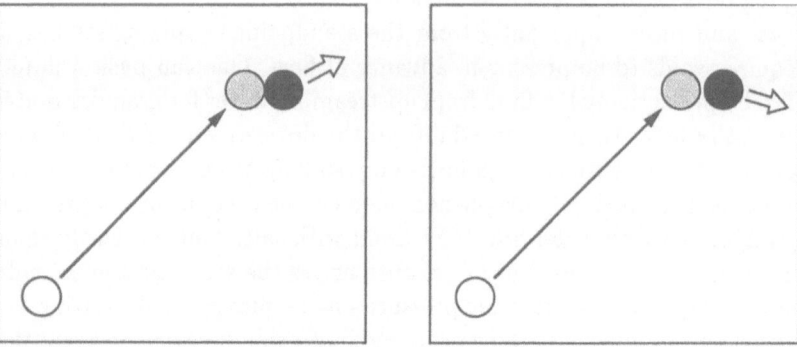

Fig. 8. — Billiards touching (left), slightly separated (right)

Fig. 9. — Axial segregation in long rotated cylinder (from [22], with permission)

The most celebrated example of this behavior is the so-called "Brazil-nut" effect, whereby large particles (Brazil-nuts) rise to the top of a granular gemisch (a bowl of mixed nuts). Multiple mechanisms have been proposed for this effect. According to the *'percolation'* mechanism, small particles can squeeze into small voids below a large particle, but the reverse cannot occur – and as a result large particles tend to rise. According to a *convection* argument [17], large particles can rise with the mean flow of granular material, but can be too large to fit into a narrow down-welling region near a boundary. Still other segregation effects are explained by neither of these models. For example in a rotating cylinder containing dissimilar solids, the solids are experimentally observed to arrange themselves into alternating bands along the axial direction (Fig. 9). Here the mechanism is believed to be due to differences in angles of repose of the two materials. Experiments have revealed that one speed of rotation of the cylinder (where the difference in angles of repose is large) two granular materials segregate into axial bands, while at another speed (where the difference is small), the same two materials mix.

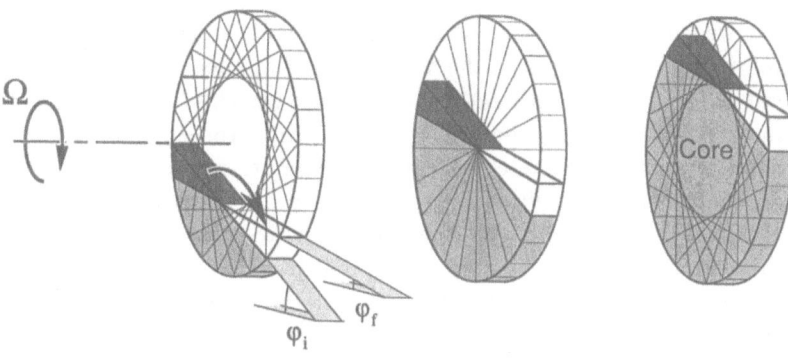

Fig. 10. — Geometric model of avalanche in rotating disk. Notice that irrespective of fill level, the effect of avalanches is to transport an uphill wedge (dark grey) of material to a downhill wedge (light grey). Global mixing depends on intersections between wedges associated with successive avalanches.

Mixing

It is perhaps not surprising that the first views – and more significantly, the vocabulary – of solid mixing were based on analogies with mixing of fluids. Thus, for example, Lacey [23], speaks of "convective mixing", the transfer of groups of particles from one location to another; "dispersive mixing", the distribution of particles over a freshly developed surface; and "shear mixing", the configuring of planes within the mass of material being mixed. Analogies help, but spurious analogies can prevent progress.

As with fluids problems, a geometrical view helps both to simplify modeling and to clarify understanding of essential mechanisms. In fact, for solids problems a geometrical view is in many ways more powerful than in the fluids case and it is surprising that it was not discovered before.

Consider a prototypical granular mixing system: a rotating 2D drum partially filled with granular particles. If the rotation speed, ω, is sufficiently slow, the angle of the free surface (φ in Fig. 10) grows until a discrete avalanche occurs, and φ relaxes from its pre-avalanche angle, φ_i, to a new angle, φ_f. The mere fact that the free surface angle changes from φ_i to φ_f allows us to construct a geometrical model. These two angles define two wedges, shaded in grey and white in the figure. As the avalanche occurs, material in the uphill wedge flows to fill the downhill wedge, as indicated by the bold arrow. Maps are present here, though in a different setting than in the fluids case. There are two types of maps in this case: a coarse map defining the gross motion of material during an avalanche — a wedge goes into a new wedge — and a finer map describing the detailed motion within the wedge. Thus the motion can be decomposed into two components: a geometrical component, consisting only of the transport of the wedges, and a dynamical component, consisting of a complex rearrangement of material within the wedge as a result of this transport.

By studying the geometrical component alone, we can learn several things about the qualitative behavior of this kind of mixing device. First, however well or poorly dynamical mixing occurs within a wedge, material cannot be transported outside of the wedges during an avalanche. Thus global mixing from material in one wedge into a different wedge can only occur within quadrilateral intersections between wedges (indicated in Fig. 10). Second, the wedge model applies to any fill level, and at a fill level of 50%, the quadrilateral intersections vanish. Consequently we expect global mixing to vanish for a half-filled drum. Third, the quadrilaterals expand as the fill level diminishes, so we expect global mixing to improve for lower fill levels. Fourth, as shown in the rightmost sketch of Fig. 10, fill levels greater than 50% produce a core in the center of the granular mass. No wedges penetrate the core, so no mixing should occur there. All of these predictions are validated by experiments (see Fig. 11). In all cases, initially all of the particles in the left half of the drum are colored red, and all particles on the right are colored blue, and the snapshots shown are taken after two complete revolutions of the drum.

We can use this model to make quantitative predictions as well. The simplest way to do this is to assume that particles within a wedge are completely randomized following each avalanche. Computationally this is accomplished by interchanging every particle within the final wedge with another particle, also within the final wedge, chosen at random. This is what was done to generate the simulations shown on the right of Fig. 11(a). In Fig. 11(b), we show mixing rates calculated for this problem. As predicted, the mixing rate goes to zero for half full drums, and increases as the fill level is reduced. Some mixing occurs for fill levels over 50%, although the mixing is substantially impeded by the core (even taking into account the fact that the mixing rate in the figure excludes the core).

Fluids comparison

We remark that the geometric viewpoint described for the drum mixer survives even if the particle characteristics are changed, provided that they are not so cohesive that distinct angles of repose become ill defined. For example, the geometrical model works for large round sugar beads equally well as for much finer, cubic salt grains; in fact, it survives when both materials are combined [28]. In this respect, the geometric view is more robust for solids than for fluids, where a change in rheology can have drastic effects on the structure of the resulting flow (cf. Fig. 6).

Lesson 1: Nonmixing regions have causes; more energy does not imply more mixing

There are technologically relevant lessons in common between the cases of fluids mixing and solids mixing. For example, in both problems there are features that do not mix. In fluids, these are islands – or more generally isolated mixing regions [25] – in solids, these are cores. In both problems, outside of these regions, mixing can be good, and an understanding of the causes of these impediments to mixing can be used to avoid them. To see this in the solids case, we note that the cause of the core is that mixing avalanches do not penetrate into this region. By including baffles, we can deliberately cause avalanches to penetrate the core, and thereby involve it in mixing. The mechanism at work is subtle. In Fig. 12, we show a sequence of sketches depicting what occurs as a tumbler with protruding baffles is slowly rotated. Initially, avalanches penetrate part way into the baffle, but the granular flow is eventually impeded by the upper corner of the baffle. Because granular flows

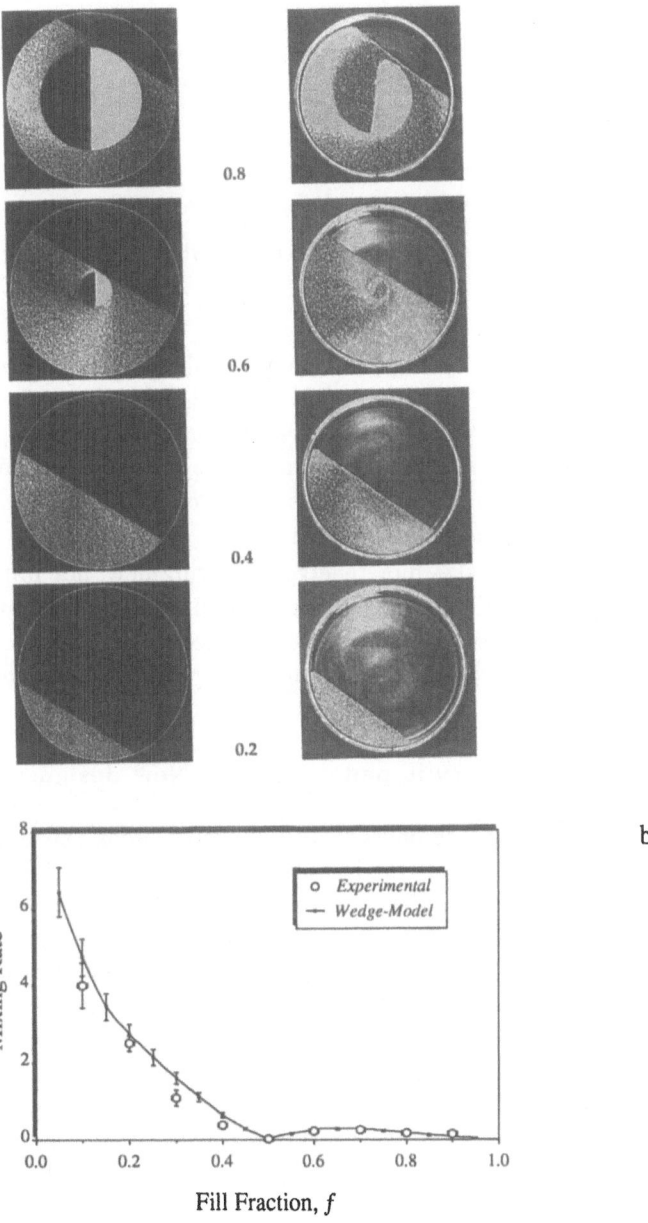

Fig. 11. — Comparisons between wedge model and experiments. (a) Side-by-side comparisons for different fill levels; simulations based on model on left, experiments on right. In all cases, blue grains are initially on the left and red grains are on the right, and photos are taken after 2 complete revolutions. (b): Mixing rate calculations from model (open symbols) and experiments (filled symbols). For details, see [24].

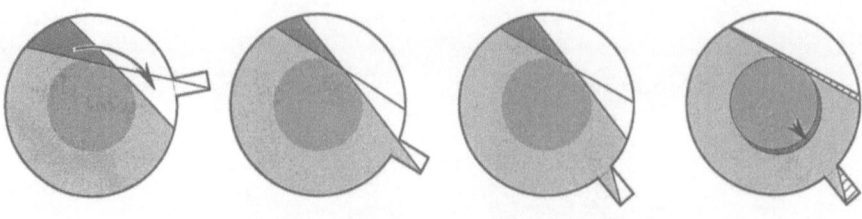

Fig. 12. — Schematic of mixing sequence in slowly rotating disk with protruding baffle.

can support a load, the cavity within the baffle remains for some period as the drum continues to rotate. After a certain point, however, the angle made by the baffle becomes sufficiently steep that the grains near the cavity collapse, and the supported material likewise settles. As this happens, the core shifts downward, and this shifting permits material on the outer edges of the core to enter the mixing zone. In this way, the core can be steadily eroded.

As in the fluids case, more energetic stirring does not improve mixing. For the problem shown in Fig. 9, large ($\sim$3 mm) and small ($\sim$0.5 mm) glass beads at a low rotation rate of (less than about 5 rpm) mix readily, but axial segregation is seen for larger rotation speeds. So here again more energy does not imply better mixing – for granular flows, in fact, the contrary is seen.

Lesson 2: Symmetry impedes mixing - New designs suggested

This brings us to a second similarity between fluids and solids, namely that symmetries interfere with mixing in both problems. To see this, consider two different baffle designs. First, a drum mixer with only one protruding baffle as shown in Fig. 12, and second a mixer with two baffles placed on opposite sides of the drum (Fig. 13(a)). While it might be thought that two baffles would accomplish twice the erosion of the core, in fact the symmetric placement of baffles results in the core shifting in one direction during one half of the rotation cycle and in the opposite direction during the second half cycle. Consequently the core undergoes no net translation, and steady erosion does not occur. The same applies to any symmetric placement of baffles. The same message appears if we destroy the chiral symmetry of the baffle, as shown in Fig. 13(b) – here clockwise rotation accomplishes core erosion; counter-clockwise rotation does not. These effects have been confirmed experimentally [28].

Lesson 3: Steady flows are poor mixers

Other lessons from fluid analysis carry over to the solids case as well. Steady flows in fluids are poor mixers. For the solids case, a similar message holds. Consider the drum mixer spun at a faster rate than before, so that the surface layer flows continuously. Here the flow is steady, and we expect different behavior from the

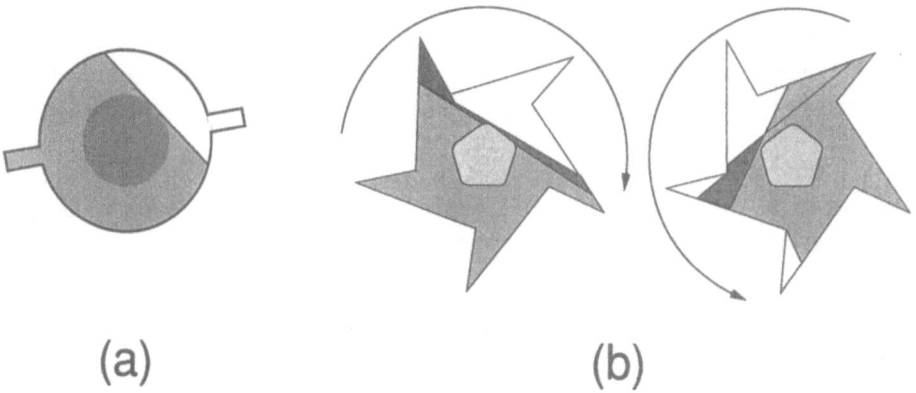

Fig. 13. — Schematic of mixing with (a) symmetric baffle placement; (b) with *rotationally asymmetric* baffles. Experimentally, core is not eroded for configuration shown in (a); core is eroded for *clockwise* rotation of blender in (b), but not for *counter-clockwise* rotation.

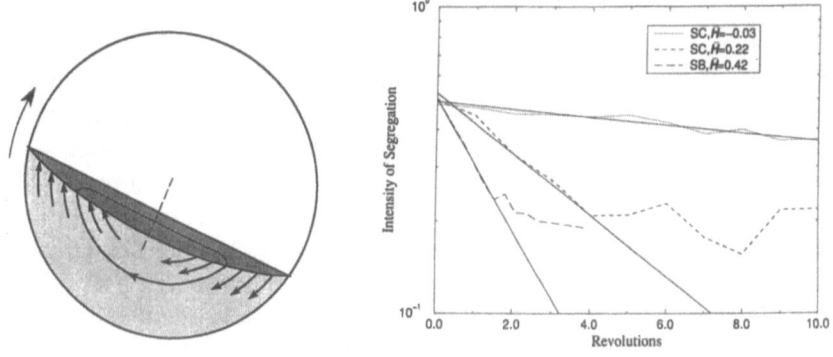

Fig. 14. — (a) Schematic view of the flow in disk mixer in cascading flow regime; (b) intensity of segregation (a measure of mixedness) vs. number of revolutions from continuum model for 3 different fill levels. Solid lines are best fits; dashed lines from experiments (SC are salt crystals; SB are sugar balls). For details, see [27].

time-periodic avalanching regime. In the continuous regime, one can break the problem in two: a bulk region, which rotates as a solid body with the drum, and a surface layer, where mixing can occur. A map can be used here as well by assuming that flow down the surface layer merely transports particles from an uphill position to a symmetrically placed downhill position [26]. In this respect, the model is similar to the avalanche model, although a continuous model can be employed to improve the description and predictive power.

A computation of the mixing rate reveals that mixing occurs more than an order of magnitude more slowly in the continuous case than in the discrete, avalanching case: it takes more *rotations* to achieve the same mixing in the continuous flow regime. A discrete avalanching mechanism is more efficient for mixing than the steady, continuous flow mechanism. In Fig. 14, we show a schematic of a mixing model used to study continuous mixing flow in a tumbler.

Another example is the case of 3D cylinders, where one may want to enhance axial mixing as well as radial mixing. In the slow rotation regime, radial mixing (possibly excepting the core for mixers filled beyond 50%) may be accomplished by avalanching. Under such conditions the axial mixing is very slow since it is essentially diffusive. Substantially faster overall mixing can be achieved by a combination of rotation and "wobbling" of the axis of the cylinder. Wobbling creates avalanches along the axial direction [28]. The best mixing occurs when the mass transported by the radial and axial avalanches are nearly equal.

5.   Concluding remarks

Let us contrast the current pictures and understanding of mixing of fluids and solids. For fluids mixing, *basic* understanding at a *continuum level* is firmly established: Navier-Stokes equations provide a first principles description valid on macroscopic scales for most problems. Such an understanding is lacking in the case of granular materials. Shortcomings of continuum descriptions manifest themselves on macroscopic scales, particle segregation being an instance where physical meso-scale processes are not well understood. The segregation problem remains largely open, as the follwing example reveals. Consider a small glass vial containing a mix of large orange sugar beads (diameter ~3 mm) and small blue salt grains (diameter ~0.5 mm). This is the so-called Brazil nut problem; from what is understood one may expect that vertical shaking will lead to the large beads rising to the surface. In fact, they do not: they mix quite readily. On the other hand, horizontal shaking causes the large beads to rapidly accumulate on the surface. By contrast a large massive object will actually sink on horizontal shaking. It is thus apparent that even the limited models that have been put forward may have easily generated counter-examples.

Another point of comparison is provided by changes in material properties. A change in the nature of a fluid is reflected in the nature of the constitutive equation – for example going to a Maxwell model or any type of viscoelastic model. [29] A problem can thus be easily formulated, though perhaps not easily solved (even numerically). No such a route is possible in the case of granular materials. Constitutive equations have not been successfully developed – certainly not to the level encountered in fluids. In some restricted contexts, it may indeed be possible to generate flow and constitutive equations. It is in general unclear which material properties may be generic, which are specific to a problem, and when very minor changes have large effects. Taking the example of segregation as a case in point, existing models consider only the size of the particles into account. Yet the *shape* is also clearly important: the angle of repose of round particles is substantially shallower than angular ones, and this is known to generate segregation on its own. Fig. 15 shows an example of radial segregation of round vs. angular particles in the rotating drum. Here the mechanism is very easily observed (and has been confirmed numerically): the round particles roll easily across the free surface and accumulate along the downhill edge of the container, to be covered by periodic avalanches of the angular particles. It is hard to conceive of constitutive equations capable of capturing such effects.

The situation is, however, not completely bleak for solids. In fact (and somewhat paradoxically) incorporating new materials properties may be easier for solids than for fluids. Let us assert first that a first principles description – particle dynamics

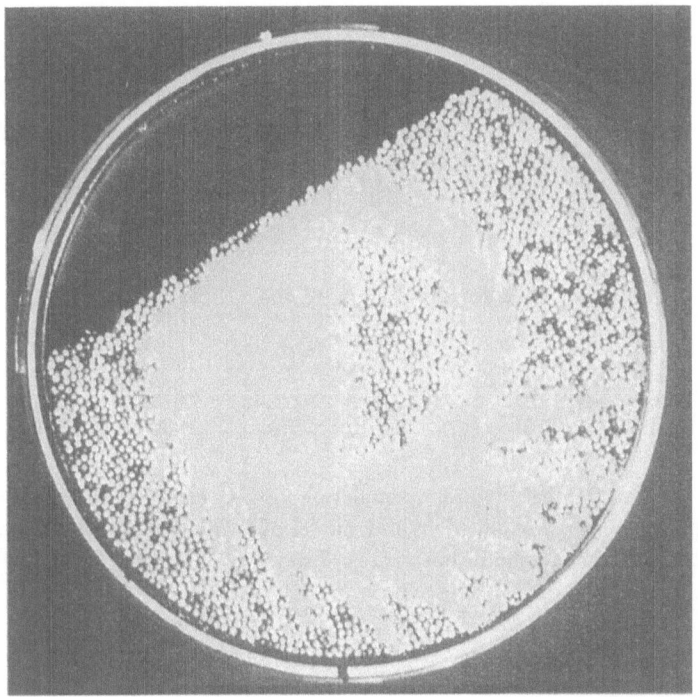

Fig. 15. — Radial segregation of sugar balls (mean diameter 3 mm) and salt crystals (mean diameter 0.5 mm) in slowly rotated disk tumbler.

– is arguably better for (spherical) solids than for fluids; granular flows sem to be tailor-made for a discrete approach. Two types of approaches seem possible: *particle dynamics* and *lattice models*. As an example of the capabilities of a particle dynamics approach consider again the problem of the rotating tumbler. The flow is simulated by computationally following each particle and determining all forces and torques (Fig. 16). For slow flows, where particles are nearly always in contact with one another, so-called *soft particle techniques* are called for. Here one allows particles to deform slightly (typically by ≪1% of a particle diameter) and one takes the normal force between particles to obey a viscously damped spring model. Tangential forces on the other hand obey a stick-slip law, with Coulombic friction applied once a maximum tangential force is exceeded. The numbers of particles that can be feasibly simulated in this way is limited: this is in principle an N2 problem which by judicious coding (i.e. breaking the problem spatially into cells so as to reduce the number of necessary computations) can be reduced to $N \ln(N)$. The number of particles which can be simulated can be increased – either in the avalanching or in the continuous flow regime – by applying the prior observation that only particles in a surface layer, typically of depth ~5 particles, can shear, and all other particles must rotate as a rigid body. This 'hybrid' model introduces minor errors in materials which settle during rotation; nevertheless it is an effective tool to permit simulations of 10^4 - 10^6

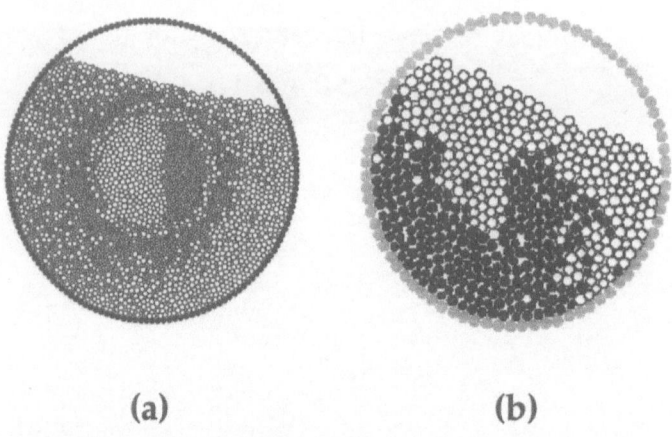

(a)   **(b)**

Fig. 16. — (a) Full particle dynamic simulation of two complete revolutions of tumbler using 600 particles. (b) Simplified hybrid model of two complete revolutions using 6000 particles. Computational time in both cases: 7 days. (Courtesy of JJ McCarthy)

particles in sensible amounts of time. The simulation shown in Fig. 16(a) does not include this simplification, and uses 600 particles; Fig. 16(b) on other hand, took the same amount of computational time using 6000 particles.

A promising alternative approach to granular flows involves lattice models [30]. Lattice models represent the ultimate contrast to continuum approaches: here spatial variations are explicitly discrete, particles can only reside on a fixed lattice, and rules of motion from one lattice site to an adjacent one are prescribed. These models, which neglect microscopic physical interactions, can be surprisingly good. As an example, in Fig. 7(c) we plot the behavior of a lattice model meant to mimic a shaken container partially filled with monodisperse particles [31]. Here the initial effect of shaking is to introduce voids into the lattice; thereafter particles settle to fill the voids according to a stochastic law. The inset above the particles in Fig. 7(c) shows the vertical displacements of marker particles averaged over 300 simulated shakes (for details see [31]). The solid curve shows good agreement with prior experimental results [32, 16].

It is apparent that mixing offers a fertile ground for theory and basic experimentation, with advances in theory likely translating into advances in practice. Understanding of mixing of fluids is more advanced than mixing of solids. Mixing of fluids is far from being closed and important problems remain; advances in the last decade, however, have been major. Mixing of solids is now where mixing of fluids was a decade ago: problems are open and successful efforts will have maximum impact. Virtually all avenues for attack have a place; geometrical ideas, continuum descriptions, and discrete particle simulations all lead to successes, with none dominating the picture. This blend offers exciting opportunities in years ahead.

References

[1] Moreover, for the problems which we discuss, careful experiments in vacuum duplicate key effects seen in the presence of ambient air.

[2] O. Reynolds, "Study of fluid motion by means of coloured bands", Nature **50**, 161 (1894).

[3] From S. Corrsin, *Lecture Notes on Introductory Fluid Mechanics* (The John Hopkins University, Baltimore, 1966).

[4] R.S. Spencer and R.M. Wiley, "The mixing of very viscous liquids", J. Colloid Sci., **6**, 133 (1951).

[5] S. Smale, "Differentiable dynamical systems", Bull. Amer. Math. Soc., **73**, 747 (1967).

[6] W-L Chien, H. Rising, and J.M. Ottino, "Laminar mixing and chaotic mixing in several cavity flows", J. Fluid Mech. **170**, 355 (1986).

[7] T. Theodorsen, "The structure of turbulence", *50 Jahre Genszchichtsforschung: Ludwig Prandtl*, (Friedr. Viegeg & Sohn, Braunschweig, 1955), pp. 55-62.

[8] O. Reynolds, "On the dynamical theory of incompresible viscous fluids and the determination of the criterion", Phil. Trans. R. Soc. London. Ser. A **186**, 132 (1895).

[9] J.G. Franjione and J.M. Ottino, "Symmetry Concepts for the Geometric Analysis of Mixing Flows", Phil. Trans. Roy. Soc. Lond., **338**, 301 (1992).

[10] J. M. Ottino, C.W. Leong, H. Rising, and P.D. Swanson, "Morphological structures produced by mixing in chaotic flows", Nature, **333**, 419 (1988).

[11] J.M. Ottino, "Unity and diversity in mixing: Stretching, diffusion, breakup and aggregation in chaotic flows", Phys. Fluids A, **3**, 1417 (1991).

[12] S.C. Jana, M. Tjahjadi, and J.M. Ottino, "Chaotic Mixing of Viscous Fluids by Periodic Changes of Geometry; The Baffle-Cavity System", AIChE Journal, **40**, 1769 (1994).

[13] T.C. Niederkorn and J.M. Ottino, "Mixing of Viscoelastic Fluids in Time-Periodic Flows", J. Fluid Mech. **256**, 243 (1993).

[14] H.K. Pak, E van Doorn & R.P. Behringer, "Effects of ambient gases on granular materials under vertical vibration", Phys. Rev. Lett. **74**, 4643 (1995).

[15] H.K. Pak and R.P. Behringer, "Bubbling in vertically vibrated granular materials", Nature **371**, 231 (1994); "Surface waves in vertically vibrated granular materials", Phys. Rev. Lett. **71** 1832 (1993); Y-H. Taguchi, "New origin of a convective motion: elastically induced convection in granular materials," Phys. Rev. Lett. **69** 1367 (1992).

[16] E.E. Ehrichs, H.M. Jaeger, G.S. Karczmar, J.B. Knight, V.Y. Kuperman, and S.R. Nagel, "Granular convection observed by magnetic resonance imaging," Science **267**, 1632 (1995).

[17] J.B. Knight, H.M. Jaeger, and S.R. Nagel, "Vibration-induced size separation in granular media: the convection connection," Phys. Rev. Lett. **70**, 3728 (1993).

[18] H. Hayakawa, S. Yue, and D.C. Hong, "Hydrodynamic description of granular convection," Phys. Rev. Lett. **75**, 2328 (1995).

[19] M. Bourzutschky and J. Miller, " 'Granular' convection in a vibrated fluid," Phys. Rev. Lett. **74** 2216 (1995).

[20] R.P. Behringer, "The dynamics of flowing sand," Nonlinear Science Today **3**, 1 (1993); J. Duran, J. Rajchenbach and E. Clément, "Arching effect model for particle size segregation," Phys. Rev. Lett. **70**, 2431 (1993); C-H Liu, S.R. Nagel, D.A. Schechter, S.N. Coppersmith, S. Majumdar, O. Narayan, and T.A. Witten, "Force Fluctuations in Bead packs," Science **269**, 513 (1995); S.S. Manna and D.V. Khakhar, "Packing of mutually interacting powder particles under gravity", in *Nonlinear Phenomena in Materials Science*, G. Ananthakrishna, L. P. Kubin, and G. Martin, eds., (Transtech, Switzerland, in press).

[21] O. Reynolds, "On the dilatancy of media composed of rigid particles in contact. With experimental illustrations." Philo. Mag. **20**, 469 (1885); "Experiments showing dilatancy, a property of granular material, possibly connected with gravitation," in *Papers on Mechanical and Physical Subjects* (Cambridge University Press, 1901), Vol. II pp. 217-27.

[22] K.M. Hill and J. Kakalios, "Reversible axial segregation of binary mixtures of granular mixtures", Phys. Rev. E **49** R3610 (1994).

[23] P.M. Lacey, "Developments in the theory of particle mixing", J. Appl. Chem. **4**, 257 (1954).

[24] G. Metcalfe, T. Shinbrot, J. McCarthy, and J.M. Ottino, "Avalanche Mixing of Granular Solids," Nature **374**, 39 (1995).

[25] L. Bresler, T. Shinbrot, G. Metcalfe, and J.M. Ottino, "Isolated Mixing Regions: Origin, Robustness, and Control", Chem. Eng. Sci. **52**, 1623 (1997).

[26] R. Hogg and D.W. Fuerstenau, "Transverse Mixing in Rotating Cylinders", Powder Tech. **6**, 139 (1972).

[27] D.V. Khakhar, J.J. McCarthy, T. Shinbrot, and J.M. Ottino, "Transverse Flow and Mixing of Granular Materials in a Rotating Cylinder", Phys. Fluids **9**, 31 (1997).

[28] J.J. McCarthy, T. Shinbrot, G. Metcalfe, J.E. Wolf, and J.M. Ottino, "Mixing of Granular Materials in Slowly Rotated Containers", AIChE J. **42**, 3351 (1996).

[29] R.B. Bird, R.C. Armstrong, and O. Hassager, *Fluid Mechanics: Dynamics of Polymeric Liquids* (Wiley, New York, 1987) Vol. I.

[30] N.B. Ouchi and H. Nishimori, "Modeling of wind-blown sand using cellular automata," Phys. Rev. E **52**, 5877 (1995); H. Caram and D.C. Hong, "Random walk approach to granular flows," Phys. Rev. Lett. **67**, 828 (1991); R.S. Anderson and K.L. Bunas, "Grain size segregation and stratigraphy in aeolian ripples modelled with a cellular automaton," Nature **365**, 740 (1993).

[31] T. Shinbrot, D.V. Khakhar, J.J. McCarthy, and J.M. Ottino, "A simple model for granular convection" (under review 1996).

[32] J.B. Knight, E.E. Ehrichs, V.Y. Kuperman, J.K. Flint, H.M. Jaeger, and S.R. Nagel, "An experimental study of granular convection," Phys. Rev. E **54**, 5726 (1996).

Scale Covariance and Geometry in Turbulent Combustion

ALAIN POCHEAU

IRPHE, UMR 6594 CNRS, Universités Aix-Marseille I & II
Centre de Saint-Jérôme, S.252, 13397 Marseille Cedex 20, France

---

The statistical behaviour of turbulent propagating interfaces is addressed in scale-invariant regimes of front-turbulence interaction. Following scale invariance, statistical laws should rely on *no* characteristic scale and *no* characteristic scale ratios. Accordingly, their form should be invariant by *global* as well as by *local* changes of scales. This symmetry, the scale-covariance, is implemented theoretically on the relation between normal and average front velocity. It is then evidenced experimentally by studying, at any scale, the equivalent relation between front roughness and turbulence intensities. Scale-covariant laws are further used to derive the evolution of front geometry from scale to scale. The asymptotic features shown at large scales and large turbulence intensities reveal that, although the most turbulent fronts look fractal in a classical scale analysis, they actually stand at a significant distance from the fractal regime. This questions the relevance of fractal analysis to these multi-scale objects and elucidates the origin of the paradoxical variation of their "apparent" fractal dimensions with turbulence intensity.

1. Introduction

The propagation of interfaces in a turbulent medium involves many interesting features. On one hand, it is mostly important in a number of practical processes, ranging from the engine industry to the propagation of epidemic or forest fires. On the other hand, it provides geometric objects, the interfaces, which may help in investigating turbulent mixing through a geometric support.

Mixing: Chaos and Turbulence, edited by Chaté *et al.*
Kluwer Academic / Plenum Publishers, New York 1999.

The present work is devoted to the study of premixed flames propagating in turbulent gases [1, 2, 3]. A tomographic cut of such flames, displayed in Fig. 1, shows front wrinkles at many different scales. These may be attributed to turbulence since they no longer appear in laminar flow. They increase the front surface and thus its efficiency in burning the medium, insofar as the normal velocity is unchanged [4]. Here, turbulence has thus increased the mean velocity of the front, as it does for the mean diffusivity of a fluid. In order to understand this enhancement of the mean transport property of the medium by turbulence, two different points of view may be considered.

The first approach, largely implemented in the past decade, focuses attention on the objects, the fronts, and aims at describing their statistical features with respect to control parameters [5]. It is especially supported by the concept of fractal which leads to a characterization of front geometry by a number: the fractal dimension. There, only scale-similar forms are considered, irrespective of the way they have been generated.

We propose to change the spirit of analysis by addressing not the objects themselves but the physical process responsible for their formation. This entails considering neither the effects of the process (here the front geometry or the mean front velocity), nor its cause (here the turbulent flow) but the *relation* between them. Accordingly, this approach may be termed "causal" as opposed to the more descriptive fractal analysis. It implies studying the structure of *laws* instead of the structure of *objects* and thus addressing the relations between measurements rather than the measurements themselves. In the present case, it will lead us to go beyond fractal analysis by considering, not only the large scales or the large mixing regimes to which fractal analysis is restricted, but also the remaining small scales or weak mixing regimes where information relevant to the physical process may also be obtained.

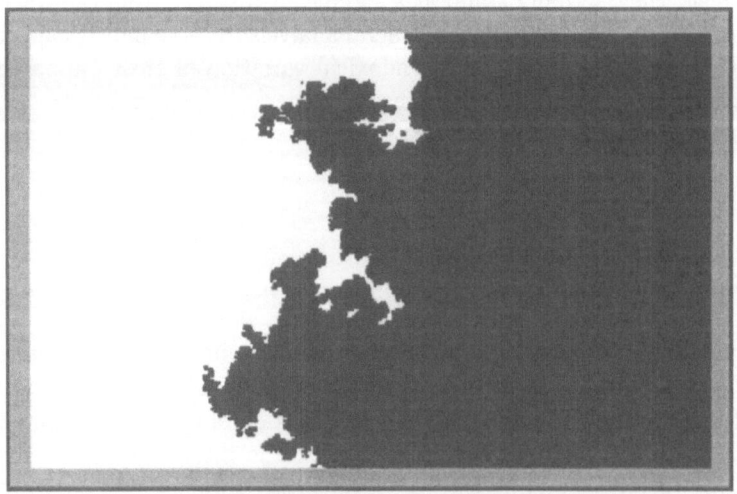

Fig. 1. — Tomographic cut of a turbulent flame front: $u'/U_N = 1.30$.

Regarding turbulent combustion, more than one century of studies have pointed out a large number of phenomena involved in flame propagation: hydrodynamic or thermo-diffusive instabilities, modification of normal front velocity by flow stretch or front curvature, cusp generation in finite-time, sensitivity to noise ...[6, 7]. All these aspects must be taken into account in the statistical relation sought between the turbulent flow and the turbulent front but, needless to say, directly solving this phenomenology lies beyond the present analytical technics. This, however, does not mean that our goal is definitely hopeless. Especially, we stress that, in the regimes where statistical symmetry is satisfied, the phenomenology of the system may actually be by-passed to work out the few possible relations compatible with the symmetry.

In the present case, this symmetry will be the scale-invariance of the system. It implies that the laws governing turbulent combustion should not rely on *any* characteristic *scale* and *any* characteristic *scale ratio*. Their form should then be invariant for global as well as for local changes of scales. These symmetries, the global and local *scale-covariance*, will be addressed in Sec. 2 and evidenced experimentally in Sec. 3. They will enable us in Sec. 4 to follow the construction of front geometry in scale space from an euclidean regime at low scale to a seemingly fractal regime at large scale. This deterministic evolution of the mean front geometry will finally lead us to relativize the relevance of fractal analysis in this system and to elucidate the origin of the scale-dependent "fractal dimensions" displayed by turbulent fronts.

2. From scale-invariance to scale-covariance

We wish to analyze turbulent combustion within a geometrical family of scales $\{L_i\}_{i=0,...,I}$, where $L_i/L_0 = a^i$. Here, L_0 denotes the smallest resolved scale, L_I the integral scale of turbulence and a the elementary scale-ratio of observation. Three assumptions or modelizations will be made on the front, the flow, and their interaction :

1. The normal front velocity U_N is a constant.
   This assumption is valid insofar as the wrinkling scales are much larger than the front thickness δ [6]. This will be the case in the present experiment except at cusp singularities.

2. The turbulence intensities follow a scale power-law.
   This is not a fundamental assumption but only a modelization of turbulence intensities which will be convenient in the following for practical purpose. Labelling $u'_{i,j}$ the turbulence intensity in the scale range $[L_i, L_j]$ and K its scale exponent, it reads:

$$u'_{i,i+1}/u'_{o,1} = (L_i/L_0)^K \tag{1}$$

Relation (1) implies the absence of any characteristic scale in the flow, a property not valid in the following experiment due to the proximity of the dissipative flow scale. In spite of this disagreement, we shall nevertheless invoke it as a well fitting relation of turbulence intensities in our scale range. It will then provide us with useful means for modelling the evolution of turbulence intensities with scales.

As the range of wrinkling scales is more dissipative than inertial and because of the feedback from combustion on turbulence, the classical value $K = 1/3$ of the Kolmogorov theory is not expected. In fact, the requirement of scale-covariance will give rise to a larger value $K = 1/2$ [8] which is actually consistent with that given by the viscous damping of turbulence intensity in non-reactive flows [2].

3. The interaction between the flow and the front is local in scale space.

This assumption implies that the wrinkles in between the range $[L_i, L_j]$ are produced by the vortices of this scale range only. In particular, they depend on no other turbulence intensity than that involved in this scale range, i.e. $u'_{i,j}$. This assumption is valid in our experiment insofar as one neglects cusp formation and as far as front velocities and turbulence intensities have similar magnitude. The latter property is actually ensured a posteriori by the cumulative effect of front wrinkles in scale space.

2.1.   Scale-invariant regime of turbulent combustion

In scale space, turbulent combustion may be viewed as an interaction between the scales of the flame and the scales of the flow. The former scales involve the flame thickness δ and the transit time of gases through the front τ_c. The latter scales involve the eddy sizes L_i ranging from the Kolmogorov scale L_η to the integral scale L_I, and their corresponding turn-over times τ_i [8]. Hereafter, we shall restrict ourselves to the so-called flamelet regime where the scales of the front are much smaller than those of the flow [9]:

$$\delta \ll L_\eta < L_i \tag{2}$$

$$\tau_c \ll \tau_\eta < \tau_i \tag{3}$$

It must be stressed that this regime is not purely academic since it is largely encountered in internal combustion engines for instance [9].

According to relations (2-3), any vortex of the turbulent flow sees the flame as an infinitely small interface, $\delta/L_i \ll 1$, with infinitely fast kinetics, $\tau_c/\tau_i \ll 1$. Applied to the search of regular laws, this suggests that the conditions of interaction between the vortices and the flame are the same for all vortices, whatever their size. Following this idea, we may then expect that the wrinkling process behaves statistically the same way for all vortices, irrespective of their scales, and is thus statistically *scale-invariant*. In the remainder of this work, our goal will consist in extracting as much information as possible from this property.

2.2.   Scale-covariant laws

Consider a front of normal velocity U_N propagating at a mean velocity U_T within a window W enclosing a turbulent flow of intensity u' (Fig. 2a). The increase of velocity U_T/U_N is a priori a function of the ratio u'/U_N, of the scales of the flow non-dimensionalized by those of the front and of other numbers specifying the physical regime such as the Reynolds number of the flow at the integral scale L_I, the Lewis

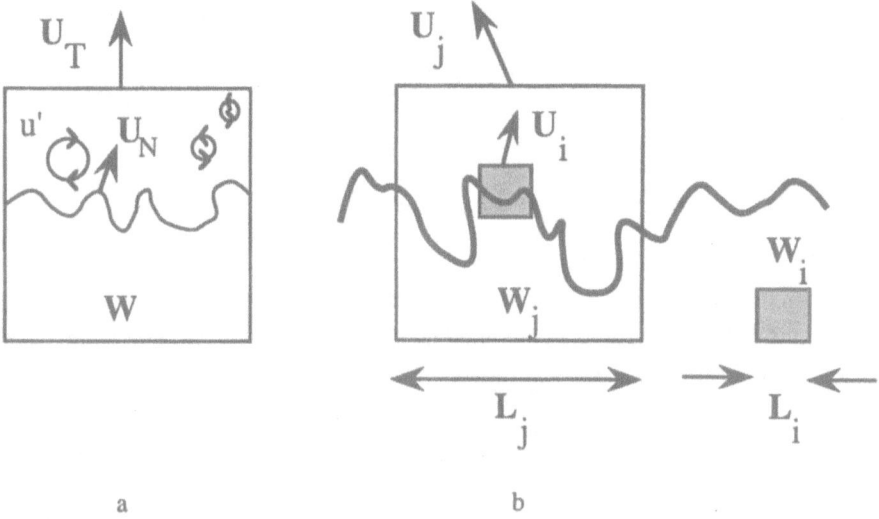

Fig. 2. — Sketch of a turbulent front : a) usual representation; b) generalization to various resolution scales L_i and observation scales L_j.

number of the reacting species, etc... :

$$\frac{U_T}{U_N} = \mathcal{L}\left(\frac{u'}{U_N}, \left(\frac{\delta}{L_i}\right)_{i=0,...,I}, \left(\frac{\tau_c}{\tau_i}\right)_{i=0,...,I}, \text{Re, Le}\right) \tag{4}$$

However, in the flamelet regime, all the quantities δ/L_i, τ_c/τ_i approach zero. Invoking a regular limit of the function $\mathcal{L}(.)$ [10], they can therefore be left out of the description. This, in particular, implies that the Reynolds number Re should be omitted in relation (4) since it is related to the ratio of scales L_I/L_η: Re = $(L_I/L_\eta)^{(1+K)}$. Note that its finite value is nevertheless implicitly taken into account in a value other than 1/3 of the exponent K of relation (1). Finally, considering the Lewis number as a fixed implicit parameter of the law, we obtain the following simplification of relation (4) :

$$\frac{U_T}{U_N} = \mathcal{L}\left(\frac{u'}{U_N}\right) \tag{5}$$

According to relation (5), law $\mathcal{L}(.)$ does not involve explicit scales. This important property means that the *same* relation would have been obtained if the analysis had been performed in another scale range. It thus implies that law $\mathcal{L}(.)$ is *scale-covariant*.

In order to formalize this property, let us observe the front within another window W_j of size L_j and at another resolution scale L_i (Fig. 2b). The mean velocity of the window W_j co-moving with the front then corresponds to the turbulent velocity U_j at scale L_j. Similarly, the mean velocity U_i of windows W_i of size L_i following the front, plays the role of a mean normal velocity at scale L_i. Both are respectively the analogues of the turbulent velocity U_T and of the normal velocity U_N previously considered. Invoking the locality of the front-turbulence interaction in scale space, U_j and U_i are thus related to the turbulence intensity $u'_{i,j}$ by a law $\mathcal{L}_{i,j}(.)$ which generalizes law $\mathcal{L}(.)$ to the scale range $[L_i, L_j]$:

$$\frac{U_j}{U_i} = \mathcal{L}_{i,j}\left(\frac{u'_{i,j}}{U_i}\right) \tag{6}$$

Scale-covariance then requires:

$$\mathcal{L}_{i,j}(.) = \mathcal{L}(.) \quad \forall i, \forall j \tag{7}$$

Its physical interpretation is the following : as soon as one rejects the intrinsic scales of both the front (δ, τ_c) and the flow (L_η, τ_η) as relevant variables of the physical process, all the resolution scales L_i and the observation scales L_j become physically equivalent. Then the wrinkling laws $\mathcal{L}_{i,j}(.)$ should not make a distinction between them. In other words, they must be scale-covariant.

2.3.   Global and local scale-covariance

The requirement of scale-covariance (7) implies that *any* change of scales $L_i \to L_{i'}$, $L_j \to L_{j'}$, should preserve the wrinkling law.

Firstly, this includes changes preserving scale ratios: $L_{j'}/L_{i'} = L_j/L_i$. This simply corresponds to a change of unit, i.e. $L_k \to \gamma L_k$ with constant γ, and thus, in logarithmic coordinates, to a uniform translation in scale space (Fig. 3a). Covariance with respect to these changes means that the value of scale is physically *irrelevant* so that the system does not single out - nor rely on - any specific one. We call the covariance with respect to these restricted scale changes "global covariance".

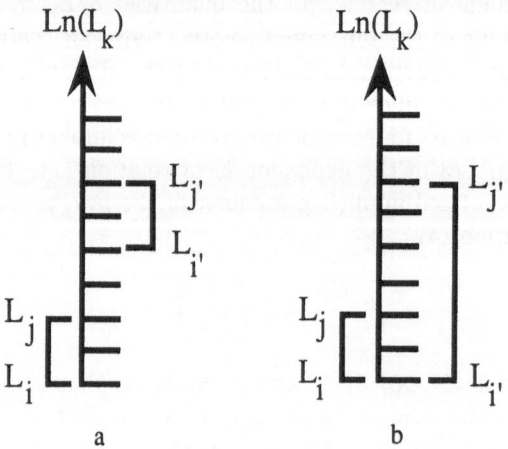

Fig. 3. — Sketch of scale-covariance : the wrinkling law keeps the same expression in the scale ranges $[L_i, L_j]$ or $[L_{i'}, L_{j'}]$: a) global covariance: scale ranges have the same size: $L_j/L_i = L_{j'}/L_{i'}$; b) local covariance: scale ranges have different sizes: $L_j/L_i \neq L_{j'}/L_{i'}$

However, the scale-covariance involved in (7) means something more: it also includes the invariance of law with respect to scale changes modifying scale ratios: $L_{j'}/L_{i'} \neq L_j/L_i$. In scale space, these changes imply not only a translation but also a stretch of scale range (Fig. 3b). Covariance with respect to them means that the size of scale range is physically *irrelevant* so that the system does not single out - nor rely on - any specific *scale ratio*. We call this symmetry "local covariance".

For functions of a single scale, global covariance is sufficient for selecting power laws [1]. Then local covariance is redundant and thus usually overlooked. The present case is, however, different since the function $\mathcal{L}_{i,j}(.)$ depends on an argument $u'_{i,j}/U_i$ which includes a variable monitored by the sub-scales $L_k < L_i$: the normal velocity U_i. It is thus an implicit function of many scales which, for this reason, is *not* prescribed to a simple power law. Here, global covariance actually stands as a weaker constraint than local covariance so that the latter has to be explicitly implemented [1].

2.4. Scale-covariant wrinkling laws

Global covariance of $\mathcal{L}(.)$ means that this law involves no explicit length scale. As this is validated, within relation (6), by any algebraic form of $\mathcal{L}(.)$, this requirement is actually fulfilled by a large number of laws proposed in turbulent combustion [11]. However, those which do not satisfy local scale-covariance make a distinction between various sizes of scale ranges and thus, for instance, between different definitions of the integral scale of turbulence L_I: $\mathcal{L}_{0,I}(.) \neq \mathcal{L}_{0,I'}(.)$. This would be a way of differentiating scales that we reject in this system.

A more detailed analysis enables us to determine the family of fully scale-covariant laws [1]. It relies on the fact that the law $\mathcal{L}_{i,j}(.)$ linking the turbulent velocities U_j and U_i in the scale range $[L_i, L_j]$ may be derived by composing the laws $\mathcal{L}_{k,k+1}(.)$ relating the sub-ranges velocities U_{k+1} and U_k, step by step, from $k = i$ to $j - 1$: $\mathcal{L}_{i,j}(.) = \mathcal{L}_{j-1,j}(.) \circ \ldots \circ \mathcal{L}_{i,i+1}(.)$. As scale-covariant laws $\mathcal{L}(.)$ must remain unchanged by this kind of integration in scale space, they should thus correspond to a fixed point of renormalization by scale change : $\mathcal{L}(.) = \mathcal{L}(.) \circ \ldots \circ \mathcal{L}(.)$ [12]. This constraint provides a suitable criterion for selecting them [1].

We shall focus attention hereafter on a particular scale- covariant law, $\mathcal{L}(x)^2 = 1 + \beta x^2$ or, from relations (6) and (7):

$$U_j^2 = U_i^2 + \beta u'^2_{i,j} \quad \forall i \, , \, \forall j \tag{8}$$

We may check its invariance by integration through the following equivalences :

$$U_{i+1}^2 = U_i^2 + \beta u'^2_{i,i+1} \qquad \forall i \tag{9}$$

$$\Longleftrightarrow \quad U_j^2 = U_i^2 + \beta \sum_{k=i}^{k=j-1} u'^2_{k,k+1} \qquad \forall i \, , \, \forall j \tag{10}$$

$$\Longleftrightarrow \quad U_j^2 = U_i^2 + \beta u'^2_{i,j} \qquad \forall i \, , \, \forall j$$

Here, the latter equivalence makes use of the additivity of the squares of turbulence intensities with respect to scale ranges, which follows from the Parseval relation or from energetic considerations. The equivalence between relations (8) and (9) implies that using the same law (8) in either small scale ranges $[L_i, L_{i+1}]$ or large scale ranges $[L_i, L_j]$ yields no internal contradiction. This validates this particular form (8) of scale-covariant laws.

3. Experimental evidence of scale-covariance

Below, we confront below scale-covariant law (8) to experiment [2].

3.1. Experimental set-up and data processing

The experimental set-up is made of a closed combustion chamber provided by a piston at one end, a grid in the middle, a matrix of sparks at the other end and three optical windows on the sides (Fig. 4). The chamber is initially filled with premixed propane-air gases seeded with oil particles. Then the piston is pushed suddenly to compress part of the gases through the grid. This produces turbulent motions within which a flame front generated by spark ignition propagates.

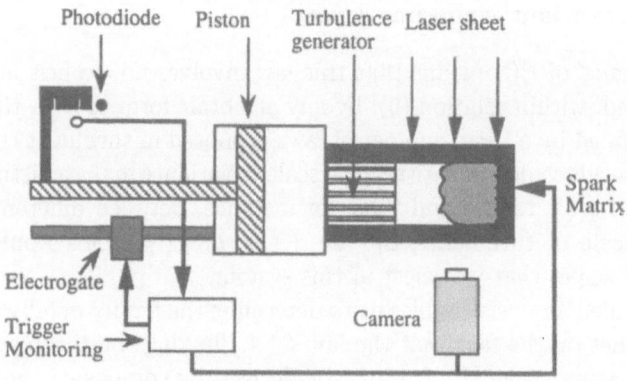

Fig. 4. — Sketch of the experimental set-up.

Vizualization of the flame front is provided by tomography [13] (Fig. 1). A plane laser sheet illuminates the chamber along the mean direction of propagation. It is diffused by oil particles towards a camera. As these are burnt by the flame, observation of the laser sheet in a transverse direction reveals a bright zone not yet burnt, a dark zone already burnt and the front cut in between. These images are registered on a CCD camera and then processed on a computer. Their resolution, 120μm, is typically five times smaller than the lowest noticeable wrinkling scale.

The normal velocity U_N is varied by changing the mixture composition. As the piston is pushed at the highest available rate, the turbulence intensity u' is changed by monitoring its decay through an ignition delay. The measurements of u' by anemometry laser doppler [14] fit with an exponential decay involving a characteristic time scale $\tau = 136$ms. Compared to τ, the propagation time (around 10ms) is then short enough for preventing noticeable further decay of u' until front observation.

Combustion of the medium induces pressure and temperature rises and adiabatic compression of the fresh gases. These effects are included in the calculation of U_N according to the empirical laws determined in laminar experiments [15]. In contrast, these effects are not used to renormalize u' so that one might question the relevance of its value at the observation time. However, since the thermodynamic conditions were the same for each flame at its observation time, one may guess that the modifications of u' from ignition to observation were the same for all. For this reason, the value of u' presumably remains relevant to the turbulent flow at the observation time, up to an assumed constant prefactor.

The ranges of parameters scanned in our study are: combustion volume $10 \times 6 \times 6\text{cm}^3$, normal velocity at observation time ($T \simeq 520K$, $p \simeq 7.2\text{atm}$) $0.40\text{ms}^{-1} \leq U_N \leq 0.86\text{ms}^{-1}$, integral scale of turbulence $L_I = 5.2\text{mm}$, $0.30 \leq u' \leq 0.93\text{ms}^{-1}$ and $0.7 \leq u'/U_N \leq 2.2$.

Image processing involves the measurement of the roughness $\Lambda_{i,S}$ of the front cut at increasing resolutions L_i, the scale L_S denoting here the size of the chamber. Then surface roughness $R_{i,S}$ is deduced by the relation $R_{i,S}^2 = 2\Lambda_{i,S^2} - 1$. This relation is valid at first order in $\Lambda_{i,S} - 1$, at large values of $\Lambda_{i,S}$ where, following the quasi-fractal nature of the front, surface roughness and linear roughness are expected to be proportional and, finally, in intermediate regimes [21]. Surface roughness $R_{i,j}$ at various resolutions L_i and within different windows W_j of size L_j, are deduced from the series $\{R_{i,S}\}_{i=0,...,I}$ by a property inherent to the definition of roughness, the multiplicativity: $R_{i,j} = R_{i,S}/R_{j,S}$.

Surface roughnesses drive an enhancement of front velocity which may be derived by considering the volume burnt by a flame per unit time. At a given resolution scale L_i, it corresponds to the product of the normal velocity U_i by the front surface S_i seen at this resolution. Equating this determination for two different scales L_i and L_j gives $U_iS_i = U_jS_j$. Since the ratio of surfaces S_i/S_j is simply the surface roughness $R_{i,j}$, one obtains :

$$R_{i,j} = U_j/U_i \quad \forall i \, , \, \forall j \tag{11}$$

According to (6), they are thus related by the law $\mathcal{L}_{i,j}(.)$ to a so-called mixing variable $m_{i,j}$:

$$m_{i,j} = u'_{i,j}/U_i \tag{12}$$

$$R_{i,j} = \mathcal{L}_{i,j}(m_{i,j}) \quad \forall i \, , \, \forall j \tag{13}$$

Turbulence intensities $u'_{i,j}$ are calculated according to the power law (1) for a given value of K by distributing the turbulence intensity u' on the scale range $[L_\eta, L_i]$, the Kolmogorov length L_η referring here to the fresh gases. Normal velocities U_i are deduced from roughness measurements by $U_i = R_{0,i}U_0$ using the fact that, for a scale L_0 smaller than the smallest wrinkling scale, $U_0 = U_N$.

Since both roughness $R_{i,j}$ and mixing variables $m_{i,j}$ may be determined at all scales, a comparison between the various laws $\mathcal{L}_{i,j}(.)$ can be made. It is performed below in two ways: firstly by fixing the lowest scale L_i at the value L_0, secondly by taking for the scale L_j the scale immediately larger than L_i.

3.2.   Integral analysis

We consider here the scale ranges $[L_0, L_j]$ at a fixed L_0 but a variable L_j. This entails including in our description the effects of more and more vortices or equivalently, taking the integral scale at larger and larger values. The measurements $(m_{0,j}, R_{0,j})$ are made, for each mixture composition, on a dozen flames and, for each flame, on 20 scales L_j in between L_0 and L_I ($L_I/L_0 = 44$ with an elementary scale-ratio $a = 1.2$).

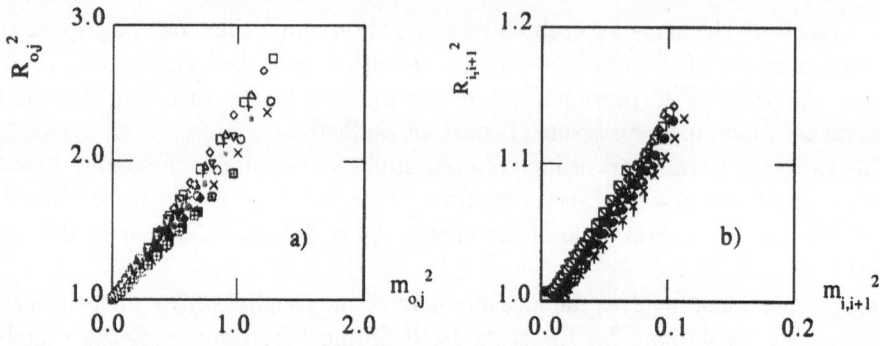

Fig. 5. — Plots of roughness $R_{i,j}$ versus mixing variables $m_{i,j}$. a) integral analysis: scale ranges $[L_0, L_j]$; b) local analysis: scale ranges $[L_i, L_{i+1}]$.

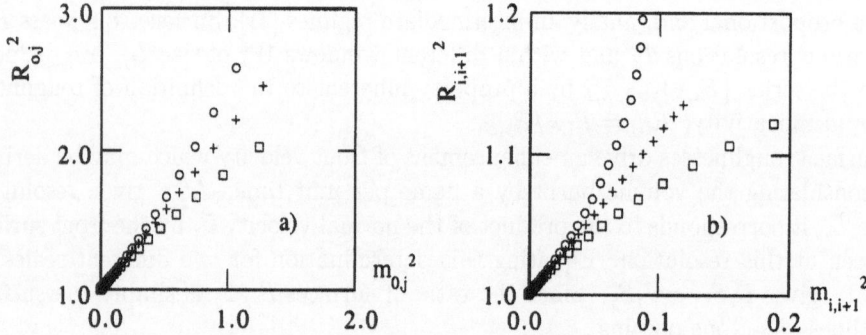

Fig. 6. — Theoretical wrinkling laws (14) (crosses), (16) (circles), (17) (squares) within the experimental scale range $[L_0, L_I]$ for $a = 1.2$ and $u'/U_N \leq 1.3$. Plots (a) and (b) as in Figs. 5. In (b), the data relevant to relations (16) and (17) even extend beyond the plotted range.

They are plotted in Fig. 5a in quadratic coordinates $(m_{0,j}^2, R_{0,j}^2)$ for $K = 1/2$ and for a given mixture composition.

Fig. 5a shows some scatter around a mean straight line. This suggests that, to our experimental uncertainty in u' and U_N, the plots are linear for each flame. This is confirmed by the fact that the slope dispersion is much smaller on each flame (4%) than within the flame set (18%).

Since, at a given L_j, the values of both $m_{0,j}$ and of $R_{0,j}$ vary with the turbulence intensity undergone by the flame, another piece of information can be extracted from Fig. 5a: all the laws $\mathcal{L}_{0,j}(.)$ are actually linear in quadratic coordinates, with the same slope $\beta = 1.14 \pm 0.21$:

$$R_{0,j}^2 = 1 + \beta m_{0,j}^2 \quad \forall j \tag{14}$$

In particular, this implies that the laws $\mathcal{L}_{0,j}(.)$ are the same. This property corresponds to a covariance by scale change and, more precisely, to a local scale-covariance since the change of L_j modifies the size of the scale ranges studied. On the other hand, the quadratic expression of the laws (14) agrees with the expected scale-covariant forms (8). This suggests that scale-covariance should extend to any other scale change.

3.3.   Local analysis

We now consider consecutive scales $[L_i, L_{i+1}]$ for various scales L_i. This entails isolating the effect of each vortex family of scale L_i and thus to scan the wrinkling process from small scales to large scales.

The results are plotted in Fig. 5b for the flames analyzed in Fig. 5a and still with $K = 1/2$ and quadratic coordinates $(m_{i,i+1}^2, R_{i,i+1}^2)$. Here again, a line with a small scatter is observed. As in the previous analysis, this implies that the laws $\mathcal{L}_{i,i+1}(.)$ are linear in quadratic coordinates with the same slope $\beta' = 1.18 \pm 0.17$. In addition β and β' appear to be equal to our level of accuracy:

$$R_{i,i+1}^2 = 1 + \beta m_{i,i+1}^2 \quad \forall i \tag{15}$$

Here too, the fact that laws $\mathcal{L}_{i,i+1}(.)$ are the same means that there is scale-covariance. More precisely, this covariance corresponds to a global scale-covariance since the size of the scale ranges studied are the same: $L_{i+1}/L_i = a \; \forall i$. In addition, the equality between these laws (15) and the former laws (14) shows that scale-covariance extends to various kinds of scale changes, as expected from the analytic property of the quadratic laws investigated in Sec. 2.4. This equality, although apparent over the flame set by comparison between the plots of Fig. 5a and 5b may even be better evidenced by comparing the data from flame to flame. The variation of their slope in one (β^*) or the other (β'^*) analysis are then much smaller than that displayed in average: $|(\beta^* - \beta'^*)/\beta^*| < 5\%$.

3.4.   Sensitivity of scale-covariance

Owing to the experimental uncertainty and to the assumptions inherent to our description, one may fear that the above evidence of scale covariance may be only an artifact of data processing. In particular, one might say that our turbulence intensities are not large enough to evidence a non-linearity of the wrinkling law in quadratic coordinates.

To answer these concerns, we first notice that the change of representation from integral analysis to local analysis is actually non-linear and non-local, owing to the use of the normal velocity U_i in the mixing variable $m_{i,i+1}$ (12). For this reason, the plot of Fig. 5b is definitely *not* an enlargement of that of Fig. 5a. Obtaining another line, moreover the same as in Fig. 5a, is therefore actually surprising. This statement is supported by the comparison of the behaviour of the following different laws: the Clavin-Williams law (16) [16], the scale-covariant law (14) and the Yakhot law (17) [17] :

$$R_{0,j} = 1 + \frac{\beta}{2} m_{0,j}^2 \quad \forall j \tag{16}$$

$$R_{0,j}^2 = \exp(\beta m_{0,j}^2 / R_{0,j}^2) \quad \forall j \tag{17}$$

As may be seen in Fig. 6a, these three laws are actually tangent at the origin $(0,1)$ and, within the present experimental range, close to one another in integral analysis. However, turning to the local analysis produces dramatic changes of both the Clavin-Williams law and of the Yakhot law but, on the opposite, leaves the scale-covariant law unchanged (Fig. 6b) (Fig. 5b). This behaviour shows that the non-linear change of representation applied when going from the integral analysis to the local analysis enhances the differences between non-covariant laws and covariant ones. In particular, whereas our data agree equally well with any of the three laws (14)(16)(17) in integral analysis, they definitely lie much farther from the non-covariant laws (16)(17) in local analysis than allowed by our experimental uncertainties. The sensitivity of scale-covariance thus accurately selects a covariant law in our experiment, without need of larger turbulence intensities.

This sensitivity calls for more clarified detail in our data processing. In order to reduce noise, we have actually deduced the measurements of roughness from a polynomial fit $P(.)$ of degree 4 of the series $(R_{i,S})$ as a function of scales L_i/L_S in logarithmic coordinates: $\ln(R_{i,S}) = P[\ln(L_i/L_S)]$. The non-linearity of $P(.)$ was expected to reproduce the evolution from the euclidean regime at low scale to the "fractal" regime at large scale. However, it cannot correspond exactly to that which would be produced by a strictly scale-covariant wrinkling law (15). This means that the fitting functions are intrinsically non-covariant so that some disagreement with scale-covariance may be further expected. They are actually evidenced in Fig. 5a,b by some curvatures of the curves and some shifts with respect to the origin (0,1). The important thing, however, is that these disagreements are slight despite the large sensitivity to non-covariance. They thus reveal that, although the polynomials used to model the scale evolution of roughness correspond to relations a priori far from satisfying scale-covariance, data systematically fit to the polynomial which most closely approaches this property. Beyond the evidence of some non-covariant artifacts induced by data processing, our measurements thus actually point to scale-covariance in the present experiment.

4.   Scale construction of turbulent fronts

We shall now use the scale-covariant law (15) to derive the evolution of the front geometry in scale space [3].

4.1.   Deterministic construction of front geometry in scale space

Each scale range $[L_i, L_{i+1}]$ increases front roughness by a factor $R_{i,i+1}$. The way this factor changes with the scale L_i indicates how the front geometry is built up in scale space. For instance, the euclidean and the fractal regimes correspond to a factor $R_{i,i+1}$ constant with scale whereas the transition from one to the other involves some variations of it.

For simplifying the notations, we denote $\rho_i = R_{i,i+1}^2$. Using relations (1)(15), the link between roughness and front velocities (11) and the definition of mixing variables (12), we obtain the following series:

$$\frac{\rho_{i+1} - 1}{\rho_i - 1} = \frac{\rho_\infty}{\rho_i} \tag{18}$$

where $\rho_\infty = a^{2K}$. Its solution is :

$$\rho_i = \frac{1 + \sigma_0 \rho_\infty^{i+1}}{1 + \sigma_0 \rho_\infty^i} \tag{19}$$

where σ_0 is given by the initial condition ρ_0:

$$\sigma_0 = \frac{\rho_0 - 1}{\rho_\infty - 1} \qquad (20)$$

These relations show that the iteration of the roughnesses converges, in the limit of an *infinite* number of scales, towards an attracting value ρ_∞, provided that $\sigma_0 \neq 0$, i.e. $u' \neq 0$. Each scale then brings about the same additionnal front roughness: the front has reached a fractal regime. Its fractal dimension D corresponds to the dimension of the euclidean surface plus the ratio of the increase of roughness to the increase of length scale, in logarithmic coordinates:

$$D = 2 + \lim_{i \to \infty} \ln(R_{i,i+1})/\ln(L_{i+1}/L_i) \qquad (21)$$

As, in this limit regime, $R_{i,i+1} = \rho_\infty^{1/2} = a^{2K}$, we obtain:

$$D = 2 + K \; ; \; i \to \infty \qquad (22)$$

This prediction derived from scale-covariance agrees with that found by other arguments [7, 18]. It is compared in Fig. 7a to the behaviour of one of the most turbulent fronts ($u'/U_N = 1.85$). Crosses correspond to the iteration (19) for an *infinite* number of scales, F to its fixed point, i.e. to the fractal regime, and circles to the experimental points. In agreement with (19), the iteration of experimental points with scale shows a decrease of their distance to F and a near stagnation in the vicinity of both the euclidean ($\rho_i = 1$) and the fractal ($\rho_i = \rho_\infty$) regimes. On the other hand, it reveals that the roughness brought about by each scale of the turbulent flow is actually *not* a constant, even at large scales. This means that this front, although largely turbulent, is, strictly speaking, *not a fractal*. It actually displays "fractal dimensions" varying with scales, as the turbulent jets studied by Dimotakis and Catrakis [19]. Following these authors, it might thus be termed "scale-dependent fractal".

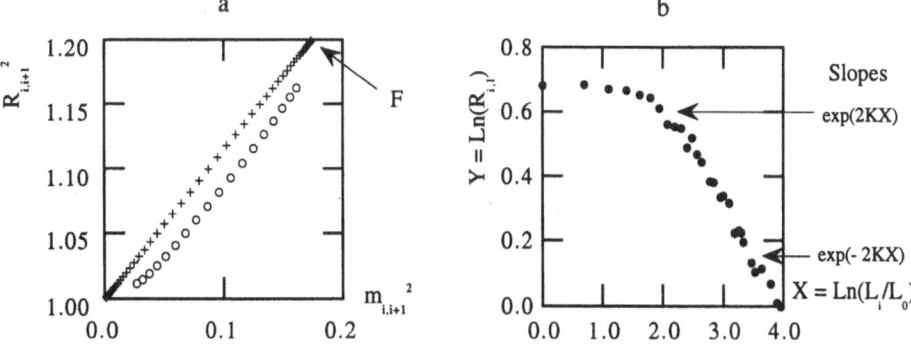

Fig. 7. — a) Geometry formation in scale space, $a = 1.2$, $K = 1/2$. Circles: experimental front $u'/U_N = 1.85$. Crosses: simulation of fractal construction for $\beta = 1.15$ and for an *infinite* number of scales. F: fractal regime. b) Classical scale analysis of the front studied in a). The evolution of curvature according to (25) and (28) shows the difficulty of appreciating the distance to a pure fractal regime.

4.2. "Apparent" fractal dimensions

We confront the scale dependence of the "fractal dimensions" of the above turbulent front to the conclusions that might be drawn from a classical scale analysis of this object. We then seek its origin within the framework of scale-covariance.

Classical scale analysis relates the front roughness $R_{i,I}$ measured at increasing resolution scales L_i to the scale ratio L_i/L_0, in logarithmic coordinates. The corresponding curve $Y(i) = \ln(R_{i,I})$ vs $X(i) = \ln(L_i/L_0)$ is plotted in Fig. 7b for the front studied in Fig. 7a. It shows a zero slope domain (the euclidean regime) followed by an increasing slope domain (the transition from the euclidean to the fractal regime) and ends up with an apparently constant slope zone corresponding to fractal behaviour. The latter evidence apparently contradicts the conclusion drawn in Sec. 4.1. It shows that the classical scale analysis might be misleading in regard to the determination of the scale-similarity of forms.

To better clarify this disagreement, we consider below local box-counting dimensions $D(i)$ at scale L_i:

$$D(i) = 2 + \ln(R_{i,i+1})/\ln(L_{i+1}/L_i) \tag{23}$$

and determine their scale-evolution according to the scale-covariant iteration (18). We notice that this will provide the local slope of the curve $(X(i), Y(i))$ according to $dY/dX(i) = 2 - D(i)$ with $D(i) = 2 + 1/2 \ln(\rho_i)/\ln(a)$.

1. Quasi-euclidean regime

   We define this regime by the condition $\sigma_0 \rho_\infty^i \ll 1$. This corresponds to the vicinity of the resolution scale L_0. Relation (19) yields:

   $$\rho_i \approx 1 + (\rho_\infty - 1)\sigma_0 \rho_\infty^i \tag{24}$$

   As $\rho_\infty - 1 \approx 2K \ln(a)$ is much smaller than one here, this implies :

   $$D(i) \approx 2 + K\sigma_0 \rho_\infty^i \tag{25}$$

   and, since $\ln(\rho_\infty^i) = 2KX(i)$,

   $$dY/dX \approx -K\sigma_0 \exp(2KX) \tag{26}$$

   The curve $(X(i), Y(i))$ thus shows a slope decreasing exponentially from zero at $L_i = L_0$. This implies an exponential growth of its curvature: we find the bent part of the transition from the euclidean to the fractal regime.

2. Quasi-fractal regime

   We define this regime by the condition $\sigma_0 \rho_\infty^i \gg 1$. It corresponds to the vicinity of the largest observed scale L_I and to the largest turbulence intensities. Relation (19) yields :

   $$\rho_i \approx \rho_\infty [1 - (\rho_\infty - 1)\sigma_0^{-1} \rho_\infty^{-(i+1)}] \tag{27}$$

   As $\rho_\infty - 1 \approx 2K \ln(a) \ll 1$, this implies :

   $$D(i) \approx D - K\sigma_0^{-1} \rho_\infty^{-(i+1)} \tag{28}$$

or :

$$dY/dX \approx 2 - D + \frac{K}{\sigma_0 \rho_\infty} \exp(-2KX) \tag{29}$$

In this large scale regime, the slope of the curve $(X(i), Y(i))$ thus reaches its asymptotic value $2 - D$ in an exponential manner. This implies, for this curve, an exponentially small curvature which makes the distance to a constant slope regime, i.e. a fractal regime, especially hard to evidence. This is responsible for the difficulty in distinguishing pure scale-similarity from an approximate one in the classical representation $(X(i), Y(i))$. We stress that this might lead one to *abusively* consider the most turbulent fronts as fractal here.

On a more general ground, relation (28) shows that, according to the scale-covariance of the wrinkling law, the local box-counting dimension $D(i)$ varies with scale. This corroborates the observation of "scale-dependent fractal" in this system and shows that this feature may simply follow from scale-covariance.

4.3.   An artifact of finite size: the variation of "fractal dimensions" with turbulence intensity

Classical scale analysis leads one to attribute a fractal dimension to the experimental fronts insofar as their curve $(X(i), Y(i))$ displays an apparently constant slope. Figure 8, shows the measurements of these numbers for the experimental fronts observed at various mixing variables $m = u'/U_N$. They are not constant and show an increase with the turbulence intensity, as already found in other studies [20]. These features are somewhat surprising since the fractal regime is actually expected in the limit of large turbulence intensity : if its vicinity is thought to be sufficiently approached at finite u'/U_N so that fronts can be considered as fractal, is it meaningful that their fractal dimensions vary so much and are so distant from the expected value $2 + K = 2.5$?

To elucidate this paradox, we calculate the evolution of the apparent fractal dimension at the integral scale L_I, $D(I)$, according to the scale-covariant behaviour (19). Since $R_{0,I}^2$ is the product of ρ_i from $i = 0$ to $I - 1$, iteration of (19) yields $R_{0,I}^2 = (1 + \sigma_0 \rho_\infty^I)/(1 + \sigma_0)$ and, by elimination of σ_0:

$$\rho_I = \frac{1 + f(\rho_\infty)(R_{0,I}^2 - 1)}{R_{0,I}^2} \tag{30}$$

where $f(\rho_\infty) = \rho_\infty + (\rho_\infty - 1)/(\rho_\infty^I - 1)$. As $\rho_\infty - 1 \ll 1$, $\rho_\infty^I \gg 1$ and $R_{0,I}^2 = 1 + \beta(u'/U_N)^2$, one obtains:

$$D(I) = 2 + \frac{K}{\ln(\rho_\infty)} \ln[\frac{1 + \rho_\infty \beta m^2}{1 + \beta m^2}] \tag{31}$$

Relation (31) evidences a variation of $D(I)$ with $m = u'/U_N$. This may be interpreted, at fixed scale L_I and fixed normal velocity U_N, as a variation of $D(I)$ with the turbulence intensity u': $\partial D(I, m)/\partial m \neq 0$. Within our framework, this simply appears as a corollary of the dependence of $D(i)$ with scale at fixed turbulence intensity: $\partial D(i, m)/\partial i \neq 0$ (28).

As shown in Fig. 8, taking for β and K the values selected by scale covariance ($\beta = 1.15$ and $K = 1/2$), yields a good agreement between relation (31) and the curve obtained experimentally. This corroborates our analysis of the "apparent" fractal dimensions in this system. According to it, the evolution of these dimensions with $m = u'/U_N$ can be traced back to the fact that they are calculated at a finite scale, the integral scale L_I of turbulence, instead of at an infinite scale. This therefore corresponds to an artifact due to the finite system size.

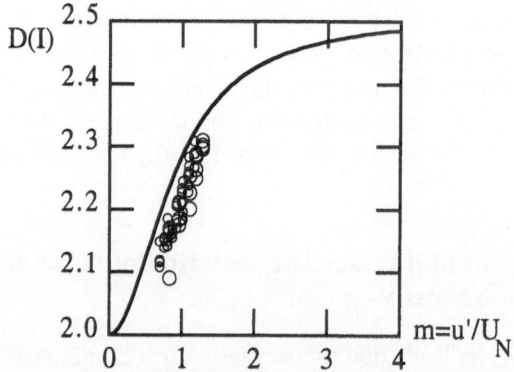

Fig. 8. — Fractal dimension $D(I)$ measured at the integral scale of turbulence L_I. Circles: experimental points; solid line: evolution of $D(I)$ with m according to scale-covariance (31) with $\beta = 1.15$, $\rho_\infty = a^{2K}$, $a = 1.2$, and $K = 1/2$.

5.   Conclusion

We have studied turbulent combustion in a scale-invariant regime of front-turbulence interaction by focusing attention on statistical laws rather than on the statistical forms of fronts. This has led us to evidence a scale symmetry of laws, scale-covariance, more extended than the scale-symmetry of forms, i.e. scale-similarity. The main advantages of scale-covariance upon scale-similarity are the following :

Scale-covariance reveals an unexpected universality in this stochastic system: the euclidean regime and the fractal regime are physically equivalent since both of them, as well as their transition, are described by the same statistical law. Along the same lines, the weak and strong mixing regimes (called wrinkled and corrugated regimes in combustion) also appear as referring to the same laws. This differs with the picture derived from scale-similarity where the weak mixing regime is overlooked since not displaying scale-similar forms. By contrast, scale-covariance emphasizes its relevance by showing that it contains as much information regarding statistical laws as the large mixing regime and, furthermore, provides the additional advantage of better scanning the transition from euclidean to fractal forms.

On another level, the scale-representation inherent to scale-covariance evidences the distance to the fractal regime much better than that used in classical scale analysis. In particular, it reveals that the most turbulent fronts are abusively qualified as fractal since they are actually not scale-similar. This means that their apparent fractal dimensions do not refer to an asymptotic fractal regime and should thus be considered with care. In particular, they show a paradoxical evolution with turbulence intensity, which simply traces back to an artifact due to the finite size of the system.

Since scale-covariance relies on the scale-invariance of physical processes, it should be valid in a number of systems presently studied under the heading of "scale-similarity". Given its larger advantages, it might be worth focussing attention on it, as explored, for instance, in fully developped turbulence [22]. This would make it possible to determine not only the characteristics of the forms generated in scale-invariant systems, but also the shape of the laws which actually govern them.

References

[1] A. Pocheau, Phys. Rev. E, **49** 1109 (1994); A. Pocheau, Europhys. Lett., **20** 401 (1992) A. Pocheau, C. R. Acad. Sci., Ser.II, **315** 21 (1992)

[2] A. Pocheau and D. Queiros-Condé, Phys. Rev. Lett., **76** 3352 (1996)

[3] A. Pocheau and D. Queiros-Condé, Europhys. Lett., **35** 439 (1996)

[4] The front normal velocity keeps the same, except at scales of the order of the flame thickness.

[5] Dynamics of Fractal Surfaces, F. Family and T. Vicsek Eds., (World Scientific, Singapore, 1991); K.R. Sreenivasan, Ann. Rev. Fluid Mech., **23** 539 (1991); M. Kardar, G. Parisi and Y.-C. Zhang, Phys. Rev. Lett., **56** 889 (1986)

[6] F.A. Williams, *Combustion theory*, 2nd Edition, (The Benjamin Cummins Publishing Company, Menlo Park, 1985)

[7] P.Clavin, in *Theory of Flames in Disorder and Mixing*, E.Guyon et al. Eds, NATO Advanced Study Institute,Series E: Applied Science, Vol.152, 293 (Plenum, New-York,1987); P.Clavin, in *Fluid Dynamical Aspects of Combustion Theory*, M. Onofri and A. Tesei eds., Pitman Reserach Notes in mathematics Series, Vol.223, 43 (Longman Scientific & Technical, Essex, 1991)

[8] Since $K = 1/2$ is still smaller than one, we note that the turn-over times $\tau_i = L_i/u'_{i,i+1}$ increase with L_i.

[9] R. Borghi, Prog.Energy Combust. Sci., **14** 245 (1988); J. Lasheras, Lecture given at the NATO Advanced Studies Institute "Mixing, Chaos, and Turbulence", Cargèse, France, 1996

[10] G.I. Barenblatt, *Similarity, Self-similarity and Intermediate Asymptotics*, (Consultants Bureau, New York, 1979)

[11] O.L. Gülder, XXIII Symposium (International) on Combustion, The Combustion Institute, 743 (1990)

[12] This is similar to the requirement of transitivity for the Lorentz transformation in the search for mechanical covariance.

[13] L. Boyer, Comb. Flame **39**, 321 (1980)

[14] A. Floch, M. Trinité, F. Fisson, T. Kageyama, C. Kwon, and A. Pocheau, in *Dynamics of Deflagrations and Reactive Systems*, A.L. Kuhl et al. Eds, Progress in Astronautics and Aeronautics **131**, 378 (1989)

[15] M. Metgalchi and J.C. Keck, Combust. Flame **48** 191 (1982)

[16] P. Clavin and F.A. Williams, J. Fluid Mech. **90** 589 (1979)

[17] V. Yakhot, Combust. Sci. Tech. **60** 191 (1988)

[18] N. Peters, XXI Combustion Symposium, The Combustion Institute, 1231 (1986); A. Kerstein, Combust. Sci. Tech. **60** 441 (1988)

[19] P.E. Dimotakis and H.J. Catrakis, this Volume; H.J. Catrakis and P.E. Dimotakis, this Volume; H.J. Catrakis and P.E. Dimotakis, J. Fluid. Mech. **317** 369 (1996)

[20] F.C. Gouldin, Combust. Flame **68**, 249 (1987); J. Mantzaras, P.G. Felton, and F.V. Bracco, Combust. Flame, **77** 295 (1989)

[21] A. Pocheau, in preparation

[22] A. Pocheau, Europhys. Lett. **35**, 183 (1996); B. Dubrulle and F. Graner, J. Phys. II (France), **6** 7976 (1996)

Part III

Mixing and Dispersion in Geophysical Contexts

Dynamics of Lagrangian Tracers in Barotropic Turbulence

A. Provenzale[1], A. Babiano[2], and A. Zanella[1]

[1] *Istituto di Cosmogeofisica, Corso Fiume 4, I-10133 Torino, Italy*
[2] *Laboratoire de Météorologie Dynamique de l'Ecole Normale Supérieure*
24 rue Lhomond, 75231 Paris Cedex 05, France

We discuss the dynamics of passively advected tracers in barotropic turbulence. After introducing the relevant fluid dynamical equations, we consider (a) the properties of Lagrangian advection inside coherent vortices; (b) the presence of anomalous dispersion laws generated by the complex structure of geostrophic turbulence; and (c) the dynamics of non neutrally-buoyant particles.

1. Introduction

A classic approach to the study of particle transport in turbulent flows is based on the use of stochastic models, see e.g. [1]. In this framework, the motion of a passive particle, advected by a two-dimensional turbulent velocity field, may be described by the Langevin equation (see e.g. [2])

$$d\mathbf{r} = (\mathbf{u} + \mathbf{u}') \, dt \tag{1}$$

$$d\mathbf{u}' = -\frac{1}{T_L}\mathbf{u}'dt + K_v^{1/2} \, d\xi \tag{2}$$

where $\mathbf{r}(t) \equiv (x, y)$ is the position of the particle at time t, $\mathbf{u} = (u, v)$ is the mean flow, $\mathbf{u}' = (u', v')$ is a random velocity and ξ is a gaussian white noise process with zero mean and variance $< d\xi \cdot d\xi >= 2dt$. In Eq.(2), T_L is the Lagrangian decorrelation time and K_v is related to the variance of $\mathbf{u}'$.

Mixing: Chaos and Turbulence, edited by Chaté *et al.*
Kluwer Academic / Plenum Publishers, New York 1999.

Parallel to the stochastic approach, in the last ten years much attention has been devoted to the issue of chaotic advection; namely, to the fact that even very simple Eulerian flows may generate chaotic and unpredictable particle trajectories. In the case of two dimensional, divergence-free flows, the equations of motion for an advected particle may still be written in the form (1), where $\mathbf{u'} \equiv 0$ and the velocity field $\mathbf{u} = (u, v)$ is now given by

$$u = -\frac{\partial \psi}{\partial y} \; ; \; v = \frac{\partial \psi}{\partial x} \; . \tag{3}$$

The stream function ψ plays the role of the Hamiltonian, i.e.,

$$\frac{dx}{dt} = -\frac{\partial \psi}{\partial y} \; ; \; \frac{dy}{dt} = \frac{\partial \psi}{\partial x} \tag{4}$$

and x and y play the role of a couple of canonically conjugate variables.

If the stream function is time independent, system (4) is integrable and advected particles move along isolines of constant ψ. In this case, no chaotic motion is possible. On the other hand, if ψ does depend on time (e.g., in a time-periodic manner), system (4) is no longer integrable and chaotic particle trajectories are possible and usually observed, see e.g. [3] and [4] for reviews.

Most studies on chaotic advection have considered a simple form for the stream function, such as a travelling wave [5], an array of regular convection cells [6], an oscillating jet stream [7], or simple models of oceanic mesoscale motions [8] and have hypotized a simple, periodic time evolution of ψ. This leads to a kinematic problem where the Eulerian aspects are trivial, and the attention is focussed on the Lagrangian dynamics. In these systems, regular and anomalous particle dispersion have been measured, and some of the properties of particle motion are quite similar to those obtained with a stochastic approach. Thanks to the simplicity of the Eulerian flow (in particular, to its time-periodic nature), for these systems several rigorous results have been obtained, see e.g. [9].

As mentioned above, in the stochastic approach described by Eq. (1,2) and in the simple models of chaotic advection, the spatio-temporal behaviour of the Eulerian field is taken to be rather simple. This has to be contrasted with the extremely rich structure of real geophysical flows, that are characterized by the presence of intense coherent vortices, jets, waves, and by an irregular (or turbulent) time evolution. Due to this, it is not clear whether the simplified approaches mentioned above may really capture the relevant aspects of transport in highly structured turbulent flows.

For these reasons, in the last few years we have conducted a series of numerical experiments on transport in simple turbulent flows dominated by coherent structures. One of the main goals of this type of work is to define the Eulerian parameters that characterize the topology of the turbulent flow, and to identify the main effects the various structural domains of the flow may have on the dynamics of the advected particles. We are thus led to study chaotic advection in systems with non-periodic, irregular time evolution and complex spatial structure. A simple prototypical example of this type of flow is barotropic turbulence in rotation-dominated systems.

2.   The quasi-geostrophic approximation

The starting point is provided by the three-dimensional Euler equations for an incompressible fluid in a rotating frame, written as (see e.g. [10, 11, 12, 13])

$$\frac{\partial \mathbf{v}}{\partial t} + (\mathbf{v} \cdot \nabla_{3D})\mathbf{v} = -\frac{1}{\rho}\nabla_{3D}p - \nabla_{3D}\Phi - 2\Omega \times \mathbf{v} \tag{5}$$

$$\nabla_{3D} \cdot \mathbf{v} = 0 \tag{6}$$

where $\mathbf{v} = (u, v, w)$ is the (3D) fluid velocity, ρ is density, p is pressure, Ω is the angular velocity of the rotating frame, Φ is the gravitational potential and ∇_{3D} is the three-dimensional gradient operator. The centrifugal term $-\Omega \times (\Omega \times \mathbf{r})$ has been included in the gravitational potential, $\Phi \to \Phi - |\Omega \times \mathbf{r}|^2/2$.

In the case of the Earth, we define a local Cartesian coordinate frame where z points upward along the local vertical direction, x points Eastward and y points Northward. The origin of this coordinate frame is fixed at the reference longitude ϕ_0 and latitude θ_0. The projection of twice the angular velocity of the Earth along the local vertical direction is $f = 2\Omega \sin \theta$. This is, of course, a function of latitude. If the meridional extent of the motion under study is not large, i.e., if $\theta - \theta_0$ is small, and if we are far from the equator, we may expand f around θ_0 and write $f = f_0 + \beta y$ where $f_0 = 2\Omega \sin \theta_0$ and $\beta = 2\Omega \cos \theta_0/R_E$, here R_E is the radius of the Earth.

In the rotating frame, we now consider the dynamics of a shallow fluid layer with average depth D and free surface $z = h(x, y, t) = D + \eta(x, y, t)$ over a bottom topography defined by $z = h_B(x, y)$. The total depth of the fluid at a point is given by $H(x, y, t) = D + \eta(x, y, t) - h_B(x, y)$. We assume in general $|\eta| \ll D$ and $|h_B| \ll D$. Both η and h_B are assumed to have zero mean. Calling L a typical horizontal scale of the flow, we note that in the case of large-scale motions in the ocean and the atmosphere one has that

$$\delta = D/L \ll 1. \tag{7}$$

By expanding Eq.(5) in powers of δ, and retaining only terms of first order in δ, we obtain the shallow-water approximation

$$\frac{\partial u}{\partial t} + u\frac{\partial u}{\partial x} + v\frac{\partial u}{\partial y} - fv = -g\frac{\partial \eta}{\partial x} \tag{8}$$

$$\frac{\partial v}{\partial t} + u\frac{\partial v}{\partial x} + v\frac{\partial v}{\partial y} + fu = -g\frac{\partial \eta}{\partial y} \tag{9}$$

and the hydrostatic balance

$$\frac{\partial p}{\partial z} = -g\rho \ . \tag{10}$$

In this approximation, the horizontal velocity $\mathbf{u} = (u, v)$ is independent of the vertical direction z. The vertical velocity w is given by

$$w(x, y, z, t) = (h_B - z)\left(\frac{\partial u}{\partial x} + \frac{\partial v}{\partial y}\right) + u\frac{\partial h_B}{\partial x} + v\frac{\partial h_B}{\partial y} \tag{11}$$

with the kinematic condition at the surface

$$w(x, y, z = D + \eta, t) = \frac{\partial \eta}{\partial t} + u \frac{\partial \eta}{\partial x} + v \frac{\partial \eta}{\partial y} \ . \tag{12}$$

Equations (8,9) have an important Lagrangian invariant, namely, the shallow-water potential vorticity

$$q_{sw} = \frac{\zeta + f}{H} = \frac{\zeta + f}{D + \eta - h_B} \tag{13}$$

that obeys the equation

$$\frac{Dq_{sw}}{Dt} = \frac{\partial q_{sw}}{\partial t} + u \frac{\partial q_{sw}}{\partial x} + v \frac{\partial q_{sw}}{\partial y} = 0 \ . \tag{14}$$

The quantity ζ is the vertical component of relative vorticity,

$$\zeta = \frac{\partial v}{\partial x} - \frac{\partial u}{\partial y} \ . \tag{15}$$

The shallow water equations describe both fast and slow motions, provided the fluid layer is shallow compared to the horizontal scale of the motion. We are now interested in obtaining an equation for motions that are slow compared with the Earth rotation period and are thus dominated by the Coriolis force. To this end, one has to introduce the Rossby number

$$\epsilon = \frac{U}{fL} \tag{16}$$

where U is a typical horizontal velocity.

By writing Eq. (8,9) in adimensional form and expanding it in powers of the Rossby number, at first order in ϵ one obtains the barotropic quasi-geostrophic (QG) approximation

$$\frac{\partial q}{\partial t} + [\psi, q] = 0 \tag{17}$$

where the symbol $[A, B]$ is the Jacobian, $[A, B] \equiv \frac{\partial A}{\partial x} \frac{\partial B}{\partial y} - \frac{\partial B}{\partial x} \frac{\partial A}{\partial y}$, and the quasi-geostrophic potential vorticity q is defined in term of the adimensional variables as

$$q = \epsilon^{-1} + \omega - \gamma^2 \psi + h_0 \tag{18}$$

where we have omitted the term βy due to the variation of the Coriolis parameter with latitude.

In Eq. (18), the adimensional variables ω, h_0 and ψ are related to the dimensional ones (ζ, h_B and η) by

$$\zeta = \frac{U}{L}\omega \tag{19}$$

$$h_B = h_0 D\epsilon \tag{20}$$

$$\eta = \epsilon\gamma^2 D\psi \tag{21}$$

where $\gamma = L/L_R$ and L_R is the Rossby deformation radius, $L_R = (gD)^{1/2}/f_0$. Note that Eq. (17,18) could have been obtained by expanding the shallow-water potential vorticity conservation law in powers of ϵ, and retaining only terms $O(\epsilon)$.

In the barotropic quasi-geostrophic approximation, the adimensional free surface ψ plays the role of the stream function. The adimensional horizontal velocities u and v are given by $u = -\partial\psi/\partial y$ and $v = \partial\psi/\partial x$ and the relative vorticity is given by $\omega = \nabla^2\psi$, where $\nabla^2 = \partial_x^2 + \partial_y^2$ is the horizontal Laplacian. For simplicity, in the above expressions we have indicated the adimensional horizontal velocities and positions with the same symbol used for the dimensional ones. Note that, in the QG approximation, the horizontal velocity is non-divergent.

The term ϵ^{-1} in Eq.(18), being constant, may be omitted from the equations of motion. In the case of a flat bottom, $h_0 = 0$, one obtains the equation

$$\frac{\partial}{\partial t}(\nabla^2\psi - \gamma^2\psi) + [\psi, \nabla^2\psi] = 0 \; . \tag{22}$$

This equation is known as the Charney equation in geophysical fluid dynamics, where it describes what is sometimes called geostrophic turbulence, or as the Hasegawa-Mima equation in plasma physics, where ψ plays the role of the magnetic field. In the next section we discuss some of the properties of this model equation, and consider in detail the case $\gamma^2 = 0$.

3. Dynamics of barotropic turbulence

The Charney equation (22) has two important quadratic invariants, namely the total (average) energy

$$E = \frac{1}{2}\frac{1}{(2\pi)^2} \int \left[(\nabla\psi)^2 + \gamma^2\psi^2\right] dx dy \tag{23}$$

and the (average) potential enstrophy

$$Z = \frac{1}{2}\frac{1}{(2\pi)^2} \int \left(\nabla^2\psi - \gamma^2\psi\right)^2 dx dy . \tag{24}$$

The conservation of these two quantities determines the existence of a direct cascade of enstrophy, from large to small scales, and of an inverse cascade of energy, from small to large scales, contrarily to what happens in 3D turbulence where energy cascades from large to small scales.

In almost all natural situations and in numerical simulations, a small viscous damping is usually present, as well as, eventually, a forcing term. The typical viscous term, due to either molecular viscosity, bottom friction or an eddy viscosity parameterizing unresolved small-scale motions, has the form

$$(-1)^{n-1}\nu_n\nabla^{2n}\nabla^2\psi \tag{25}$$

where $n = 1$ for standard Newtonian viscosity, $n = 0$ for bottom friction, and $n > 1$ for the so-called hyperviscosity. The dynamical equation thus takes the form

$$\frac{\partial}{\partial t}(\nabla^2\psi - \gamma^2\psi) + [\psi, \nabla^2\psi] = (-1)^{n-1}\nu_n\nabla^{2n}\nabla^2\psi + F \tag{26}$$

where F indicates an external forcing, to be better specified in the following. Figures 1a and 1b show the vorticity field at time $t = 20$ obtained from two numerical

a   b

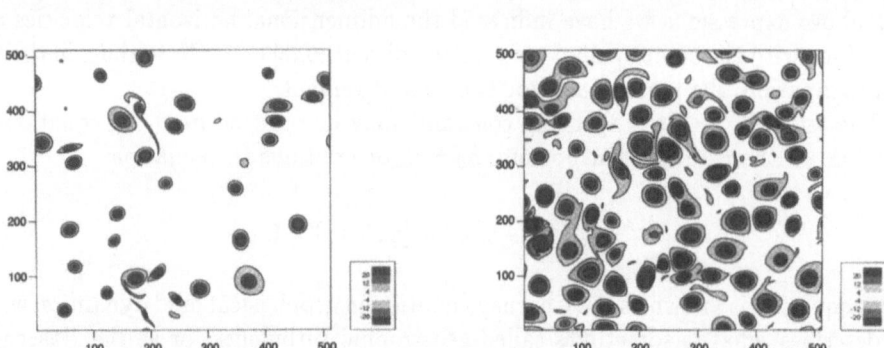

Fig. 1. — Vorticity field at time $t = 20$ obtained by a pseudo-spectral numerical simulation of freely decaying turbulence, described by the Charney equation (26). Panel (a) is for pure 2D turbulence ($\gamma^2 = 0$), panel (b) for $\gamma^2 = 100$. The two simulations have the same initial conditions.

simulations of freely-decaying turbulence (i.e., $F = 0$) with $\gamma^2 = 0$ and $\gamma^2 = 100$ respectively. The case $\gamma^2 = 0$ corresponds to considering motions on scales smaller than the Rossby deformation radius. Conventionally, this case is called 2D turbulence.

A pseudo-spectral code with resolution 512^2 grid points in the square domain $(0, 2\pi)^2$ and standard 2/3 spectral dealiasing has been used to evolve the vorticity field. We have used an hyperviscosity with $n = 2$ and $\nu_2 = 10^{-7}$. For the case with $\gamma^2 = 0$, the initial state is a narrow-band gaussian random vorticity field as given in [14] with energy spectrum peaked at wavenumber $k_0 = 30$. The total initial energy is $E_{in} = 0.5$, this value fixes the meaning of the time scale. The same initial vorticity field has been used for the simulation with $\gamma^2 = 100$, being interpreted as potential vorticity in this case.

In both cases, the evolved field is characterized by the presence of strong coherent vortices, that form after the time of maximum energy dissipation rate and dominate the dynamics of the system. Even these simple models are thus capable of producing intense coherent vortices, and may be used as idealized approaches for studying the role of coherent structures in real geophysical flows. The main difference between the two cases illustrated in Figs. 1a and 1b is that for $\gamma^2 = 100$ the shielding effect generated by the finite value of the Rossby deformation radius strongly decreases the interaction between vortices that are separated by more than about $d_R \approx 2\pi/\gamma$. For this reason, the vortices tend to increase until they reach a size of order d_R, while the further evolution is much slower, see e.g. [15]. In the case $\gamma^2 = 0$, on the other hand, the interactions extend to large distances, the vortices are characterized by a broad distribution of size and circulation and no typical length-scale is observed. At late times, the merging of same-sign vortices lead to the presence of a vortex dipole

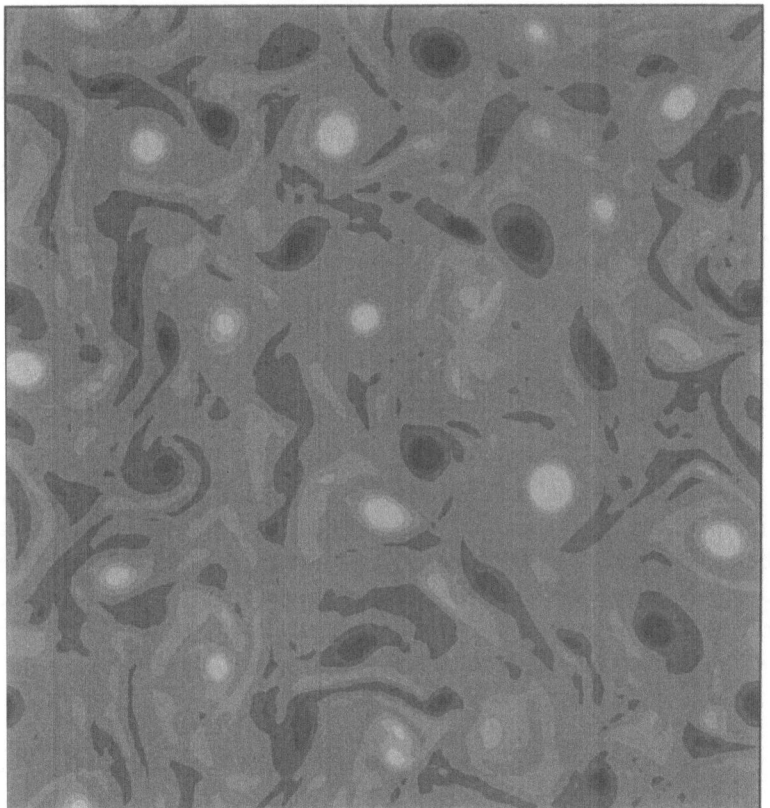

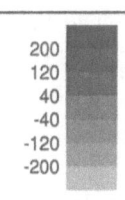

200	
120	
40	
-40	
-120	
-200	

Fig. 2. — Vorticity field obtained by a pseudo-spectral simulation of forced and dissipated 2D turbulence. The system is in a statistically stationary state.

at the largest scales, that decays slowly due to dissipation.

From the above discussion, it is clear that freely decaying turbulence is not in a state of statistical equilibrium. In real geophysical flows, however, the instabilities of the large-scale flow provide a source of energy for the mesoscale turbulence, and vortices are continuosly generated. In order to have a statistically stationary turbulent flow, forcing and dissipation must be balanced such that the total energy and enstrophy of the system do not show any temporal trend.

Simulations of forced, statistically stationary 2D turbulence have been discussed by [16, 17, 18], to which we refer for details. In the following, we discuss the results of numerical simulations obtained with a pseudo-spectral code where the dissipation is introduced as both an higher-order hyperviscosity ($n = 8$) acting at small scales and a term acting at the largest scales, obtained by using $n = -1$ in Eq. (25), in order to dissipate the energy piled up by the inverse cascade. The forcing is obtained by keeping fixed the amplitude of the energy spectrum at a given wavenumber; this choice attempts to simulate the effect of an energy and enstrophy reservoir representing e.g. the baroclinic instability of a large scale flow.

Figure 2 shows a typical vorticity field at statistical equilibrium, obtained with a pseudo-spectral simulation with resolution 128^2 grid points in the square domain $(0, 2\pi)^2$. Standard spectral dealiasing has been used. The forcing wavenumber has been chosen to be $\mathbf{k}_F = (10, 0)$. An initial random field has been evolved until the statistical equilibrium was reached. At equilibrium, the non-dimensional kinetic energy is $E_{av} = 53.5$ and the enstrophy is $Z_{av} = 2600$, giving a non-dimensional Eulerian integral time scale $T_E = 0.14$. Note the presence of many coherent vortices, that are continuously generated and destroyed in the course of time. At variance with what happens for freely decaying turbulence, in this type of forced turbulence the vortices have a narrow distribution of size, centered on the forcing scale, and their number is approximately constant in time.

4. Topology of 2D turbulence

In past years, many efforts have been devoted to developing a simple parameterization of the complex topological structure of 2D turbulence. In particular, the simple distinction between "vortices" and outside "homogeneous turbulence" has turned out to be too simplistic. In an attempt of providing an appropriate parameterization, Okubo [19] and Weiss [20], have proposed a partition of the turbulent field based on the local value of the velocity Jacobian $[u, v]$, see also [21, 18, 22, 23]. This criterion is based on considering the quantity $Q(x, y, t) = 4\lambda^2 = s^2 - \omega^2$ where $\pm\lambda(x, y, t)$ are the eigenvalues of the velocity Jacobian. Here

$$s^2 = s_n{}^2 + s_s{}^2 \tag{27}$$

where $s_n = \partial_x u - \partial_y v$ and $s_s = \partial_x v + \partial_y u$ are the normal and shear components of strain respectively.

The quantity Q defines the basic topology of the turbulent field as it measures the relative contribution of the squared strain s^2 with respect to the squared vorticity ω^2. Regions dominated by rotation are called elliptic and are characterized by $Q < 0$; regions dominated by strain and deformation are called hyperbolic and are characterized by $Q > 0$.

In the past years, various authors have studied the Okubo-Weiss criterion. In particular, Basdevant and Philipovitch [22] have shown that this criterion is strictly justified only in a small portion of the turbulent field. In a different study, Larcheveque [21] has shown that by applying the divergence operator to the incompressible Navier-Stokes equation in the limit of vanishing viscosity, or for a statistically stationary situation where forcing and damping balance each other, one gets

$$\nabla^2 p = -\frac{1}{2}Q \tag{28}$$

where p is pressure. This diagnostic equation is valid for all time and it provides a strict link between the value of Q and the local nature of the flow field.

Based on the value of Q, we thus identify (a) the vortex cores, which are characterized by strong negative values of Q; (b) the strain cells surrounding the vortices, characterized by large positive values of Q; and (c) the background where Q fluctuates between small positive and negative values. Depending on the sign of Q, the background field may be further divided into elliptic or hyperbolic patches. Figure 3

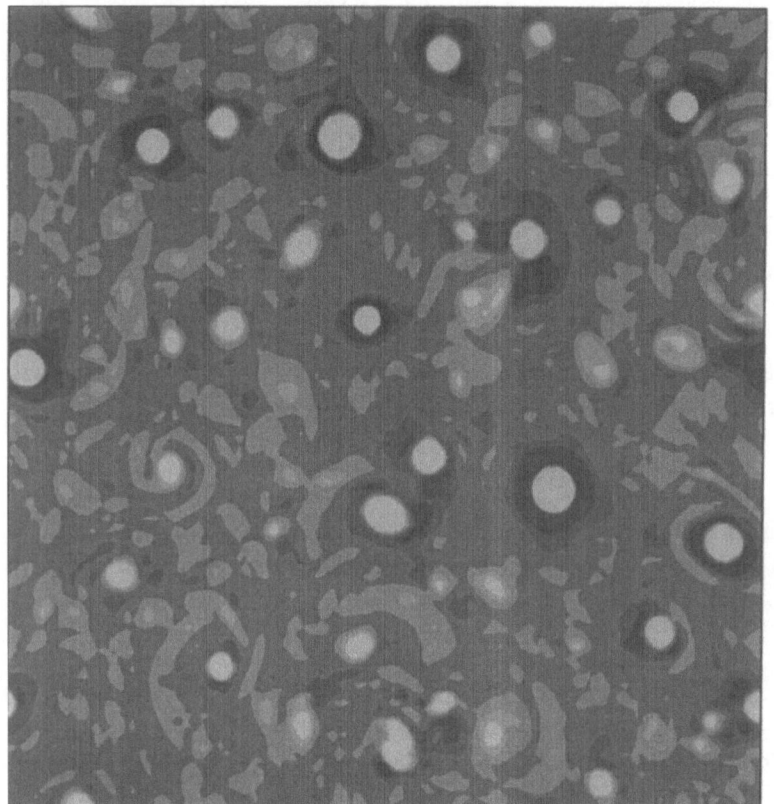
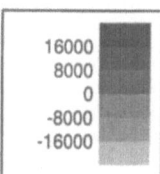

Fig. 3. — Spatial distribution of the Okubo-Weiss parameter Q for the vorticity field shown in Fig. 2.

shows the Q field corresponding to the vorticity field shown in Fig. 2. Note the presence of extended strain cells around coherent vortices.

Before closing this section, we mention that Hua and Klein [24] have recently proposed a new parameterization that takes into account the time evolution of the velocity and vorticity field in 2D turbulence. Work is now in progress to verify whether this approach may provide better results than the standard Okubo-Weiss parameterization.

5.   Particle dynamics in coherent vortices

The presence of coherent vortices heavily affects the dynamics of passively advected tracers. In particular, vortices act as barriers to particle exchanges between the inside and outside of a vortex and enhance coherent transport of particles trapped in the vortex cores, see e.g. [18, 25, 23].

Particles initially seeded inside a coherent vortex remain there for long times, and are ejected mainly during strong, inelastic vortex-vortex interactions. Particles initially seeded outside coherent structures tend not to enter the vortex cores. This impermeability of coherent vortices induces strong inhomogeneities in the particle distribution even for long times, and, possibly, significant differences between Eulerian and Lagrangian averages over times that are much longer than the typical eddy turnover time.

Further, passively advected particles behave quite differently depending on whether they are inside or outside coherent vortices. Babiano et al [25] have shown that particles inside coherent vortices undergo regular trajectories, characterized by a null Lagrangian Lyapunov exponent, while particles outside coherent vortices undergo chaotic advection and are characterized by a positive Lagrangian Lyapunov exponent.

To study the Lagrangian dynamics in a coherent vortex, we consider the case of a large, isolated vortex centered at the point (x_c, y_c) with a non-circular profile given by

$$\omega(R, \phi) = A e^{-\frac{1}{2}(\frac{R}{\Sigma(\phi)})^4} \tag{29}$$

where R is the distance from the center of the vortex, $R^2 = (x - x_c)^2 + (y - y_c)^2$, and $\phi = \tan^{-1}[(y - y_c)/(x - x_c)]$ is the angular coordinate. The function $\Sigma(\phi)$ determines the non radially-symmetric shape of the vortex. In particular, we choose

$$\Sigma(\phi) = \Sigma_0 + \Delta \cos(k\phi) \tag{30}$$

where k is the azimutal wavenumber. Figure 4a shows the field obtained with a vortex centered in (π, π), with $\Sigma_0 = \pi/2$, $\Delta = 0.08\pi$ and $k = 3$. The initial energy of the field is $E(0) = 1$ and the resolution is 256^2. The white line indicates the initial distribution of 5,000 passive particles that are to be advected by the velocity field of the vortex. The field is let to undergo freely decaying evolution, we use hyperviscosity with $n = 2$ and $\nu_2 = 10^{-8}$. The vortex ejects vorticity filaments and tends, over long times, toward a more circularly symmetric state. Figure 4b shows the vorticity field at time $t = 40$; the white dots mark the positions of the 5,000 advected particles at this time.

From Fig. 4b, it is clear that in the vortex core, no radial mixing of the particles has taken place. The initial straight line of particles has been distorted into a spiral, due to the presence of differential rotation inside the vortex. Beyond the vortex edge, particles are advected chaotically and they wander along the separatrices between the vortex in the basic cell $(0, 2\pi)^2$ and its periodic images. In this case, full mixing has taken place.

Before making any further progress, one has to properly define the vortex edge. This issue has been discussed at length in several applications, such as the study of the stratospheric polar vortex, see e.g. [26]. The edge of the vortex may be defined as the region where the gradient of relative (or potential) vorticity is maximum, or as the isoline $Q = 0$ closer to the vortex center, or as the isoline of the maximum kinetic energy. These two latter definitions give similar results in most occasions. Figure 5 shows a section of the vorticity (solid line), kinetic energy (dotted line), and Okubo-Weiss parameter Q (dashed line) for the vortex shown in Fig. 4b. In this work, we define the vortex edge as the region where the kinetic energy is maximum and the Okubo-Weiss parameter is zero; this is clearly visible in Fig. 5 (grid points number 70 and 180, approximately).

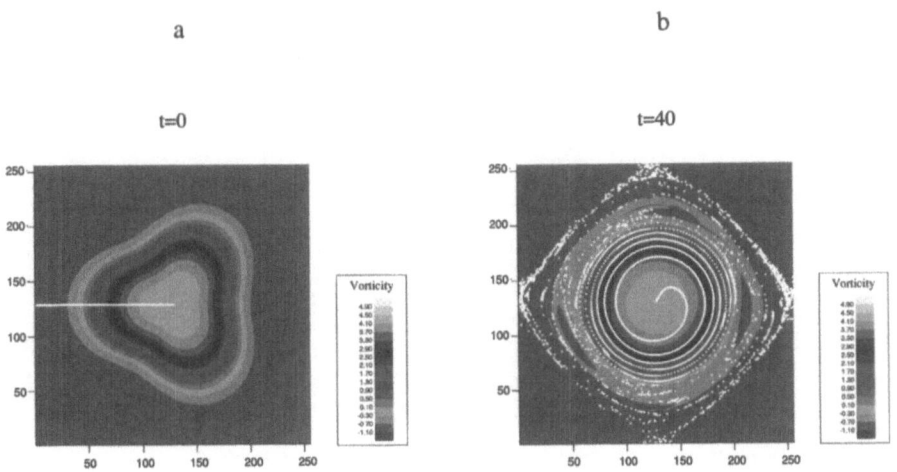

Fig. 4. — Vorticity field and passive particle distribution from a pseudo-spectral numerical simulation of a freely decaying isolated vortex with initially non-circular form. Here $\gamma^2 = 0$. Panel (a) shows the initial condition and panel (b) the evolved state at $t = 40$.

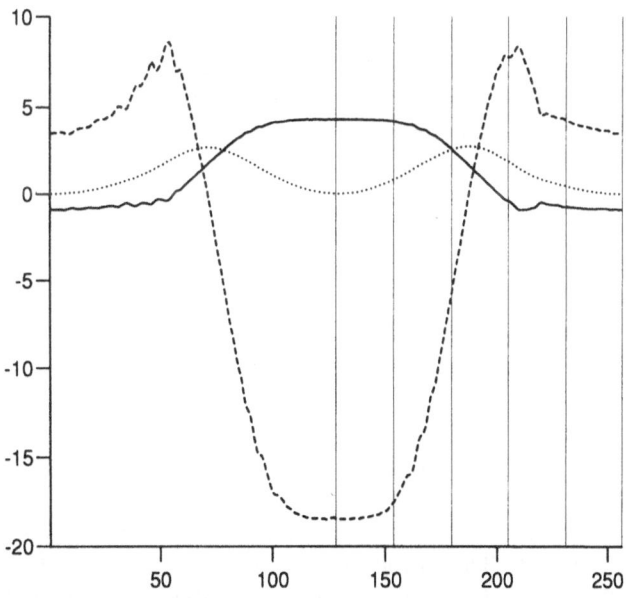

Fig. 5. — Profile of the vorticity (solid line), kinetic energy (dotted line), and Okubo-Weiss parameter Q (dashed line) for the vortex shown in Fig. 4b. The thin vertical lines indicate the edges of the five regions over which we compute relative radial dispersion.

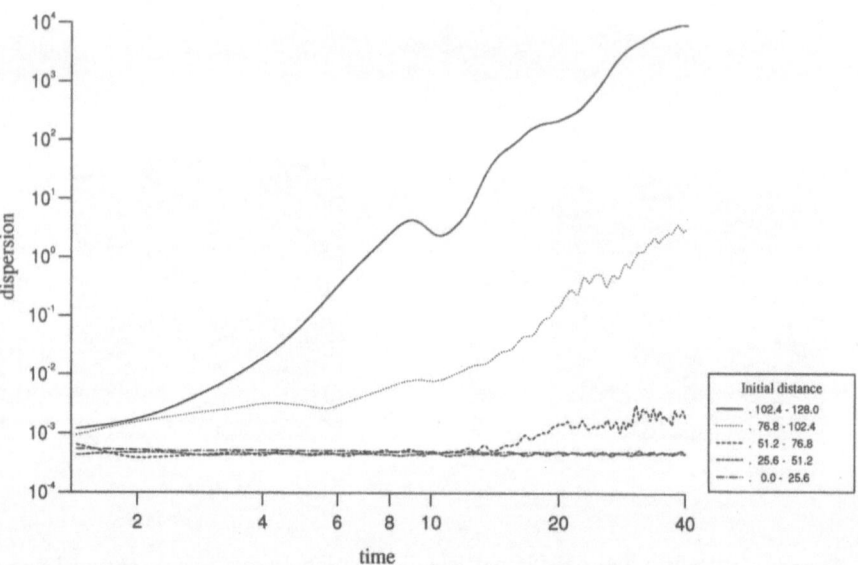

Fig. 6. — Relative radial dispersion for the five groups of advected particles discussed in the text. The legend indicates the initial distance of the particles from the center of the vortex. Dispersion is computed in units of grid spacings.

To make a more quantitative study, we compute the relative radial dispersion of the passively advected particles, as a function of their distance from the vortex center. The 5,000 passive particles have been initially seeded from the center of the vortex to the edge of the simulation domain, as shown in Fig. 4a. The thin vertical lines in Fig. 5 indicate the five different regions (containing 1,000 particles each) over which we compute the dispersion. The first two regions are completely inside the vortex core, while the last two are outside the vortex. Region 3 is approximately on the vortex edge.

Figure 6 shows the curves of relative radial dispersion, for the five regions mentioned above. Relative radial dispersion is herein defined as

$$D_R^2(t) = \frac{1}{M/2} \sum_{j=1,3,5,\ldots}^{M} (R_j(t) - R_{j+1}(t))^2 \qquad (31)$$

where j is the label of the particle and $R_j(t)$ is the distance of particle j from the vortex center at time t. The initial distance between the particles in each couple, $D_R(0)$, is 128/5000 grid intervals. Note that the sum in Eq. (31) runs on $j = 1, 3, 5, \ldots$; i.e., the average is over the set of $M/2$ couples of neighbouring particles in a given (initial) region.

From Fig. 5, it is clear that the particles inside the vortex (regions 1 and 2) or on its edge (region 3) undergo almost no relative radial dispersion, consistent with the results discussed by Babiano et al [25] indicating regular Lagrangian motion inside vortex cores. By contrast, particles that are outside the vortex undergo significant dispersion. In no case, we have observed particles entering the vortex from the outside.

The impermeability of vortex cores to inward particle fluxes is observed also during strong vortex-vortex interactions such as the merging of same-sign vortices. Figure 7 shows the time history of a vortex merging in pure 2D turbulence. Two equal vortices with initially circular shapes undergo a merging event, illustrated in the left panels. Passive particles have been initially seeded inside the two vortex cores; the two colors mark the particles belonging to one or the other vortex. The right panels of Fig. 7 show an enlargement of the particle distribution in the central region. The two particle distributions start wrapping around each other, undergoing filamentation, but the two populations do not mix, consistent with the lack of radial mixing inside vortex cores. A detailed study of the merging of two equal, same-sign vortices in 2D turbulence has been discussed by Nielsen et al. [27].

A similar situation is observed also in the presence of a finite Rossby deformation radius. Figure 8 shows the evolution of the same initial condition (that has now to be interpreted in terms of potential vorticity) for a case with $\gamma^2 = 30$. Note that, in this case, a thin filament of vorticity precedes the bulk of each vortex, and that both vortices are strongly deformed, consistent with the fact that a finite shielding distance inhibits (or slows down) the tendency toward forming circularly symmetric shapes. Even in this case, though, the particles of the two vortices do not mix, keeping track of their origin.

To further illustrate this behaviour, in Fig. 9 we show the evolution of two same-sign vortices with different initial circulation and size. The weaker vortex has amplitude 2/3 of the stronger one, and twice its size. The resolution is 128^2 and $\gamma^2 = 30$. The stronger vortex is smaller than the Rossby deformation radius, and due to its larger amplitude is quite robust. During the evolution, the weaker vortex is sheared around the stronger one, and most of its particles are trapped around the core of the final structure generated by the merging. The particles of the stronger vortex remain at the center of this structure, forming its core. Also in this case, the core of the stronger vortex does not mix with external material.

As a general conclusion of this section, we may state that the cores of coherent vortices are strongly impermeable to inward particle fluxes, and tend not to exchange material with the outside, while they exist. A similar situation has been observed for the winter polar vortex in the antarctic stratosphere [26].

6.   Absolute dispersion in 2D turbulence

In this section we briefly discuss the properties of absolute dispersion in the bulk of the turbulent flow, trying to understand the role played by the ensemble of coherent structures and by the complex topology of 2D turbulence. For this discussion, we consider only the case $\gamma^2 = 0$.

Absolute dispersion is defined as

$$A^2(t; t_0) = \frac{1}{M} \sum_{j=1,2,3,\ldots}^{M} (\mathbf{r}_j(t) - \mathbf{r}_j(t_0))^2 \qquad (32)$$

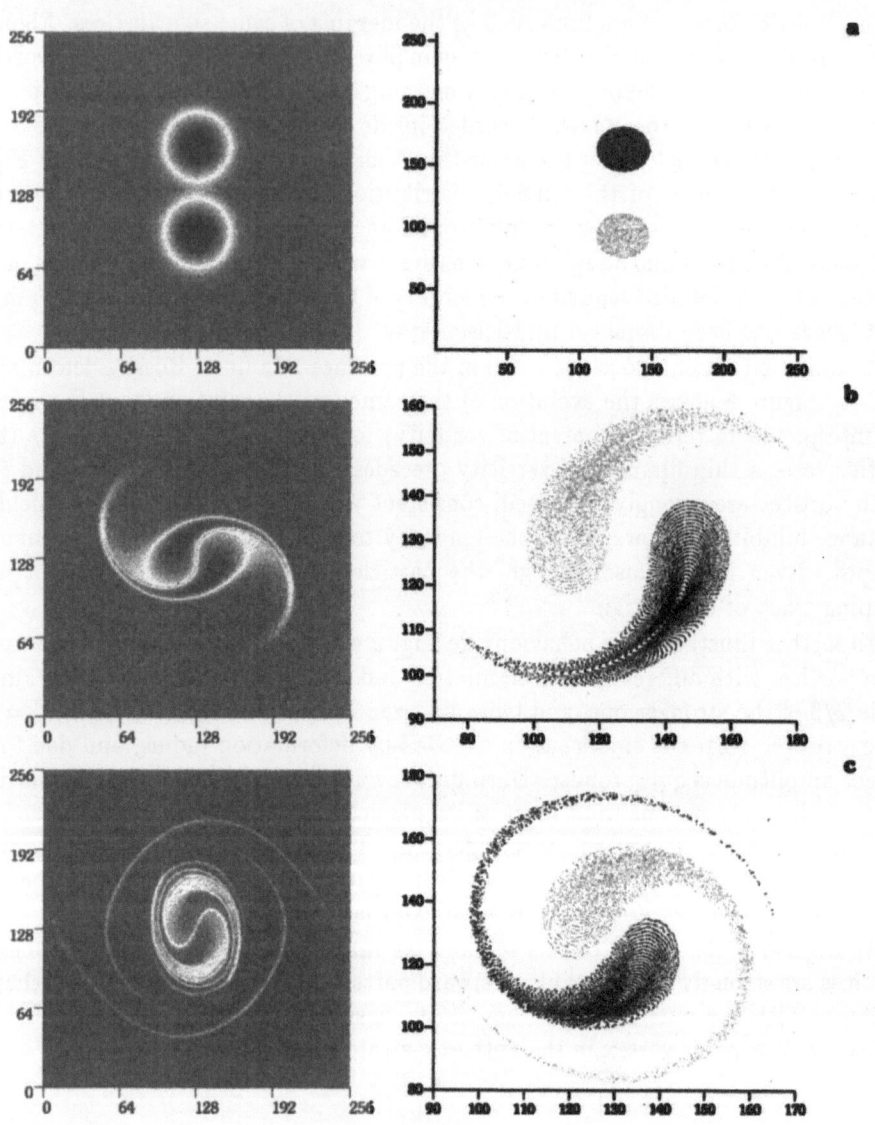

Fig. 7. — Merging of two equal, same-sign vortices in freely decaying 2D turbulence with $\gamma^2 = 0$. The left panels show the vorticity field and the right panels the particle distribution.

where $\mathbf{r}_j(t) = (x_j(t), y_j(t))$ is the position of the j-th particle at time t and the average is over the set of M particles. If the turbulent flow is statistically stationary, as in the case of the simulations of forced and dissipated 2D turbulence mentioned above, the properties of dispersion depend only on $t - t_0$. For simplicity, from now

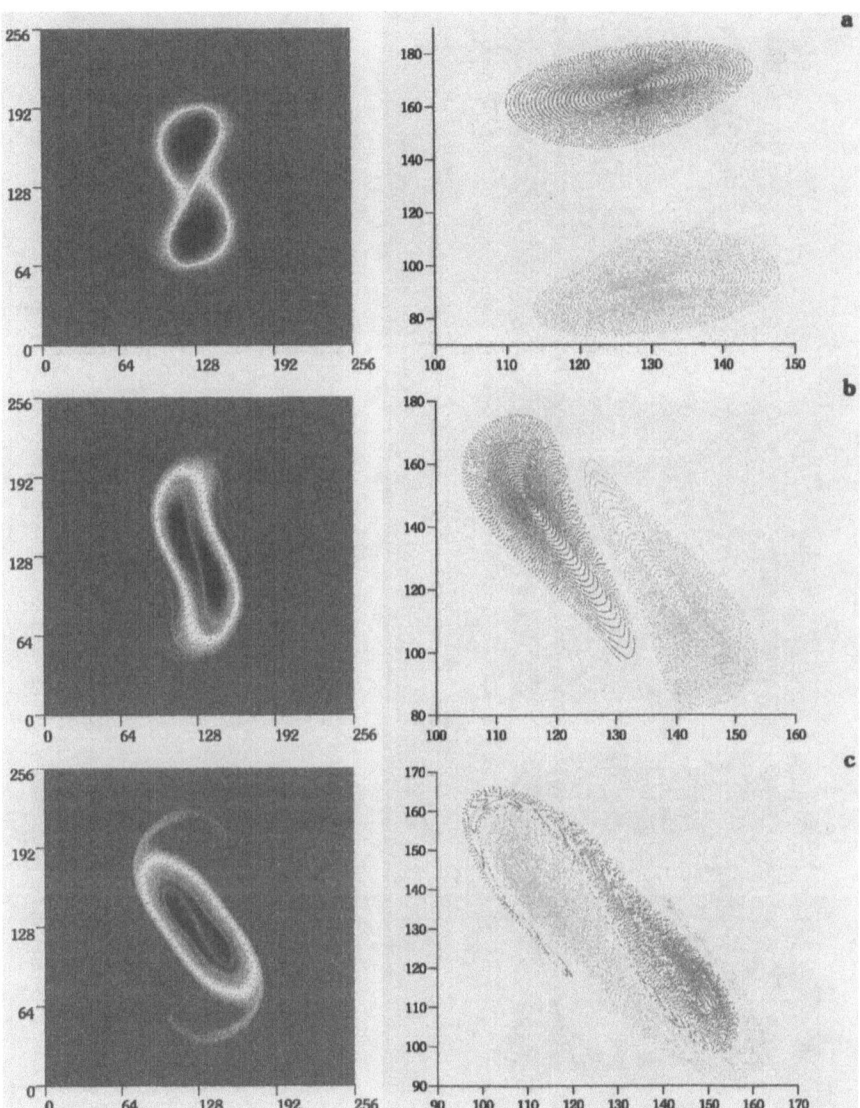

Fig. 8. — Merging of two equal, same-sign vortices in freely decaying turbulence with $\gamma^2 =$ 30. The left panels show the vorticity field and the right panels the particle distribution.

on we define $t_0 = 0$; i.e., we fix the origin of time at the beginning of the dispersion process.

The absolute dispersion (32) has two important limits. For $t \rightarrow 0$, particle motion

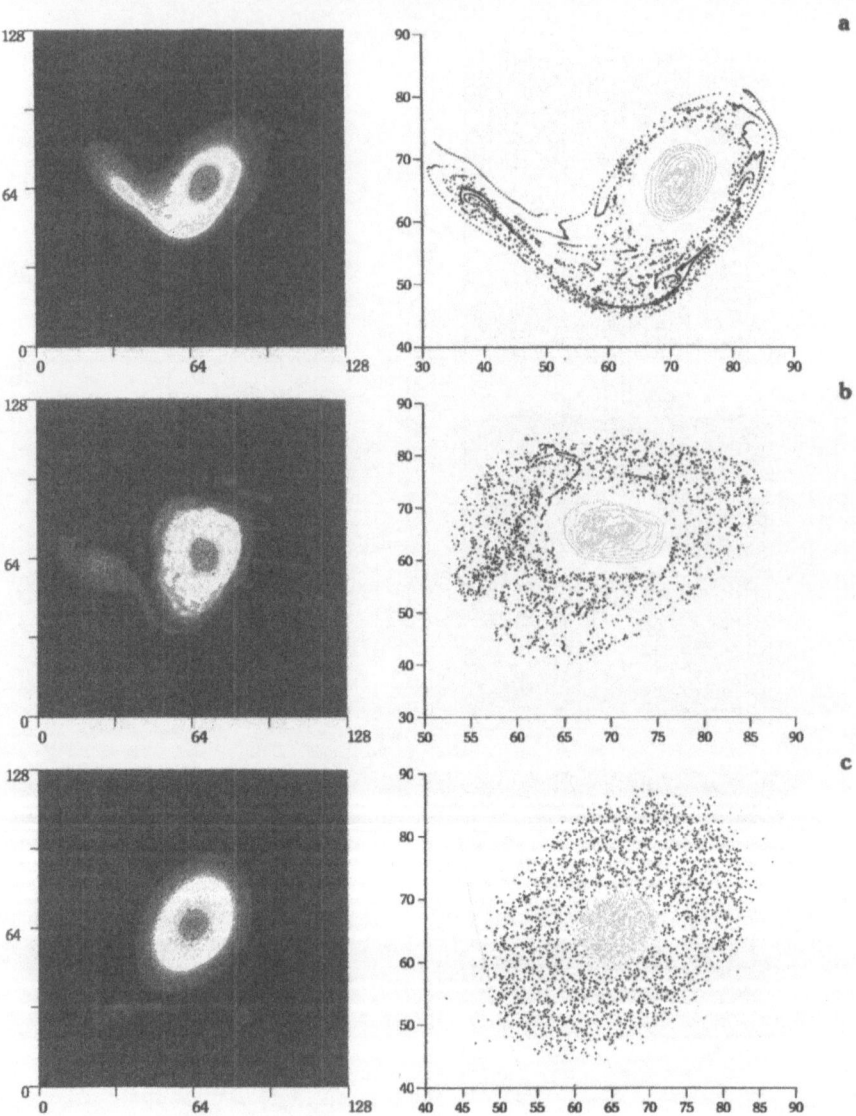

Fig. 9. — Merging of two same-sign vortices with different circulation and size in freely decaying turbulence with $\gamma^2 = 30$. The left panels show the vorticity field and the right panels the particle distribution.

is ballistic and one obtains

$$A^2(t) = 2E_{av}t^2 \tag{33}$$

where E_{av} is the average kinetic energy of the advected particles. For large times

$t \to \infty$, particle motion becomes brownian-like

$$A^2(t) = 2Kt \tag{34}$$

where K is the absolute dispersion coefficient. In general, the ballistic regime (33) is observed for times much shorter than the Lagrangian decorrelation time T_L, while the brownian regime is reached only for times much longer than T_L. The Lagrangian decorrelation time is defined as

$$T_L = \int_0^\infty C(\tau) \, d\tau \tag{35}$$

where

$$C(\tau) = \lim_{T \to \infty} \frac{1}{T\sigma^2} \int_0^T \mathbf{u}(t+\tau) \cdot \mathbf{u}(t) \, dt \tag{36}$$

is the autocorrelation function; the velocity $\mathbf{u}$ has zero mean and variance σ^2. For $t \gg T_L$, particle velocities are uncorrelated in time; this is at the origin of the brownian-like dispersion.

In numerical simulations of forced and dissipated 2D turbulence, both the ballistic and the brownian limits of absolute dispersion have been observed. In real flows, the large-time brownian behaviour is observed only when there are no large-scale inhomogeneities in the velocity field. In most geophysical flows, the presence of large-scale gyres, currents and the finite size of the basin in the case of the ocean often forbids the detection and/or the existence of a brownian regime.

In addition to the two asymptotic limits, the motion of freely drifting surface floats (drifters) and of subsurface floats in the ocean have often indicated the existence of an anomalous power-law dispersion regime on times of the order of T_L, i.e., $A^2(t) \propto t^\alpha$ with α between 1 and 2, see e.g. [28] and references therein. This is particularly significant, as in most geophysical flows T_L is of the order of the time scale of interest. For $t \approx T_L$, the particle velocities are correlated in time and the brownian motion approximation is not adequate; the concept of fractional brownian motions [29] has sometimes been advocated for describing the intermediate regime of anomalous dispersion.

In an attempt to better understand the origin and the properties of the intermediate regime, Elhmaidi et al [18] and Provenzale et al [23] have studied the properties of single-particle dispersion in 2D turbulence. In this framework, it has been shown that, starting from a uniform initial distribution of advected particles, there is often the appearance of an anomalous dispersion regime characterized by

$$A^2(t) \propto t^{5/4}. \tag{37}$$

By using conditional averages over the ensemble of advected particles, this regime has been shown to be associated with particle advection in hyperbolic regions (both the straining cells and hyperbolic patches in the background). This regime lasts for about one decade in time, on times around T_L. The fact that this regime is often visible also in unconditioned averages over the whole set of advected particles is due to the fact that hyperbolic regions occupy the largest portion of the turbulent field. A second anomalous dispersion regime, $A^2(t) \propto t^{5/3}$, associated with advection in strongly elliptic patches, has also been detected, even though in a much less clear way than the 5/4 regime. Anomalous dispersion also appears in the dynamics of points vortices [30].

More recently, evidence for a 5/4 anomalous dispersion law on times of the order of T_L has been detected in the study of the trajectories of SOFAR floats at 700 m below the surface, in highly energetic regions close to the Gulf Stream in the Western North Atlantic [31]. Also in this case, the 5/4 anomalous dispersion seems to be associated with motions in proximity of strong coherent structures.

7.   Dynamics of non neutrally-buoyant particles

Up to this point, we have considered only the motion of advected particles that have (a) the same density of the fluid and (b) an infinitesimally small size. When one of these assumptions is violated, a more complete Lagrangian advection equation has to be used for what we may call an "impurity."

The equations of motion for a small spherical impurity advected by a turbulent Eulerian velocity field may be written in terms of Newton's second law of motion as

$$m_p \frac{d^2\mathbf{r}}{dt^2} = \mathbf{F} \tag{38}$$

where $\mathbf{F}$ represents all the forces that the fluid exherts on the advected impurity with mass m_p. The detailed form of $\mathbf{F}$ is quite complicated (see e.g. [32], and it contains (1) a term representing the forces that the fluid would exhert on a fluid particle placed in the position of the impurity; (2) a Stokes drag term forcing the advected impurity to have the same velocity of the fluid at that point; (3) an added-mass term that describes the effects of the induced dipolar flow around the advected particle; (4) a Basset history term that depends on the past differences between Eulerian and Lagrangian accelerations; (5) the Faxen corrections that take into account the curvature of the velocity field at the particle position; (6) a buoyancy term acting along the local vertical direction, and (7) a Coriolis term in the case of rotating systems.

In the following, we consider motions on a horizontal plane; the buoyancy term can thus be discarded. We also discard the Faxen corrections, that are usually of limited relevance, and the Basset history term. For simplicity, here we also discard the added mass term. In this simplified form, Eq. (38) becomes

$$\frac{d^2\mathbf{r}}{dt^2} = \delta\left[\frac{\partial\mathbf{u}}{\partial t} + (\mathbf{u}\cdot\nabla)\mathbf{u}\right] - \tau_s^{-1}\left(\frac{d\mathbf{r}}{dt} - \mathbf{u}\right) - 2\mathbf{\Omega}\times\left(\frac{d\mathbf{r}}{dt} - \delta\mathbf{u}\right) \tag{39}$$

where $\mathbf{u}(\mathbf{r},t)$ is the fluid velocity at the position of the impurity, $\mathbf{r}(t)$, $\mathbf{\Omega}$ is the angular velocity of the reference frame, $\delta = \rho_f/\rho_p$ is the ratio of the density of the fluid to the density of the impurity, and

$$\tau_s = \frac{9}{2}\left(\frac{a}{L}\right)^2 Re \ , \tag{40}$$

where Re is the Reynolds number, L is a typical scale of the flow and a is the radius of the advected impurity. The standard behaviour of advected fluid particles, described by the equation $d\mathbf{r}/dt = \mathbf{u}$, is recovered as a singular limit when $\delta = 1$

and $a/L \to 0$. Note that, at variance with the classic approach based on equating Lagrangian and Eulerian velocities at a point, the motion of an advected impurity described by Eq. (39) defines a four-dimensional, dissipative dynamical system even for a divergence-free, two-dimensional velocity field. Chaotic motion of the advected impurities is thus possible also for stationary, 2D velocity fields.

The motion of advected impurities in simple, kinematic velocity fields without the Coriolis term has been thoroughly studied, see e.g. [32, 33, 4, 34]. Consider for example the case of a simple, stationary stream function given by [4]

$$\psi(x, y) = A(\cos x + \cos y) \tag{41}$$

and $\Omega = 0$. In this flow, light particles, characterized by $\delta > 1$, tend to the centers of advection cells, that are neutrally stable fixed points for the fluid particles. Heavy particles with $\delta < 1$, by contrast, wander chaotically throughout the flow and undergo a brownian dispersion process over long enough times, concentrating along the separatrices between the advection cells. A similar behaviour has been observed also for time-dependent velocity fields [4] and in the case of evolving 2D turbulence, where light particles concentrate inside coherent vortices and heavy particles wander chaotically in the background turbulence [35]. Clearly, this is very different from the behaviour of fluid particles. It is interesting to note that even particles that are neutrally buoyant ($\delta = 1$), but have finite size, may behave quite differently from fluid particles.

The presence of a Coriolis term does complicate the picture further. As shown by Tanga et al [36], the motion of an impurity along a curved trajectory, such as near a stationary vortex, is affected by two important forces: (a) the centrifugal force that pushes heavy particles to escape from the vortex and light particles to enter the vortex, and (b) the Coriolis force that deflects the particle trajectory. These forces act jointly to push heavy particles outside cyclonic vortices; the two forces compete in the case of anticyclonic vortices. In this latter case, the Coriolis force pushes the heavy particles toward the center of the vortex while the centrifugal force pushes them outside. For a fast enough rotation, i.e., for weakly nonlinear vortices in strongly rotating systems, and/or, for a small enough Rossby number, the Coriolis force dominates and the heavy particles tend to concentrate inside the cores of stationary anticyclonic vortices, where large concentrations are obtained quite rapidly. This mechanism has been recently suggested to play an important role for the formation of planetesimals in the solar nebula, see [37, 38, 36].

In the case of evolving 2D turbulence, the motion of advected impurities is more complicated, but it maintains some of the properties observed for the simpler Eulerian flows [35]. In particular, in the presence of a strong enough rotation, heavy particles tend to concentrate inside the cores of anticyclonic vortices, even when these latter are non stationary and undergo turbulent evolution. Figure 10 shows the distribution of 8,000 advected impurities in a pseudo-spectral simulation of forced and dissipated 2D turbulence, with the same characteristics as the simulation shown in Fig. 2. The particles, with $\delta = 0.8$ and $\tau_s = 0.01$, have been initially seeded uniformly in the square domain $(0, 2\pi)^2$. The rotation rate is $\Omega = 160$ (in adimensional units). After a short time, a large fraction of the particles concentrate in the cores of anticyclonic vortices, analogously to what has been found for stationary velocity fields.

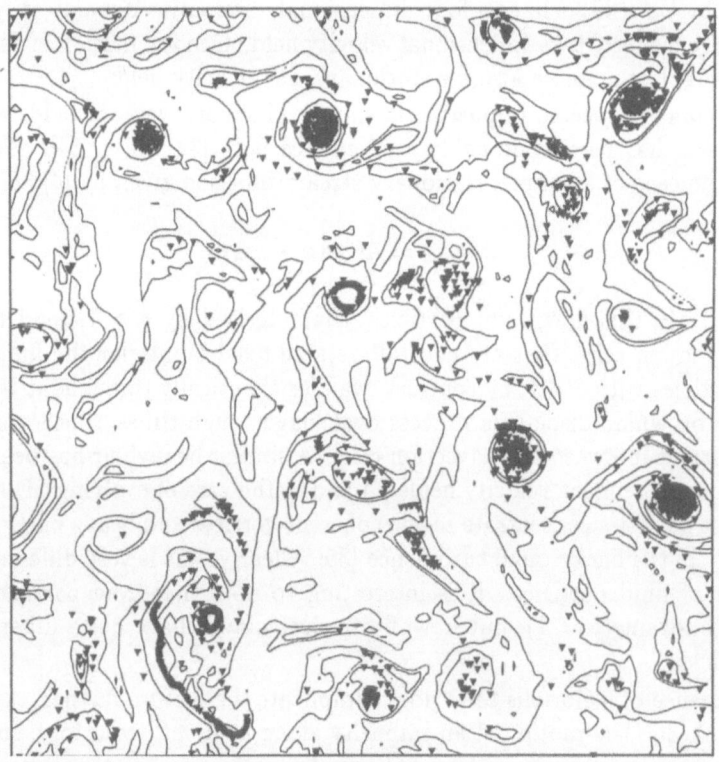

Fig. 10. — Distribution at time $t = 6$ of $8,000$ heavy impurities that are passively advected by forced and dissipated, rotating 2D turbulence. The thin solid lines indicate the isolines of the vorticity field. The impurities were initially distributed uniformly in the simulation domain.

8. Conclusions

In this work, we have reviewed some of the available results on the dynamics of Lagrangian tracers in barotropic flows, and we have discussed some new results on the advection of passive particles inside coherent structures.

The first general conclusion is that coherent vortices do strongly influence the dynamics of advected tracers. In particular, the vortex cores trap passive particles for long times, and are characterized by a strong impermeablity to inward particle motion. Inside the vortices, radial mixing is minimal. These properties have to be taken into account when studying tracer transport in flows characterized by the presence of coherent structures.

The complex topology of geostrophic turbulence has then been shown to induce anomalous dispersion of advected tracers on time scales of the order of days to weeks for the atmosphere and for the upper ocean, on slightly longer times for the deep ocean. The understanding of these dynamics, and their parameterization in terms of Eulerian quantities, is still quite fragmentary. On the other hand, a better intuition of these processes would be extremely valuable for properly interpreting the available Lagrangian data, and for modelling sub-grid scale processes in atmospheric and oceanic general circulation models. Analogously, these results should be taken into account for developing more complete stochastic models.

Finally, we have discussed how non neutrally-buoyant particles, or particles with finite size, may behave quite differently from fluid particles. Again, this fact should be taken into account when interpreting the results provided by Lagrangian tracers.

Acknowledgments

It is a pleasure to acknowledge the Organisers of the Advanced Study Institute "Mixing: Chaos and Turbulence" for the stimulating atmosphere they have been able to create during the School, and for their kind insistence in pretending a written version of our lectures. We are grateful to C. Basdevant, A. Bracco, B. Lien Hua, J. C. McWilliams, F. Paparella, M. Schmidt, P. Tabeling and J. B. Weiss for useful comments and interesting discussions.

References

[1] A.S. Monin and A.M. Yaglom, *Statistical Fluid Mechanics: Mechanics of Turbulence* (The MIT Press, Cambridge, 1971).

[2] A. Griffa, in *Stochastic Modelling in Physical Oceanography*, R. Adler, P. Muller, and B. Rozovsky eds. (Birkhauser, Boston, 1996).

[3] J.M. Ottino, *The Kinematic of Mixing: Stretching, Chaos, and Transport* (Cambridge University Press, Cambridge, 1989).

[4] A. Crisanti, M. Falcioni, G. Paladin, and A. Vulpiani, Riv. Nuovo Cimento **14**, 1 (1991); A. Crisanti, M. Falcioni, A. Provenzale, P. Tanga, and A. Vulpiani, Phys. Fluids **4**, 1805 (1992).

[5] J.B. Weiss and E. Knobloch, Phys. Rev. A **40**, 2579 (1989).

[6] T.H. Solomon and J.P. Gollub, Phys. Rev. A **38**, 6280 (1988).

[7] D. del Castillo Negrete and P.J. Morrison Phys. Fluids **5**, 948 (1992).

[8] R.M. Samelson, in *Stochastic Modelling in Physical Oceanography*, R. Adler, P. Muller, and B. Rozovsky eds. (Birkhauser, Boston, 1996).

[9] S. Wiggins, *Chaotic Transport in Dynamical Systems* (Springer, New York, 1992).

[10] J. Pedlosky, *Geophysical Fluid Dynamics* (Springer, New York, 1987).

[11] J. Pedlosky, *Ocean Circulation Theory* (Springer, Berlin, 1996).

[12] M. Ghil and S. Childress, *Topics in Geophysical Fluid Dynamics: Atmospheric Dynamics, Dynamo Theory, and Climate Dynamics* (Springer, New York, 1987).

[13] D.J. Tritton, *Physical Fluid Dynamics* (Clarendon Press, Oxford, 1988).

[14] J.C. McWilliams, J. Fluid Mech. **219** 361 (1990).

[15] V.D. Larichev and J.C. McWilliams, Phys. Fluids A **3**, 938 (1991).

[16] A. Babiano, C. Basdevant, B. Legras, and R. Sadourny, J. Fluid Mech. **183**, 379 (1987).

[17] A. Babiano, C. Basdevant, P. Le Roy, and R. Sadourny, J. Fluid Mech. **214**, 535 (1990).

[18] D. Elhmaidi, A. Provenzale, and A. Babiano, J. Fluid Mech. **257**, 533 (1993).

[19] A. Okubo, Deep-Sea Res. **17**, 445 (1970).

[20] J. Weiss, Physica D **48**, 273 (1991).

[21] M. Larcheveque, C. R. Acad. Sci. Paris, Series II, **311**, 33 (1990).

[22] C. Basdevant and T. Philipovitch, Physica D **73**, 17 (1994).

[23] A. Provenzale, A. Babiano, and B. Villone, Chaos, Solitons & Fractals **5**, 2055 (1995).

[24] B.L. Hua and P. Klein, "An exact criterion for the stirring properties of nearly two-dimensional turbulence", preprint 1997.

[25] A. Babiano, G. Boffetta, A. Provenzale, and A. Vulpiani Phys. Fluids **6**, 2465 (1994).

[26] F. Paparella, A. Babiano, C. Basdevant, A. Provenzale, and P. Tanga, J. Geophys. Res. **102**, 6765 (1997).

[27] A.H. Nielsen, J.J. Rasmussen, and M.R. Schmidt, "Vortex merging and corresponding energy cascade in two-dimensional turbulence", preprint 1997.

[28] A. Provenzale, A.R. Osborne, A.D. Kirwan, and L. Bergamasco, in *Nonlinear Topics in Ocean Physics*, Proceedings of the CIX Course of the Internation School of Physics "E. Fermi", Varenna 1988, A. R. Osborne Ed. (North-Holland, Amsterdam, 1991).

[29] B.B. Mandelbrot, *The Fractal Geometry of Nature* (Freeman, San Francisco, 1982).

[30] J.A. Viecelli, Phys. Fluids A **2**, 2036 (1990); J.B. Weiss, A. Provenzale, and J.C. McWilliams, Phys. Fluids A **10**, 1929 (1998).

[31] V. Rupolo, B.L. Hua, A. Provenzale, and V. Artale, J. Physical Ocean. **26**, 1591 (1996).

[32] M.R. Maxey, and J.J. Riley, Phys. Fluids **26**, 883 (1983).

[33] M.R. Maxey, Phys. Fluids **30**, 1915 (1987).

[34] P. Tanga and A. Provenzale, Physica D **76**, 202 (1994).

[35] A. Provenzale, A. Babiano, O. Piro, and P. Tanga, "Dynamics of passive impurities in 2D turbulence", preprint 1997.

[36] P. Tanga, A. Babiano, B. Dubrulle, and A. Provenzale ICARUS **121**, 158 (1996).

[37] P. Barge and J. Sommeria, Astron. & Astrophys. **295**, L1 (1995).

[38] F.C. Adams and R. Watkins, Astrophys. J. **451**, 314 (1995).

Transport, Stirring, and Mixing in the Atmosphere

Peter Haynes

Centre for Atmospheric Science [*],
Department of Applied Mathematics and Theoretical Physics
University of Cambridge, Cambridge, CB3 9EW, England

1. Introduction

1.1. Outline

Our understanding of atmospheric transport has increased considerably over the last twenty years or so. There have been important theoretical advances, often prompted by enormous improvements in the observations, both of dynamical quantities and of chemical species, and in numerical models. The combined use of numerical models and observations has been particularly important, especially in the last decade. At the same time, there has been much interest in the application of results of dynamical systems theory to fluid dynamics under the heading of 'chaotic advection'. [See e.g. the textbook by Ottino [73].] Such results are of undoubted qualitative relevance to atmospheric flows, but nonetheless there are relatively few specific examples where insight from chaotic advection has lead to concrete, quantitative insight into atmospheric transport.

This article will review our current understanding of atmospheric transport, stirring and mixing and identify areas where insight from theories of chaotic advection may be quantitatively useful. The article is intended to be of interest to a reader who has some knowledge of dynamical systems and chaotic advection and wishes to learn more about possible applications to the real atmosphere.

The remainder of this section will summarize relevant aspects of our knowledge of the observed state of the atmosphere. Then in §2 some of the ideas underpinning our understanding of large-scale atmospheric dynamics will be reviewed, focussing

---

(*) The University of Cambridge Centre for Atmospheric Science is a joint initiative of the Department of Applied Mathematics and Theoretical Physics and the Department of Chemistry

Mixing: Chaos and Turbulence, edited by Chaté *et al.*
Kluwer Academic / Plenum Publishers, New York 1999.

on those that have particular relevance to characterising the transport, stirring and mixing problem in the atmosphere. The two important features that emerge are (i) that the stable stratification tends to inhibit vertical motion, so some important aspects of atmospheric transport may be captured by considering only quasi-horizontal motion along stratification surfaces (in fact, as explained in §2, isentropic surfaces) and (ii) that in many cases the large-scale flow gives the dominant contribution to advection. (i) and (ii) together mean that two-dimensional chaotic advection provides a useful paradigm for certain aspects of atmospheric transport. §3 considers observations and models of quasi-horizontal transport, stirring and mixing along isentropic surfaces, which is the part of the overall transport, stirring and mixing problem to which the application of chaotic advection ideas seems most promising. §4 considers some of the transport, stirring and mixing issues to which vertical transport and vertical structure are central. §5 summarises and identifies some areas where application of the mathematical of chaotic advection and dynamical systems may give greater understanding of the real atmosphere.

1.2.   Observational background

We shall be primarily concerned here with transport, stirring, and mixing due to the large-scale flow in the troposphere and stratosphere. The troposphere is the part of the atmosphere below about 10km (higher in the tropics, lower at the poles) where the temperature decreases with height. The stratosphere is the part of the atmosphere above the troposphere, and below about 50km, in which the temperature increases with height. (The temperature variation with height has important implications for the stability to vertical displacements - see later.) Books such as James (1994) [49], Holton (1992) [44], and Andrews et al. (1985) [2] are dynamical texts (the last is primarily concerned with the stratosphere) which also give useful summaries of the observed state of the atmosphere.

It is perhaps useful to point out at the beginning that in many respects the atmosphere is relatively well-observed. [See, for example, [49], §2.2.]. The requirements for reliable weather forecasts have lead to an extensive network of observing stations, giving improved knowledge of the winds, temperatures etc. in the troposphere and lower stratosphere. These observing stations are, of course, concentrated in the continental areas of the Northern Hemisphere and coverage in the Southern Hemisphere in particular is extremely sparse. More uniform geographical coverage has been provided by a number of satellite-based instruments used to measure temperature, water vapour, ozone and other quantities. This satellite data has not had quite the impact on weather forecasts that one might have expected, since in the troposphere the vertical resolution of the data tends to be rather poor and the presence of clouds also causes problems. However satellite data has had an enormous impact on our knowledge of the stratosphere.

Besides 'operational' satellites, which make measurements on a routine basis, there have been a number of important satellites that operate for a restricted period. For example, the Upper Atmosphere Research Satellite (UARS), launched in 1991, has measured a large range of different trace chemicals, such as nitrous oxide (N_2O), nitric acid (HNO_3) and chlorine monoxide (ClO), as well as ozone (O_3), water vapour (H_2O), temperature, and aerosol. See, for example, the recent issues of Journal of Atmospheric Sciences (volume 51, number 20) or Journal of Geophysical Research (volume 101, number D6), devoted to UARS observations and their interpretation.

The advantage of satellite data is that the coverage is quasi-uniform over the globe. The disadvantage is that the spatial resolution tends to be rather coarse. At the opposite extreme is data obtained from aircraft or balloons, often during special observational campaigns such as the Antarctic Airborne Ozone Experiment (AAOE), the Arctic Airborne Stratospheric Expedition (AASE), and their successors. This data tends to have very spatial high resolution, but is obtained only in a few locations, leaving a non-trivial problem about how best to interpret it, given that it is a restricted sample of chemical and dynamical fields that vary in three spatial dimensions and in time.

One important problem in numerical weather prediction has been to devise ways of effectively using observational data, obtained on an irregular space and time grid, in a numerical model where the corresponding grid is regular. A number of different techniques have been applied to use observational data to provide initial conditions for such numerical models. One approach has been 'analysis' where the observational data is interpolated in space and time and combined with fields predicted by the model in an integration started at an earlier time. [See e.g. Daley, 1991 [22].] This usually has to be followed by another procedure called 'initialisation' before integration of the model can proceed. A different technique is to insert data in a continuous fashion into a model as it is integrated forward in time. This procedure is called assimilation. Development and improvement of assimilation methods is a significant area of current research [see e.g. the literature survey by Courtier et al., (1993) [21]]. Existing methods routinely used for numerical weather prediction models (which normally focus on the troposphere and lower stratosphere) have been successfully applied to models which aim to represent the full troposphere-stratosphere system [92]. There is also the prospect that, in future, assimilation techniques may allow much better use of chemical observations [34].

Historically, many observations of the atmosphere or the ocean have provided measured temperature or density, but not velocity. Velocity therefore has to be deduced indirectly. A crude approach is to build up a pressure field via the hydrostatic approximation and then to infer the velocity field from geostrophic balance. (See §2 for precise definitions of these terms.) Both 'analysis' and assimilation provide alternative, more accurate, approaches to this problem. Both give systematic ways of using temperature data (and what ever other data is available, including perhaps some direct velocity observations) together with a dynamical model, to infer velocity fields. Such velocity fields have recently been used with much success to perform various numerical experiments to give greater insight into observed chemical tracers, e.g. associated with one of the airborne expeditions, or with the observations from UARS. (The term 'chemical tracer' is used to mean a chemical species that is conserved following the fluid motion for more that one or two eddy turnaround times, i.e. for more than a few days in the cases of interest here. One might say that, at leading order, the distribution of such a species was determined by transport. Many chemical species in the troposphere and stratosphere have a very short photochemical lifetime and their distributions are determined, at leading order, not by transport, but by local photochemical equilibrium.)

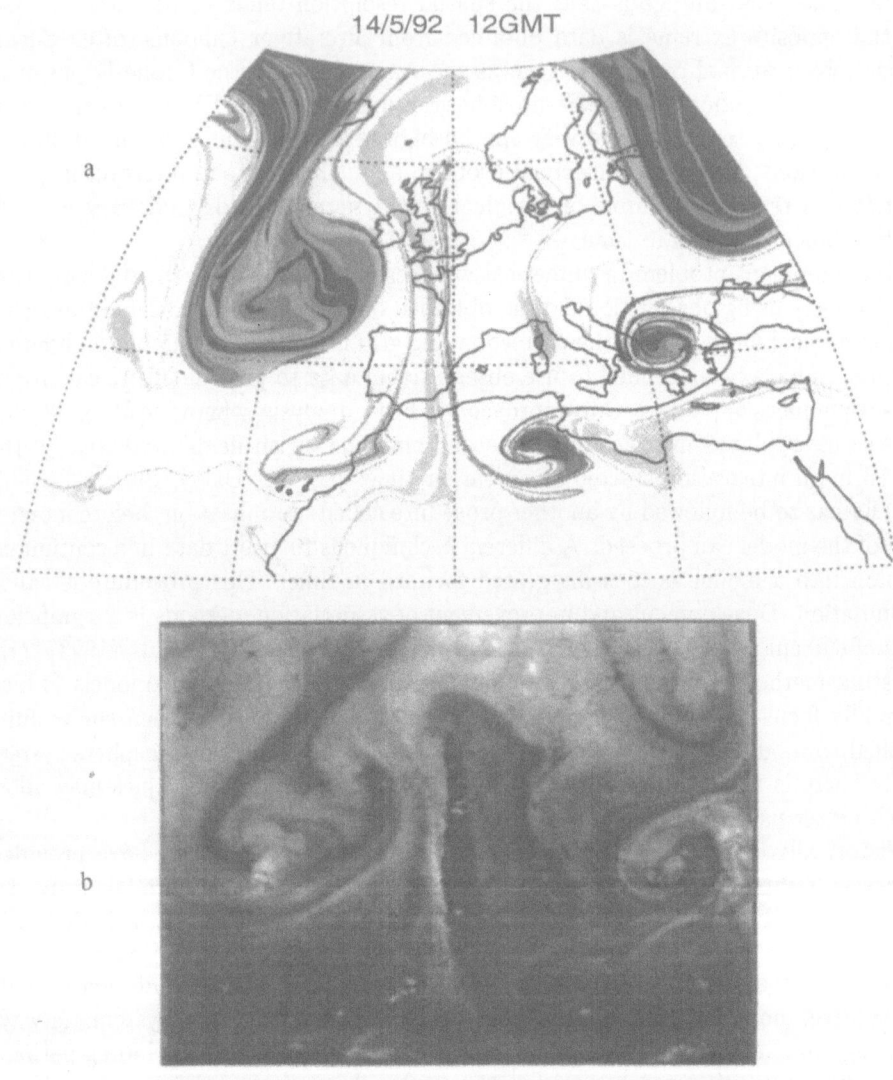

14/5/92   12GMT

a

b

Meteosat water vapour image 14 May 1992 12:00UT

Fig. 1. — (a) Passive tracer on 320-K isentrope for May 14 1992, at 1200UT, predicted using a high-resolution Lagrangian tracer-advection algorithm, using winds from meteorological data. The tracer contours were initialized five days earlier to coincide with contours of a dynamical tracer, potential vorticity, that was itself calculated from meteorological data. For more details about potential vorticity, see §2.3. (b) Meteosat 5.7 to 7.1 μm water vapour image. The dark regions signify high radiance values and, in effect, low moisture content in the upper troposphere and lower stratosphere. (Both (a) and (b) taken from [4]

Some aspects of the numerical experiments mentioned above will be discussed in §3, but a first example is given in Figure 1, taken from [4]. Figure 1a shows the predicted contours of a tracer on a quasi-horizontal surface, at a level of about 10km above the surface of the Earth. This tracer might represent water vapour and for comparison a direct satellite measurement of water vapour is also shown in Figure 1b. The latter has rather coarse horizontal and, most importantly, vertical resolution and this, plus a number of other factors, makes it difficult to quantify the precise relation between the fields shown in Figures 1a and 1b. Nonetheless, the same gross structures are present in both and it is believed that Figure 1a gives a useful impression of the fields of water vapour and other tracers at scales that are well below the resolution of present day observational techniques.

2.   Large-scale atmospheric dynamics

Some readers may wish to skip dynamical details and should proceed directly to §2.5 at this point.

2.1.   The equations

The full equations of motion for the atmosphere describe a vast range of different sorts of flow, including sound waves, three-dimensional turbulence, convection, gravity waves, large-scale weather systems in the troposphere and their stratospheric analogues. The emphasis here is on the last two, i.e. on the flow on space scales of a hundred kilometres or more and time scales of a day or more. This sort of flow has a vortex-like character and is sometimes described as 'slow' motion in order to distinguish it from 'fast' sound waves and gravity waves. It is the transport associated with 'slow' motion that largely determines the global scale distributions of quantities such as ozone and water vapour, as well as a whole host of other chemical species that play a leading role in the chemical-climate system.

Understanding of the large-scale 'slow' flow and the transport associated with it is aided by a number of mathematical (and hence conceptual) simplifications. One such simplification, which is particularly useful when considering transport, arises from the fact that the atmosphere is stably stratified. A new vertical coordinate may be chosen that neatly takes account of some of the effects of the stable stratification and is discussed below in §2.2. The stable stratification, plus the fact that the atmosphere is rotating rapidly relative to an inertial frame, also allows the derivation of an approximate set of governing equations that gives a much simplified view of the dynamics.

2.2.   Stable stratification and potential temperature

If we are concerned with large-scale motion in the atmosphere, on scales of a few tens of kilometers or more, i.e. safely larger than the scale of clouds, then the atmosphere may be regarded as stably stratified. By this we mean that the density gradient is such that if a fluid parcel is displaced in the vertical, the density of its surroundings and the corresponding pressure forces are such as to return it towards its original position [see e.g.[44], §2.7]. Stable stratification therefore implies inhibition of vertical mixing and vertical transport. [For a recent review see [33].]

In an incompressible stratified fluid it is the density that is the appropriate measure of stratification. Fluid parcels conserve their density and if the density decreases with height then the fluid is stably stratified. Often, e.g. in laboratory experiments, density variations are associated with temperature variations and it is temperature increase with height that implies stable stratification. The atmosphere is composed of compressible gas and neither temperature nor density is a relevant measure of stratification. Instead the relevant measure is the density that a fluid parcel would have if it were moved adiabatically (i.e. without irreversible thermodynamic change) to a reference pressure, p_* say. For a perfect gas this measure is simply a function of the entropy, conveniently chosen to be the potential temperature θ, where

$$\theta = T(p/p_*)^{-\kappa}$$

where T is the temperature, p is the pressure and κ is the ratio of specific heat at constant volume to that at constant pressure (equal to 5/7 for a diatomic gas). [The potential temperature is the temperature that a parcel of air would have if it were brought adiabatically, i.e. without change of entropy, to a reference pressure p_*].

If the fluid is stably stratified then the potential temperature must increase monotonically upwards and it may therefore be used to replace geometric height as a vertical coordinate. The corresponding coordinate surfaces are surfaces of constant potential temperature, or equivalently, constant entropy. These surfaces are often referred to as isentropic surfaces and the coordinates as isentropic coordinates. The vertical velocity in such a coordinate system is proportional to the material rate of change of potential temperature and therefore to the rate of change of entropy. Viewing the motion in isentropic coordinates therefore has the advantage of focussing attention on whether or not air parcel motion is adiabatic versus diabatic (i.e. undergoing irreversible thermodynamic changes).

In the atmosphere there are a number of different physical mechanisms by which diabatic changes may occur. The two that are most important from the point of view of this paper, which focusses attention on the upper troposphere and the stratosphere, are (i) radiative transfer and (ii) molecular dissipation occuring in patches of three-dimensional turbulence. In the middle and lower troposphere one would also have to take account of diabatic effects due to moist processes and due to boundary-layer turbulence etc. With regard to (ii) note that there is no strong evidence in the upper troposphere and lower stratosphere for an ubiquitous background field of three-dimensional turbulence, presumably because of the stabilising effect of stratification. However, there is evidence for intermittent patches of turbulence, almost certainly associated with the breakdown of inertio-gravity waves that are excited by flow over topography, frontal systems, convection etc. (see, e.g., [24], [9], [88] and references therein).

There is a useful division of time scales between an advective time scale along isentropic surfaces, which might be typically three or four days, and the time scale for diabatic effects, which might typically be ten to twenty days. Equivalently one might consider the relevant velocities, which along isentropic surfaces are about 10 ms^{-1} and across isentropic surfaces are less than 10^{-3}ms^{-1}, giving a ratio of about 10^{-4}. This should be contrasted with the natural geometric aspect ratio between vertical and horizontal length scales of about 10^{-2} or so. There is a real sense in which diabatic vertical velocities are weak. Thus, viewed in isentropic coordinates,

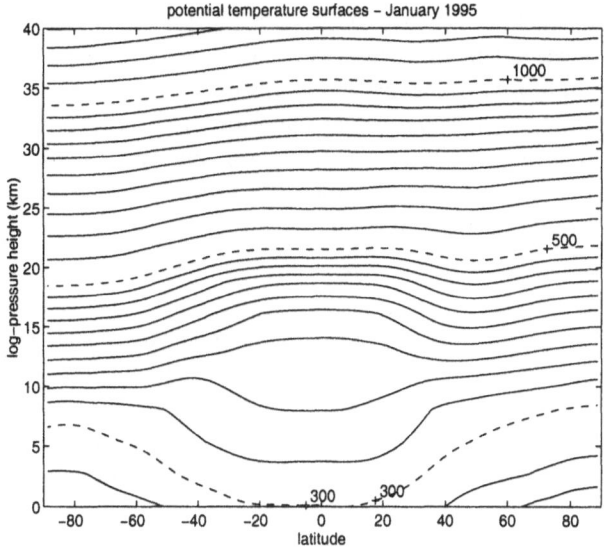

Fig. 2. — Contours of potential temperature averaged longitudinally and over month of January 1995, as a function of latitude and log-pressure height, calculated using UK Meteorological Office data. The latter is defined to be log (1000mb / pressure) multiplied by 7km and is a reasonable approximation to geometric height. Contours are at intervals of 20K, up to 500K, then 50K above that. The 300K, 500K and 1000K contours are dashed. The contours show the average configuration of potential temperature surfaces, often know as 'isentropic' surfaces.

fluid transport can usefully be divided into rapid motion along isentropic surfaces and much slower motion across isentropic surfaces. This leads to considerable conceptual and practical advantages to using isentropic coordinates when considering atmospheric tracer transport. Isentropic coordinates have been exploited formally to good effect in mathematical formulations of the tracer transport problem (see e.g. [2], Chapter 9). Here we shall be content to note some of the qualitative insights that arise from their use. See also [47] and [46] for further details and discussion. The basic point is that, if we are interested in transport on timescales of a few days, then we may neglect diabatic changes and consider only transport along isentropic surfaces. The full three-dimensional transport problem has therefore been reduced to the two-dimensional problem of transport along isentropic surfaces by the horizontal velocity field. It is this aspect of the transport problem that will be discussed in §3. For reference, Figure 2 shows the configuration of isentropic surfaces, labelled by potential temperature, in the lowest 40 km of the atmosphere. Note that many such surfaces, particularly in lowest 10km, cover quite a broad range of pressures (and of geometric heights).

2.3. Balanced dynamics and the quasi-geostrophic equations

Returning to questions of dynamics, the derivation of a simplified form of the governing equations that applies to the 'slow' motion when the fluid is stably stratified and rapidly rotating is a standard exercise in geophysical fluid dynamics, and cov-

ered in textbooks such as those by Gill (1982) [37] or Pedlosky (1987) [74]. The
reader who is interested in mathematical details is recommended to consult such
textbooks. Here only a brief motivation is given.

One key to finding a suitable set of equations is to use one of the interesting
properties of a stratified fluid, that there is a a quantity called potential vorticity
that is materially conserved when the flow is frictionless and adiabatic. The potential
vorticity P is generally defined by

$$P = \zeta_a . \nabla \theta / \rho$$

where ζ_a is the absolute vorticity, i.e the vorticity measured with respect to an
inertial frame, so that $\zeta_a = \nabla \times \mathbf{u} + 2\mathbf{\Omega}$, where $\mathbf{u}$ is the velocity with respect to the
rotating Earth and $\mathbf{\Omega}$ is the rotation rate of the Earth, ρ is the density and θ, as
before is the potential temperature. [In fact, θ may be replaced by any quantity χ
that (a) is itself materially conserved in adiabatic frictionless motion and (b) such
that $\nabla \chi . (\nabla \rho \times \nabla p) = 0$ where p is pressure.]

At first sight the material conservation of potential vorticity may seem no more
than a curiosity. It allows the possibility of constructing a materially conserved
quantity from dynamical quantities such as velocity and temperature and hence a
way of visualising the motion by displaying time sequences of the P field. But, of
course, the same could be done using observed fields of any chemical tracer. The
special property of potential vorticity is that for the 'slow' motion it satisfies an
invertibility principle. That is to say, knowledge of the potential vorticity field, plus
certain other quantities, specifically the potential temperature field on boundaries,
plus some knowledge of the 'average' stratification, perhaps measured by the total
mass distribution between isentropic surfaces, instantaneously determines all other
flow quantities such as velocity field, temperature and pressure. The instantaneous
relations between potential vorticity and the other flow quantities follow through
certain 'balance' conditions, that eliminate the 'fast' gravity-wave and sound-wave
motion. This gives a complete description of the 'slow' fluid motion: a single prognostic equation describes the advection of the potential vorticity field by the fluid
velocity field (and perhaps the changes in potential vorticity through the effects of
friction or diabatic heating) and at each instant the velocity field (and other flow
quantities) are determined by the potential vorticity. [For a more detailed account,
see e.g.[48], [58].] Potential vorticity (hereafter PV) is sometimes described as an
'active' tracer (to be contrasted with a 'passive' tracer) since its redistribution by
transport affects the structure of the flow.

In general there is no closed-form mathematical representation of the invertibility
principle (and hence of a system of equations for the 'slow' motion). However, approximate forms may may be written down in a straighforward mathematical way
when at least one of two important dimensionless parameters, the Rossby number
(which measures the ratio between a typical rate of change and the rotation rate)
or the Froude number (which measures the ratio between typical velocities and
the gravity wave speed) are small. One much-used example that is valid at small
Rossby number is the quasi-geostrophic system, which describes the time evolution
of a flow that is close to geostrophic balance, i.e with the dominant balance in the
horizontal momentum equation between the Coriolis force and the pressure gradient.
Geostrophic balance therefore gives the necessary instantaneous balance conditions
between different flow quantities. Again, see textbooks such as those by Gill and

Pedlosky for derivations of the quasi-geostrophic system. Here we simply note the form of the equations and some of the implications. For simplicity we will work in a Cartesian coordinate system where x, y and z are distances in the eastward, northward and vertical directions respectively. For realistic atmospheric applications there are advantages to using pressure p in place of z as a vertical coordinate, but that is not necessary here. The assumption that the flow is close to geostrophic balance means that we can define a leading order 'geostrophic' flow field with horizontal components (u_g, v_g) given by $u_g = -\partial\psi_g/\partial y$ and $v_g = \partial\psi_g/\partial x$, where ψ_g is the quasi-geostrophic streamfunction equal to $\tilde{p}/f_0\rho_0$. Here f_0 is the local value of the Coriolis parameter (twice the local vertical component of the Earth's rotation), and ρ_0 is an average density. The perturbation density $\tilde{\rho}$ may be deduced from the hydrostatic equation as $\tilde{\rho} = -f_0\rho_0 g^{-1}\partial\psi_g/\partial z$, where g is the gravitational acceleration. [The terms 'perturbation pressure' and 'perturbation density' effectively mean the horizontally varying parts of the total pressure and density fields respectively.] There is a single predictive (or ' prognostic') equation for the flow in the interior (away from horizontal boundaries) of the form

$$\frac{D_g q_g}{Dt} = 0 \qquad (1)$$

where D_g/Dt denotes the time derivative following the 'geostrophic' flow and q_g is the quasi-geostrophic potential vorticity (hereafter QGPV) defined in terms of ψ_g by

$$q_g = \frac{\partial^2\psi_g}{\partial x^2} + \frac{\partial^2\psi_g}{\partial y^2} + \frac{\partial}{\partial z}\left\{\frac{f_0^2}{N(z)^2}\frac{\partial\psi_g}{\partial z}\right\} + \beta y. \qquad (2)$$

$N(z)$ is a ' background' buoyancy frequency associated with the stable stratification, and β is the latitudinal gradient of the Coriolis parameter. The βy term an approximation to the latitudinal variation of the Coriolis parameter.

Note that here we have the structure identified earlier, an equation (1) which represents the material conservation of PV (or an approximation to it) and an invertibility principle (2) which allows calculation of all other flow quantities, given q_g. Recall that knowledge of ψ_g determines the leading-order velocity field and also the density field.

The calculation of ψ_g from the QGPV field q_g requires the solution of an elliptic equation, so that there is a global dependence of other flow quantities on q_g and in addition there is a requirement for boundary conditions on ψ_g. The relevant boundary condition at 'side' boundaries is that of no normal flow. At 'top' and 'bottom boundaries the natural condition is on the vertical velocity, but the leading order flow is horizontal. In fact the leading-order terms in the density equation may be used to deduce the leading-order vertical velocity, w and in terms of the quasi-geostrophic streamfunction it follows that

$$w = \frac{g}{N^2\rho_0}\frac{D_g\tilde{\rho}}{Dt} - \frac{f_0}{N^2}\frac{D_g}{Dt}\frac{\partial\psi}{\partial z} \qquad (3)$$

Note that if w is known on the upper and lower boundaries, then (3) represents another prognostic equation, for $\partial\psi/\partial z$.

The form of the quasi-geostrophic equations has many similarities with the more widely known equations for two-dimensional vortex dynamics (hereafter 2DVD), which take the form

$$\frac{Dq}{Dt} = 0, \tag{4}$$

where the velocity $\mathbf{u} = (u, v)$, may be defined in terms of a streamfunction ψ, with $u = -\partial\psi/\partial y$ and $v = \partial\psi/\partial x$. The relation between the vorticity q and the streamfunction ψ is that

$$q = \frac{\partial^2\psi}{\partial x^2} + \frac{\partial^2\psi}{\partial y^2}. \tag{5}$$

Given suitable boundary conditions on ψ, (5) may be regarded as an invertibility principle for the two-dimensional flow in the same way that (2) is an invertibility principle for the quasi-geostrophic flow.

2.4. Quasi-geostrophic motion

What is the nature of quasi-geostrophic motion and, hence, what its stirring and mixing properties? The type of dynamics described by (4) and (5) is familiar from many studies of 2DVD and two-dimensional turbulence. For example, in a flow where vorticity is concentrated into localised coherent vortices (point vortex dynamics is an extreme idealisation of such a flow, but flows with this character are more generally observed in e.g. later stages of 2-D turbulent 'spin-down' experiments), each vortex moves with a velocity that is determined by the strengths and relative positions of the other vortices. This is an indication of the non-local dependence of ψ on q implied by (5). Far vortices are less important than near vortices in this respect; the velocity field associated with a given vortex decays inversely with distance from that vortex.

Given the close similarity between the quasi-geostrophic equations (1) and (2) and those for 2DVD (4) and (5), one might expect quasi-geostrophic flows to evolve in a similar manner. A coherent vortex flow analogous to that described above would in the QG system again have each vortex moving horizontally under the influence of the others, but the vortices might be on different levels and the decay of the velocity field due to a particular vortex would be as the inverse square of the distance.

The similarity between the two systems is one of the reasons why 2DVD and two-dimensional turbulence have received a great deal of attention in the literature, even though two-dimensionality is almost impossible to achieve in a homogeneous fluid, except under conditions where there is a very strong geometric constraint, e.g. in soap films [20]. There have now been many analytical and numerical studies of quasi-geostrophic flows, e.g. simple point vortex configurations [100], coherent vortex evolution including effects of background shear and strain and interactions between vortices [65], [98], [27], and in turbulent spin-down experiments [63], [64], [28].

It seems clear that almost all phenomena encountered in 2DVD have analogues in the QG system. Nonetheless there are differences between the two systems and these may be particularly important when one is considering specific applications to the atmosphere. The extra space dimension in the QG system allows some new phenomena. For example, interactions between coherent vortices at different levels but similar horizontal positions are possible in the QG system but have no analogue in 2DVD.

One important question that is of current research interest concerns spatial isotropy. From (2) one can infer a useful rough guide to length scales for quasigeostrophic flows, that the ratio of horizontal to vertical scales is about N/f_0, often referred to as Prandtl's ratio. However that does not mean that there is necessarily exact three-dimensional isotropy if the z coordinate is suitably scaled. For one thing, it is only the horizontal components of the gradient of the streamfunction that determine the velocity field. The recent simulations of [64] suggest that the manifestation of breakdown of isotropy in a spin-down experiment may be the formation of tall vortex structures. However, detailed understanding of the mechanisms that lead to the formation of these structures is still lacking. (And rather contradictory results have been obtained by [28].

The quasi-geostrophic system is no longer regarded as suitable for detailed numerical calculation of atmospheric and oceanic flows, e.g. in numerical weather prediction, even for flows that are dominated by 'slow' motion. However, it remains an extremely useful conceptual model on which to base understanding of 'slow' atmospheric and oceanic flows and with which to interpret observed flows and realistic numerical simulations, in particular because it includes an explicit and easily understood invertibility principle.

In what follows we shall concentrate on transport, stirring and mixing of passive tracers. But it should be plausible from what has been said already that considering the transport of PV itself can give important insights into the dynamics. See e.g., [47], [58], [84], the latter focussing on the oceanic case.

2.5. Implications for the transport and mixing problem

The close relation between stratified, rotating flows and 2DVD has important implications for how stirring and mixing occurs. One might say that we have two extreme paradigms for stirring and mixing. One is chaotic advection, where flow that varies smoothly in space and time, and in particular has a well-defined length scale, leads to small-scale structure in advected tracer fields. Since the advection is dominated by the large scales, the eddy turnover time remains the same at all scales. The other extreme is that of three-dimensional turbulence where eddy turnover time decreases as scale decreases. It follows that, at least above the Kolmogorov scale where viscous dissipation becomes important, the eddies of a given length scale make a substantial contribution to stirring and mixing at that scale.

2DVD, and, by implication, rotating stratified flow in the atmosphere and in the ocean appear to be closer to the first paradigm, where large-scale advection dominates. Exactly why this is the case is a subtle issue. For example, as noted by Pierrehumbert et al. (1995) [80], spectral arguments for quasigeostrophic dynamics and 2DVD suggest that all scales of the flow contribute equally to local strain rates. Pierrehumbert et al. [80] gain some insight into why this is not the case in practice by constructing analogues to 2DVD systems in which there is a different relation between the materially conserved quantity and the streamfunction. One interesting example that can also be motivated by certain atmospheric flows [50] [43] is where the materially conserved quantity corresponds to a first spatial derivative, rather than a second spatial derivative of the streamfunction. It turns out that this is relevant to quasi-geostrophic flows in which there is constant interior QGPV, but where there are lower (or upper) boundary density gradients and it is the boundary density anomalies that control the flow, through (3) with $w = 0$. In this case when a lower boundary density anomaly reduces in scale the induced vorticity increases. Such flows are characterised by the formation of filamentary structures in the boundary temparature field that frequently roll up into small vortices. Indeed it is useful to constrast the canonical example of the roll-up of a shear layer in the two cases. In conventional 2DVD the shear layer (equivalent to a filament of anomalous vorticity) is unstable and rolls up into vortices that, in the absence of dissipation, are connected by thin filaments of anomalous vorticity. These filaments do not themselves roll up into vortices. The reason appears to be that the filaments are being stretched out in the strain field due to the vortices, which has the effect of inhibiting vortex roll-up [29]. In surface temperature dynamics, on the other hand, the filaments connecting the primary vortices themselves roll up into vortices. The reason is that as the surface temperature filaments are stretched out the vorticity that they induce increases and this means that the strain field due to the primary vortices is ineffective in inhibiting the roll-up of such filaments. Nice examples contrasting the two cases are shown by Held et al. (1996) [43].

It therefore appears that the phenomenon of stabilization of vortex filaments by the strain field due to the large-scale flow means that in practice small-scale vorticity features are usually, although not always, passive and it is large scales that dominate the velocity field. The chaotic-advection paradigm for transport, stirring and mixing is therefore potentially relevant to 2DVD and to quasigeostrophic dynamics.

2.6. Rossby waves and instabilities

The close analogy between 2DVD and quasigeostrophic flows implies the possibility that some atmospheric and oceanic flows will have the character of interactions between small numbers of coherent vortices. In such cases the sort of chaotic advection studies pioneered by Aref (1984) [5] might be relevant. However, more often atmospheric flows are dominated by strong background shear flow in the longitudinal direction. Disturbances of such flow away from longitudinal symmetry, either by waves due to external forcing or by the spontaneous growth of disturbances through hydrodynamic instability, are crucial for effective transport, stirring and mixing.

The relevant wave propagation mechanism for these flows is associated with gradients of vorticity or potential vorticity. When fluid parcels are moved in a direction perpendicular to such gradients, the vorticity field is changed and a pattern of circulations is set up that tends to restore the fluid parcels to their original position. This mechanism is often referred to as the 'Rossby restoration mechanism' and gives rise to the propagation of 'Rossby waves'. The mechanism is analogous to, but different in important details from, that which operates in a stably stratified fluid and gives rises to the propagation of internal gravity waves. Figure 3 shows in a schematic way how the mechanism works in a simple case. Rossby waves can propagate in the horizontal and in the vertical and therefore provide an important mechanism for communicating the effects of forcing to parts of the flow that are far from the forcing location. For example, it has been suggested that the effects of tropical heating are communicated to midlatitudes through Rossby wave propagation and thereby affect the month-to-month variability of mid-latitude weather systems. Similarly, Rossby waves that are generated in the troposphere, e.g. by geographical variations in topography and heating, propagate up into the stratosphere, where they have important implications for transport, stirring and mixing.

The alternative idea, that flow asymmetries arise through instability, is most rele-

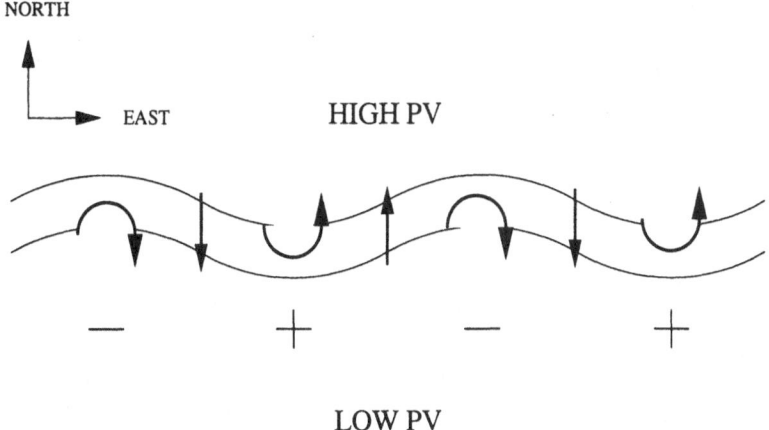

NORTH

EAST   HIGH PV

LOW PV

Fig. 3. — The propagation mechanism for Rossby waves: In the background flow there is a north-south gradient of potential vorticity, so that potential vorticity contours generally run east-west. Here a wave pattern of north-south displacements has been applied so that the potential vorticity contours, shown as wavy lines, are disturbed from their background configuration. This implies a pattern of relative vorticity anomalies of alternating positive and negative sign (shown by the +s and −s). This relative vorticity field has a corresponding velocity field shown by the bold arrows. Note that the positive northward velocities are displaced one quarter of a wavelength to the west of the positive northward displacements (and correspondingly for negative velocities and negative displacements). This implies that the pattern of displacements will propagate to the west, as a Rossby wave.

vant in the troposphere. The longitudinally symmetric flow that might be expected to arise through the action of differential heating may be argued to be unstable, through a mechanism known as baroclinic instability, so that what is achieved in practice is a flow with strong longitudinal asymmetries that can act to transport tracers in the latitudinal direction. In the stratosphere baroclinic instability is much less effective and the main mechanism causing flow asymmetry is the upward propagation of large-scale Rossby waves mentioned earlier. Both baroclinic instability, and another relevant hydrodynamic instability, barotropic instability, which is more closely analogous to the shear instability familiar from 2D vortex dynamics, may be understood in terms of Rossby wave dynamics. Both instabilities involve the disturbance of the PV field by the flow and the resulting change in the flow field due to the changes in PV. Again James (1994) [49] and Andrews et al. (1985) [2] provide useful background material on Rossby wave dynamics in the atmosphere and [49] discusses tropospheric baroclinic instability More details of the physical mechanisms for Rossby waves and for barotropic and baroclinic instabilities are provided by Hoskins et al. (1985) [48].

Of course, longitudinal asymmetry of the flow is not enough, by itself, to ensure effective transport. The latitudinal particle excursions associated with an asymmetric flow may be entirely reversible and have no permanent effect on tracer fields. There has been considerable emphasis on this fact in the literature on stratospheric transport, partly for the historical reasons that the simplest mathematical models often assumed that disturbances to the flow were steady linear waves and that attention was often focussed on deriving parametrizations of transport for use in two-dimensional (height-latitude) models. When disturbances have non-trivial time dependence they are usually effective in transporting tracers and we shall ignore this issue from now on. A careful discussion is given in [2], Chapter 9.

3.   Quasi-two-dimensional transport

3.1.   Introduction

Restricting interest to advection along isentropic surfaces means that the transport problem is two-dimensional. This is a considerable simplification and, of course, allows comparison with studies of chaotic advection based on two-dimensional velocity fields. In this section we review the structure of two-dimensional isentropic transport in the stratosphere that has been revealed over the decade or so by new tracer observations. We also review some of the techniques using wind fields constructed from observations that are now used to gain further insight into transport. We then consider how this picture of transport is consistent with that predicted by chaotic advection studies and how ideas from chaotic advection can be used to give quantitative insight into observed flows.

3.2.   Transport by breaking Rossby waves – the surf zone

One approach to recognising the spatial structure of transport, stirring, and mixing along isentropic surfaces is to consider the distribution of PV, which, as noted earlier, is itself a tracer. PV fields calculated for large-scale atmospheric flows typically show two very different sorts of behaviour. The difference is particularly clear in the winter stratosphere, where the flow is dominated by a strong westerly (eastward) vortex around the high-latitude 'polar night'. This vortex is disturbed by Rossby waves that are generated in the troposphere and propagate up into the stratosphere.

In one part of the flow, generally at higher latitudes and corresponding to the main vortex, the PV contours are displaced from latitude circles and are distorted, but remain quasi-circular. This behaviour is that which is implicit in linear wave theory and can be regarded as a manifestation of the upward Rossby-wave propagation. At lower latitudes, on the other hand, PV contours are stretched out by the flow and strongly deformed away from a quasi-circular shape. This behaviour is outside the scope of linear wave theory and it has been argued by McIntyre and Palmer (1983, 1984) [60] [61] that it is a manifestation of the breaking of Rossby waves. Indeed McIntyre and Palmer named the region outside the main vortex – where PV contours are strongly stretched and deformed – the 'stratospheric surf zone', in order to emphasise the analogy with surface gravity waves breaking on an ocean beach. It is clear that there will be effective stirring and mixing of chemical tracers, as well as PV, in the surf zone.

The flow studied by McIntyre and Palmer was in the middle stratosphere, at around 30km. But we now realise that that the stirring properties of the flow are highly inhomogeneous in latitude at almost all levels in the stratosphere, with the location of the regions of strongest and weakest stirring varying substantially with height.

The distinctive surf-zone, main vortex structure is illustrated in Figure 4, which shows particles that are advected by the flow in a high-resolution, single-layer, numerical simulation of the type pioneered by Juckes and McIntyre (1987) [51], [McIntyre and Pinhey, personal communication; see also [62]]. In such simulations the disturbances are forced in-situ by a large-scale vorticity forcing, but the effect is rather similar to the arrival in a given layer between two isentropic surfaces, say in the middle stratosphere, of Rossby waves that have propagated up from below. High-resolution three-dimensional simulations (e.g. [55], [38], have confirmed this similarity.

The particles shown in Figure 4 are shaded to mark their initial latitude, with darker shades corresponding to higher latitudes. The flow is on the surface of a sphere and is viewed from above the North pole. The initial flow takes the form of a strong anti-clockwise vortex at high latitudes, with weaker clockwise flow at low latitudes. This longitudinally symmetric flow is subsequently disturbed by a large-scale forcing, centred at midlatitudes. The particles are stirred up by the flow, but the pictures clearly show the strong inhomogeneity of the stirring. The largest stirring region, the 'surf zone', is

at mid and low latitudes. The high-latitude vortex persists as a coherent material entity. There is evidence of stirring within the vortex, but there appears to be a strong barrier to transport associated with the vortex edge. In this particular simulation the barrier is perfect and no particles leak from the interior of the vortex to the surf zone or vice versa.

3.3. Tracer edges

We have seen how in the flow depicted in Figure 4 the strong stirring is localised to the surf zone. Simple consideration of the effect of localised stirring on a tracer field with an initially smooth distribution shows that, across the edges of the stirring region, a sharp jump in the value of the tracer will tend to form. Figure 5 gives a schematic illustration of how this occurs. If the pattern of stirring in the real atmosphere is similarly localised, then we might expect to see evidence for sharp jumps in the values of tracers, both chemical tracers and PV at the vortex edge, as the poleward boundary of the stirring region.

Figure 6 [which is a reconstructed version of Plate 5 in Tuck (1989)] shows tracer data from the AAOE (Antarctic Airborne Ozone Expedition), measured on the ER-2 aircraft. The resolution of the data is limited by the response time of the various instruments and the aircraft speed, but is certainly better than 10km or so. The tracers shown are N_2O , which generally decreases poleward in the stratosphere, NO_y , which generally increases poleward, and ozone, which also increases poleward, at least in early and midwinter and in midlatitudes. (By NO_y we mean a group of reactive nitrogen compounds. The total of the members of this group is relatively long-lived in the lower stratosphere, but the individual members are not, since there is rapid photochemical conversion from one species to another within the group.)

There is a clear jump in the value of these tracers at about 59 deg S. Unfortunately it is not possible to measure PV directly, so there are no corresponding measurements to verify that there is a jump in PV colocated with that in the chemical tracers. It is possible to construct PV fields from large-scale meteorological analyses, but such field will have very coarse resolution compared to the aircraft measurements. However, simple mathematical models of a dynamical structure incorporating a PV edge suggest that there will be a jump in the wind shear across the PV edge. The

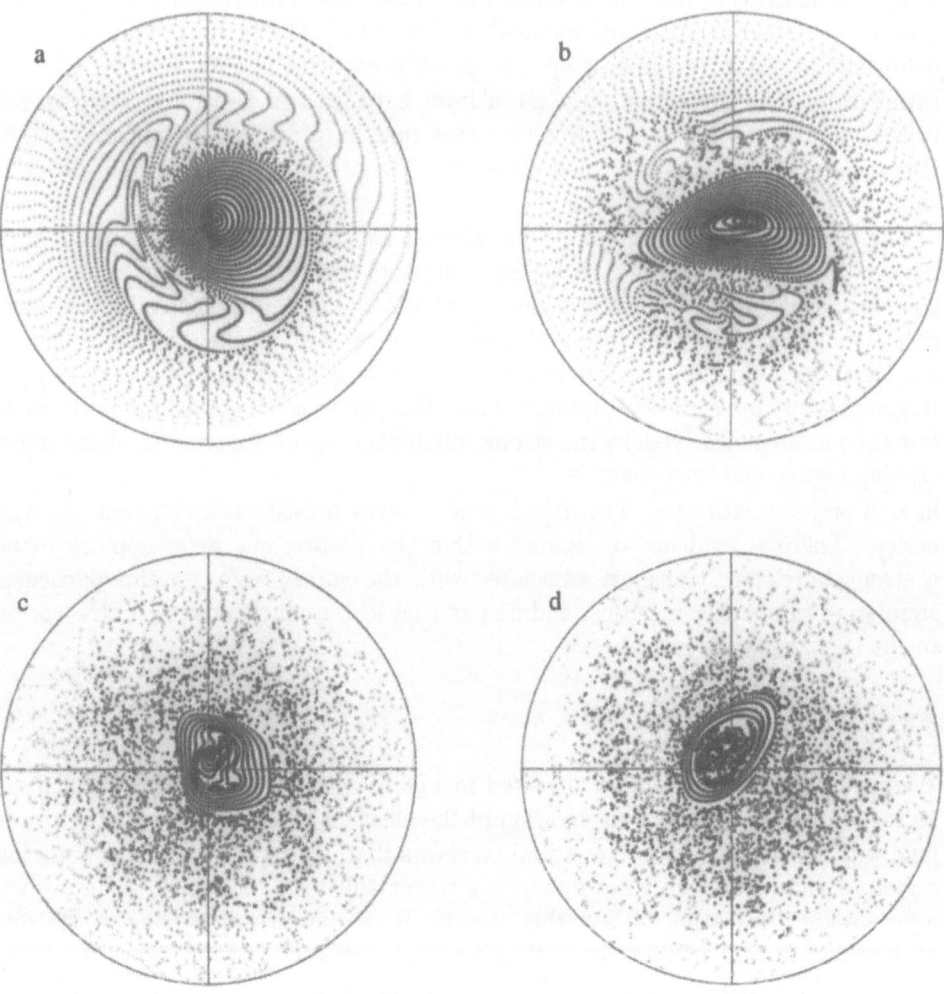

Fig. 4. — Tracer particles advected in a dynamically consistent numerical simulation of single-layer flow on a sphere. The flow is viewed from above the North pole. The particles are shaded according to their initial latitude, with darker shading corresponding to higher latitude. The flow consists of a strong cyclonic (anti-clockwise) vortex at high latitudes, disturbed by a large-scale forcing centred at mid latitudes. The amplitude of the forcing is smoothly increased from zero at the beginning of the simulation and is smoothly decreased and then increased again, with a period of 20 days, as the simulation progresses. Particles are shown at (a) 10 days, (b) 20 days, (c) 40 days, and (d) 80 days. It may be seen that there is a systematic adjustment of the particles to a statistically quasi-steady state where they are apparently well-mixed outside the vortex and inside the vortex, but there is no exchange across the vortex boundary.

aircraft measurements indeed show that there the shear changes sign from north of 58 deg S to south of 60 deg S. This is at least an indication that there is a PV edge at this location.

There are a number of other features seen in Figure 6 that deserve explanation.

EFFECT OF LOCALISED STIRRING

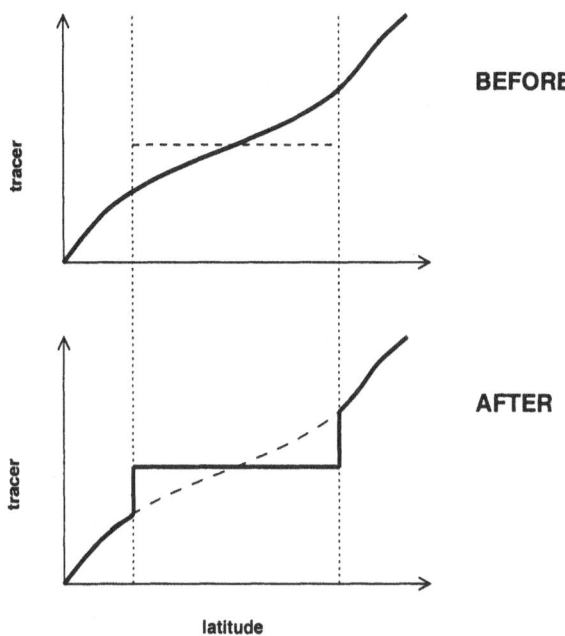

Fig. 5. — Effect of localized stirring on a tracer distribution. The variation with latitude of initial tracer distribution is shown by the solid curve. The tracer is assumed subsequently to be stirred (and perfectly mixed) over a limited range of latitudes, bounded by the finely dashed lines. The effect of the localised stirring and the resultant mixing is not only to make the tracer constant in this region, but also to produce sharp tracer gradients at the edges of the region.

Firstly there is another striking jump in the values of some tracers, particularly NO_y and O_3 , at around 65 deg S. Note that here there is apparently no jump in N_2O , nor is there any striking change in the wind shear. The accepted explanation for this feature is that it marks the edge of the chemically perturbed region, defined not by any change in transport regime, but by whether air parcels have encountered polar stratospheric clouds. The latter form when temperatures are sufficiently low and there is no reason why the boundary of this region should coincide with the PV edge. The polar stratospheric clouds allow chemical reactions that decrease NO_y and increase ClO , but do not affect N_2O . The ClO can, under appropriate conditions, catalyse the destruction of ozone. There is another interesting feature at around 62 deg S. This is interpreted by Tuck (1989) [95] as being associated with a filament of low ozone air originally from inside the chemically perturbed region, that is partially mixed with its surroundings so that, in particular, the high NO_y introduced by mixing has destroyed all the ClO . This is a reminder that in the stratosphere (and elsewhere in the atmosphere) stirring and mixing may have important dynamical consequences.

The schematic picture of the effect of localised stirring shown in Figure 5 suggests that, if the strong stirring associated with Rossby wave breaking is confined to midlatitudes, there should be a sharp edge in tracers not only on the poleward side,

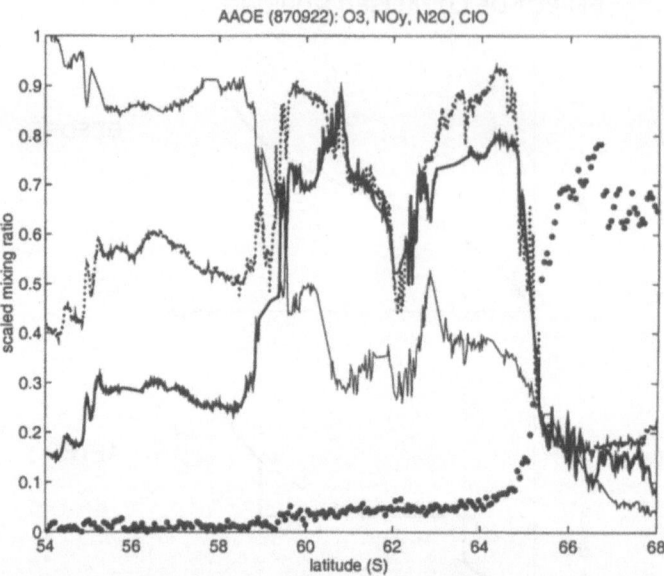

Fig. 6. — [After Tuck (1989), Plate 5]. Variation of mixing ratios of ozone (O_3) (dotted curve), nitrous oxide (N_2O) (fine solid curve), 'odd nitrogen' (NO_y) (heavy solid curve) and chlorine monoxide (ClO) (heavy dots), measured along flight track of ER-2 aircraft on 22 September 1987 during AAOE. The mixing ratios have had a constant value subtracted and then multiplied by a suitable factor to allow all to be shown together. What is plotted is, for O_3, 0.0005 times (mixing ratio (ppbv) - 300), for N_2O, 0.008 times (mixing ratio (ppbv) - 100) for NO_y, 0.1 times (mixing ratio (ppbv) - 2), and for ClO, mixing ratio (ppbv). See text for further details and explanation.

but also on the equatorward side. There is indeed evidence for strong latitudinal tracer gradients between the tropics and subtropics in the stratosphere from aerosol observations e.g. [94] and from observations of chemical tracers e.g. [69]. Figure 7 has been constructed following Figure 5 in the latter paper and shows a composite of lower stratospheric aircraft measurements of NO_y and ozone (O_3) . Both NO_y and O_3 may be regarded as conserved tracers in the lower stratosphere. The fact that the aircraft measurements are from a number of different heights, and that the concentrations of both NO_y and O_3 have strong vertical gradients, means that latitudinal contrasts are difficult to distinguish when the concentrations are plotted individually. However, the ratio of the two concentrations (which may itself be regarded as the concentration of a conserved tracer) has a much weaker vertical gradient and when, as in Figure 7, it is plotted against latitude it shows clearly that there are strong latitudinal gradients centred at about 18 deg S and 15 deg N.

3.4.   Use of observed velocity fields for transport studies

As noted earlier, one of the important developments that has complemented the improved observations of chemical tracers has been the construction of velocity datasets for the stratosphere, using assimilation techniques. These velocity datasets have

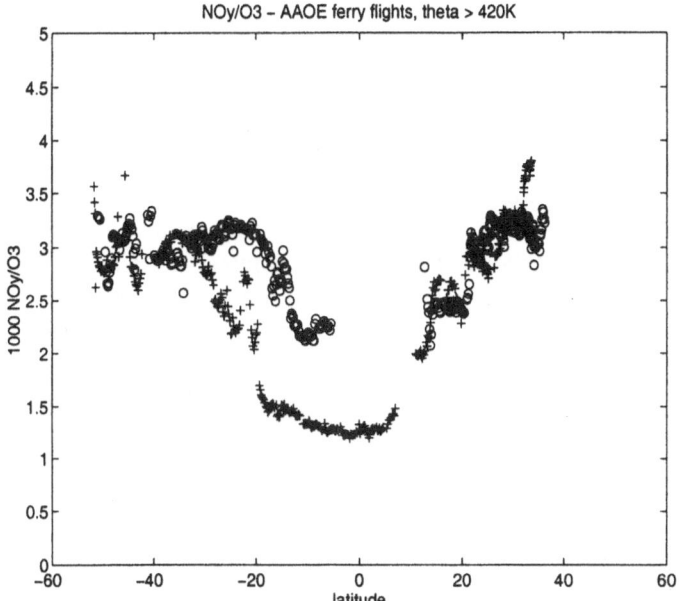

Fig. 7. — [After [69], Figure 5.] Ratio of mixing ratios of NO_y and O_3 versus latitude for AAOE ferry flights from NASA Ames to Punta Arenas. Points marked o are for the southward flights (12, 14, 15 August 1987) and points marked + for the northward flights (29 September, 1 and 3 October 1987). Only values corresponding to potential temperature greater than 420K have been plotted.

been exploited in a number of different techniques for tracer studies.

One technique is simply to use the velocity fields to follow trajectories of air parcels. Trajectory studies have a long history in dynamical meteorology but the improved velocity datasets together with increases in computing power have allowed a very large number of trajectories to be followed simultaneously. The so-called 'domain filling trajectory' approach has been used by Fisher et al. (1993) [35] and Sutton (1995) [91] to follow the three-dimensional trajectories of air parcels on seasonal time scales, but has also been used by Bowman (1993) [12] to follow particle trajectories restricted to isentropic surfaces, to gain insight into transport on shorter timescales.

A second technique is to use the velocity fields in a tracer model, in which the tracer equations are integrated using finite-difference, finite-element or spectral methods. Most such studies to date have restricted the problem to two dimensions by considering the evolution of the tracer on an isentropic surface. For example, Chen et al. (1994) [19] have used this approach to study transport into and out of the tropics on different isentropic surfaces.

The third technique in common use is to use the velocity fields to follow the evolution of a material contour. The contour is defined by a finite set of particles, but the algorithm used allows redistribution of the particles around the contour, or the introduction of new particles on the contour, as the integration proceeds, so as best to represent the contour. Once again the problem is usually restricted to two dimensions by using winds on a particular isentropic surface. This approach is generally referred to as 'contour advection' or in its more sophisticated versions 'contour advection with surgery' (hereafter CAS) [72], [102]. The 'surgery' refers to the fact that as the contour becomes more and more contorted, as it inevitably does in a flow that is stirring effectively, it may be necessary to remove parts of the contour, e.g. associated with thin filaments, in order to keep the number of particles needed to define the contour to within the limits necessary for practical computation. Algorithms for effective representation of contours and effective application of surgery were first developed in the context of solving the two-dimensional vorticity equation for flows in which the vorticity field is piecewise constant (and therefore defined entirely by the contours dividing vorticity of different values) (see e.g.[26]).

All of these techniques have limitations. They are usually applied to an initial-value problem, and the picture of the tracer field obtained for a particular date and time will depend both on the initial condition and the time that has elapsed since the initial condition was applied. For example, CAS calculations are typically initialised by placing contours to coincide with contours of a tracer for which large-scale observations are available. This is usually PV, calculated using wind and temperature from meteorological datasets, but chemical tracers such as ozone, for which there are satellite·data, have also been used (A. Mariotti, personal communication 1997). The large-scale observations typically have useful resolution only at scales of a few hundred kilometers or more. Integrating forward in time from the initial condition, the contour is deformed by the flow and typically develops filamentary structure at increasingly small scales. At the end of the integration, typically ten to twenty days, the resulting contour is taken, in some sense, to give a fine-resolution view of a tracer field for which only coarse-resolution data was available initially. The precise relation between the predicted contour and contours of a real tracer is a subtle one. Any small-scale filaments that are represented are only those that have been produced from large-scale tracer features over the integration period. Small-scale filaments that persisted for the period of the integration would not be captured (assuming that they were below the useful resolution of the observations used for the initialisation). Small errors in the observed velocity fields will lead to corresponding errors in the positions of filaments that will often be as large, or larger than, the width of the filaments themselves. It is therefore clear that any simple quantitative measure of the correlation between CAS predictions and high-resolution tracer measurements is unlikely to show strong agreement.

Nonetheless CAS, and the other techniques mentioned above, have significantly enhanced useful interpretation of tracer observations. For example, CAS has been used to follow the evolution of filaments of vortex air that have peeled off into the surf zone and has shown convincing qualitative agreement with filamentary structures seen in aircraft data (see, for example, Figures 1 and 2 of [103]. Such studies have been important in supporting the idea that the structures seen in the aircraft data can be explained as due to stirring by the large-scale flow on isentropic surfaces. It seems that the success of CAS here is due to the fact that the strongest contrasts in tracer are associated with filaments of vortex air that are drawn away into the surf zone by relatively infrequent erosion events. These are exactly the sorts of structures, arising over a period of a few days from the contrast in tracer across the vortex edge (that is, of course, captured in the large-scale observations), that CAS is best able to capture. It is also the case that techiques such as CAS have given a qualitative sense of the tracer structures that would be observed if the high resolution aircraft measurements gave global coverage, or if satellite measurements were at much higher spatial resolutions. But once these qualitative insights have been gained, one might legitimately look for more quantitative insights from such techniques.

A more quantitative application of CAS has been to quantify transport, e.g. across the boundary of the vortex. Waugh et al. (1994) [103] have performed calculations in which a contour is initialised to coincide with a PV contour in the neighbourhood of the vortex edge. The subsequent advection leads to the formation of filaments on this contour, which are stretched out into the surf zone. The area associated with such filaments is interpreted as a measure of the transport out of the vortex into the surf zone and is quantified by applying the surgery algorithm with a rather coarse cut-off scale. This leads to the removal of filaments and a reduction in the area enclosed by the contour. A similar approach has been used by Waugh (1996) [101] to quantify transport across the subtropical barrier.

3.5. Relevance of chaotic advection

It is perhaps appropriate at this stage to note the predictions of theories of chaotic advection and their possible relevance to atmospheric flows. The first requirement for the usefulness of the chaotic advection paradigm is that small-scale structure arises in tracer fields through the action of velocity fields that have simple large-scale spatial structure. It has already been argued from basic dynamical considerations in §2 that this is the case in the atmosphere, at least on large enough scales. There is also useful evidence from CAS calcuations that, at least in the stratosphere, it is indeed the large-scale part of the velocity field that dominates. For example, Waugh and Plumb (1994) [102] compare results of contour advection calculations with velocity fields at different spatial resolutions and show that there is rather good agreement between results with velocity fields that are resolved at say, 3 deg latitude × 3.6 deg longitude and 10 deg latitude × 12 deg longitude.

Much of the work on chaotic advection has concerned two-dimensional, non-divergent flows, i.e. Hamiltonian dynamical systems [5], and this may be argued to be relevant to the real atmosphere on the following basis. The potential relevance of two-dimensional flows has already been justified by arguing that restriction to isentropic surfaces is relevant, perhaps for twenty days or so in the lower stratosphere, less in the middle and upper stratosphere. Whilst the horizontal flow on isentropic surfaces is not exactly non-divergent, it is non-divergent to some reasonable approximation. (For example, for flows in which the Rossby number Ro is small, the divergence is $O(\text{Ro})$ smaller than a naive estimate of velocity gradients.) There have, of course, been a great many studies of chaotic behaviour in Hamiltonian dynamical systems, and much of the structure found, chaotic regions, in which there is mixing, bounded by regions of invariant tori, i.e. surfaces made up of trajectories, which act as transport barriers, is generic. Two studies that are particularly motivated by atmospheric and oceanic flows are those of [77] and [23]. Given the structure of the wintertime stratospheric flow, as revealed by numerical simulation and observations, it is not surprising that it has been suggested that the surf-zone should be regarded as a chaotic mixing region, with the vortex edge as a set of invariant tori (e.g. [76], [75], [78]).

Much of the study of two-dimensional Hamiltonian systems has been based on

time-periodic flows, since this allows, through the use of Poincaré sections, application of the theory of area-preserving maps. As noted by Wiggins (1988), flows that
are not time-periodic are ubiquitous in practice and one of the continuing challenges
in this subject is to develop methods for analysing transport in non-periodic flows
(such as those in the atmosphere). One possible approach to quantifying atmospheric
transport, that does not depend on time-peridodicity is to use finite-time Liapunov
exponents as a diagnostic of stretching [76]. The finite-time Liapunov exponent $\lambda(t)$
is defined for a fluid particle by assigning an infinitesimal fluid line element $\delta\mathbf{x}$ to
that particle and following the stretching of that line element. Then

$$\lambda(t) = \frac{1}{t}\log\{\frac{|\delta\mathbf{x}(t)|}{|\delta\mathbf{x}(0)|}\}. \tag{6}$$

Note that $\lambda(t)$ is not independent of the initial orientation of the line element, but in
a stretching flow might be expected to become independent of the initial orientation
after a few stretching time scales. For an ensemble of fluid particles in a mixing region
of a simple chaotic advection flow, the distribution of the $\lambda(t)$ will, with increasing
time, narrow about the value of the Liapunov exponent (the latter defined in the
usual way for chaotic dynamical systems). However, at any finite time there will be
a spread of values of the $\lambda(t)$.

Pierrehumbert's idea is that transport barriers (or tori) will be manifested as
regions where the $\lambda(t)$ are anomalously low. In [76] examples are shown where low
values of the $\lambda(t)$ for a simple chaotic map reveal the presence of invariant tori
embedded within a mixing region. The advantage over Poincaré sections is that
only a relatively short time window is needed to identify such features (and, in
particular, there is scope for application to non-periodic flows). Diagnostics based
on finite-time Liapunov exponents have been applied to the atmosphere using wind
fields from large-scale observations [76] and also from a general circulation model
[78] to advect particles and stretch line elements.

Another diagnostic, which is closely related to finite-time Liapunov exponents,
calculates the change in length of a finite material contour rather than the changes
in lengths of an ensemble of material line elements. The evolution of the contour
length may be followed simply by tracking large numbers of particles or using the
CAS technique. The stretching of the contour is clearly equivalent to a weight
average of finite-time Liapunov exponents over a set of particles distributed along
the contour. If a set of contours is considered, anomalously low stretching of a subset
may indicate that the subset is embedded in a transport barrier.

In performing such calculations the clearest results are obtained if the contours
are initialised in a configuration that is aligned with the flow. The reason for this
is that, if a material contour is placed in a flow, its length initially increases rapidly
with time whether or not the flow is 'stretching' in the sense of a chaotic mixing
flow. Such stretching is just that which would be seen if the contour were placed in a
simple linear steady shear. During this initial adjustment material lines tend to align
with the flow and eventually, e.g. for a material line in a chaotic mixing region, the
initial linear growth is swamped by the exponential growth associated with chaotic
stretching. The practical approach that has often been taken to minimize the effects
of this initial linear growth is to choose the initial configuration of the contours
with PV contours (determined from observed data or from a numerical model as
appropriate). The idea is that, since PV is approximately materially conserved,

the potential vorticity contours are aligned with the flow in the required manner. A chemical tracer could be used instead of PV. Initialisation along instantaneous streamlines is another possibility.

Such contour-lengthening calculations were first performed by Bowman (1993) [13] and Norton (1994) [72], both using single-layer dynamical models of the stratospheric circulation, and Pierce and Fairlie (1993) [75] using velocity fields from a coarse resolution three-dimensional model. All these investigations showed clear differences in stretching rates between contours embedded in the vortex and those in the 'surf-zone'. Important applications to observed atmospheric flows have made by Chen (1995) [18], for example, who used to contour stretching rates to identify in an objective manner the presence of a transport barrier at the edge of the Antarctic polar vortex and Waugh et al. (1994) [103] who showed that large stretching rates sometimes occur inside the Arctic polar vortex, indicating that there may be effective mixing within the vortex, even if there is a strong transport barrier at its edge.

Another important transport barrier is believed to exist on isentropic surfaces that intersect the tropopause, i.e. the notional, quasi-material boundary separating stratosphere and troposphere. On such isentropic surfaces the tropopause is marked by a sharp jump in PV, just as is the case for the stratospheric vortex edge. Indeed it has been argued that the tropopause may form through a similar mechanism to the vortex edge, i.e. through inhomogeneous eddy stirring [61], [1], [41]. The eddies thought to be important in this case are the 'synoptic' eddies, with scales of a few thousand kilometers that are responsible for day-to-day and week-to-week variations in the weather at midlatitudes. The presence of a transport barrier in this region is required by the strong chemical contrasts between troposphere and stratosphere. If there were no barrier then these strong contrasts could not persist. However it has been remarkably difficult to distinguish such a barrier using the tracer-advection techniques outlined above. Bithell and Gray (1997) [10] report that contour advection calculations do not show a clear minimum in stretching rates at the expected location of the tropopause. Pierrehumbert and Yang (1993) [78] calculated finite-time Liapunov exponents for a large ensemble of particles on a relevant isentropic surface, but no clear minima in stretching rates are seen in the expected locations. One problem may be that both such diagnostics are, by their very nature, Lagrangian. For this reason they may fail when the transport barrier being considered is rather leaky. For example, if a material contour is placed in the location of a partial barrier, but then subsequently portions of that contour are transported rather rapidly into mixing regions, then the stretching rate will be substantial. Furthermore, there is unlikely to be a strong minimum in stretching as the initial position of the contour is varied. Quantifying the effectiveness of the hypothesised transport barrier at the tropopause remains an important problem if we are to understand fully the processes that control chemical distributions in this part of the atmosphere.

4.   Three-dimensional effects

4.1.   Beyond isentropic transport

Focussing on transport, stirring and mixing along isentropic surfaces has proved a productive approach and, as noted above, potentially allows exploitation of the large body of theory concerning chaotic advection by two-dimensional time-dependent flows. However, there are many aspects of the atmosphere that cannot be explained by considering isentropic transport alone. Vertical transport across isentropic surfaces or the presence of vertical structure must be included in models if the important physical mechanisms are to be captured.

Here we consider two important ways in which these extra effects play a crucial role. The first concerns the way that cross-isentrope and along-isentrope transport interact to set up the observed variation of chemical tracers in height, latitude and longitude. The second concerns the importance of the formation of small vertical scales in tracer fields as a route to dissipation and mixing.

4.2.   The structure of transport in the height-latitude plane

We can now put together a simple picture of the structure of the atmospheric transport in height and latitude. This is depicted in the schematic shown in Figure 8. The inhomogeneity of the transport along isentropic surfaces noted in the previous section suggests that the winter stratosphere should, on each isentropic surface, be divided into 'vortex', 'surf-zone' and 'tropical' regions, separated by (imperfect) barriers to transport. There is evidence from e.g. contour advection studies [18] that below about 16km (corresponding roughly to the 400K isentropic surface) the transport barrier associated the polar vortex itself weakens and there is free stirring of air parcels between mid and high latitudes. McIntyre (1995) [59] calls this the 'sub-vortex' region. In the summer hemisphere of the stratosphere there is no analogue of the vortex, surf-zone structure, since the characteristics of Rossby waves are such that they cannot propagate up into the easterly winds that are present there. However, there is some stirring in the lower stratosphere due to the upward penetration of the effect of the tropospheric synoptic-scale eddies. Finally, as noted earlier, in each hemisphere there is a transport barrier associated with the tropopause.

Superimposed on this picture of transport along isentropes there must be cross-isentropic transport, associated with diabatic heating or cooling. Such heating or cooling is in reality longitudinally varying, but it turns out that we can gain a reasonable qualitative sense of transport if we simply consider the longitudinally averaged cross-isentropic transport, corresponding to the longitudinally averaged diabatic heating or cooling. There is not space here to explain why this is the case (see e.g. [106] [Chapter 6], [2] [Chapter 9]). Nor is there space to explain how this 'diabatic circulation' arises, except to note that much of it arises not from latitudinal variations in e.g. solar heating, but from momentum transport associated with dissipating Rossby waves and gravity waves. (See e.g.[2] [Chapter 7], [39], [46].) In the troposphere and in the lower stratosphere the diabatic vertical velocities are generally upward at low latitudes and downward at high latitudes. A number of different quantitative estimates of the diabatic circulation have been made over the past fifteen years or so. For a recent calculation of the diabatic circulation

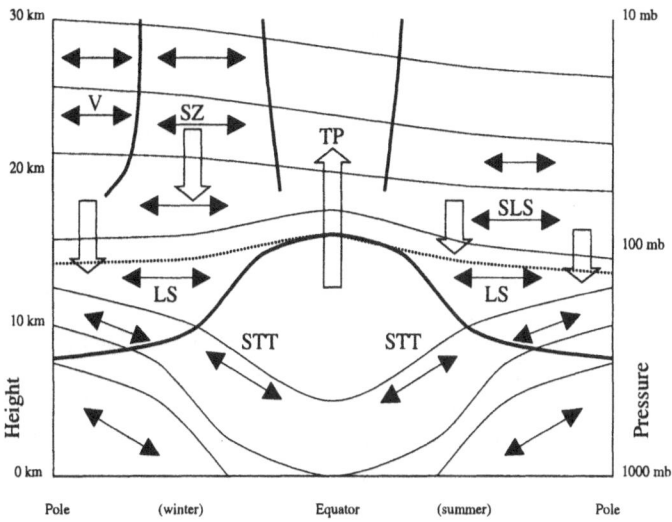

Fig. 8. — Schematic of the dominant pattern, under solstice conditions of large-scale transport in the troposphere and lower stratosphere . The solid quasi-horizontal lines are isentropic surfaces. Stirring along isentropic surfaces is indicated by quasi-horizontal solid arrows. Diabatic transport across isentropic surfaces is indicated by vertical unfilled arrows. The important transport barriers, namely the vortex edge, the two tropical edges and the tropopause, are denoted by thick solid lines. The lower stratosphere may be divided into six regions, the winter polar vortex (V), the surf zone (SZ), the tropical pipe (TP), the summer lower stratosphere (SLS) and, in each hemisphere, the lowermost stratosphere (LS), meaning that region which lies above the tropopause but is accessible from the troposphere along isentropic surfaces. Each region is shown on the schematic picture. Also marked in each hemisphere is the subtropical upper troposphere (STT), that part of the troposphere which is well away from the ground and accessible from the stratosphere along isentropic surfaces. The upper boundary of the lowermost stratosphere is the isentropic surface that just grazes the tropopause in the tropics and is indicated by the dotted quasi-horizontal line.

in the stratosphere [85], in the troposphere [52]. [In the troposphere the upward motion at low latitudes is associated with the well-known Hadley circulation. The corresponding upward motion in the stratosphere is sometimes referred to as part of the 'stratospheric Hadley circulation'. But this is potentially misleading. The stratospheric transport is much weaker and the mechanisms for the two circulations are believed to be very different.]

The combination of upwelling, plus relative isolation from mid-latitudes, in the tropics in the lower stratosphere means that it is now common to talk about a 'tropical pipe' [81]. Chemical tracers with sources in the troposphere, such as N_2O , are taken up through the lower stratosphere in the tropical pipe and gradually leak out into the extratropics. [This is part of the reason why, as noted in §3, N_2O tends to decrease poleward on each isentropic surface. In addition, the higher latitude downwelling tends to bring low values of N_2O down from above, since N_2O has a photochemical sink in the upper stratosphere. The same holds for other tracers with similar distributions of sources and sinks.]

Dividing up the atmosphere in the manner shown in Figure 8, and ideas such as the 'tropical pipe' might seem simplistic to an outsider, but they are providing a very useful conceptual framework on which to base studies of transport and within which to interpret chemical measurements. For example, a number of recent papers have used different combinations of observational data and simple transport models to estimate the timescales for air parcels to leak into or out of the tropics (e.g. [67], [68], [99], [42].

4.3.   Vertical dispersion

Many of the earliest models that included transport in an attempt to explain chemical observations were one-dimensional in space, representing vertical but not horizontal variation. In part this was because many of the early chemical observations were of the form of vertical profiles and therefore contained information only about vertical variation. The transport parametrization in such models was diffusive. Of course, in the light of the picture of the circulation described above, the shortcomings of such a transport parametrization are clear. The next generation of chemical models typically resolved both height and latitude variations and included a more physically plausible representation of transport. Again some simplification was sought and it was argued that along-isentrope transport is dominated by the effects of eddies (i.e. circulation features with a strong longitudinal variation) which, in a longitudinally averaged, view lead to dispersion along isentropes and might be represented by an eddy diffusion. The cross-isentrope transport, on the other hand, was argued to be dominated by advection by the longitudinally averaged circulation and might therefore be represented by an advective transport. (See, e.g., [96], [53], [106], [83]). This is close to the picture outlined above, except that the strong spatial inhomogeneity of the stirring along isentropic surfaces observed in reality suggests that approximating it by eddy diffusion may be seriously flawed.

The two ingredients of latitudinal diffusion (on a latitudinally confined domain) and vertical advection by a latitudinally varying velocity field correspond exactly to those in the classical fluid-dynamical problem of shear dispersion in a pipe or channel. As shown by G.I. Taylor, the net effect is diffusive dispersion, now referred to as Taylor dispersion, along the pipe. It is similarly the case that, under certain conditions, the vertical transport of globally averaged quantities can be described by a downgradient diffusion [45], [54], [82]. This result gives some basis for interpreting the results from the early one-dimensional chemical models. The vertical diffusion in those models may be argued to capture the effect of the large-scale circulation, including both the eddies and the longitudinal mean circulation, on global average values of tracers. Plumb [81] refers to these as 'global-diffuser' models. But, of course, it is not at all clear whether such models can be usefully used to interpret chemical profiles taken at individual horizontal locations.

The exact mathematical correspondence between the Taylor problem and that of vertical advection plus latitudinal diffusion in a longitudinally averaged chemical transport model makes it inevitable that in the corresponding globally averaged problem the vertical transport is diffusive. What is in some ways more interesting is to consider whether the diffusive vertical transport persists in more general circumstances, for example, when the latitudinal transport is not diffusive, that might be relevant to the real atmosphere rather than just to idealised models. In the classical pipe problem the cross-pipe diffusion means, in effect, that molecules of tracer follow a random walk in the cross-pipe direction. As they do so, they sample what is, in effect, a random time series of along-pipe velocity and therefore also follow a random walk in the along-pipe direction. On time scales longer than the decorrelation time for the along-pipe velocity experienced by each molecule, equivalent to the cross-pipe diffusion time, this random walk in the along-pipe direction is diffusive.

What is required for diffusive vertical dispersion of an ensemble of particles is thus that each particle experiences a random time series of vertical velocity and that the statistical properties of the time series are identical over all particles. It is not necessarily required that horizontal transport be diffusive. Recent calculations by Sparling et al. (1997) [89] have considered the dispersion of particles across isentropic surfaces and have shown when such dispersion is likely to be diffusive.

Figure 9 is taken from [89] and shows the final positions in the height-latitude plane (with height being defined by potential temperature) of a large set of particles. The particles were all started on the 500K surface and were advected for a period of 60 days. Horizonal velocities were taken directly from large-scale meteorological datasets. Vertical velocities (relative to isentropic surfaces and hence proportional to diabatic heating rates) were calculated using relatively sophisticated radiation schemes (of the type used in atmospheric general circulation models), taking temperatures from the meteorological datasets. The period covered was 1 January 1993 to 1st March 1993 (i.e. in Northern Hemisphere winter). Since such calculations necessarily depend on the quality of the meteorological dataset and the radiative scheme, results from two independent calculations are displayed, each using different meteorological datasets and radiative schemes.

The first thing that is clear from these pictures is the large-scale latitudinal pattern of upwelling and downwelling. Particles in the tropics have ascended and particles in mid and high latitudes have, on average, descended. However, particularly for the particles in Northern Hemisphere mid-latitudes and also for those in the Southern Hemisphere extratropics, there has been considerable vertical dispersion. This dispersion results because different particles experience different time series of diabatic heating. Sparling et al. (1997) [89] show that this is primarily because of particles moving latitudinally across a strong latitudinal gradient in diabatic heating, though there is also a contribution from longitudinal and temporal variations. The largest latitudinal displacements are in the mid-latitude winter hemisphere, i.e. in the surf zone, and in the summer hemisphere, where as noted earlier, there is stirring in the lower stratosphere. The small vertical dispersion at high latitudes in the winter hemisphere is a manifestation of the trapping of particles, which therefore undergo only limited latitudinal displacement, in the polar vortex.

The time variation of the vertical dispersion within different ensembles of particles, measured by the variance of potential temperature within such ensembles, is shown in Figure 10, also taken from [89]. As is well-known from classicial work on turbulent dispersion, under many conditions the variance of displacement over an ensemble of particles increases quadratically with time at early times (the so-called 'advective' regime) and then at later times increases linearly with time (the so-called 'diffusive' regime). The transition from one regime to another occurs on a time scale roughly equal to the Lagrangian deorrelation time. The results in Figure 10 show that ensembles of particles that stay in the surf zone, or in the extratropical summer hemisphere, display this transition within the 60-day time period of the integration. However, the global ensemble, of all parcels, shows advective dispersion for the

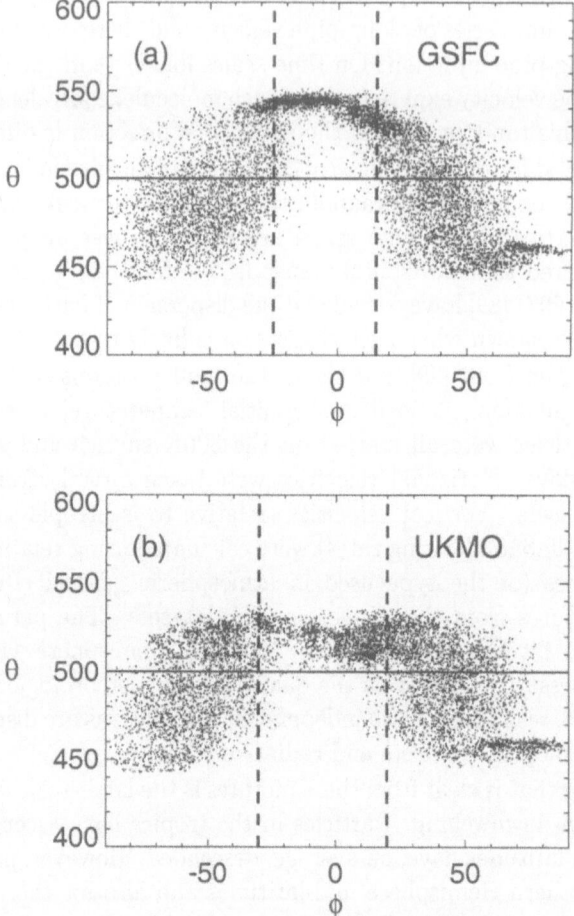

Fig. 9. — Particle positions after 2 months projected onto the θ (potential temperature) - ϕ (latitude) plane. Particles were initialised on the 500K potential surface and then advected by winds calculated from meteorological data for the period 1 Jan 1993 to 28 Feb 1993. (Vertical winds relative to isentropic surfaces, which are proportional to heating rates, were calculated using a radiative code.) Results from two independent calculations based on different datasets, Goddard Space Flight Center (GSFC) and UK Meteorological Office (UKMO), and radiative codes are displayed. The dashed lines show the approximate boundaries of the tropical subensemble of particles. [Figure reproduced from [89].]

whole time period. This is because, as could be inferred from Figure 9, some of the particles remain trapped in the tropics, others remain trapped in the vortex, etc. In the language used earlier, the decorrelation time for the vertical velocity for the global ensemble, i.e. the time taken for particles to 'forget' the vertical velocity with which they started is certainly larger than 60 days. Hence the one-dimensional 'global diffuser' models, described earlier, for the vertical variation of trace species, can be valid only on timescales substantially longer than this. On the other hand, models that distinguish between say the surf-zone and tropics, have the prospect of being valid on shorter time scales, since the decorrelation time scales within these individual regions are less than the global decorrelation time.

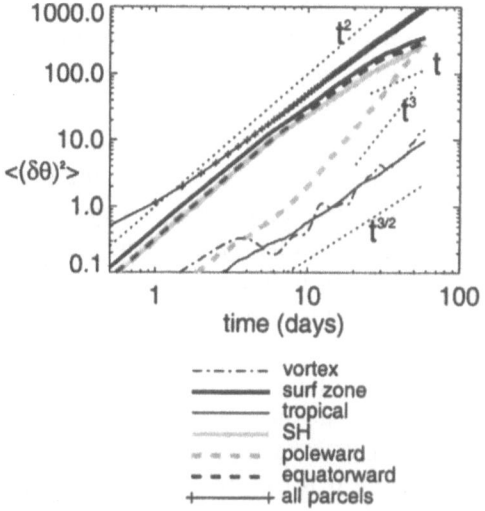

Fig. 10. — Time evolution of the potential temperature variance of different subensembles, according to the GSFC calculation. Various powers of time t are indicated on the graph to allow comparison. [Figure reproduced from [89].

Plumb (1996) has generalised the one-dimensional models to represent both a 'tropical pipe' where chemical species are upwelled from the troposphere and an extratropical region where the transport is assumed to have the same character as in the 'global-diffuser' models. The only difference is that there can be a net vertical mass flux in the extratropics (balanced by an opposing flux in the tropical pipe). This means that the vertical transport of the horizontally averaged chemical concentrations appears as a combinatin of advection and shear-diffusion. This model has the advantage that may predict different chemical distributions in tropics and extratropics, reflecting the strong differences between these regions that are observed. The validity of the diffusive approximation for vertical transport, particularly the lack of dependence of diffusive transport in the horizontal, is further discussed in Thuburn and McIntyre (1997), who derive general conditions under which the approximation is valid.

4.4. Three-dimensional routes to mixing

The second topic of this section concerns routes to mixing in the real atmosphere. We define mixing to mean the physical processes by which two distinct fluid elements, e.g. with distinct chemical characteristics, mix at the molecular level, i.e. essentially become one homogeneous fluid element. This definition is a little different to that which is sometimes used in the dynamical systems context (e.g. see [6]). In particular, under this definition mixing cannot arise purely through advection, but requires molecular diffusion to homogenize tracer concentrations at the molecular level. Of course, the stirring action of advection plays a vital catalytic role in mixing, by stretching and distorting fluid elements so that spatial gradients and hence the effect of diffusion are systematically enhanced.

The techniques, such as contour advection, described in §2 for simulating the evolution of tracer fields on isentropic surfaces have played an important role in atmospheric science by allowing visualisation of the formation, through the stirring processes, of fine-scale horizontal structure in tracer fields. However, in order to appreciate the precise significance of these fine-scale structures for mixing, it is necessary to consider the mixing processes in a little more detail.

Processes that lead to dissipation and mixing are generally quasi-isotropic in the sense that, for a structure of a given length scale they do not differentiate by direction. For example, two neighbouring temperature anomalies, one warm, one cold, separated by 100 meters, say, might be expected to affect each other through radiative transfer, but it will not matter whether the separation is in the vertical or the horizontal. Other similarly isotropic physical processes are molecular diffusion in a laminar flow and mixing due to enhanced molecular diffusion in three-dimensionally turbulent flow. In principle scale reduction in the horizontal will lead eventually to dissipation or mixing. However it is clear from observational experience that structures in the atmosphere and the ocean tend to have much smaller length scales in the vertical than in the horizontal. This suggests that mixing and dissipation are most likely to result from small vertical scales rather than small horizontal scales. Is there a systematic reduction in vertical scales, as well as in horizontal scales, due the large-scale flow? This question has recently been investigated by Haynes and Anglade (1997) and some of the key arguments in that paper are repeated again below.

Certainly the large-scale flow is not isotropic. As has been noted earlier, cross-isentrope velocities are much smaller than along-isentrope velocities. This suggests that a suitable model for analysing the scale-cascade process might take a velocity field that is purely horizontal (representing the flow along isentropic surfaces), but which varies in both the horizontal and the vertical. As has already been noted in §3, the effect of chaotic advection along isentropic surfaces is to reduce horizontal scales at an exponential rate. The extra effect that needs to be included is that of vertical variation of the horizontal flow, i.e. vertical shear.

It turns out that the relevant effects are captured by the simple linear flow with components $(\Gamma x, -\Gamma y + \Lambda z, 0)$ in the x, y and z directions respectively, where (x, y, z) are Cartesian coordinates, with z vertical. For simplicity we assume the flow to be steady and therefore take the horizontal strain Γ and the vertical shear Λ to be constants. Exponential reduction in scale in the horizontal at rate Γ is therefore accomplished by the steady horizontal strain. The equation for the evolution in this flow of a tracer with concentration $\chi(x, y, z, t)$ is

$$\frac{\partial \chi}{\partial t} + \Gamma x \frac{\partial \chi}{\partial x} + (\Lambda z - \Gamma y)\frac{\partial \chi}{\partial y} = \kappa(\frac{\partial^2 \chi}{\partial x^2} + \frac{\partial^2 \chi}{\partial y^2} + \frac{\partial^2 \chi}{\partial z^2}), \qquad (7)$$

where κ is the diffusivity, assumed constant. This equation has solutions of the form $\chi(x, y, z, t) = \text{Re}\{\hat{\chi}(t)\exp(i[kx + ly + mz])\}$ where k, l and m are functions of time satisfying the ordinary differential equations $\dot{k} + \Gamma k = 0$, $\dot{l} - \Gamma l = 0$ and $\dot{m} + \Lambda l = 0$.

It follows that if $(k, l, m) = (k_0, l_0, m_0)$ at $t = 0$ then

$$(k, l, m) = (k_0 e^{-\Gamma t}, l_0 e^{\Gamma t}, m_0 - \frac{\Lambda l_0}{\Gamma}(e^{\Gamma t} - 1)). \tag{8}$$

Reduction in the scale of variation of a tracer in this flow may be deduced by considering the time variation of the wavenumber vector (k, l, m). Note that at large times the size of the horizontal wavenumber $(k^2 + l^2)^{1/2}$ increase exponentially with time at rate Γ, indicating the expected reduction in horizontal scale associated with the steady horizontal strain flow. However, the inclusion of vertical shear means that vertical wavenumber m also increases exponentially at the same rate, implying an accompanying reduction in vertical scale cascade. The aspect ratio α, say, of horizontal to vertical scales defined by $\alpha = |m|/(k^2 + l^2)^{1/2}$ is equal, at large times, to Λ/Γ, i.e. the ratio of vertical shear to horizontal strain. The reason for this is quite clear if one considers the geometry of the flow, as depicted in Figure 11. The effect of adding the vertical shear to the horizontal strain is to cause the axis of strain to slope with height. The angle between the axis and the horizontal is the inverse tangent of Γ/Λ. Convergence of the flow at each level naturally leads to structures in tracer fields that slope at the same angle. These structures thin exponentially fast in the horizontal and therefore also in the vertical.

The implications for diffusive mixing are manifested in the time evolution of the coefficient $\hat{\chi}$ which satisfies the equation

$$\frac{d\hat{\chi}}{dt} = -\kappa(k^2 + l^2 + m^2)\hat{\chi}. \tag{9}$$

Substituting from (8) and integrating, it follows that

$$\log(\hat{\chi}(t)) - \log(\hat{\chi}(0)) =$$

$$-\kappa\{\frac{l_0^2}{2\Gamma}(1 + \frac{\Lambda^2}{\Gamma^2})(e^{2\Gamma t} - 1) + \frac{2l_0\Lambda}{\Gamma^2}(e^{\Gamma t} - 1)(m_0 + \frac{\Lambda l_0}{\Gamma}) + (m_0 + \frac{\Lambda l_0}{\Gamma})^2 t + \frac{k_0^2}{2\Gamma}(1 - e^{-2\Gamma t})\}. \tag{10}$$

The term in brackets on the right-hand side increases with time and diffusive decay becomes important when this term $\sim \kappa^{-1}$. If κ is sufficiently small then the time at which this occurs will be determined by the first term inside the bracket, which is most rapidly growing for large time. Note first that this term is independent of m_0 since, for sufficiently small diffusivity, it is not the initial vertical structure, but the small vertical structure that arises from tilting of initial horizontal structure, that controls diffusive effects. Note also that this term is increased by a factor $1 + \Lambda^2/\Gamma^2$ over the form it would have in the absence of vertical shear. The diffusive decay, then, is almost exactly as it would be in a problem with no vertical structure, with the effects of horizontal diffusion enhanced by the scale reduction associated with horizontal strain. However, the effect of the vertical shear is to augment the horizontal diffusivity κ by an amount $\kappa\Lambda^2/\Gamma^2 = \kappa\alpha^2$ due to the diffusivity acting on structure of aspect ratio $\alpha = \Lambda/\Gamma$. This augmentation of the horizontal diffusivity is closely analogous to that found by Young et al. (1982) [108] due to the effects of oscillatory vertical shear (argued to result from near-inertial waves). In the simple model reported above the vertical shear should be regarded as that associated with the large-scale flow.

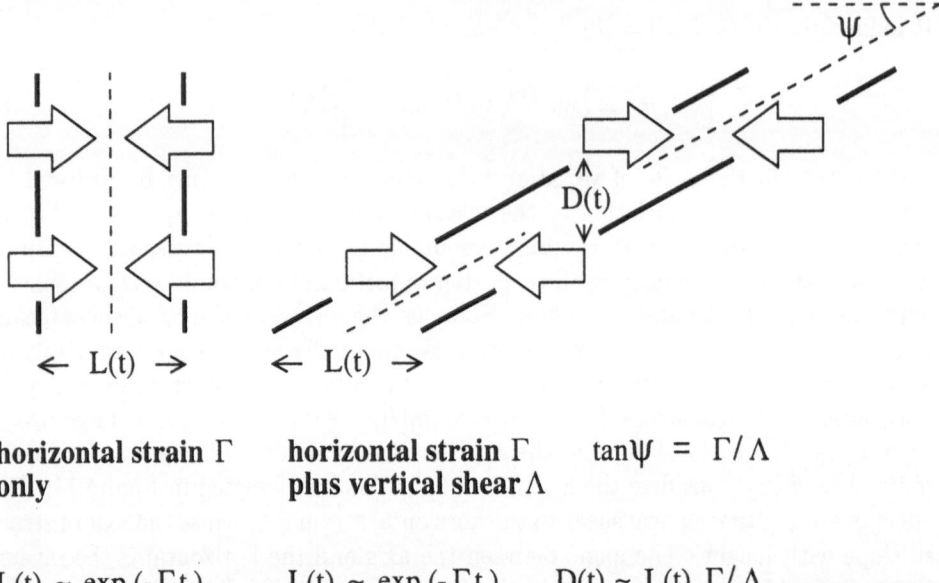

horizontal strain Γ only	horizontal strain Γ plus vertical shear Λ	$\tan\psi = \Gamma/\Lambda$
$L(t) \sim \exp(-\Gamma t)$	$L(t) \sim \exp(-\Gamma t)$	$D(t) \sim L(t)\,\Gamma/\Lambda$

Fig. 11. — Illustration of the basic mechanism by which vertical scales are reduced in a flow containing a combination of horizontal strain and vertical shear. The effect is shown of such a flow on a tracer sheet, with tracer contours represented by the bold lines. The flow could be that considered in §4.4, in which case the view is along the x-axis, which is the axis of extension for this flow. The left-hand picture shows the effect of a flow including horizontal strain alone. The horizontal thickness L of the tracer sheet is reduced in time as $e^{-\Gamma t}$, where Γ is the horizontal strain rate. The right-hand picture shows the effect when a flow in the y direction with shear Λ in z is added. The axis of the strain field is now tilted to the horizontal at angle $\tan^{-1}(\Gamma/\Lambda)$, as is the resulting tracer sheet. The horizontal thickness L and the vertical thickness D therefore both decrease as $e^{-\Gamma t}$ and stay in the aspect ratio Λ/Γ. [Figure reproduced from [40]. Further calculations in that paper show how this effect persists when the horizontal strain and vertical shear fields are time dependent.]

The assumption of steady velocity fields in the above model is, of course, a drastic one. However, Haynes and Anglade (1997) show how similar behaviour is found in more general flows, by integrating the equation

$$\frac{d\mathbf{k}}{dt} = -(\nabla\mathbf{u}).\mathbf{k} \tag{11}$$

for the gradient $\mathbf{k}$ of a conserved tracer, measured following a fluid particle, given the velocity field $\mathbf{u}$. One approach is to impose the components of the velocity gradient tensor $\nabla\mathbf{u}$, i.e. the horizontal strain and vertical shear, as random functions of time. This has been used classically to study tracer evolution in turbulent flows, but might also be considered as a simple representation of the time varying stretching and shearing encountered by a fluid element as it moves in a chaotic advection flow. A second approach is to calculate $\mathbf{u}$ from observations (from the wintertime lower stratosphere). In both cases both horizontal and vertical tracer gradients are found, on average, to increase exponentially at the same rate and to attain a characteristic aspect ratio. For observed wintertime lower stratospheric flows the aspect ratio α is about 250.

4.5. Implications for stratospheric mixing

Assessment of the strength of mixing processes in the stratosphere (and elsewhere in the atmosphere) is important not only because of fundamental interest, but also because of implications for chemical evolution. Recent investigations have suggested that there may be sensitivity of stratospheric chemical evolution in a gross sense (e.g. in the total amount of reaction product or the total amount of a given species destroyed) to the strength of mixing processes [31], [93]. In the case of [31], this sensitivity is expressed in terms of sensitivity to numerical resolution. However, any numerical model with finite resolution (however cleverly its transport scheme has been designed) is based on the implicit assumption that chemical species are well-mixed on the scale of the grid. If the grid size is larger than the scale on which mixing actually occurs then the effect is equivalent to artificially enhancing the strength of mixing processes.

The molecular mixing required for chemical reaction, for example, depends ultimately on molecular diffusion. As is well known, the effects of molecular diffusion can be enhanced by fluid-dynamical stirring. Up to now we have considered the stirring effects of chaotic-advection flows. However, as noted earlier, the lower stratospheric flow is not entirely laminar but is believed, on the basis of balloon, aircraft and radar observations, to contain patches of three-dimensional turbulence. The effect of these patches may spectacularly enhance mixing on small scales. Dewan (1981) [24] has argued that the effects of the turbulent patches will be to give a diffusive transport in the vertical, associated with a diffusivity D say, essentially because fluid parcels encountering a succession of such patches will undergo a random walk with small vertical step size. We might further hypothesize that representing this transport by downgradient diffusion also captures the mixing effects of the patches – this seems plausible if the turbulent patches become well mixed within their lifetime. (It is useful to contrast the case of the 'diabatic dispersion' discussed earlier, where the dispersive transport might under some conditions be parameterizable by downgradient diffusion, but is not likely, by itself, to lead to an equivalent molecular mixing. In the language sometimes used in other areas of fluid dynamics, diabatic dispersion leads to 'macromixing,' but not to 'micromixing'.)

One way to estimate the strength of the mixing processes and to understand their effects, therefore, is to estimate the vertical diffusivity D and to consider the interaction of the vertical diffusion with the large-scale 'chaotic-advection' flow. The arguments of [40] suggest that the latter interation will lead to an effective horizontal diffusivity $\alpha^2 D$, where α is the aspect ratio. Thus tracer evolution on a quasi-horizontal surface might be explained at leading order, by two-dimensional chaotic advection with diffusivity $\alpha^2 D$. If a typical stretching rate is S, then the minimum horizontal scales seen in such flows, i.e. the horizontal scales at which mixing occurs, are expected to be of order $\alpha(D/S)^{1/2}$.

A number of estimates of the vertical diffusivity D are available from calculations based on radar data. Some of these (e.g. [36]) are based on hypothesised relations between turbulent energy dissipation rates and vertical eddy diffusivities. That of Woodman and Rastogi (1984) [107] is based on the model posed in [24]. However most of these calculations seem to give a value of around 0.2 m^{2}s^{-1}. A lower bound on D is certainly the molecular diffusivity, which is about $1.5 \times 10^{-4}\text{m}^2\text{s}^{-1}$. These two values for D, together with the Haynes and Anglade (1997) [40] estimate of 250 for the aspect ratio α and a stretching rate S of about 0.3 days^{-1}, imply minimum horizontal scales of about 50km and 2km respectively. As noted earlier, there is considerable evidence from aircraft data of small-scale horizontal variation of chemical tracers and it does not seem to be the case that this variation is strongly suppressed below scales of 50km (recall Figure 6). The aircraft data is, indeed, itself a potential source of information about mixing processes and two recent pieces of work, by Waugh et al. (1997) [104] and Balluch and Haynes (1997) [7], have attempted to estimate D from this data. Both conclude, in broad terms, that D is about $10^{-2}\text{m}^2\text{s}^{-1}$, i.e. considerably less than the radar estimates. This is a troubling discrepancy that it seems important to resolve. Progress is likely to be complicated by the fact that the mixing due to patches of turbulence is almost certainly highly variable in time and space. The value for D of $10^{-2}\text{m}^2\text{s}^{-1}$, which would imply a minimum horizontal scale of about 12km, might represent the 'average' effect of mixing processes. Balluch and Haynes (1997) [7] show there is 'believeable' variation in the chemical tracers down to scales of order 1km and suggest that these might be associated with air parcels that have encountered relatively few turbulent patches and hence relatively weak mixing.

5. Discussion

5.1. Summary

It has been argued in the above that transport in the atmosphere is dominated by advection by the large-scale flow and that, at least for time scales of a few days, the transport is, quasi-horizontal, along isentropic surfaces. Thus two-dimensional chaotic advection is an appropriate paradigm in the sense that complicated spatial structure arises through the action of flow that has a rather simple space and time variation. One might describe the flow as layerwise two-dimensional chaotic advection. The structure of the transport on each isentropic surface shows the same sort of strong spatial inhomogeneity, with mixing regions and transport barriers, that is now seen as characteristic of 2-D chaotic advection.

If the description of transport is to be any better than semi-quantitative, then the third space dimension must be taken into account. In particular, if transport on longer time scales is of interest then it is essential to take account of vertical advection across isentropic surfaces. As is well known from the work of [25] and others, in three dimensions there may be chaotic advection without any time-dependence in the flow. However, it is not very clear studying three-dimensional steady flows will give much insight into the real atmosphere. A more interesting possibility is to add weak vertical advection to otherwise two-dimensional chaotic advection flows. Work on three-dimensional time-dependent flows (e.g.[32], see also [17]) suggests that there may be interesting effects in such flows. Whether these effects are relevant to transport in realistic atmospheric flows has yet to be determined. If realistic mixing processes are of interest, then the vertical variation of the flow must be taken into account, since this results in a reduction in vertical scale, as well as in horizontal scale, of tracer fields and it is the vertical-scale reduction that is most significant for mixing.

The remainder of this section discusses some examples of topics in atmospheric transport, stirring and mixing that seem particularly interesting to the author, both in their importance for understanding the atmosphere and in their relation to problems addressed by theoretical work on chaotic advection.

5.2. Dynamical consistency

One aspect of chaotic advection that many working on atmospheric (or indeed oceanic) flows find troubling is the fact that almost all problems are posed kinematically, i.e. the velocity field is taken as a given function of space and time, without any requirement for dynamical consistency. As has been noted in §2, the evolution of large-scale atmospheric and oceanic flows may be understood in terms of an advected scalar (PV) whose distribution at any instant determines the flow at that instant. If a flow is to be time-periodic, for example, then clearly the PV field also has to be time-periodic. It is not clear that this is consistent with the existence of chaotic advection regions in the flow unless the PV (and hence the flow) satisfies certain strong constraints. For example, Brown and Samelson (1994) [16] have suggested that it is necessary for the potential vorticity to be constant in the chaotic advection regions. One way round this problem is to consider special flows where the potential vorticity (perhaps in a certain localised region) happens to be a passive tracer, e.g. the nonlinear Rossby-wave critical layer flow [70, 71]. While such special cases are useful in justifying the results of kinematic studies as, in some sense, dynamically realisable, they do not, of course, provide any insight into what actually happens in the much larger class of problems where PV is active, rather than passive.

More generally, the properties of the 'inversion' operator appearing in e.g. equations (2) and (5) mean that small-scale features in vorticity or potential vorticity have a relatively weak influence on the flow. Thus the stirring and mixing of vorticity or potential vorticity at small scales is likely to be much as predicted for stirring and mixing of passive tracer by chaotic advection studies, provided that the structure of the large-scale flow is similar. This is nicely discussed and illustrated by Pierrehumbert (1991) [77] who shows results from an initial-value dynamical simulation and notes how the structures arising in the small-scale vorticity field correspond in some detail to those predicted by a kinematic model.

One way to assess which parts of the large body of results for two-dimensional time-periodic flow carry over to the dynamically consistent case is to consider the effects of more general time-dependence (the idea being that the requirement of dynamical consistency will almost always disrupt time-periodicity). There has already been some progress in analysing transport, for example, in the quasi-periodic case (i.e. two or more periods in irrational ratios) (see e.g. [30]) and even with the aperiodic case [56].

Take, for example, the interesting issue of the 'stickiness' of the regular trajectories that border a chaotic region (e.g.[66] sections II.D.4 and IX.C.3). These have implications both for transport and for stirring and mixing. For the former, the 'stickiness' implies long-term correlations in velocities that may lead, e.g. to algebraic tails in the probability distributions for time of escape from certain regions, perhaps in turn leading to the possibility of anomalous diffusive effects. For the latter, it implies that the distribution of the cumulative stretching experienced by fluid elements has a larger than expected tail on the low stretching side [see [76] and references). This might, for example, have implications for the diffusive decay of tracer variance. If 'stickiness' is a property that only holds for periodic velocity fields, or some other relatively small class of velocity fields, then it and the associated transport, stirring and mixing behaviour, might not be observed for dynamically consistent flows.

5.3.   Microstructure

Mathematical models for the tracer microstructure that results from chaotic advection are important for interpretation of small-scale tracer measurements. They are also a useful preliminary step towards understanding chemical evolution or the effects of other processes that depend nonlinearly on tracer concentrations.

One rather crude measure of microstructure is the tracer spectrum. Recent calculations have shown that the horizontal spectra of tracers measured during the various aircraft ozone campaigns has a slope somewhere between -2 and -1.5 [90] [8]. Theories of tracer behaviour in a large-scale flow fields suggest that in the forced-dissipative case the slope should be -1 (the 'Batchelor' spectrum) [e.g. [79] and references] and indeed this has been verified in numerical simulations by Antonsen et al. (1996) [3]. Bacmeister et al. (1996) [8] have suggested that in order to explain the steeper slope of the observed spectra it may be necessary to take account of both advection by the large-scale flow and the advective effect of gravity waves on small-vertical-scale tracer features produced by the large-scale flow. Ngan and Shepherd (1997) [71], on the other hand, have argued that the form of the spectrum may be due to the fact that the dominant tracer features are near discontinuities associated with the vortex edge (and filaments drawn away from it) and that these near discontinuities remain spatially isolated, because dissipative processes are strong enough to remove filaments before they become space-filling. The Saffman (1971) model [86], predicting a -2 slope for the specrum is therefore relevant. Of course, the important scientific question here is what physical processes need to be invoked to explain the spectrum? The Bacmeister et al. (1996) [8] suggestion requires more than large-scale advection and simple dissipation, the Ngan and Shepherd (1997b) suggestion [71] does not.

Different approaches to quantifying microstructure have, of course, been suggested. Pierrehumbert (1995) [79] discusses a number of relevant issues, including the possibility, argued by Varosi et al. (1991) [97], that chaotic advection flows may lead to multifractal scaling of tracer gradients. Pierrehumbert does not find such scaling in his numerical simulations and notes the need for further examination of this issue. He does find scaling consistent with tracer fields that are filamentary (i.e., locally one-dimensional) with appreciable spatial correlations across filaments. Certainly filamentary structure is one of the striking features of visualizations of tracers in chaotic-advection flows (and is in strong contrast to the appearance of tracers in fully three-dimensional turbulent flow.) Balluch and Haynes (1997) [7] give a heuristic argument for this locally one-dimensional structure, based on the fact that the direction (but not the magnitude) of the tracer gradient at some spatial location and time is determined only over the recent history (a few stretching times) and therefore varies only on the large scale. They note that the analogous structure for 'layerwise two-dimensional chaotic advection' consists of tracer sheets rather than tracer filaments and exploit this in their assessment of the effects of vertical mixing. It might be that this approach could be incorporated into a more sophisticated representation of microstructure.

5.4. Transport barriers

The partial barriers to transport between tropics and surf zone in the lower stratosphere, or at the tropopause, play an important role in determining the distribution of chemical tracers. Partial barriers to transport play similarly important roles in the ocean. For example, it has been noted by Bower et al. (1985) [11] that at upper levels there are strong water mass contrasts across the Gulf stream suggesting that it acts as a barrier to transport. At lower levels, on the other hand, there is little contrast, presumably because the eddy motion associated with the Gulf Stream acts as an efficient 'blender'.

Both atmospheric and oceanic problems have already stimulated kinematic studies of transport associated with disturbed jets [87] [30] [23]. How can such studies be exploited or extended? What one might seek is some insight into what controls the leakiness of barriers in realistic flows. Consider, for example, a two-dimensional time-periodic flow where there may be perfect barriers to transport. What ingredient of the real atmospheric flow would it be most relevant to add to such a model in order to give realistic leakiness? Adding weak vertical advection is one possibility. Another might be more complex time-dependence. But in the latter case is it weak random fluctuations that are most important or occasional large-amplitude disturbances? One practical approach to addressing these questions for the Southern Hemisphere polar vortex has been made by Bowman (1996) [15], who uses observed winds on isentropic surfaces to study the effect on the transport barrier at the vortex edge of applying different time and space filters to the wind fields.

If, in model simulations, such barriers are too 'leaky', tracer distributions are likely to be incorrectly predicted. In the case of tracers such as ozone and water vapour, which play a crucial role in radiative transfer, such errors might feed back on the simulation of the circulation itself.

5.5. Conclusion

There is no doubt that the development of chaotic advection as a sub-discipline of fluid dynamics has given important stimulus to understanding transport, stirring and mixing in atmospheric flows. It is also the case that there is an appealing analogy between the structures typically generated in chaotic advection models and those that have been inferred from new atmospheric observations and numerical simulations over the last decade or so. However, it still seems arguable that the insight provided by the chaotic advection theories is only qualitative. The challenge in the future is to determine to what extent the mathematical apparatus that has been constructed to study nonlinear dynamical systems, in particular chaotic advection, can be used to give hard, quantitative insight into transport, stirring and mixing in realistic atmospheric flows. The examples given above are some suggestions of where progress would certainly be significant and might also be feasible.

Acknowledgments

I am grateful to Emily Shuckburgh and Jacques Vanneste for helpful comments on early drafts, also to Michael McIntyre and Nicholas Pinhey for allowing me to show their results in advance of publication. Nicholas Pinhey also gave valuable assistance in producing some of the figures. I acknowledge research support from the UK NERC (grant GR9/02013 and through UGAMP), from the UK EPSRC (grant 65314) and from the European Community. Finally, I congratulate the organisers on an enjoyable and educational Cargèse School on Mixing: Chaos and Turbulence.

References

[1] Ambaum, M., 1997: Isentropic formation of the tropopause. J. Atmos. Sci., 54, 555–568.

[2] Andrews, D. G., Holton, J. R., Leovy, C. B., 1987: Middle Atmosphere Dynamics. Academic Press, 489pp.

[3] Antonsen, T. M., Fan, Z., Ott, E., Garcia-Lopez, E., 1996: The role of chaotic orbits in the determination of power spectra of passive scalars. Phys. Fluids, 8, 3094–3104.

[4] Appenzeller, C., Davies, H. C., Norton, W. A., 1996: Fragmentation of stratospheric intrusions. J. Geophys. Res., 101, 1435–1456.

[5] Aref, H., 1984: Stirring by chaotic advection. J. Fluid Mech., 143, 1–21.

[6] Aref, H., 1991: Stochastic particle motion in laminar flows. Phys. Fluids A, 3, 1009–1016.

[7] Balluch, M. G., Haynes, P. H., 1997: Quantification of lower stratospheric mixing processes using aircraft data. J. Geophys. Res., 102, 23487–23504.

[8] Bacmeister, J. T., Eckermann, S. D., Newman, P. A., Lait, L., Chan, K. R., Loewenstein, M., Proffitt, M. H., Gary, B. L., 1996: Stratospheric horizontal wavenumber spectra of winds, potential temperature, and atmospheric tracers observed by high-altitude aircraft. J. Geophys. Res., 101D, 9441–9470.

[9] Barat, J., 1982: Some characteristics of clear-air turbulence in the middle stratosphere. J. Atmos. Sci., 39, 2553–2564.

[10] Bithell, M. and L.J. Gray, 1997. Contour lengthening rates near the tropopause. Geophys. Res. Lett., 24, 2721–2724.

[11] Bower, A. S., Rossby, H. T., and Lillibridge, J. L., 1985: The Gulf Stream - barrier or blender? J. Phys. Oceanogr., 15, 24-32.

[12] Bowman, K. P., 1993: Large-scale isentropic mixing properties of the Antarctic polar vortex from analyzed winds. J. Geophys. Res., 98, 23013–23027.

[13] Bowman, K. P., 1993: Barotropic simulation of large-scale mixing in the Antarctic polar vortex. J. Atmos. Sci., 50, 2901–2914.

[14] Bowman, K. P., 1996: Rossby wave phase speeds and mixing barriers in the stratosphere. Part I: observations. J. Atmos. Sci., 53, 905–916.

[15] Bowman, K. P., 1996: Rossby wave phase speeds and mixing barriers in the stratosphere. Part I: observations. J. Atmos. Sci., 53, 905–916.

[16] Brown, M. G., Samelson, R. M., 1994: Particle motion in incompressible, inviscid vorticity-conserving fluids. Phys. Fluids, 6, 2875–2876.

[17] Cartwright, J. H. E., Feingold, M., Piro, O., 1996: Chaotic advection in three-dimensional unsteady incompressible laminar flow. J. Fluid Mech., 316, 259–284.

[18] Chen, P., 1994: The permeability of the Antarctic vortex edge. J. Geophys. Res., 99, 20563–20571.

[19] Chen, P., J. R. Holton, A. O'Neill and R. Swinbank, 1994. Isentropic mass exchange between the tropics and extratropics in the stratosphere, J. Atmos. Sci., 51, 3006-3018.

[20] Couder Y., and Basdevant C., 1986: Experimental and numerical study of vortex couples in two-dimensional flows. J. Fluid Mech., 173, 225-251.

[21] Courtier, P., Derber, J., Errico, R., Louis, J.-F. and Vukićević, T. 1993: Important literature on the use of adjoint, variational methods and the Kalman filter in meteorology. Tellus, 45a, 342-357.

[22] Daley, R., 1991: Atmospheric Data Analysis. Cambridge, Cambridge University Press, 457pp.

[23] del Castillo-Negrete, D., Morrison, P. J., 1993: Chaotic transport by Rossby waves in shear flow. Phys. Fluids, A5, 948–965.

[24] Dewan, E. M., 1981: Turbulent vertical transport due to thin intermittent mixing layers in the stratosphere and other stable fluids. Science, 211, 1041–1042.

[25] Dombre, T., Frisch, U., Greene, J. M., Henon, M., Mehr, A., Soward, A. M., 1986: Chaotic streamlines in the ABC flows. J. Fluid Mech., 167, 353–391.

[26] Dritschel, D. G., 1989: Contour dynamics and contour surgery: numerical algorithms for extended, high-resolution modelling of vortex dynamics in two-dimensional, inviscid, incompressible flows. Computer Phys. Rep., 10, 77–146.

[27] Dritschel, D. G., de la Torre, M., 1996: The instability and breakdown of tall columnar vortices in a quasi-geostrophic fluid. J. Fluid Mech., 328, 129–160.

[28] Dritschel, D. G., Ambaum, M. H. P., 1997: A contour-advective semi-Lagrangian numerical algorithm for simulating fine-scale conservative dynamical fields. Q. J. Roy. Meteorol. Soc., 123, 1097–1130.

[29] Dritschel, D. G., Haynes, P. H., Juckes, M. N., Shepherd, T. G., 1991: The stability of a two-dimensional vorticity filament under uniform strain. J. Fluid Mech., 230, 647–665.

[30] Duan, J., Wiggins, S., 1996: Fluid exchange across a meandering jet with quasiperiodic variability. J. Phys. Oceanog., 26, 1176–1188.

[31] Edouard, S., Legras, B., Lefèvre, F., Eymard, R., 1996: The effect of small-scale inhomogeneities on ozone depletion in the Arctic. Nature, 384, 444–447.

[32] Feingold, M., Kadanoff, L. P., Piro, O., 1988: Passive scalars, three-dimensional volume-preserving maps, and chaos. J. Statist. Phys., 50, 529–565.

[33] Fernando, H. J. S., 1991: Turbulent mixing in stratified fluids. Ann. Rev. Fluid Mech., 23, 455–493.

[34] Fisher, M., Lary, D. J., 1995: Lagrangian four-dimensional variational data assimilation of chemical species. Q. J. Roy. Meteorol. Soc., 121, 1681–1704.

[35] Fisher, M., O'Neill, A., Sutton, R., 1993: Rapid descent of mesospheric air into the stratospheric polar vortex. Geophys. Res. Lett., 20, 1267–1270.

[36] Fukao, S., Yamanaka, M. D., Ao, N., Hocking, W., Sato, T., Yamamoto, M., Nakamura, T., Tsuda, T., Kato, S., 1994: Seasonal variability of vertical eddy diffusivity in the middle atmosphere. 1. Three-year observations by the middle and upper atmosphere radar. J. Geophys. Res., 99, 18973–18987.

[37] Gill, A. E., 1982: Atmosphere-Ocean Dynamics. Academic Press, 662pp.

[38] Haynes, P. H., 1990: High-resolution three-dimensional modelling of stratospheric flows: quasi-two-dimensional turbulence dominated by a single vortex. In: Topological Fluid Mechanics, ed. H. K. Moffatt, Tsinober; Cambridge University Press, 345–354.

[39] Haynes, P. H., Marks, C. J., McIntyre, M. E., Shepherd, T. G., Shine, K. P., 1991: On the "downward control" of extratropical diabatic circulations by eddy-induced mean zonal forces. J. Atmos. Sci., 48, 651–678.

[40] Haynes, P. H., Anglade, J., 1997: The vertical-scale cascade of atmospheric tracers due to large-scale differential advection. J. Atmos. Sci., 54, 1121–1136.

[41] Haynes P.H. and Scinocca, J.F. 1998: Formation of the tropopause through the action of baroclinic eddies. (In preparation.)

[42] Hall, T. M., Waugh, D. W., 1997: Tracer transport in the tropical stratosphere due to vertical diffusion and horizontal mixing. Geophys. Res. Lett., 24, 1383–1387.

[43] Held, I. M., Pierrehumbert, R. T., Garner, S. T., Swanson, K. L., 1995: Surface quasi-geostrophic dynamics. J. Fluid Mech., 282, 1–20.

[44] Holton, J. R., 1992: An introduction to dynamic meteorology (3rd edition). San Diego, Academic Press, inc., 511pp.

[45] Holton, J. R., 1986: A dynamically based transport parameterization for one-dimensional photochemical models of the stratosphere. J. Geophys. Res., 91D, 2681–2686.

[46] Holton, J. R., Haynes, P. H., McIntyre, M. E., Douglass, A. R., Rood, R. B., Pfister, L., 1995: Stratosphere–troposphere exchange. Revs. Geophys., 33, 403–439.

[47] Hoskins, B. J., 1991: Towards a PV-θ view of the general circulation. Tellus, 43AB, 27–35.

[48] Hoskins, B. J., McIntyre, M. E., Robertson, A. W., 1985: On the use and significance of isentropic potential-vorticity maps. Q. J. Roy. Meteorol. Soc., 111, 877–946.

[49] James, I. N., 1994: Introduction to circulating atmospheres. Cambridge, Cambridge University Press, 422pp.

[50] Juckes, M. N., 1994: Quasigeostrophic dynamics of the tropopause. J. Atmos. Sci., 51, 2756–2779.

[51] Juckes, M. N., McIntyre, M. E., 1987: A high resolution, one-layer model of breaking planetary waves in the stratosphere. Nature, 328, 590–596.

[52] Karoly, D. J., McIntosh, P. C., Berrisford, P., McDougall, T. J., Hirst, A. C., 1997: Similarities of the Deacon cell in the Southern Ocean and Ferrel cells in the atmosphere. Q. J. Roy. Meteorol. Soc., 123, 519–526.

[53] Mahlman, J. D., Andrews, D. G., Hartmann, D. L., Matsuno, T., Murgatroyd, R. G., 1984: Transport of trace constituents in the stratosphere. In: Dynamics of the Middle Atmosphere, ed. J. Holton, T. Matsuno; Tokyo, Terrapub, Dordrecht, Reidel, 387–416.

[54] Mahlman, J. D., Levy II, H., Moxim, W. J., 1986: Three-dimensional simulations of stratospheric N_2O: predictions for other trace constituents. J. Geophys. Res., 91, 2687–2707.

[55] Mahlman, J. D., Umscheid, L. J., 1987: Comprehensive modeling of the middle atmosphere: the influence of horizontal resolution. In: Transport Processes in the Middle Atmosphere. ed. G. Visconti, R. R. Garcia; Dordrecht, Reidel, 251–266.

[56] Malhotra, N. and Wiggins, S. 1997: Geometric structures, lobe dynamics and Lagrangian transport in flows with aperiodic time dependence. J. Nonlin. Sci. (submitted).

[57] McIntyre, M. E., 1990: Middle atmosphere dynamics and transport: some current challenges to our understanding. In: Dynamics, Transport and Photochemistry in the Middle Atmosphere of the Southern Hemisphere, ed. A. O'Neill, Dordrecht, Kluwer.

[58] McIntyre, M. E., 1992: Atmospheric dynamics: some fundamentals, with observational implications. In: Proc. Internat. School Phys. "Enrico Fermi", CXV Course, ed. J. C. Gille, G. Visconti; Amsterdam, Oxford, New York, Toronto, North-Holland, 313–386.

[59] McIntyre, M. E., 1995: The stratospheric polar vortex and sub-vortex: fluid dynamics and midlatitude ozone loss. Phil. Trans. Roy. Soc. London, 352, 227–240.

[60] McIntyre, M. E., Palmer, T. N., 1983: Breaking planetary waves in the stratosphere. Nature, 305, 593–600.

[61] McIntyre, M. E., Palmer, T. N., 1984: The "surf zone" in the stratosphere. J. Atm. Terr. Phys., 46, 825–849.

[62] McIntyre, M. E., Pinhey, N., 1998: Permeability of the stratospheric vortex edge: on the possible effects of inertia waves. J. Geophys. Res., (to be submitted).

[63] McWilliams, J. C., 1989: Statistical properties of decaying geostrophic turbulence. J. Fluid Mech., 198, 199–230.

[64] McWilliams, J. C., Weiss, J. B., Yavneh, I., 1994: Anisotropy and coherent vortex structures in planetary turbulence. Science, 264, 410–413.

[65] Meacham, S. P., Pankratov, K. K., Shchepetkin, A. F., Zhmur, V. V., 1994: The interaction of ellipsoidal vortices with background shear flows in a stratified fluid. Dyn. Atmos. Oceans, 21, 167–212.

[66] Meiss, J. D., 1992: Symplectic maps, variational principles and transport. Rev. Mod. Phys., 64, 795–848.

[67] Mote, P. W., Rosenlof, K. H., McIntyre, M. E., Carr, E. S., Gille, J. C., Holton, J. R., Kinnersley, J. S., Pumphrey, H. C., Russell, J. M., Waters, J. W., 1996: An atmospheric tape recorder: the imprint of tropical tropopause temperatures on stratospheric water vapor. J. Geophys. Res., 101, 3989–4006.

[68] Minschwaner, K., Dessler, A. E., Elkins, J. W., Volk, C. M., Fahey, D. W., Loewenstein, M., Podolske, J. R., Roche, A. E., Chan, K. R., 1996: The bulk properties of isentropic mixing into the tropics in the lower stratosphere. J. Geophys. Res., 101, 9433–9439.

[69] Murphy, D. M., Fahey, D. W., Proffitt, M. H., Liu, S. C., Chan, K. R., Eubank, C. S., Kawa, S. R., Kelly, K. K., 1993: Reactive nitrogen and its correlation with ozone in the upper troposphere and lower stratosphere. J. Geophys. Res., 98D, 8751–8773.

[70] Ngan, K., Shepherd, T. G., 1997: Chaotic mixing and transport in Rossby-wave critical layers. J. Fluid Mech., 334, 315–351.

[71] Ngan, K., Shepherd, T. G., 1997: Comments on some recent measurements of anomalously steep N20 and O3 tracer spectra in the stratospheric surf zone. J. Geophys. Res., 102, 24001–24006.

[72] Norton, W. A., 1994: Breaking Rossby waves in a model stratosphere diagnosed by a vortex-following coordinate system and a technique for advecting material contours. J. Atmos. Sci., 51, 654–673.

[73] Ottino, J. M., 1989: The kinematics of mixing: stretching, chaos and transport. Cambridge Univ. Press, 378pp.

[74] Pedlosky, J., 1987: Geophysical Fluid Dynamics (2nd edition). New York, Springer-Verlag, 710pp.

[75] Pierce, B. R., Fairlie, T. D. A., 1993: Chaotic advection in the stratosphere: implications for the dispersal of chemically perturbed air from the polar vortex. J. Geophys. Res., 98, 18589–18595.

[76] Pierrehumbert, R.T., 1991a: Large-scale horizontal mixing in planetary atmospheres. Phys. Fluids, A3, 1250-1260.

[77] Pierrehumbert, R.T., 1991b: Chaotic mixing of tracer and vorticity by modulated travelling Rossby waves. Geophys. Astrophys. Fluid Dynamics, 58, 285-319.

[78] Pierrehumbert, R. T., Yang, H., 1993: Global chaotic mixing on isentropic surfaces. J Atmos Sci, 50, 2462–2480.

[79] Pierrehumbert, R. T., 1995: Tracer microstructure in the large-eddy dominated regime. In: Chaos applied to fluid mixing, ed. H. Aref, M. S. El Naschie; Pergamon/Elsevier, 347–365.

[80] Pierrehumbert, R. T., Held, I. M., Swanson, K. L., 1995: Spectra of local and non-local two-dimensional turbulence. In: Chaos applied to fluid mixing, ed. H. Aref, M. S. El Naschie; Pergamon/Elsevier, 367–371.

[81] Plumb, R. A., 1996: A "tropical pipe" model of stratospheric transport, J. Geophys. Res., 101, 3957-3972.

[82] Plumb, R. A., Ko, M. K. W., 1992: Interrelationships between mixing ratios of long-lived stratospheric constituents. J. Geophys. Res., 97, 10145–10156.

[83] Plumb, R. A., Mahlman, J. D., 1987: The zonally averaged transport characteristics of the GFDL general circulation/transport model. J. Atmos. Sci., 44, 298–327.

[84] Rhines, P. B., 1986: Vorticity dynamics of the oceanic general circulation. Ann. Rev. Fluid Mech., 18, 433–497.

[85] Rosenlof, K. H., 1995: Seasonal cycle of the residual mean meridional circulation in the stratosphere. J. Geophys. Res., 100, 5173–5191.

[86] Saffman, P. G., 1971: On the spectrum and decay of random two-dimensional vorticity distributions at large Reynolds number. Stud. Appl. Math., 50, 377–383.

[87] Samelson, R., 1992: Fluid exchange across a meandering jet. J. Phys. Oceanog., 22, 431–440.

[88] Sato, T., Woodman, R., 1982: Fine altitude resolution observations of stratospheric turbulent layers by the Arecibo 430MHz radar. J. Atmos. Sci., 39, 2546–2552.

[89] Sparling, L.C., Kettleborough, J.A., Haynes, P.H., McIntyre, M.E., Rosenfeld J.E., Schoeberl M.R. and Newman, P.A., 1997: Diabatic dispersion in the lower stratosphere. J. Geophys. Res., 102, 25817–25829.

[90] Strahan, S. E., Mahlman, J. D., 1994: Evaluation of the SKYHI general circulation model using aircraft N_2O measurements 2. Tracer variability and diabatic meridional circulation. J Geophys Res, D99, 10319–10332.

[91] Sutton, R., 1994: Lagrangian flow in the middle atmosphere. Q. J. Roy. Meteorol. Soc., 120, 1299–1321.

[92] Swinbank, R., O'Neill, A., 1994: A stratosphere-troposphere data assimilation system. Mon. Wea. Rev., 122, 686–702.

[93] Tan, D. G. H., Haynes, P. H., MacKenzie, A. R., Pyle, J. A., 1998: Effects of fluid-dynamical stirring and mixing on the deactivation of stratospheric chlorine. J. Geophys. Res., 103, 1585–1605.

[94] Trepte, C. R., Hitchman, M. H., 1992: Tropical stratospheric circulation deduced from satellite aerosol data. Nature, 355, 626–628.

[95] Tuck, A. F., 1989: Synoptic and chemical evolution of the Antarctic vortex in late winter and early spring, 1987. J. Geophys. Res., 94, 11687–11737.

[96] Tung, K. K., 1982: On the two-dimensional transport of stratospheric trace gases in isentropic coordinates. J. Atmos. Sci., 39, 2330–2355.

[97] Varosi, F., Antonsen, T. M., Ott, E., 1991: The spectrum of fractal dimensions of passively convected scalar gradients in chaotic fluid flows. Phys. Fluids (A), 3(5), 1017–1028.

[98] Viera, F., 1995: On the alignment and axisymmetrization of a vertically tilted geostrophic vortex. J. Fluid Mech., 289, 29–50.

[99] Volk, C. M., Elkins, J. W., Fahey, D. R., Salawitch, R. J., Dutton, G. S., Gilligan, J. M., Proffit, M. H., Loewenstein, M., Podolske, J. R., Minschwaner, K., Margitan, J. J., Chan, K. R., 1996: Quantifying transport between the tropical and midlatitude lower stratosphere. Science, 272, 1763–1768.

[100] Walsh, D. and Pratt, L.J., 1995: The interaction of a pair of point potential vortices in uniform shear. Dyn. Atmos. Oceans, 22, 135–160.

[101] Waugh, D. W., 1996: Seasonal variation of isentropic transport out of the tropical stratosphere. J. Geophys. Res., 101, 4007–4023.

[102] Waugh, D. W., Plumb, R. A., 1994: Contour advection with surgery: a technique for investigating finescale structure in tracer transport. J. Atmos. Sci., 51, 530–540.

[103] Waugh, D. W., Plumb, R. A., Atkinson, R. J., Schoeberl, M. R., Lait, L. R., Newman, P. A., Loewenstein, M., Toohey, D. W., Avallone, L. M., Webster, C. R., May, R. D., 1994: Transport of material out of the stratospheric Arctic vortex by Rossby wave breaking. J. Geophys. Res., 99, 1071–1078.

[104] Waugh, D. W., Plumb, R. A., Elkins J. W., Fahey, D. W., Boering K. A., Dutton, G. S., Keim, E., Gao, R. -S., Daube, B. C., Wofsy, S. C., Loewenstein, M., Podolske, J. R., Chan, K. R., Profitt, M. H., Kelly, K. K., Newman, P. A., Lait, L. R., 1997: Mixing of polar air into middle latitudes as revealed by tracer-tracer scatter plots. J. Geophys. Res., 102, 13119-13134.

[105] Wiggins, S., 1988: Stirred but not mixed. Nature, 333, 395–425.

[106] WMO, 1985: Atmospheric ozone 1985: Assessment of our understanding of the processes controlling its present distribution and change. (World Meteorological Organization Global Ozone Research and Monitoring Project Report No. 16) (3 volumes). Geneva, World Meteorological Organisation, 1095+86pp.

[107] Woodman, R. F., Rastogi, P. K., 1984: Evaluation of effective eddy diffusive coefficients using radar observations of turbulence in the stratosphere. Geophys. Res. Lett., 11, 243–246.

[108] Young, W. R., Rhines, P. B., Garrett, C. J. R., 1982: Shear-flow dispersion, internal waves and horizontal mixing in the ocean. J. Phys. Oceanogr., 12, 515–527.

The Ozone Hole

BERNARD LEGRAS

*Laboratoire de Météorologie Dynamique du CNRS, Ecole Normale Supérieure
24 rue Lhomond, 75005 Paris, France*

We present an overview of the processes leading to the ozone hole in the lower stratosphere with an emphasis on the role of transport and mixing. It is shown that the polar vortex behaves as a chemical container owing to the existence of a barrier to transport along its edge, only broken by sparse filamentary structures. Mixing is governed by chaotic layerwise motion down to horizontal scale of about 10 km. Difficulties in modelling non diffusive mixing may lead to serious underestimates of the ozone depletion.

1.   Chlorine and ozone

It is now widely admitted that the destruction of stratospheric ozone is due to the the release of man-made products in the atmosphere [13]. The most common species of chlorofluorocarbons (CFC) are very stable molecules which reside one century in the atmosphere before being destroyed by photochemistry or dissolved in the ocean. In contrast, the inorganic chlorine, mainly under the form HCl, has much larger sources in marine aerosols and volcano but is easily dissolved in atmospheric water and washed out so that it generates a negligible input to the stratosphere The 2.5 increase of stratospheric chlorine over the last 20 years is thus entirely due to organic chlorine among which the natural sources (of CH_3Cl) accounts for less than 20% of the total [14].

The air is entrained within the stratosphere above the large cumulus clouds that form in the equatorial region. For the portion which reaches 20 km and above, photochemical conversion effectively breaks CFC molecules and liberates chlorine atoms which soon recombine to produce HCl and $CLONO_2$. Since water is nearly

Mixing: Chaos and Turbulence, edited by Chaté *et al.*
Kluwer Academic / Plenum Publishers, New York 1999.

absent in the stratosphere, these molecules are now stable and disappear mainly by reinjection into the troposphere.

Although ozone is increasingly present in polluted areas, still 90 % is contained in the stratosphere between 15 km and 50 km with a maximum in concentration of nearly 10^{19} molecules m^{-3} at 25 km that is 2 ppmv (parts per million in volume) . Stratospheric ozone is produced by UV photolysis of molecular oxygen (wavelength < 240 nm) releasing two free atom of oxygen which combine in turn with two other oxygen molecules to form ozone. Ozone itself is easily photolysed by UV (wavelength < 320 nm) and reacts with atomic oxygen, returning to molecular oxygen. This set of reactions, known as the Chapman cycle is summarized as

$$
\begin{aligned}
O_3 + h\nu &\rightarrow O_2 + O \quad (\lambda < 320\text{nm}) \\
O + O_3 &\rightarrow 2O_2 \\
O_2 + h\nu &\rightarrow O + O \quad (\lambda < 240\text{nm}) \\
O + O_2 &\rightarrow O_3
\end{aligned}
$$

The first and fourth reaction are the most common, so that ozone acts as an effective absorber converting UV radiation into heat[1]. When the whole cycle is considered, it predicts that an equilibrium is reached depending on the solar flux. Below 30 km, the photochemical lifetime of odd oxygen (O_3 + O) is, however, of several weeks, large enough to be importantly influenced by transport. Indeed, while ozone is mainly generated over the tropics, the observed total ozone at 60° is almost the double than at the tropics and there is more ozone at high latitudes in the northern hemisphere in January than in July.

Chlorine under the forms HCl and CLONO$_2$ does not react directly with ozone. It is first necessary to transform it into reactive radicals, Cl, ClO or Cl$_2$O$_2$. The simplest catalytic cycle involving a free oxygen atom produced by photodissociation of O$_2$ is

$$
\begin{aligned}
Cl + O_3 &\rightarrow ClO + O_2 \\
ClO + O &\rightarrow Cl + O_2 \\
\text{Net}: O_3 + O &\rightarrow 2O_2
\end{aligned}
\tag{1}
$$

This cycle is similar to two other natural cycles, where Cl is replaced by NO or OH. Using this observation, Molina and Rowland [9] pointed out the possible destructive role of chlorine. This reaction, however, requires an important energy input to photodissociate O$_2$ in large quantity, available only at low latitudes, and alone cannot account for a massive destruction of ozone. Therefore, in 1985, it was a surprise to discover that ozone was disappearing in the least expected region and time, above Antarctica in winter, and that the destruction was almost total at the stratospheric level where ozone is normally most abundant.

---

[1] The radiative equilibrium in the stratosphere is governed by UV absorption by ozone and IR emission by carbon dioxide

2. Chemistry of the ozone hole

As we shall see shortly, the story is somewhat intricate and appears as a conspiracy between chemistry, microphysics and hydrodynamics where transport and mixing processes play an essential role. Let us begin by the chemical part. The most effective cycle in the polar region was described by Molina and Molina in 1987 [10]:

$$
\begin{aligned}
ClO + ClO + M &\rightarrow ClOOCl + M^* \\
ClOOCl + h\nu &\rightarrow Cl + ClOO \\
ClOO + M &\rightarrow Cl + O_2 + M^* \\
2(Cl + O_3 &\rightarrow ClO + O_2) \\
Net : 2O_3 &\rightarrow 3O_2
\end{aligned}
\tag{2}
$$

The slowest reaction is the first one which determines the kinetics of the whole cycle. Therefore, ozone destruction is basically given by

$$
\frac{\partial}{\partial t}[O_3] = -k[ClO]^2
\tag{3}
$$

One denotes usually as $[Cl]_x = [ClO] + [Cl] + 2[ClOOCl]$ the concentration in active chlorine in the atmosphere (by opposition with the reservoir species HCl and $ClONO_2$). Under normal circumstances, this concentration is less than 0.1 ppbv (part per billion in volume). With such value, the cycle would take more than 5000 days to destroy ozone assuming the absence of regeneration. It is therefore a negligible effect. The concentrations reached within the polar stratosphere are, however, up to 3 ppbv which leads to a total destruction of ozone within less than one week. A second cycle where one of the ClO is replaced by BrO in (2) plays a significant role in ozone destruction. As we shall see shortly, the two cycles behave very differently with respect to mixing.

In order to account for so large a concentration in active chlorine, it is necessary to convert most of chlorine from the reservoir species to the active form. The process involves heterogeneous reactions at the surface of ice crystals. Since the stratosphere is very dry, water seldom condensates and form clouds. However, when the temperature falls below 195K, a condition met in winter above polar regions, ice clouds can form. Some chemical reactions which are forbidden or ineffective within the gas are then possible on the surface of ice crystals. They transform chlorine into its active form while the other product of the reaction, HNO_3, incorporates and is being trapped within the crystal. For instance, one gets

$$
ClONO_2 + HCl \rightarrow Cl_2 + HNO_3 \,.
\tag{4}
$$

The trapping of nitrated components is completed by the reaction

$$
N_2O_5 + H_2O \rightarrow 2HNO_3 \,,
\tag{5}
$$

which is very important because it eliminates also the radicals NO, NO_2 and NO_3 (which are in equilibrium with N_2O_5). These radicals usually limit the presence of active chlorine by the recombination reaction

$$
ClO + NO_2 \rightarrow ClONO_2 \,.
\tag{6}
$$

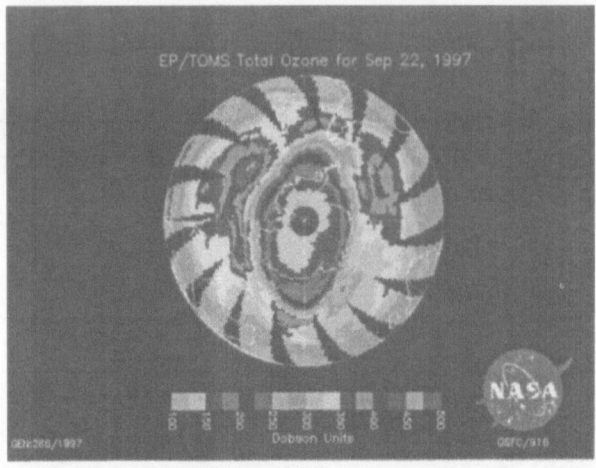

Fig. 1. — Chart of total ozone integrated on the vertical over the South hemisphere on September 22,1997. Units in DU. 1 DU is equivalent to a layer of 0.01 mm of ozone at 0°C and sea surface pressure, or to $2.7\ 10^{22}$ molecules.m^{-2}. The measurements are done by the Total Ozone Mapping Spectrometer (TOMS) on board Earth-Probe satellite of NASA. The very small value ($<$ 150 DU) of the ozone column within the polar vortex over Antarctica is typical of the ozone hole when ozone has almost entirely been removed from the lower stratosphere.

The elimination of nitrated compounds is completed if the crystals are big, for temperature below 185K, and the clouds persist long enough to leave time for sedimentation, at a falling rate of few hundreds of m per day, to wash out the stratosphere. These conditions are precisely those which are met during winter in the Antarctic stratosphere at the altitude (20 km) where ozone is normally most abundant. In August, when the polar stratosphere, denitrified and enriched in chlorine, receives afresh solar radiations, the process of ozone destruction is engaged without limit until total destruction, as seen in fig. (1).

3. Transport in the stratosphere

Within this process, fluid dynamics is responsible of the existence and the stability of the polar vortex. The stratospheric circulation is basically dominated by the thermal wind balance which combines the balance of horizontal pressure gradient by the Coriolis force with the hydrostatic equilibrium in the vertical. Hence, a vertical wind gradient results from the horizontal temperature gradient, rotated by 90° in the northern hemisphere and by −90° in the southern hemisphere. In the winter hemisphere, ozone heating establishes a temperature gradient towards the equator which produces strong cyclonic winds blowing eastward. They are particularly intense near the polar-night terminator where the absorption of short-wave radiation vanishes. In the summer hemisphere, the high altitude solar flux is greater at the pole than at the equator. This establishes a temperature gradient towards the pole and generates anticyclonic winds blowing westward.

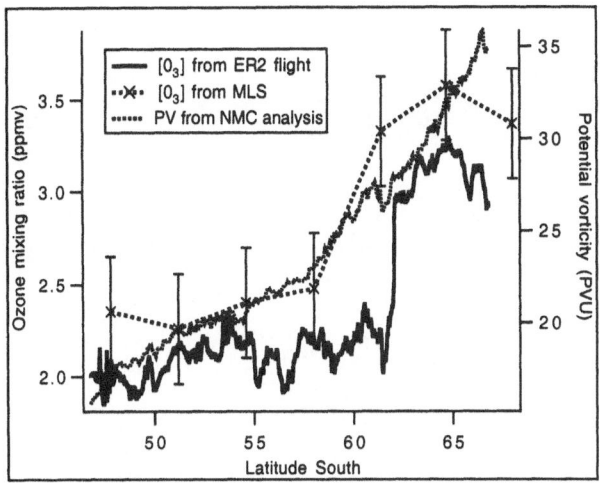

Fig. 2. — In solid line, the ozone mixing ratio measured on board the ER-2 aircraft flighing southward on a meridional track on August 10, 1994. The aircraft was flying at approximately a constant potential temperature level 470 K. Units are ppmv. At this time of the year the ozone hole is not formed and the vortex is still rich in ozone. The transition between subtropical air to polar air near 62S occurs very abruptly in less than 5 km. As the spatial resolution of 3.8 km is limited by the response time of the instrument, the existence of an even sharper boundary cannot be excluded. The dashed curve shows the satellite measurements provided by the Microwave Limb Spectrometer (MLS) on board the Upper Atmosphere Research Satellite. The satellite track is close to but does not exactly coincide with the track followed by the aircraft. The comparison illustrates that satellite measurements reach a fairly good agreement with in situ data but are far from resolving the edge of the vortex. The dotted line shows the potential vorticity profile from the operational weather analysis of the National Meteorological Center. Units are PVU with 1 PVU = 10^{-6} $K.m^2.s^{-1}.kg^{-1}$). From dynamical arguments (see the text) this curve is expected to exhibit a jump on the edge of the vortex. The absence of this feature indicates that meteorological analysis do not resolve the vortex edge.

Under strict radiative equilibrium, the air parcels experience no net heating and the potential temperature is conserved[2]. There is no vertical motion across the potential temperature surfaces and by continuity no meridional motion. However, the observed temperature depart from the radiative equilibrium. At the top of the stratosphere near 50 km, above the winter pole, the temperature easily exceeds the thermal equilibrium by more than 50K. The net cooling above the pole has to be balanced by adiabatic compression and heating by a slow descending motion [12].

The dynamical properties of the polar vortex largely contribute to the ozone hole

---

[2] The potential temperature is defined as the temperature obtained by adiabatic compression to a reference pressure p_0. Using the perfect gas law, one gets

$$\theta = T \left(\frac{p_0}{p}\right)^{R/C_p} .$$

Since θ grows monotonically with altitude in astable atmosphere and is conserved under adiabatic motion, it is often used as a vertical coordinate.

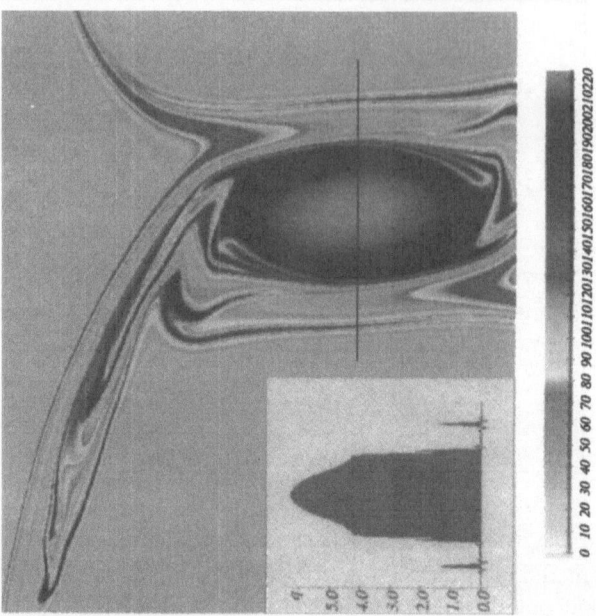

Fig. 3. — Vorticity field of a vortex submitted to a uniform adverse shear at a late stage after the onset of stripping. Filaments surround the vortex and a sharp boundary develops as seen in the displayed vorticity cross section.

because it creates a large-scale container which maintains the polar air isolated from the mid-latitudes. In-situ chemical measurements made by research aircraft flying in the lower stratosphere show a very well defined vortex edge characterized by a very sharp variation of concentrations over horizontal distances of a few kilometers (see figure 2). This chemical edge is a manifestation of the vertical stratification turned into horizontal gradient by the descent inside the vortex.

The very cause of the containment and the sharpness of the edge lies in quasi-horizontal motion. Owing to the strong stratification of the stratosphere where Richardson number is always larger than 100, the air motion takes place basically within quasi-horizontal surfaces which slide one upon the other. These surfaces are isentropic surfaces, usually labelled with the potential temperature θ. Near 20 km, the adiabatic approximation remains valid for about 2 weeks during which the particles remain within less than 1 km from the original θ-surface.

Under the same approximation, the flow possesses a particle invariant, the Ertel potential vorticity $P = (2\vec{\Omega} + \vec{\nabla} \times \vec{u}) \cdot \vec{\nabla}\theta/\rho$ which is a generalization of the conservation of vorticity in two-dimensional Euler equation to three-dimensional flows. The conservation of P is an exact result for any flow in the absence of heating and body forces. It does not rely on the motion being layerwise but is particularly useful in this case as it allows to draw an analogy with two-dimensional turbulence which has been well studied, both numerically and experimentally. This type of flow is known to easily generate long live vortices which trap the fluid particles inside[3]. However, the flow is not fully isolated but is submitted to the deformation of external straining field produced by other vortices (or any imposed non-uniform flow).

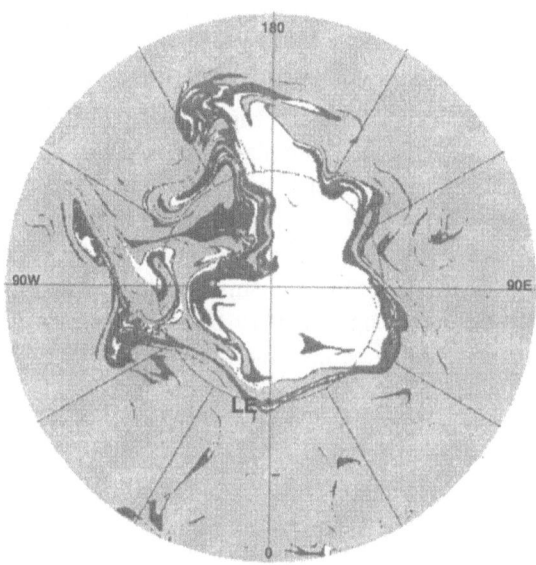

Fig. 4. — Reconstruction of the potential vorticity on the isentropic surface $\theta = 375\mathrm{K}$ on January 8, 1992 at 1200 UT using contour advection with surgery calculations initiated on January, 1st. The indicated station is Lerwick at 60.1N and 1.2E. The contours are for $P = 9$ and 11 PVU.

The superposition of a vortex having closed streamlines and an external straining field (i.e., $u = \gamma x - \Omega y, v = -\gamma y + \Omega x$) yields a total stream function field having : (i) two critical points, often located inside the vortex, (ii) and associated separatrices passing through these critical points, capable of transporting fluid to large distances from the vortex into the background flow. Figure 3 shows a vortex within a row submitted to an adverse shear. The filaments around have been produced by shredding vorticity from the vortices in the row and by stretching and folding the bits around. The irreversible removal of vorticity affects only the periphery of the vortex while the core undergoes a reversible deformation. The dividing flow near the critical points works like a knife peeling an apple; it generates the sharp boundary seen in the profile of fig. 3. For some combinations of γ and Ω, the vortex may be destabilized, but only rarely external fluid penetrates within an established vortex.

Figure 4 shows that potential vorticity in the stratosphere closely follows the behaviour of two-dimensional turbulence. This figure is obtained as a two-stage reconstruct, since direct data processing would not allow this kind of resolution. The first stage of the reconstruction is the so-called analysis provided by operational weather forecast models shown in fig. 5 at a different level. These models, which are based on a comprehensive description of atmospheric processes, integrate numerical equations for about 10^7 variables representing an atmospheric state. The observations provided by the meteorological network are used to constrain the evolution of the system. In spite of the increasing use of satellites, there are much less observations than internal degrees of freedom, by about a factor 10, which are unevenly distributed

(³) In particular, the stratosphere above 20 km is only observed by teledetection with a crude resolution and there are large areas, especially over the Pacific and Southern Hemisphere which are almost void of in situ measurements. Therefore, the information contained in fig. 5 is to a large extend a result of the interpolation and extrapolation of data by the model. Still, the accuracy of this analysis is insufficient to resolve the edge of the vortex and the filaments (see also fig. 2). The second stage for obtaining fig. 4 is based on the assumption that potential vorticity is transported as a passive scalar on surfaces of constant potential temperature. It is also assumed that it is sufficient to know the large-scale component of the wind, which is provided by the operational analysis, and that the small-scale non analysed fluctuations can be neglected. The rationale is that velocity fluctuations over a distance r scale as r for layered turbulence, unlike three-dimensional turbulence for which they scale as $r^{1/3}$. Hence, the local strain is governed by the large-scale structures as long as three-dimensional turbulence can be neglected. The cutoff scale below which three-dimensional mixing cannot be neglected is estimated by P. Haynes in this issue to be of the order of 10 km. It is likely, however, that vertical mixing cannot be characterized by a single number as there are numerous evidences that it is sparse and depends on the regime of the flow.

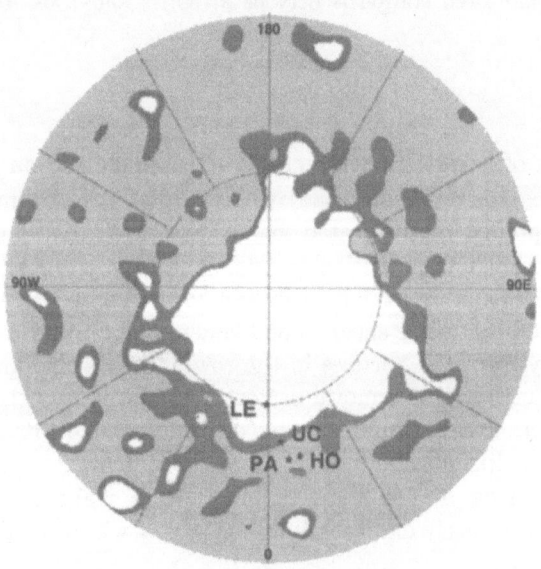

Fig. 5. — Analysed potential vorticity on the isentropic surface $\theta = 450K$ on January 8, 1992 at 1200 UT from European Center for Medium Range Weather Forecast analysis. The contours are for $P = 11$ and 14 PVU.

(³) One may wonder what is the true number of degrees of freedom of the atmosphere. Some strangely small dimensions have sometimes appeared in the literature due to undiscriminating use of box counting algorithm but there is a general belief that this number is huge.

Figure 4 is obtained by integrating in time the transport equation

$$\frac{D}{Dt}P + \vec{v} \cdot \vec{\nabla}P = 0, \tag{7}$$

on the surface $\theta = 375K$ (about 13 km), using the analysed horizontal wind $\vec{v}$ on this surface and starting from the analysed distribution of potential vorticity (which has the same resolution as fig. 5) 10 days earlier[8]. The vorticity field is discretized as nodes on a small number of contours. As the integration proceeds, renoding increases and relocates the nodes, in order to maintain a very high accuracy [1, 17, 11]. This method has been used over the recent years to predict the location of filamentary structures. The predictions generally agree fairly well with aircraft measurements[18] in the northern hemisphere. Here we present (cf fig. 6) a vertical sounding of ozone which exhibits a laminae associated with the crossing of the filament shown at the bottom of fig. 4. The agreement with observation is degraded in regions where uncertainties on the wind field are large. This is the case of most of the southern hemisphere and also of the tropical region where in addition one has to take into account significant vertical motion at the bottom of the stratosphere and to consider the role of badly analysed gravity waves.

The crucial role of the polar vortex in the ozone hole is to create an isolated container which allows the polar air that does not mix with mid-latitude air to reach the very low temperatures necessary for heterogeneous chemistry to take place and maintains the activated chlorine separated from the mid latitude air containing nitrated species able to deactivate it. In the Antarctic, the polar vortex is stable enough to persist until late spring as it does not usually disappear before mid-November. Therefore, activated air undergoes a prolonged exposure to solar light which yields to total destruction of ozone. The lowest ozone concentrations are usually recorded in late October before the vortex breaks and mixes with subtropical air. Breaking leads to i) a dissemination of the ozone loss, ii) a supplementary destruction of ozone as chlorine encounters ozone, iii) and a rapid halt of the process as chlorine reacts with nitrated radicals and returns to the inactive reservoir species. Before the final breaking, significant amount of polar air may have moved to mid-latitudes within filaments shredded from the vortex.

The Arctic situation is very different. During winter the vortex is much more disturbed by planetary waves associated with orography and land-sea contrast and it hardly persists after mid-March. Owing to the sparse distribution of low temperatures, denitrification proceeds only to a limited extend. Although a large amount of the polar air is activated during winter, a lot of chlorine returns to the reservoir species before having a chance to destroy ozone. Natural variations of ozone in the Arctic within one season or from one year to the other, are also larger than in the Antarctic and can mask a partial depletion. Therefore, it is necessary to compare the observed evolution of ozone with that calculated by models in the absence of chemistry to establish that ozone is actually destroyed in the Arctic. During the recent winters 1994-1995 and 1995-1996, the temperatures remained unusually cold in the stratosphere , leading to a destruction of about 30 % of the total ozone within the vortex [7, 4, 5].

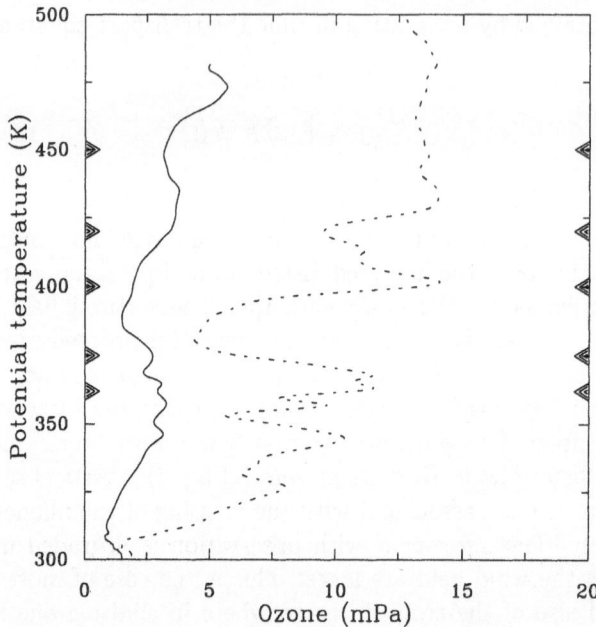

Fig. 6. — Ozone partial pressure profile measured by a probe carried under a balloon launched from Lerwick on January 8, 1994 at 11TU. Units are in mhPa. The whole structure of the vortex is sloping southward by about 500 km in the horizontal for 5 km in the vertical direction. Soon after entering the stratosphere, the rising balloon samples the rich in ozone filament of polar air shown near Lerwick. Between 370K and 400K, over about 3 km, the balloon crosses the pool of poor in ozone subtropical air lying between the filament and the body of the vortex. This is seen as a strong laminae in the profile. After entering the vortex, the balloon crosses near 420K a region of partially mixed air which is partly seen at lower level in fig. 4. At 450K, the balloon is well inside the vortex (see fig. 5)

4. Sensitivity to mixing

The mechanisms of ozone destruction has been reproduced in numerical models. State-of-the-art three dimensional chemistry and transport models (CTM) include a very comprehensive chemistry with about 50 chemical species and 150 chemical and photolytic reactions. The species react together and are transported using the analysed three-dimensional winds. The predictions of the models are in fairly broad agreement with the observations (see Ref [6] for a detailed account) but some quantitative discrepancies are apparent. For instance, the models reproduce the extension of the ozone hole in the Antarctic but fail by about 50 % to account for the smaller destruction of ozone during Arctic winter. Several possible causes have been proposed for this discrepancy championing the role of new chemical mechanisms such as heterogenous reactions on liquid aerosols. A detailed discussion of these issues is out of the scope of this paper. Here we rather concentrate on the role of the lack of resolution. Owing to the large number of transported species and reactions in a CTM, only very coarse resolution can be afforded. A typical advanced CTM has 30 levels in the vertical and a resolution of 200 km in the horizontal; and many use lower resolution. It is therefore necessary to consider the role of unresolving the smaller scales.

Grossly speaking, mixing occurs on quasi-horizontal isentropic layers down to a scale of about 10-20 km or less at which three dimensional motion begins to dominate. The horizontal mixing is chaotic with the local strain being dominated by large-scale structures. Mixing by three dimensional motion is dominated by local structures and is much faster, leading rapidly to the molecular scale at which chemical species really mix. Hence, a filament of polar air transported towards mid-latitudes preserves its chemical character until its width reaches the crossover scale to three dimensional regime after which it rapidly mixes with subtropical air. Deactivation of chlorine by reaction 4 ensues, halting the destruction of ozone. In a coarse resolution CTM, the filament undergoes earlier mixing, leading to premature deactivation of chlorine and to underestimate of ozone loss. Along the edge of the polar vortex, where steep concentration gradients tend to form, excessive diffusion in CTM enhances mixing across the edge, leading again to spurious deactivation. This limiting process due to lack of resolution takes place where ozone destruction is normally most effective because the insolation is large only at the periphery or outside the vortex. Haynes in this issue and ref. [15] discuss this effect in more details. It has been proposed [16] to correct from resolution effects in CTM by decreasing the reaction rate of the deactivation reaction.

A second type of sensitivity to mixing arises from the nonlinearity of reaction (3). Let us consider the example of a circular patch of ClO which evolves in the horizontal plane under pure diffusion ν. The self-similar solution is

$$[ClO] = \frac{a_0^2}{a^2}[ClO]_0 \exp\left(-\frac{r^2}{a^2}\right), \tag{8}$$

with $a^2 = a_0^2 + \nu t$. In the case of *fast* chemistry, when $[O_3]/k[ClO]_0^2 \ll a^2/\nu$, O_3 is fully destroyed in the area covered by ClO. Hence, the loss varies as νt. In the case of a *slow* chemistry, that is when $[O_3]/k[ClO]_0^2 \gg a^2/\nu$, O_3 is only partly destroyed and the loss varies as $ka_0^2\nu^{-1}\ln(a^2/a_0^2)$. The rationale is that when ClO is diluted two times it pollutes twice the initial domain but its concentration is divided by 2, so the destruction rate is divided by 4. Therefore, according to the initial conditions, the ozone loss can either increase or decrease as the diffusivity increases. This effect has been studied in ref. [2] using a simplified ozone chemistry and a high resolution transport model. It is found that ozone depletion in the Arctic falls in the case of slow chemistry and is highly sensitive to dilution of ClO as long as the cycle (2) dominates. The estimate of the error committed by CTM is of the order of the observed discrepancy. On the contrary, the Antarctic depletion is not sensitive to dilution, since a broad part of the polar air is almost fully depleted, and there is plenty of available ClO. The depletion is then basically proportional to the area of the vortex. The cycle involving BrO + ClO is not sensitive to dilution because active bromine is well mixed. Its contribution to ozone depletion depends on the average concentration of ClO and not on its variance.

5.   Discussion

The destruction of ozone in the polar stratosphere requires a combination of chemical, microphysical and hydrodynamical processes which is now firmly established.

Transport and mixing is one of the main pieces of the puzzle as it determines to a large extend how the chemical species are distributed and can react together. Strangely enough, the dynamics establishes some barriers to transport which confine chemical species in domains where different conditions and chemistry prevail. The detailed and quantitative understanding of the low stratosphere is, however, far from being completed. Ozone disappears at mid-latitude at a rate which is not accounted by models. The exchanges between the upper troposphere and the stratosphere, in particular in the tropical region, are still poorly understood. The role of gravity waves in transport and mixing is hardly known. As the spatial and temporal resolution of observations will increase, it is likely that the study of small-scale distributions will exhibit further discrepancies with models and that atmospheric data will offer an application field to elaborate theoretical tools.

References

[1] Dritschel, D., Contour dynamics and contour surgery: numerical algorithms for extended, high-resolution modelling of vortex dynamics in two-dimensional, inviscid, incompressible flows. *Computer Phys. Rep.* **10**, 3 (1989), 77–146.

[2] Edouard, S., Legras, B., Lefè,vre, F., and Eymard, R. The effect of small-scale inhomogeneities on ozone depletion in the Arctic. *Nature* **384** (1996), 444–447.

[3] Elhmaï,di, D., Provenzale, A., and Babiano, A. Elementary topology of two-dimensional turbulence from a Lagrangian viewpoint and single-particle dispersion. *J. Fluid Mech.* **257** (1993), 533–558.

[4] Goutail, F., Pommereau, J., Phillips, C., Sarkissian, A., , Lefè,vre, F., Kyro, E., Rummukainen, M., Ericksen, P., Andersen, S., Kaastad-Hoiskar, B., Braathen, G., V., D., and Khatatov, V. Total ozone depletion in the Arctic during the winters of 1993-94 and 1994-95. *J. Atmos. Chem.* (1997), in press.

[5] Hansen, G., Svenø,e, T., Chipperfield, M., Dahlback, A., and Hopps, U.-P. Evidence of substantial ozone depletion in winter 1995/1996 over Northern Norway. *Geophys. Res. Lett.* **24**, 7 (1997), 799–802.

[6] Lefèvre, F., Brasseur, G. P., Folkins, I., Smith, A. K., and Simon, P., Chemistry of the 1991-1992 stratospheric winter: Three-dimensional model simulations. *J. Geophys. Res.* **99**, D4 (1994), 8183–8195.

[7] Manney, G., Froidevaux, L., Waters, J., Santee, M., Read, W., Flower, D., Jarnot, R., and Zureck, R., Arctic ozone depletion observed by UARS MLS during the 1994-95 winter. *Geophys. Res. Lett.* **23**, 1 (1996), 85–88.

[8] Mariotti, A., Moustaoui, M., Legras, B., and Teitelbaum, H., Comparison between vertical ozone soundings and reconstructed potential vorticity maps by contour advection with surger. *J. Geophys. Res.* **102**, D5 (1997), 6131–6142.

[9] Molina, M., and Rowland, F., Stratospheric sink for chlorofluoromethanes : chlorine atom catalysed destruction of ozone. *Nature* **249** (1974), 810.

[10] Molina, M. J., and Molina, L. T., Production of Cl_2/O_2 from the self-reaction of the ClO radical. *J. Phys. Chem.* **91** (1987), 433–436.

[11] Norton, W. A., Breaking Rossby waves in a model stratosphere diagnosed by a vortex-following coordinate system and a technique for advecting material contours. *J. Atmos. Sci.* **51**, 4 (1994), 654–673.

[12] Rosenfield, J., Newman, P., and Schoeberl, M., Computation of the diabatic descent in the stratospheric polar vortex. *J. Geophys. Res.* **99** (1994), 16,677–16,689.

[13] Rowland, F., Stratospheric ozone depletion. *Ann. Rev. Phys. Chem.* **42** (1991), 731.

[14] Solomon, S., Progress towards a quantitative understanding of Antarctic ozone depletion. *Nature* **347** (1990), 347.

[15] Tan, D., and Haynes, P., The effects of fluid-dynamical stirring and mixing on the deactivation of stratospheric chlorine. *submitted to J. Geophys. Res.* (1997).

[16] Thuburn, J., and Tan, D. G., A parameterization of mixdown time for atmospheric chemicals. *J. Geophys. Res.* **102**, D11 (1997), 13,037–13,049. CGAM, Reading.

[17] Waugh, D., and Plumb, R., Contour advection with surgery: A technique for investigating fine scale structure in tracer transport. *J. Atmos. Sci.* **51**, 4 (1994), 530–540.

[18] Waugh, D., Plumb, R., Atkinson, R., Schoeberl, M., Lait, L., Newman, P., Lowenstein, M., Toohey, D., Avallone, L., Webster, C., and May, R., Transport of material out of the stratospheric Arctic vortex by Rossby wave breaking. *J. Geophys. Res.* **99**, D1 (1994), 1071–1088.

The effect of Schmidt Number on Stratified Entrainment

ALINE J. COTEL

Department of Mechanical and Industrial Engineering
University of Manitoba, Winnipeg, Manitoba, Canada, R3T 5V6.

———————

From a review of the literature, including a new theory on vortex persistence, the effect of Schmidt (or Prandtl) number on stratified entrainment is discussed. In contrast to unstratified turbulence, stratified turbulence entrainment depends on diffusive properties of the stratifying agent. Stratification inhibits the large scale motions of the flow which would otherwise dominate entrainment, so that smaller scale processes become important. At sufficiently large Richardson number, the diffusion of solutal (or heat) at small scales determines the entrainment rate. Thus stratification can act as a probe of the turbulence, suppressing normally dominant features to reveal weaker, more subtle effects.

1. Introduction

In unstratified turbulence, the entrainment process is controlled by engulfment of the large scale vortices (Roshko 1976 [1]). Because these vortices are essentially inviscid, the entrainment rate is independent of Reynolds number. Therefore the spreading angles of shear layers, jets, and the like are constant for all Reynolds numbers above a few thousand, at the mixing transition where the small scale turbulence develops.

While the entrainment rate is independent of molecular properties, the amount of molecular-scale mixing within the turbulent layer does depend on the molecular diffusivity. Thus it depends (at fixed Schmidt or Prandtl number) on the Reynolds number. However, the viscosity is not important, so the true dependence is on the Péclet number, the product of Schmidt and Reynolds numbers. This can be

Mixing: Chaos and Turbulence, edited by Chaté *et al.*
Kluwer Academic / Plenum Publishers, New York 1999.

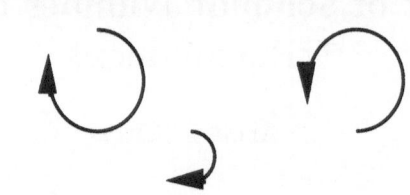

Fig. 1. — Turbulence below a stratified interface in the limit of large Ri.

understood in terms of Taylor layers, or 'flame sheet' (Broadwell and Breidenthal 1982 [2], Broadwell and Mungal 1991 [3]). Some mixed fluid resides with diffusive Taylor layers, the thickness of which depends on the Péclet number. So unlike entrainment, unstratified mixing depends on molecular diffusivity. The entrainment and mixing processes are related but not identical.

However, when the fluid is stratified, the engulfment by large scale vortices is inhibited. Consider for simplicity a thin stratified interface with turbulence beneath it as sketched in figure 1. Suppose the turbulence obeys a Kolmogorov spectrum. One can then define an eddy Richardson number, based on the potential energy of pulling a tongue of buoyant fluid across the interface a distance equal to the eddy diameter, and based on the kinetic energy of the eddy . The eddy Richardson number is a measure of the engulfment ability at the stratified interface. Only if it is less than unity can the corresponding eddy entrain at the interface (Breidenhtal 1992 [4]). If the eddy Richardson number of the largest eddies exceeds unity, then such eddies can neither complete a rotation, pull down a tongue of buoyant fluid, nor engulf it. Only smaller eddies can participate in entrainment. Presumably, the eddy size dominating the entrainment process would be the largest one which satisfies this eddy Richardson number criterion. Therefore the entraining eddy scale is that distinguished by the condition that its eddy Richardson number is unity, sometimes called the Ozmidov scale (Turner 1973 [5]).

In the limit of extreme stratification, so that even the smallest eddies at the Kolmogorov microscale have an eddy Richardson number of unity or greater, then engulfment is not possible by any existing eddy, even the smallest ones. The stratified interface is then essentially flat. It seems clear that in this limit diffusion must become important to the entrainment process, since advective turbulent fluid motions can not transport fluid across a flat interface. At least in this limit, diffusion of the stratifying agent (salt or heat for exemple) across the interface controls the entrainment rate.

Therefore, one expects diffusion to become important for entrainment at least at large Richardson number (Ri). Precisely where does this first happen? Is it at the appearance of a flat interface? As will be shown below, diffusion can become important even at lower Richardson number, when the interface is still highly convoluted.

2. Turner's experiments

The first evidence that diffusion can affect stratified entrainment for non- flat interfaces was the classical stirring grid experiments of Turner (1968) [6]. In a water tank, he measured the entrainment rate of an oscillating grid when the stratifying agents were either salt or heat. With salt stratification, the entrainment rate was proportional to $Ri^{-3/2}$ for Ri greater than one. With heat stratification, the results were identical for Ri less than about seven. However, for Ri greater than seven the heat-stratified entrainment was proportional to Ri^{-1}. The entrainment rate now depends on the diffusivity of the stratifying agent. If the agent is salt, then the dimensionless number is the Schmidt number. If the agent is heat, then it is the Prandtl number.

3. Kolmogorov nibbling of Taylor layers

Turner's discovery reveals several important clues of the physics. First, the diffusion transition occurs at a Richardson number of seven, which is about the Prandtl number of water. Second, the entrainment rates on either side of the diffusivity transition appear to be independent of Reynolds number, although a thorough experimental examination of this assumption has not yet been done, to the author's knowledge. Third, the entrainment rates on both sides of this transition still depend strongly on Richardson number. Consequently, the interface can not be flat. Diffusion is controlling the entrainment process even though the interface is highly contorted by turbulent motions. Finally, the transition is smooth and the entrainment rate is continuous across it.

A proposed theory attempts to account for all the evidence listed above (Breidenthal 1992 [4], Cotel 1995 [7], Cotel & Breidenthal 1996 [8]). The entrainment rate can be defined in terms of the ratio of the relevant length and time scales of the entrainment process. Since diffusion is important, the most logical explanation is that the Taylor layers have somehow become involved in the entrainment process. Because the impinging eddies are nonstationary or nonpersistent, the relevant large eddy size which defines the Taylor layer is the rebound eddy of Linden (Linden 1973 [9], Cotel & Breidenthal 1996 [8]). This Taylor layer thickness sets a length scale and is equal to $(D\tau_r)^{1/2}$ where D is the stratifying agent diffusivity and τ_r is the time scale based on the rebound eddy.

Since the largest eddies are presumed to be ineffective at entrainment for Richardson numbers greater than unity, there are only two distinguished eddy sizes remaining: the Ozmidov scale and the Kolmogorov microscale. It can not be the former, since this would lead to a Reynolds number dependence on the entrainment rate, in conflict with the assumption mentioned above. Thus the relevant time scale must be that of the Kolmogorov microscale associated with the rebound vortex: τ_{λ_0}. It is encouraging that this choice eliminates the Reynolds number dependence. The entrainment velocity is then:

$$W_e = \frac{(D\tau_r)^{1/2}}{\tau_{\lambda_0}} \; . \tag{1}$$

Physically, the picture is of diffusive Taylor layers, associated with the rebound

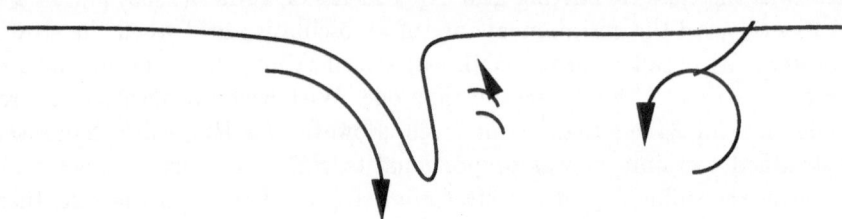

Fig. 2. — Kolmogorov nibbling of Taylor layers.

vortex, which are nibbled on by the Kolmogorov microscale vortices (Figure 2). These little vortices have an eddy Richardson number less than one, provided the interface is not flat, so they are capable of engulfing buoyant fluid. From their viewpoint, the interface in not thin, but rather thick, the Taylor layer.

4.   Flat interface case

When the Ozmidov scale becomes as small as the smallest eddies, i. e. the Kolmogorov microscale, then the interface becomes effectively flat. The entrainment rate must be the ratio of a diffusion length scale and an eddy time scale, both of which depend on the choice of eddy size. There are only two possible choices of distinguished eddy sizes, the largest and the smallest.

The new model proposes that under some conditions the choice is the largest eddy scale, while under other conditions, it is the smallest. The parameter which determines which is called the 'persistence'. It is a measure of the stationarity of the largest eddies with respect to the interface. The persistence is defined to be the number of rotations a vortex makes during the time interval it moves a distance equal to its own diameter with respect to the interface. If the vortex makes many rotations while remaining in one place, the persistence parameter T is large compared to one. More usually, however, the vortex only makes about one rotation during this time, as it convects itself around. Then $T \sim 1$.

The model proposes that if T is sufficiently large, the largest eddies control the flux at a flat interface; if T is small, then the smallest eddies. The critical value of T between the two limits is set by the ratio of velocities of the largest and smallest eddies. Thus $T_{crit} = Re^{1/4}$ (Landau & Lifshitz 1959 [10]). The entrainment velocity is then for the persistent case,

$$W_e = \frac{(D\tau_D)^{1/2}}{\tau_D} \text{ , and} \tag{2}$$

$$W_e = \frac{(D\tau_{\lambda_0})^{1/2}}{\tau_{\lambda_0}} . \tag{3}$$

for the non persistent case.

5. Heat transfer

The implications of this result are counter-intuitive. It says that the fine scale turbulence has no effect on the surface flux if the large eddies are sufficiently stationary. By analogy, the surface flux could be the heat transfer at a solid wall. Indeed, experiments of wall heat transfer confirm that fine scale, freestream turbulence has no effect on wall heat transfer for a forced, laminar boundary layer (Edwards and Furber 1956 [11], Schlichting 1960 [12], Dowling et al. 1991 [13]). Turbulence only enhances the heat flux if the large eddies are moving, as within a turbulent boundary layer.

6. Stratified entrainment model for $Sc \geq 1$

Figure 3 is an entrainment diagram showing the different entrainment regimes as a function of the Reynolds and Richardson numbers for Schmidt number (Sc) greater than one (Cotel 1995 [7], Cotel and Breidenthal 1996 [8]), as discussed in part in the previous sections. The double lines represent a sudden transition between entrainment regimes, and the single line a smooth transition. The Roman numerals represent the regime number, and the entrainment rate is proportional to the indicated value for each regime.

7. Stratified entrainment model for $Sc \ll 1$

Salt diffusion in the ocean has a Schmidt number of about 600, and aqueous heat diffusion has a Prandtl number of about 7. In the atmosphere the Prandtl numbers is about one. The Prandtl number is much less than one for substances like liquid metals. While there may be relatively few flows with Schmidt (or Prandtl) numbers much less than unity, this limit is nonetheless considered for completeness and as a consistency check for the entire model. Figure 4 is a diagram of the theoretical entrainment rate for $Sc \ll 1$. It differs from figure 3 in three respects. First, Regime III has disappeared. This is because a vortex must rotate at least once during the time it moves its own diameter with respect to the interface. Consequently, the minimum possible value of the persistence parameter T is unity, and Sc is assumed to be less than one in figure 4. Second, the boundary between flat and non-flat regimes is shifted slightly. Consider a trajectory in figure 4 of constant (but large) Re and increasing Ri. As Ri is increased, the size of the eddies which are just stratified is reduced; i.e., the Ozmidov scale declines. Eventually Ri is such that eddies only slightly larger than the Kolmogorov microscale are just stratified. If Sc is very much less than unity, the time for molecular diffusion across such a stratified eddy will be much smaller than the convective flux. As a result, the diffusive flux at the stratified interface will be greater than the convective flux. The entrainment rate becomes independent of the convective motions, so that Ri drops out of the problem, just as in the case of the flat interface. Third, the dependence on Schmidt number is different for the flat interface, Regimes VII and VIII, because the thickness of the

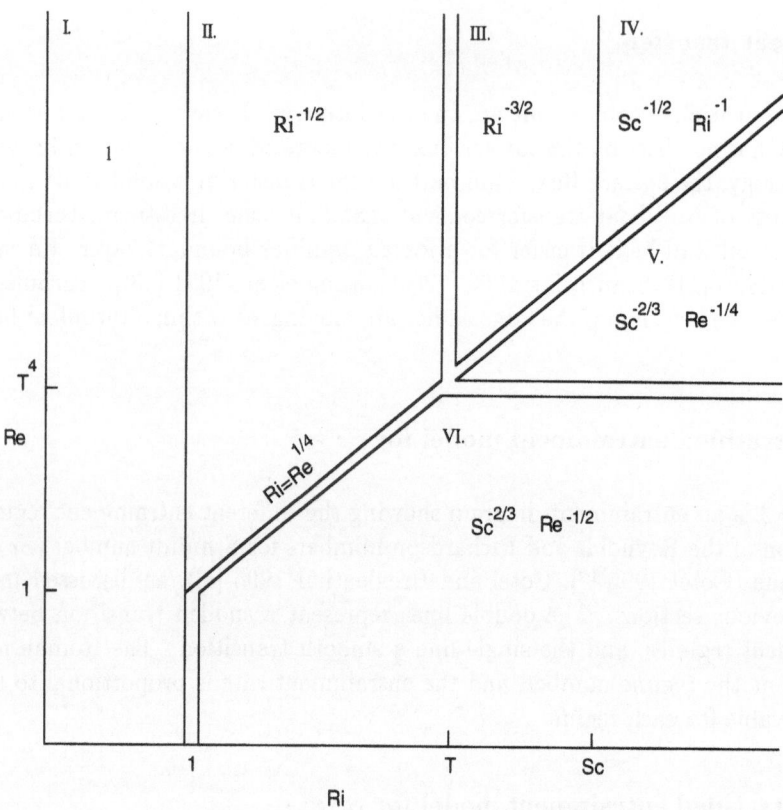

Fig. 3. — Stratified entrainment diagram for $Sc \geq 1$

diffusive interface is no longer small in comparison to the velocity boundary layer there. Therefore, the entrainment velocity is proportional to the Schmidt number to the -1/2 power.

8.   Comparison with experiments

Fernando's results [14] are consistent with Regime IV if his definition of Richardson is changed to the definition of Ri used here. In addition, the magnitude of Fernando's entrainment rate is in approximate agreement with the $Sc^{-1/2}$ coefficient in Regime IV. The non- averaged measurements of Noh and Fernando [15] are also consistent, both with Regime IV and the sharp transition to Regime V. An apparent exception is the oscillating grid experiment of Krylov and Zatsepin [16], who simultaneously measured the heat and salt entrainment rates when the stratification was mainly due to salt. Surprisingly, they found that the addition of salt increased the heat entrainment rate at a fixed Richardson number. However, their result is at least qualitatively consistent with the model. This can be seen by imaging that the salt

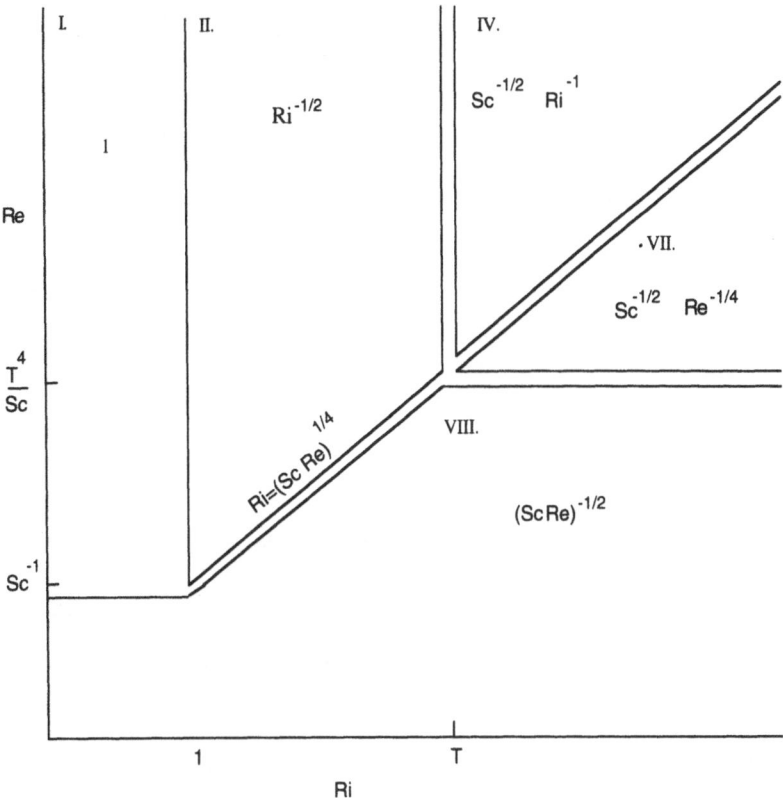

Fig. 4. — Stratified entrainment diagram for $Sc \ll 1$

contribution to the stratification is increased while holding the overall stratification and hence Ri constant. According to the model, the tongues of the rebounding eddies will have Taylor layers subject to nibbling by the Kolmogorov microscale eddies. The temperature Taylor layer will be much thicker than that of salt as a result of the large difference in Prandtl and Schmidt numbers. As the stratification component due to heat is reduced, the ability of a Kolmogorov microscale eddy to entrain a parcel of buoyant fluid from the heat Taylor layer increases. It is as if the effective Richardson number is reduced.

Heat transfer measurements (Elias 1930 [17], Edwards and Furber 1956 [11], and Kestin et al. 1961 [18]) over a wide range of Prandtl numbers satisfy the Colburn relation (Colburn 1933 [19]), which is in approximate agreement with Regime V (Bejan 1984 [20]). However, the effective velocity in the model goes as $Re^{-1/4}$, while in the Colburn formula, it is proportional to the turbulent skin friction. The skin friction does not obey a simple power law, so there is a small difference between the two relations. The measurements may not be accurate enough to resolve the difference. These same measurements are also in agreement with Regimes VI, VII, and VIII.

9.   Conclusion

It was previously thought that the main parameter controlling stratified entrainment was the Richardson number. But from the literature and using new concepts for stratified entrainment, it is proposed that other parameters are also of primary importance, such as the Reynolds, Schmidt numbers, and the newly proposed persistence parameter. The effect of the Schmidt number is particularly subtle since it represents the effects of diffusion even in the case of a convoluted interface. The importance of the Schmidt number is more obvious in the case of the flat interface, since the only possible transport process is diffusion. There is little data for the entrainment at flat interfaces. However, the heat transfer literature is extensive. It provides a useful test of these concepts, but more field and laboratory experiments are needed to fully verified the proposed model.

References

[1] Rosko, A. 1976. Structure of turbulent shear flows: A new look. AIAA J., 14: 1349-1357.

[2] Broadwell, J.E. and Breidenthal, R.E. 1982. A simple model of mixing and chemical reaction in a turbulent shear layer. J. Fluid Mech., 125: 397-410.

[3] Broadwell, J.E. and Mungal, M.G. 1991. Large-scale structures and molecular mixing. Phys. of Fluids A, 3(5) Part 2: 1193-1206.

[4] Breidenthal, R.E. 1992. Entrainment at thin stratified interfaces: The effect of Schmidt, Richardson, and Reynolds numbers. Phys. Fluids A, 4(10): 2141-2144.

[5] Turner, J.S. 1973. Buoyancy effects in fluids. Cambridge University Press.

[6] Turner, J.S. 1968.. The influence of molecular diffusivity on turbulent entrainment across a density interface. J. Fluid Mech., 173: 431- 471.

[7] Cotel, A.J. 1995. Entrainment and Detrainment of a Jet Impinging on a Stratified Interface, Ph.D. thesis, University of Washington.

[8] Cotel, A.J. and Breidenthal, R.E. 1996. A Model of Stratified Entrainment Using Vortex Persistence. Submitted to Applied Scientific Research.

[9] Linden, P.F. 1973. The interaction of a vortex ring with a sharp density interface: a model for turbulent entrainment. J. Fluid Mech., 60: 467-480.

[10] Landau, L.D. and Lifshitz, E.M. 1959. Fluid Mechanics. Pergamon Press, 122.Leveque, M.A. 1928. Les lois de la transmission de la chaleur par convection. Ann. Mines, 13: 201-239.

[11] Edwards, A. and Furber, B.N. 1956. The influence of freestream turbulence on heat transfer by convection from an isolated region of a plane surface in parallel flow. Proc. Inst. Mech. Eng., 170: 941.

[12] Schlichting, H. 1960. Boundary Layer Theory, 6th ed., McGraw-Hill, New York.

[13] Dowling, D.R., Buonadonna, V.R. and Breidenthal, R.E. 1990. Temperature transients in a cylindrical pressure vessel filled from vacuum. J. Thermophysics and Heat Transfer, 4(4): 504-511.

[14] Fernando, H.J.S. 1987. Turbulent mixing in the presence of a stabilizing buoyancy flux. Proceedings of the Third International Symposium on Stratified Flows, ed. E.J. List and G.H. Jirka, Caltech, 447.

[15] Noh, Y. and Fernando, H.J.S. 1993. The role of molecular diffusion in the deepening of the mixed layer. Dynamics of Atmos. and Oceans, 17: 187-215.

[16] Krylov, A.D. and Zatsepin, A.G. 1992. Frazil ice formation due to difference in heat and salt exchange across a density interface. J. Marine Systems, 3: 497-506.

[17] Elias, F. 1930. Der Warmeubergang einer geheizten Platte an stromende Luft. Abhdlg. Aerodyn. Inst. T. H. Aachen, 9; 1929 ZAMM, 9: 434-453; and 10: 1-14.

[18] Kestin, J., Maeden, P.F. and Wang, H.E. 1961. Influence of turbulence on the heat transfer of heat from plates with and without a pressure gradient. Inst. J. Heat Mass Transfer, 3: 133-154.

[19] Colburn, A.P. 1933. A method of correlating forced convection heat transfer data and a comparison with fluid friction. Trans. Am. Inst. Chem. Eng., 29: 174-210.

[20] Bejan, A. 1984. Convective Heat Transfer. Wiley: 248-256.

Dispersion at Large Péclet Number

Yves Pomeau

Laboratoire de Physique Statistique de l'Ecole Normale Supérieure
24 rue Lhomond, 75231 Paris Cedex 05, France
and
Department of Mathematics, University of Arizona, Tucson, AZ 85721, USA

This is a review of results on the dispersion at large Péclet number in flows with a stationary structure. Therein, the molecular diffusion acts like a singular perturbation: although small, in the long run it makes its effects felt at order one. The two basic situations I will consider are: 1) dispersion in spatially periodic parallel rolls; 2) the settling of small particles in periodic rolls.

The dimensionless form of the advection-diffusion equations for steady flows reads (boldface are for vectors):

$$\frac{\partial c}{\partial t}(\mathbf{r}, t) + \mathbf{u}(\mathbf{r}).\nabla c(\mathbf{r}, t) = \frac{1}{Pe}\nabla^2 c(\mathbf{r}, t) \tag{1}$$

In this equation $c(\mathbf{r}, t)$ is the concentration of a passive scalar (as temperature at small Rayleigh number or the concentration of an impurity). The molecular diffusion coefficient D (heat diffusion in the case of temperature, Fick diffusivity for chemical impurities) that should be in front the Laplacian has been replaced by the inverse of the Péclet number defined as $Pe = UL/D$, U order of magnitude of the fluid velocity, L typical length of variation of the velocity field. The physics of the large Péclet number flows can be described in a first, very rough approximation as follows: without the right hand side of (1) (= at a strictly infinite Péclet number), the field $c(\mathbf{r}, t)$ follows the flow lines, without wandering away from the flow lines it started on. In other terms, (1) can be solved formally by the method of trajectories: given an initial concentration field

$$c(\mathbf{r}, t = 0) = c_0(\mathbf{r}) \tag{2}$$

Mixing: Chaos and Turbulence, edited by Chaté *et al.*
Kluwer Academic / Plenum Publishers, New York 1999.

the solution of

$$\frac{\partial c}{\partial t}(\mathbf{r}, t) + \mathbf{u}(\mathbf{r}).\nabla c(\mathbf{r}, t) = 0 \tag{3}$$

is $c(\mathbf{r}, t) = c_0(\mathbf{R}(\mathbf{r}, t))$, where $\mathbf{R}(\mathbf{r}, t)$ is the solution of the trajectory equation:

$$\frac{d\mathbf{R}}{dt}(\mathbf{r}, t) = \mathbf{u}(\mathbf{R}(\mathbf{r}, t)). \tag{4}$$

with the initial condition $\mathbf{R}(\mathbf{r}, t = 0) = \mathbf{r}$.

This solution at infinite Péclet number has already some non trivial features, depending upon the dimension of space:

1) Two-dimensional incompressible and stationary flows (as the ones I shall consider) have a simple structure: the flow lines are either closed, or they can be open in a infinite geometry. From the point of view of the present analysis, the interesting fact is the existence of separatrices: these separatrices are special flow lines ending at hyperbolic stagnation points. The incoming (/outgoing) flow line is an unstable (/stable) manifold. These special flow lines are truly separatrices: a patch of concentration described by (3) that starts by sitting on a separatrix will be split when arriving at the fixed hyperbolic point: part of the patch will be carried along on one side of the unstable manifold, and the other part will be carried to the other side. This splitting breaks the continuity of the concentration field, that is one of the reason (presumably) why large gradients may appear in concentration field of turbulent flows.

2) Three-dimensional (3D) incompressible flows, even when stationary, have a non trivial flow line structure [1]: they are equivalent, in a rather precise sense, to Hamiltonian systems with two degrees of freedom, and may show a complicated mixture of KAM tori and of ergodic flow lines filling a finite volume. Nevertheless, for the problem of settling in a real 3D periodic flow that I shall consider, it seems that the flow lines were mostly non chaotic, either closed or winding on tori, so that the random dispersion leading ultimately to the settling was due to the small effect of molecular diffusion (and was a Péclet number effect).

3) Although this is not the main topic of this review, I make here a few comments upon real turbulent flows, that are both 3D and time dependent. A non trivial point is the possibility of splitting by separatrices (again in the absence of molecular diffusion). To understand this point, let us consider the dispersion equation for a time dependent flow:

$$\frac{\partial c}{\partial t}(\mathbf{r}, t) + \mathbf{u}(\mathbf{r}, t).\nabla c(\mathbf{r}, t) = 0 \tag{5}$$

This can still be solved formally as $c(\mathbf{r}, t) = c_0(\mathbf{R}(\mathbf{r}, t))$, where $\mathbf{R}(\mathbf{r}, t)$ is the solution of the trajectory equation:

$$\frac{d\mathbf{R}}{dt}(\mathbf{r}, t) = \mathbf{u}(\mathbf{R}(\mathbf{r}, t), t). \tag{6}$$

with the initial condition $\mathbf{R}(\mathbf{r}, t = 0) = \mathbf{r}$.

The time dependence can be formally incorporated in a fourth coordinate, say $r_4 = t$, such that the fourth component of the velocity field is just 1, and that the equation of motion for the four dimensional "position vector" $\mathbf{R}_4 = (\mathbf{R}, r_4)$ reads as:

$$\frac{d\mathbf{R}_4}{dt} = \mathbf{u}_4(\mathbf{R}_4). \tag{7}$$

the fourth (time) component of the velocity field $\mathbf{u}_4$ being obviously equal to 1. This four dimensional velocity field has a zero divergence, therefore the sum of the Lyapunov exponents along a given trajectory has to be equal to 0. As usual, because this 4D flow is autonomous, one instability exponent has to be equal to zero, the one along the trajectory itself. Generically, out of the three other exponents one at least must be positive to make the divergence equal to zero, making all trajectories exponentially unstable. Therefore, almost all trajectories tend to split exponentially, and can be considered as separatrices, even in the absence of fixed point of this 4D flow. This might explain the formation of sharp gradients of passive scalars in turbulent flows at large Péclet number. Presumably too this phenomenon is not easily reachable by perturbative expansion, because the infinite Péclet number situation yields already highly irregular solutions. Notice too that for 3D stationary flows, the situation is marginal from the point of view of the counting of the exponents, since the sum of two exponents transverse to the flow lines must be zero, which can be realized either with one positive and one negative exponent or with two conjugate purely imaginary exponents, whence the complicated generic structure of KAM tori and of ergodic region, because the KAM tori split the phase space into an outer and an inner domain.

I now review the two situations that have been studied in some details for the large Péclet number limit. First this is the problem of diffusion of a passive scalar in a periodic roll pattern, this pattern being for instance made of Rayleigh-Bénard rolls near threshold. Then, this is the problem of settling of particles in the same kind of roll structure: I intend to explain the remarkable observation that particles settle almost exclusively at the bottom of the up-going separatrices.

Although observations have been made of flow lines in 3D structure, I shall restrict myself to the problem of dispersion in the large Péclet number limit in a periodic 2D pattern of rolls between two parallel plates. These rolls draw an infinite system of equal rectangles, tiling the stripe between two parallel lines (the top and bottom plates). Every rectangle is filled with closed flow lines circling around a fixed point. Each rectangle is limited from below and from above by the plates, and is separated sidewise from its neighbours by two separatrices going from the top to the bottom plate, or conversely, depending on the sense of rotation inside each rectangle. As in Rayleigh-Bénard convection or in Taylor-Couette rolls, the sense of rotation in neighboring cells are opposite. This flow structure is non trivial from the point of view of the long term dynamics defined by the flow lines (without molecular diffusion): inside each cell, the flow lines are closed loops, and a particle starting inside will remain there forever. However the flow lines on the boundary are open and a particle starting there will stay on the boundary forever, jumping from one cell to the next. These separatrices introduce a kind of discontinuity for trajectories, because the equations for the dispersion are hyperbolic and so can have this kind of discontinuous solution. As it is well known to happen in many instances, this discontinuity does not survive once molecular diffusion is taken into account: molecular diffusion makes higher spatial derivatives appear in the equations, that are definitely incompatible with non smooth solutions. To be more concrete, consider the dispersion in periodic system of rolls in 2D. The roll structure enters as a periodic stream function $\psi(\mathbf{r})$, $\mathbf{r} = (x, y)$, x and y Cartesian coordinates. The horizontal (u) and vertical (v) components of the velocity field are:

$$u(\mathbf{r}) = \frac{\partial \psi}{\partial y}(\mathbf{r}) \tag{8}$$

$$v(\mathbf{r}) = -\frac{\partial \psi}{\partial x}(\mathbf{r}) \tag{9}$$

This velocity field is stationary and incompressible, and the flow lines are the level lines of $\psi(\mathbf{r})$. Moreover the stream function is periodic in space, along x:

$$\psi(x + \Lambda, y) = \psi(x, y) \tag{10}$$

Λ being the wavelength. For simplicity, I shall also assume that the flow lines have a mirror symmetry with respect to $x = \Lambda/2$:

$$\psi(\Lambda - x, y) = -\psi(x, y) \tag{11}$$

Furthermore, the boundary conditions are on the "horizontal" plates at $y = \pm a/2$. I shall take the usual "no-slip" condition (some details of the final result depends significantly upon this condition): $u = v = 0$ for $y = \pm a/2$. From the symmetry condition $\psi(\Lambda - x, y) = -\psi(x, y)$, the lines $x = \Lambda/2 \pm n\Lambda$ are vertical separatrices $(u(x = \Lambda/2 \pm n\Lambda, y) = 0$, n integer). I shall analyze now the solutions of the convection diffusion equation for this kind of geometry in the limit Pe $\to \infty$. This limit makes appears various time scales, all related to different processes.

The basic time scale where diffusion enters is the time needed for a particle to diffuse across an elementary cell of the structure. With the scaling leading to the dimensionless form (1) of the convection diffusion equation, this "short" time scale is Pe (to be multiplied by L/U, U typical velocity scale, to get an actual time, L typical length scale, of the same order of magnitude as Λ). Let us take a time TPe $(T \gg 1)$ much larger than this time scale. To understand how a diffusing particle jumps from cell to cell, we can assume that during this time $(T$Pe$)$ the diffusing particle has stayed inside a cell, and add a random jump of Λ in the x-direction every time the diffusing particle hits the boundary of this cell. Let $N(T)$ be this number of jumps. Thus the effective diffusion coefficient in the x-direction is $D_{\text{eff}} = \Lambda^2 N(T)/(TPe)$. $N(T)$ grows linearly with T because of the following remark: at large times, the diffusing particle will diffuse uniformly across the unit cell. Whenever it approaches the lateral boundary (that is a simple representation of the separatrix) at less than a critical distance the particle shall have a non vanishing probability of jumping across the separatrix to the next cell. This will happen with a (small) probability $(T$Pe$/N(T))$ in the course of time, but this can be estimated as well as the ratio of the "critical area" to the total area of the cell, since the cell is uniformly filled for long times. This critical area is the area of the boundary layer near the vertical separatrix where the particle has a non vanishing probability to cross the separatrix during one travel close to this separatrix. The travel time along this separatrix is 1, in our dimensionless units (it is actually of order L/U), and during this time the particle diffuses of a distance Pe$^{-1/2}$ across the separatrix (the "dimensionless" diffusion coefficient is Pe^{-1}), so the "critical area" is of order

$Pe^{-1/2}$ times the total area of the cell, which is the ratio $N(T)/(TPe)$, so that the effective diffusion coefficient is $D_{eff} = Pe^{-1/2}$. This effective diffusion coefficient is the geometric mean of the molecular diffusion coefficient D and of the (much larger) dispersion coefficient UL.

This result was predicted in [2], and the prefactors have been computed in [3, 4, 5], a non trivial but delightful exercise in boundary layer theory. It has been shown too that other intermediate time scales are relevant for describing what happens whenever the number of rolls is not large enough to make the former calculation relevant. Let us assume that a small droplet of dye is put inside a cell. This droplet will spread first inside the cell it started from, until its boundary reaches the separatrix, from where it begins to spread to the other cells. It will do so by creeping first along the separatrices and along the horizontal plates, since these ones draw a connected network. Together with this creeping flow, the dye will diffuse to the center of the cells. The previous theory is valid if the number of cells is large enough to have the creeping flow reaching the lateral boundary of the roll system after the dye has spread-for instance-to the center of the first neighboring cell. The time needed to reach by diffusion the center of a cell from its periphery is of order Pe. Therefore during this time, the number of visited cell is $(D_{eff}Pe)^{1/2}$, which has to be less or of the order of the total number of cells in order to make applicable the previous theory. This means that the total number of cells has to be at least of order $Pe^{1/4}$ to make the system large enough, that might be a severe constraint in real experiments, as the Péclet number may be huge. If this condition of large system is not satisfied, intermediate scaling [6] describe the convection diffusion process, that depend in a detailed way on the boundary conditions.

I shall now turn to the settling of particles in a system of parallel rolls. The experiments show [7] that the particles tend to settle at the bottom of the up-going separatrices, this being slightly counterintuitive, since one would expect the particles to be carried downward by the down-going separatrices. At the Cargèse conference, it was suggested too that this could be explained by inertia-like effects: because of their inertia particles do not follow exactly the flow lines. This was probably not what happened in the experiment, because the particles were quite small (less than one micrometer of diameter), but it could not be excluded in general. The explanation I shall present below is based upon an analysis of the convection diffusion equation, with negative buoyancy added. It is true that the molecular diffusion becomes very inefficient compared to other phenomena, like the inertia effect, as soon as the particles become a slightly larger than a few micrometers. The relative importance of inertia vs. molecular diffusion depends upon some fancy combination of parameters, and can be estimated as follows. The centrifugal force on a particle advected along flow lines of curvature $1/R$ is (up to multiplicative constants of order 1)

$$F_{cent} = \frac{\delta\rho\, r^3 U^2}{R} \quad , \tag{12}$$

$\delta\rho$ density difference between the particles and the fluid, r radius of the particle, and U typical flow velocity. This centrifugal force yields a drift velocity v_{drift} across the flow lines obtained by balancing this force with the Stokes drag:

$$v_{drift} = \frac{\delta\rho\, r^3 U^2}{Rr\eta} \quad , \tag{13}$$

η shear viscosity of the fluid. This yields a typical time scale for the centrifugal drift across the cell, that is obtained by assuming that the flow structure has only one length scale, L ($L \sim R$) that is run with speed v_{drift}. The ratio of this drift time to the diffusion time across the cell is:

$$\frac{t_{\text{drift}}}{t_{\text{diff}}} = \frac{D\eta r^3 U^2}{\delta\rho\, r^2 U^2} \quad , \tag{14}$$

where D is the coefficient of molecular diffusion. This can be put in a nice form by using the Einstein formula for D:

$$D = \frac{k_B\theta}{M}\frac{\rho r^2}{\eta} \quad , \tag{15}$$

with k_B Boltzmann's constant, θ absolute temperature, M mass of the particle and ρ mass density of the fluid.

From this one deduces an obviously dimensionless form for the ratio of time scales:

$$\frac{t_{\text{drift}}}{t_{\text{diff}}} = \frac{\rho v_{\text{th}}^2}{\delta\rho\, U^2} \quad , \tag{16}$$

where $v_{\text{th}}^2 = k_B\theta/M$ is the square of the thermal speed of the diffusing particle.

The calculations that follows assume that the diffusing particle is small so that the ratio $t_{\text{drift}}/t_{\text{diff}}$ is large, making thus the inertia effects negligible compared to molecular diffusion. As explained in [7], the effect of negative buoyancy, ultimately responsible of the settling of the particles to the bottom of the container, is included in the equations of convection diffusion by adding a downward constant velocity field, called $\mathbf{g}$ to emphasize its connection with the gravity:

$$\frac{\partial c}{\partial t}(\mathbf{r}, t) + \mathbf{u}(\mathbf{r}).\nabla c(\mathbf{r}, t) + \mathbf{g}.\nabla c(\mathbf{r}, t) = \frac{1}{\text{Pe}}\nabla^2 c(\mathbf{r}, t) \tag{17}$$

In the absence of any other effect, $\mathbf{g}$ would be the constant speed of sedimentation. This speed will be assumed to be much smaller than the average flow speed, otherwise the particles would settle so fast that their distribution on the bottom plate could not be influenced by the flow pattern, contrary to what is observed [7]. If one neglects again molecular diffusion [the right hand side of (17)], one gets back to the problem of convection in a constant 2D incompressible flow: the velocity field $\mathbf{w} = \mathbf{u} + \mathbf{g}$ is still incompressible if $\mathbf{u}$ is incompressible and $\mathbf{g}$ constant. The only difference with the situation at $\mathbf{g} = 0$ is that the boundary conditions for $\mathbf{w}$ are that $\mathbf{w} = \mathbf{g}$ on the bottom plate, as $\mathbf{u} = 0$ there. Most flow lines of $\mathbf{w}$ are still closed concentric flow lines, but for a thin layer of flow lines going from the top plate to the bottom plate and crossing those two plates at right angle. The thickness of this layer is

$$\epsilon_b = L\left(\frac{g}{U}\right)^2 \tag{18}$$

along the top and bottom plates, and of order $\epsilon_s = L\,(g/U)$ close to the vertical separatrix (explicit dimension are used here to make the argument more transparent-hopefully). The geometry of the flow lines of the $\mathbf{w}$ field is such that there is now on the downward going separatrix of $\mathbf{u}$ an hyperbolic fixed point of $\mathbf{w}$ at a distance ϵ_s

from the bottom plate with an in-going unstable manifold bounding the area filled with closed flow lines, and an outgoing stable manifold on the vertical separatrix pointing to the bottom plate on one side and, on the other side, to another unstable fixed point close to the top plate. The settling takes place close to the intersection of the up-going separatrix of the field $\mathbf{u}$ because this separatrix becomes a short stretch of down-going flow line of the $\mathbf{w}$ field close to the bottom plate, and because molecular diffusion brings the particles starting inside the closed lines of $\mathbf{w}$ near the outer boundary of the domain of closed flow lines. This outer boundary is the unstable manifold of the unstable fixed point of $\mathbf{w}$ close to the bottom plate. Once a diffusing particle has reached the neighborhood of this unstable fixed point, it has a finite chance to be carried along the stable manifold that crosses the bottom plate and to reach the intersection of this bottom plate with the upward going separatrix of $\mathbf{u}$, which is as well locally the downward going separatrix of $\mathbf{w}$.

It is observed in experiments that the lines drawn by the settling particles on the bottom plate are very thin. Let us estimate then this thickness (assuming that attractive microscopic forces between the particles are unimportant, a very unlikely assumption, but to take it into account would bring us too far). This thickness results from the effect of molecular diffusion during the travel along the boundary of the domain of closed flow lines (that turns also to be the unstable manifold of the fixed point of $\mathbf{w}$ close to the bottom plate). This travel time is dominated by the time spent close to the bottom plate. There the horizontal velocity is of order $U(\epsilon_b/L)$, so that the travel time is $L^2/U\epsilon_b$. During this, the particle shall diffuse by molecular diffusion of a distance of $\left(DL^2/U\epsilon_b\right)^{1/2} = L\mathrm{Pe}^{-1/2}g^{-1/2}$ across the flow lines. This defines the typical length scale for the thickness of the settling region on the bottom plate. Note that the validity of this result depends on the fact that the product of a large quantity, $1/\mathrm{Pe}$ and of a small quantity g ($=$ ratio of the settling speed to the typical flow speed) is small. Otherwise, the settling particles would spread quite uniformly on the bottom plate.

To conclude, I would like to emphasize that these questions of diffusion at large Péclet number are quite non trivial, even for simple flow structures, and without any time dependence. It might be that some insight could be gained from these considerations for the far more complex case of diffusion in real turbulent flows, where the large scale geometry is probably quite often a relevant piece of information that should not be washed out by assuming an homogeneous and isotropic turbulent flow, that is only true, if it is ever so, at small scales only.

References

[1] L. de Sèze and Y. Pomeau, J. Physique (Paris) C **5**, 95 (1978).

[2] Y. Pomeau, C.R. Acad. Sc. Paris, **301**, 1323 (1985).

[3] W. R.Young, A. Pumir, and Y. Pomeau, Phys. Fluids A **1**, 462 (1989).

[4] B. Shraiman, Phys. Rev. A **36**, 261 (1987).

[5] M. N. Rosenbluth, H.L. Berck, I. Doxas, and W. Horton, Phys. Fluids **30**, 2636 (1987).

[6] E. Guyon, J.P. Hulin, C. Baudet and Y. Pomeau, Nucl. Phys. B **2**, 271 (1987).

[7] B. Simon and Y. Pomeau, Phys. Fluids A **3**, 380 (1991).

PART IV

STATISTICAL METHODS AND MIXING

PART IV

An Introduction to Chaotic Advection

Julyan H. E. Cartwright[1,2], Mario Feingold[3], and Oreste Piro[1,4]

[1] Departament de Física, Universitat de les Illes Balears,
07071 Palma de Mallorca, Spain

[2] Centre de Càlcul i Informatització
Universitat de les Illes Balears, 07071 Palma de Mallorca, Spain

[3] Department of Physics, Ben-Gurion University,
Beer-Sheva 84105, Israel

[4] Institut Mediterrani d'Estudis Avançats (CSIC–UIB), IMEDEA,
07071 Palma de Mallorca, Spain

———————

Understanding particle advection in incompressible laminar fluid flow, apart from being of theoretical interest, holds much relevance for technological applications. Properties of emulsions, dispersion of contaminants in the atmosphere and ocean, sedimentation, and mixing, are just a few examples. Chaotic advection is the complex behaviour a passive scalar — a fluid particle, or a passively advected quantity such as temperature or concentration of a second tracer fluid — can attain, driven by the Lagrangian dynamics of the flow. The surprise is that even laminar flow at low Reynolds number is capable of producing such behaviour. The importance of chaotic advection lies not least in the enhancement of transport it produces. In this review we provide an introduction to theoretical results, numerical simulations, and laboratory experiments on chaotic advection in two-dimensional unsteady, three-dimensional steady, and three-dimensional unsteady flow.

Mixing: Chaos and Turbulence, edited by Chaté *et al.*
Kluwer Academic / Plenum Publishers, New York 1999.

1.　Chaos in laminar flow

Only recently has it been generally appreciated that high Reynolds number turbulent flow is not necessary for complex particle trajectories in fluid dynamics. Laminar flow, once thought to have simple dynamics, can, it turns out, give rise to chaotic behaviour of Lagrangian particle trajectories, even though the Eulerian velocity at any given point in space is fixed or periodic in time[1]. Laminar flow can be one-, two-, or three-dimensional, and may be steady (time independent) or unsteady (time dependent). The character of the flow depends greatly on its dimensionality. While all fluid flows naturally occur in the full three dimensional space, there are ways in which one or two space coordinates become less important than the rest. The simplest situation is that where one or two of the velocity-field components vanish, leading to two- and one-dimensional flows, respectively. A second more interesting case is where one or more of the three velocity components may be independent of the others. This category, which might be termed that of $(2+1)$-dimensional flows, comes between truly two-dimensional flows where the third velocity component is zero, and truly three-dimensional flows where all three velocity components are coupled. Regular duct flows in general are of this type, for example the eccentric helical annular mixer model of Ottino [43], and Kusch & Ottino [36]. Another example of $(2+1)$-dimensional flow is spherical Couette flow; the steady laminar flow between coaxially-rotating concentric spheres, where there is both a primary one-dimensional flow and a secondary two-dimensional flow mathematically completely independent of the primary flow, although physically induced by it. We shall return to spherical Couette flow in §5.1.

Probably the most dramatic illustration of the power of the dimensional classification of flows is its impact on the corresponding mixing properties of passive scalars. The source of the different mixing behaviour results from the difference in the topology of the particle paths (path lines) in the various cases. Passive scalars are idealized particles that are small enough not to perturb the flow, but large enough not to engage in Brownian motion (i.e., not to diffuse), and that move with the flow itself. The same physics applies to scalar or vector quantities possessed by a fluid, such as temperature, concentration of a second fluid, or magnetic field. These quantities move with the flow — are passively advected — and so may be described by the Lagrangian representation of the fluid; their dynamics is that of a fluid element

$$\dot{x} = u_x(x,y,z,t), \qquad \dot{y} = u_y(x,y,z,t), \qquad \dot{z} = u_z(x,y,z,t), \qquad (1)$$

u_x, u_y, and u_z being the Cartesian components of the velocity field $u(u_x, u_y, u_z)$. We shall be studying laminar flows where u is time periodic (unsteady) or time independent (steady), that are incompressible, that is to say $\nabla \cdot u = 0$.

The plan of the paper is as follows. We shall review the state of knowledge of advection in laminar flows at each level of the flow hierarchy. In §2 we recall the properties of one and two-dimensional flow and in §3 those of three-dimensional steady flow. We survey in §4 the types of chaotic advection to expect in three-dimensional unsteady flow using a classification by the number of fast and slow components of the velocity field. Concentrating in §5 on flows with two slow and one fast velocity components, in §5.1 we review spherical Couette flow, and in §5.2 we go on to describe

---

[1] Of course, the laminae are then neither topologically simple nor fixed in time [4].

in detail a model we (Cartwright, Feingold & Piro [14, 16, 17]) recently introduced of an experimentally realizable flow that exhibits the novel advection phenomenon in three-dimensional unsteady flow; the global *resonance-induced* dispersion of passive scalars. Finally, in §6, we draw these different threads together and discuss the possibilities for future work based on this and other unsteady three-dimensional flows.

We shall be discussing aspects of both fluid dynamics and nonlinear dynamics, and we shall draw on the terminology of both disciplines when necessary. We hope that the reader will not be confused by the use of *stagnation point* in the context of fluid dynamics, but *fixed point* in the context of nonlinear dynamics, for example. For further clarification on similar points of terminology, we suggest that the reader consult Ottino [43]. There is, however, one technique that we refer to repeatedly whose terminology we should like to clarify. A classical method, due to Poincaré, for the study of a system of ordinary differential equations describing a dynamical system such as a fluid flow, is the reduction of the dimension of the system achieved by sectioning the flow in time or space to produce a map. The dynamical behaviour of this map can then be investigated much more easily than that of the original system, in the knowledge that its limit sets are simply related to the limit sets of the flow [7]. The technical details of this procedure differ depending on whether the flow is autonomous (steady) or nonautonomous (unsteady) [44], and the terminology is not completely fixed: here we take a Poincaré section to mean any such section, and a Poincaré map to be the dynamics on the section. We talk of a stroboscopic map when we refer to the Poincaré map of a nonautonomous system obtained by sampling the flow stroboscopically once per period. The Poincaré map of an autonomous system obtained by taking the intersections of the trajectory (path line) with a fixed plane we refer to as a return map.

2. One and two-dimensional flow

One-dimensional incompressible flow at low Reynolds number, such as Poiseuille flow through a straight pipe, or Couette flow between two cylinders, is encountered at the beginning of fluid-dynamics textbooks, precisely because its dynamics, whether steady or unsteady, is very simple: one dimensionality plus incompressibility implies that the fluid velocity must be constant at any time in the direction of flow.

The case of two-dimensional flow is already a lot more interesting. Here the incompressibility of the flow implies that $\partial u_x/\partial x + \partial u_y/\partial y = 0$. In other words, there exists an exact differential $d\psi$ such that

$$\dot{x} = \frac{\partial \psi}{\partial y}, \qquad \dot{y} = -\frac{\partial \psi}{\partial x}. \tag{2}$$

ψ is known as the stream function, and we can immediately recognize the symplectic structure of Hamilton's equations of motion. Therefore, the motion of a passive scalar in a two-dimensional incompressible flow is fully analogous to the phase space Hamiltonian dynamics of a point mass, with the stream function ψ playing the rôle of the Hamiltonian. That is, two-dimensional steady incompressible flows are, in the same way as one-degree-of-freedom time-independent Hamiltonians, integrable. Two-dimensional unsteady incompressible flows, on the other hand, have the dynamics of one-degree-of-freedom time-dependent Hamiltonians, which are generically nonintegrable. In the case where the time dependence is periodic, these have stroboscopic maps that are area-preserving maps of the plane, and show the features typical of these maps, as predicted by the famous Kolmogorov–Arnold–Moser (KAM) theorem.

In an integrable Hamiltonian system, trajectories are regular. They lie on smooth curves in phase space known as *KAM*, or *invariant*, tori. The KAM theorem tells us that for sufficiently small nonintegrable perturbations of nonlinear integrable Hamiltonian systems, most of the KAM tori survive, and are gradually destroyed as the perturbation is increased. Those KAM tori that break up immediately intregrability is lost leave behind (as the Poincaré–Birkhoff theorem explains) an even number of periodic points, forming Birkhoff periodic orbits. Generically, fixed points and periodic points in area-preserving systems are either centres (elliptic fixed points) or saddles (hyperbolic fixed points). Centres and saddles alternate on the chains of periodic points in Birkhoff periodic orbits, forming structures known as island chains. Around each centre are more KAM tori interspersed with more island chains, each containing centres, around which the structure just described is repeated *ad infinitum*; the KAM torus and island chain structure exists at all scales. There is further complexity around the saddles. Generically, the stable and unstable manifolds of a saddle become tangled up, producing *homoclinic* points from the intersection of the stable and unstable manifolds of the same saddle, or *heteroclinic* points from the intersection of the unstable manifold of one saddle with the stable manifold of another. If there is one homoclinic or heteroclinic point then there must be an infinity of them, since their forward and reverse iterates are also homoclinic or heteroclinic points. The stable and unstable manifolds must oscillate more and more wildly in phase space between intersections which can be found on any section of the manifolds, forming homoclinic and heteroclinic tangles.

To study the consequences of this complex behaviour in fluid dynamics, both numerical [2, 3, 22, 34, 20] and experimental [19, 49, 51] investigations of chaotic advection in two-dimensional unsteady incompressible laminar flow have been undertaken. Many of these investigations have used as a model the unsteady Couette flow between two eccentric cylinders, often termed unsteady journal-bearing flow. This system is particularly well suited for the study of chaotic advection in two dimensions. First, for the steady problem an analytical solution of the Navier–Stokes equations is available in the creeping flow regime. Therefore, one obtains the unsteady flow by periodically switching between two different steady flows conditions. Second, an experiment that is in close correspondence with the theoretical model can be easily designed.

The geometry of steady journal-bearing flow (see figure 1) is uniquely determined by five parameters, namely, the radii of the two cylinders, R_i (inner) and R_o (outer), the corresponding angular velocities, Ω_i and Ω_o, and the eccentricity, D, that is the distance between the centres of the cylinders. However, the corresponding stream function depends on only three parameters, $r \equiv R_i/R_o$, $\omega \equiv \Omega_i/\Omega_o$ and D. In the creeping flow limit, the Navier–Stokes equation reduces to a linear biharmonic one,

$$\nabla^4 \psi = 0, \tag{3}$$

which together with the appropriate boundary conditions was solved by Wannier [54] using complex variables, and by Ballal & Rivlin [11] using bipolar coordinates.

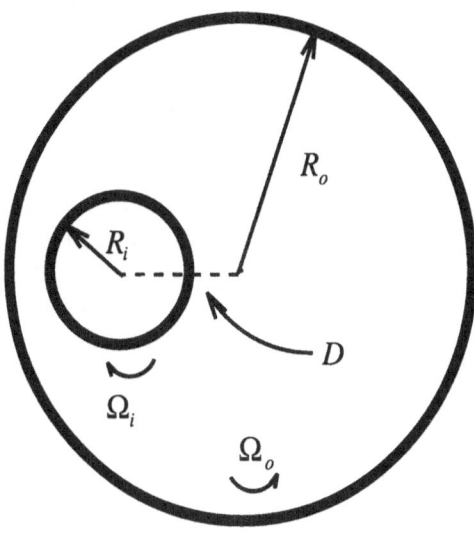

Fig. 1. — The geometry of journal bearing flow and the definitions of the parameters.

The topology of the stream lines for various values of the parameters is related to the competition between the two cylinders to impose their own flow, namely, the flow that would be obtained in the absence of the other cylinder. Due to the linearity of (3), the stream function can be expressed as a linear combination of single cylinder flow, that is,

$$\psi(x,y) = \psi_i(x,y)\Omega_i + \psi_o(x,y)\Omega_o. \tag{4}$$

The different situations can be classified in terms of the number of stagnation points that arise. These can be either enclosed by elliptical stream lines, being then stable, or alternatively, lie at the intersection of separatrix-type stream lines or endpoints of such stream lines on one of the cylinders, being then unstable. It appears that one can have at most two pairs of a stable and an unstable stagnation point for ω positive (corotating cylinders) and large enough D and at least one such pair for negative ω (counterrotating cylinders).

The simplest way to introduce time dependence in the journal-bearing stream function is to let either Ω_i or Ω_o (or both) alternate periodically between two different values. For example, we let Ω_o take the value $\Omega_{o,1}$ for time T_1 and $\Omega_{o,2}$ for time T_2 such that the period, $T = T_1 + T_2$. Whenever inertial effects can be neglected (see §5.2 for a more detailed discussion of this issue) this provides a mechanism for passive scalars that evolve along a streamline of the flow with $\Omega_{o,1}$ during T_1 to move over to a different streamline of the same flow for the next period by means of a streamline of a different flow, namely, that associated with $\Omega_{o,2}$ during T_2. The dynamics ensuing from the process of switching from one streamline to another is no longer integrable. In order to study the new type of behaviour, one strobes the motion at discrete times that are multiples of T. For $T_2 \ll T_1$, small regions of chaotic motion appear in the vicinity of the separatrix stream lines of the $\Omega_{o,1}$ flow while in the rest of the space the dynamics is regular and practically unchanged from the steady flow. However, as T_2 becomes closer in value to T_1, a gradually larger fraction of space is taken over by the chaotic motion at the expense of the regular path lines.

A significant effort has been made to understand the consequence of the chaotic dynamics on the mixing properties of journal-bearing flow. For this purpose the time evolution of continuous distributions rather than that of a single passive scalar has been studied. Such distributions may for example represent a second fluid and can be either one or two dimensional. In regions surrounding stable stagnation points the distributions rotate and at the same time are sheared by the gradient of angular velocities of the elliptical trajectories enclosing the stagnation point. The deformation process is relatively slow (being linear in time) and leads to the emergence of *whorls* [12]. On the other hand, in the chaotic regions the evolution of distributions is guided by the homoclinic or heteroclinic tangle of the stable and unstable manifolds belonging to the unstable stagnation points. In this process exponentially fast deformation occurs, generating huge *tendrils* [12]. Consequently, the mixing efficiency in the chaotic domain is significantly enhanced over that in the regular regions. However, regular path lines which are the fluid mechanical analogue of KAM tori play an additional rôle in suppressing mixing efficiency for the case of two-dimensional flows. Whenever such path lines separate two chaotic domains, they form a barrier that prevents fluid on one side from mixing with the fluid on the other. Thus, good mixing throughout the volume of the flow can occur only for large enough nonlinearities, $T_1 \approx T_2$, and appropriate values of the other parameters, where the fraction of regular path lines becomes negligible and these appear only in the form of very small elliptical islands.

Several experiments have been performed in parallel with the numerical studies. It was found [19] that using glycerine in a system where $R_i = 4.9$ cm and $R_o = 19.6$ cm ($Re \approx 0.1$) and cylinders that are about 20 cm high one obtains to a very good approximation a purely two-dimensional flow. Moreover, the agreement between experiment and numerics is almost perfect. Since diffusion was not taken into account in the theoretical model, the slow broadening of the dye lines that inevitably occurs in the experiment represents one source of discrepancy and limits the length of time for which the evolution of distributions can be studied. The extent of irreversibility in the experiment may be studied by reversing the rotation history of the cylinders. When such an experiment is performed with a distribution of passive scalars in a regular region of the flow, reversal of the rotation even after as many as 20 periods allows one to recover the initial configuration barely altered aside from some diffusive broadening and some previously absent oscillations. On the other hand, performing the reversal in a chaotic domain after only 4 periods leads to a contour that is completely different from the original one and rather complex in shape. These results may be compared with Taylor's experiments on reversibility in steady Couette flow [52].

3.   Three-dimensional steady flow

On the next level of the flow hierarchy, one encounters steady three-dimensional incompressible flow. This case however is similar to unsteady flow in two dimensions, since the return map to any section transverse to the flow can be shown to be an area-preserving map of the plane. Moreover, a steady three-dimensional incompressible flow can be mapped onto a one-degree-of-freedom time-dependent Hamiltonian system. Such mapping fails only at stagnation points of the flow (or at fixed points of the mapping in the language of dynamical systems) [6, 8, 38]. Whenever there are no stagnation points, and if the flow velocity along a particular spatial direction never vanishes, the corresponding coordinate can be used as a fictitious time [4, 38]. Arnold [5] was the first to address the problem of chaos in three-dimensional steady incompressible inviscid fluid flow. He showed that a sufficient condition for a given flow to be integrable is that its vorticity $\omega = \nabla \times u$ be nowhere parallel to its velocity u. Naturally, this lead him to conjecture that for Beltrami flows, namely, flows whose vorticity and velocity are everywhere parallel, $\omega = \lambda u$, stream lines are chaotic. In order to verify this suggestion, Hénon [31] examined the case $\lambda = 1$, that later on became known as the ABC (Arnold–Beltrami–Childress) flow,

$$\dot{x} = A \sin z + C \cos y, \tag{5}$$

$$\dot{y} = B \sin x + A \cos z, \tag{6}$$

$$\dot{z} = C \sin y + B \cos x. \tag{7}$$

The ABC flow is an inviscid flow that solves the three-dimensional Euler equation. It was subsequently studied by Dombre et al. [25], and has become the archetypal mathematical model for the study of chaos in three-dimensional steady incompressible laminar flow. Zheligovsky [56] has analyzed chaotic Beltrami flows in a sphere that represent spherical analogues of ABC flow. Other mathematical models of chaotic advection in three-dimensional steady Stokes flows within a spherical drop have been investigated by Bajer & Moffatt [9], and Stone, Nadim & Strogatz [50]. Experimentally realizable three-dimensional steady flows have also been shown to exhibit chaotic advection. In particular, the partitioned-pipe mixer model of Khakhar, Frajione & Ottino [33] has been experimentally studied by Kusch & Ottino [36]. Similarly, the twisted pipe model of Jones, Thomas & Aref [32] has led to experimental investigations by Le Guer, Castelain & Peerhossaini [37].

Since flows through pipes have wide technological and biophysical applications, it is important to enlarge our knowledge of their mixing properties. For example, in a heating system one would like to maximize the extent of transversal heat transfer between the flow around the centre of the pipe and that in the vicinity of the walls. On the other hand, in pipes that carry hot fluid towards a heat exchanger, the optimal situation occurs when there is the least mixing between the different regions of the pipe section.

The dimensionality of flows in pipes of circular section with constant radius, a, throughout their length is determined by their longitudinal shape. While in straight pipes at low Reynolds number the resulting Poiseuille flow is purely one-dimensional, in the same regime, one obtains a $(2 + 1)$-dimensional flow in curved pipes with constant curvature R. Specifically, a two-vortex secondary flow arises that moves fluid between the centre of the pipe and its walls. The longitudinal flow, which is decoupled from the transversal flow, is to first order in a/R the same as for the straight pipe. However, in many applications curved pipes have more complicated shapes than the toroidal one. The twisted pipe model [32] represents a step closer to such applications in that it leads to a truly three-dimensional flow. It consists of a sequence of consecutive half tori rotated with respect to each other by a pitch angle χ

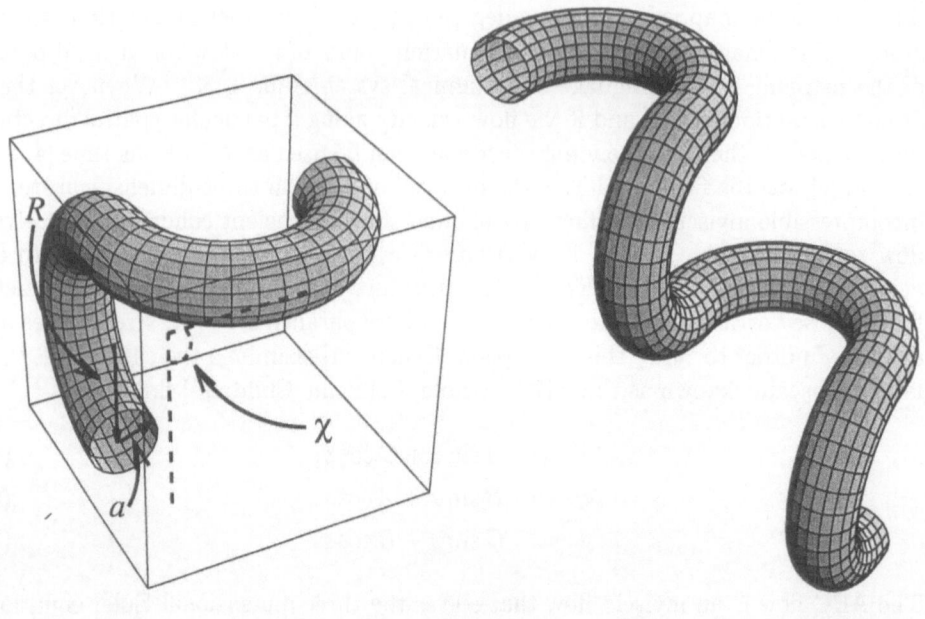

Fig. 2. — The geometry of twisted-pipe flow and of the definitions of the parameters. Left: the basic unit for a twist angle χ of 90°. Right: several units arranged into a twisted pipe.

(see figure 2). The analysis of this system is greatly simplified if one assumes that at the interface between two half tori the secondary flow adjusts itself instantaneously to the twist. While this assumption is clearly unjustified, one expects that it does not modify the qualitative features of the mixing behaviour in the system. In this approximation a passive scalar evolves following the stream lines of the half torus through which it is passing at any given time, and these stream lines are the same as for an isolated torus. The latter were first given by Dean [23, 24] and to order $O(a/R)$

$$\nabla^2 w = \frac{1}{r} \left(\frac{\partial \psi}{\partial r} \frac{\partial w}{\partial \phi} - \frac{\partial \psi}{\partial \phi} \frac{\partial w}{\partial r} \right) - C, \tag{8}$$

$$\nabla^2 \psi = \frac{1}{r} \left(\frac{\partial \psi}{\partial r} \frac{\partial w}{\partial \phi} - \frac{\partial \psi}{\partial \phi} \frac{\partial w}{\partial r} \right) \nabla^2 \psi + 2Dw \left(\frac{\sin \phi}{r} \frac{\partial w}{\partial \phi} - \cos \phi \frac{\partial w}{\partial r} \right), \tag{9}$$

where all lengths are scaled by a and the axial velocity, w, and the secondary stream function, ψ, are scaled by the average axial velocity, W, and the kinematic viscosity, ν, respectively. Moreover, r and ϕ are the polar coordinates in the pipe section such that on the wall $r = 1$, and θ is the longitudinal angle of the section with respect to the centre of the torus. Also,

$$C = -\frac{a^2}{RW\mu} \frac{\partial p}{\partial \theta}, \tag{10}$$

is the dimensionless pressure gradient, D is the Dean number

$$D = \frac{W^2 a^2}{R\nu^2}, \tag{11}$$

and

$$Re = \frac{Wa}{\nu}. \tag{12}$$

Then, to first order in D, the equations of motion for a passive scalar moving through a toroidal pipe are of the form

$$\dot{x} = \frac{\alpha}{1152}\left(h(r) + \frac{y^2}{r}h'(r)\right), \tag{13}$$

$$\dot{y} = -\frac{\alpha}{1152}\frac{xy}{r}h'(r), \tag{14}$$

$$\dot{\theta} = \frac{1}{4}\beta(1 - r^2), \tag{15}$$

where x and y are the Cartesian coordinates in the pipe section, $x^2 + y^2 = r^2$, $\alpha = DC^2$, $\beta = DC/Re$ and

$$h(r) = \frac{1}{4}(4 - r^2)(1 - r^2)^2. \tag{16}$$

Accordingly, the dynamics of the passive scalar through the twisted pipe is modelled by using (13–15) and rotating the trajectory anticlockwise by the pitch angle χ each time θ becomes an integer multiple of π. Since in each half torus the transversal flow is decoupled from the longitudinal Poiseuille flow, it is useful to replace the time variable with θ, leading to two equations of motion that depend on a single parameter, $\gamma = \alpha/\beta$. The corresponding Poincaré map for the twisted pipe model, A, is defined as the relation between the (x, y) coordinates of stream lines at $\theta = 2\pi n$, (x_n, y_n), and those at $\theta = 2\pi(n + 1)$, (x_{n+1}, y_{n+1}). This map depends on two parameters, γ and χ, and is composed of evolutions through a half torus, M_γ, and rotations through $-\chi$, R_χ, such that $A = R_\chi M_\gamma R_\chi M_\gamma$.[2] While for special values of the parameters the map A is integrable, in general it generates a mixture of regular and chaotic trajectories. Clearly, the map has three integrable cases: $\gamma = 0$, $\chi = 0$, and $\chi = \pi$. Numerical studies of A show that at finite γ, letting χ grow away from zero generates chaotic trajectories around the separatrix that divides the upper and the lower vortices. It is found that the fraction of the pipe section that is filled by chaotic stream lines grows with increasing γ and that for fixed γ it is maximal for $\chi \approx \pi/2$. A more quantitative analysis of the Poincaré map can be made using two of its symmetries. First, one can show that A is invariant under reflection with respect to the $\phi = \chi/2$ axis, and so are the corresponding section portraits. Second, the pipe wall is invariant under the map and for χ not too close to π there are two fixed points on the wall, one stable that approaches $\phi_+ = \pi/2$ as $\gamma \to \infty$ and one unstable that tends to $\phi_- = \chi - \pi/2$. For small χ, the stable and unstable manifolds of these two fixed points are responsible for the main features of the dynamics, which to a large

_____

[2] Notice however that the change of variables from t to θ has lead to a non-area preserving Poincaré map.

extent preserves the two-vortex aspect of the $\chi = 0$ integrable case. In particular, one of the manifolds from each of the fixed points lies along the border between the two regular vortices and the boundaries of the emerging heteroclinic tangle indicate the extent of the corresponding chaotic region. However, as χ grows, ϕ_+ and ϕ_- become gradually closer to each other, and as a consequence the manifolds influence the dynamics in a decreasing fraction of the pipe section.

While the global structure of the Poincaré sections obtained for the twisted pipe model is similar to that of two-dimensional time-periodic flows, for example journal-bearing flow, their physical interpretation is a lot more dramatic. Here the regular islands correspond to tubes of stream lines that do not allow the enclosed fluid or passive scalars to leave their interior throughout the entire length of the pipe. Since these tubes foliate the regular regions, passive scalars cannot move across tubes except by diffusion. On the other hand, the dynamics of stream lines belonging to the chaotic regions that connect the centre of the pipe to its walls indicate that passive scalars located in these regions, e.g. heat, are being quickly mixed. An analogous distinction between the behaviour in the regular and the chaotic regions is observed for the longitudinal dispersion. Starting with a line of passive scalars along the axis of symmetry of the Poincaré section it is found that after long enough time the dispersion in the chaotic regions becomes significantly larger than that in the regular part. Moreover, the distance in θ travelled by the passive scalar in a given time as a function of its initial position along the symmetry line is a smooth function for the part of the line lying in the regular areas of the section. On the other hand, in the chaotic intervals this function is wildly fluctuating and is expected to be fractal.

These results have found qualitative experimental confirmation in the work of Le Guer, Castelain & Peerhossaini [37] on an array of curved duct sections of square profile. For each successive section the symmetry plane of the bend is rotated by an angle of $\pi/2$ relative to the preceeding one. Water is circulated through the system at a steady rate with low enough Reynolds number to ensure laminar flow. The size and curvature of the ducts are chosen to allow the Dean vortices to develop sufficiently in the quarter turn corresponding to each bend. The flow stream lines are visualized by means of a fluorescent dye injected upstream through different arrays of needles and the Poincaré sections can be recorded by projecting a sheet of laser light transversally to the duct at equivalent locations in successive bends. Of course, only a limited number of sections can be connected and this implies that no exaustive studies of the long term dynamics can be performed. However, some basic facts such as the coexistence of regions with poor and efficient transport due to KAM tori can be easily recognized. Also, the incipient formation of fractal structure in the behaviour of the transit time across the sections has been recorded.

Another dramatic demonstration of the improvement of the transport properties due to chaotic advection with obvious importance for applications are the experiments reported by Acharya, Sen & Chang [1] and by Peerhossaini, Castelain & Le Guer [46] consisting of a heat exchanger made of a set of curved copper pipes that can be arranged both as a coil of roughly constant and small pitch producing weakly perturbed Dean vortices, and as a twisted array producing strongly chaotic stream lines. The overall heat transfer coefficient for equal Reynolds number for the in-pipe flow increases by a factor of as much as twenty percent from one configuration to the other.

4.   Classifying three-dimensional unsteady flows with Liouvillian maps

Three-dimensional unsteady incompressible flows with periodic time dependence produce stroboscopic maps in three dimensions that preserve volume. These three-dimensional volume-preserving maps do not correspond to Hamiltonian systems, which have phase spaces of even dimensionality: Hamiltonians with two or three degrees of freedom produce volume-preserving stroboscopic maps of two or four dimensions respectively. Following Liouville's theorem, we have named volume-preserving maps of three dimensions, which dimensionally interpolate between the two cases above, *Liouvillian* maps [26, 27, 28, 47, 29, 15].

To study Liouvillian maps one can employ ideas from KAM theory and imagine perturbations of integrable action–angle models where the actions are fixed and the angles change at each iteration. A perturbation away from the integrable case then causes the actions to vary slowly. For two-dimensional area-preserving maps, or for any maps derived from Hamiltonian dynamical systems, the underlying symplectic structure provides the action–angle variables on which to build. Here with Liouvillian maps such symplectic structure is absent, so we must define our variables with respect to something else. Thus we use invariant tori to define the action–angle variables.

We can consider integrable, Liouvillian maps that possess some variables, actions I, that do not change from iteration to iteration, whereas others, angles θ, vary linearly as the map is iterated([3])

$$I' = I, \qquad \theta' = \theta + \omega(I). \qquad (17)$$

We use the notation L_k^0 to represent an integrable Liouvillian map thaTt possesses a set of action–angle variables $I \in \mathbb{R}^{\daleth}$, $\theta \in \mathbb{T}^{\daleth - \daleth}$. A k-action integrable Liouvillian map L_k^0 has a family of $(3 - k)$-dimensional invariant tori, each of which is traversed by the $3 - k$ angle variables that rotate at each iteration by an amount given by the corresponding component of the $(3 - k)$-dimensional vector ω. In other words, L_k^0 maps consist of uniform translations on $(3 - k)$-tori embedded in $\mathbb{T}^{\daleth}$. We thus have four possibilities for these integrable maps:

L_0^0 Zero actions; three angles. The frequency vector ω is constant, and hence we have a uniform rotation on $\mathbb{T}^{\daleth}$.

L_1^0 One action; two angles. The motion takes place on two-tori defined by constant I, and a point on one of these tori is given by the two angles θ_1 and θ_2. Thus the motion on the tori is a uniform rotation with a frequency depending on the value of the action.

L_2^0 Two actions; one angle. The two actions parametrize a family of invariant circles (i.e., one-tori), and the angle variable rotates with a frequency $\omega(I_1, I_2)$.

L_3^0 Three actions; zero angles. Each point of space is parametrized by (I_1, I_2, I_3) and is hence invariant; the map is the identity.

([3]) Notice that at this stage it is not important whether the frequency vector is made a function of the old actions $\omega(I)$ or of the iterated actions $\omega(I')$. This will become important when we wish to perturb this map whilst retaining its Liouvillian nature: we construct the perturbation to make it automatically volume preserving.

We consider now the effects of nonintegrable perturbations on L_k^0 maps. The perturbations we wish to consider will act to couple the hitherto independent actions and angles whilst retaining the Liouvillian property of the map. Thyagaraja & Haas [53] have worked out the most general form that such perturbations can take, which in general leads to an implicit map. Below we shall discuss only the subset of perturbations that leave the map in explicit form. We believe that these explicit Liouvillian maps capture the dynamics of Liouvillian maps in general, whilst being easier to work with than implicit maps.

We deal first with the two extreme cases $k = 0$ and $k = 3$. Results of Ruelle & Takens [48] and Newhouse, Ruelle & Takens [42] about perturbing quasiperiodic flows on a three-torus indicate that small perturbations of zero-action L_0^0 maps will generically lead to completely chaotic behaviour([4]). A small nonintegrable perturbation added to three-action L_3^0 maps is a small perturbation

$$I_1' = I_1 + \varepsilon F_1(I_2, I_3), \qquad I_2' = I_2 + \varepsilon F_2(I_1', I_3), \qquad I_3' = I_3 + \varepsilon F_3(I_1', I_2'), \quad (18)$$

of the identity. In the $\varepsilon \to 0$ limit we have a set of three autonomous ordinary differential equations, which takes us back to the case of three-dimensional steady flows. In other words, the perturbation represents the *Euler map* of the flow $\dot{I} = F(I)$; the map (18) is derived from this flow by applying the forwards and backwards Euler numerical integration methods to discretize it. The dynamics of Euler maps are closely related to the flows from which they are derived [18].

For one-action L_1^0 maps, it turns out that most of the invariant two-tori are preserved in a KAM-like manner under small nonintegrable volume-preserving perturbations. However, the resonant tori satisfying the equation

$$m\omega_1(I) + n\omega_2(I) = 2\pi k, \qquad (19)$$

where k, m, and n are integers, are destroyed, and in their place appear a finite number of invariant circles, half of which are stable and the other half unstable, in the same way as invariant circles are destroyed and fixed points appear in planar area-preserving maps following the Poincaré–Birkhoff theorem. Further invariant tori surround the stable invariant circles, whilst chaos appears around the unstable ones. The preserved invariant tori act as boundaries for the chaotic layers, so a single trajectory cannot cover the whole phase space until the perturbation is large enough to destroy the final invariant torus. The whole situation is like that governed by the KAM theorem in planar area-preserving maps, except that the structures have one additional dimension; that is the invariant tori are two rather than one-dimensional, the island chains interspersed amongst them contain invariant circles, not fixed points, and the stable (elliptic) and unstable (hyperbolic) invariant circles are centres for further invariant tori and for chaos respectively [28, 15]. The existence of a KAM-like theorem in this one-action case has been proven by Cheng & Sun [21], although as Mezić & Wiggins [39] point out, here in three dimensions we have no criterion similar to the irrationality of rotation numbers in two dimensions that enables one to predict in which order the remaining invariant tori are destroyed as the nonintegrable perturbation is increased.

([4]) However, the effect of restricting the perturbations so that they retain volume conservation has not, as far as we know, been investigated.

With two-action L_2^0 maps, on the other hand, small nonintegrable volume-preserving perturbations cause behaviour rather different to that detailed above. The integrable case turns out to be a singular limit; as soon as a nonintegrable perturbation is switched on, the two-parameter family of invariant circles coalesce into invariant two-tori. We can see why this should be so using an argument based on an adiabatic approximation. Consider the perturbed two-action map

$$I_1' = I_1 + \varepsilon F_1(I_2, \theta), \qquad I_2' = I_2 + \varepsilon F_2(I_1', \theta), \qquad \theta' = \theta + \omega(I_1', I_2'). \qquad (20)$$

Suppose that ω is irrational, then when ε is small, the angle θ covers uniformly the entire interval $(0, 2\pi)$ before the actions I change significantly. The variation of I is thus sensitive only to the average of F over θ. Therefore the actions now iterate as

$$I_1' = I_1 + \varepsilon \bar{F}_1(I_2), \qquad I_2' = I_2 + \varepsilon \bar{F}_2(I_1'). \qquad (21)$$

where the bar represents the θ average. Thus the dynamics of the actions I decouples from that of the angle θ for non-resonant ω. We are left with an area-preserving map, which leads in the limit $\varepsilon \to 0$ to the Hamiltonian system

$$\dot{I}_1 = \bar{F}_1(I_2) = \frac{\partial H}{\partial I_2}, \qquad \dot{I}_2 = \bar{F}_2(I_1) = -\frac{\partial H}{\partial I_1}. \qquad (22)$$

We obtain the Hamiltonian

$$H(I_1, I_2) = \int_0^{I_2} \bar{F}_1(I) dI - \int_0^{I_1} \bar{F}_2(I) dI = \beta. \qquad (23)$$

In other words, in the $\varepsilon \to 0$ limit, the action variables evolve slowly along the level curves of (23). Including the fast motion in the θ direction, we infer that the originally invariant circles parallel to the θ axis coalesce into invariant tori Σ_β defined by the condition $H(I_1, I_2) = \beta$, the motion on which is fast in one direction and slow in the other. Resonances occur when the condition imposed above that $\omega(I)$ be irrational is not satisfied and we have

$$n\omega(I_1, I_2) = 2\pi k, \qquad (24)$$

where k and n are integers. The resonant condition defines a set of sheets that typically intersect a continuous set of invariant tori, which break down locally at the intersections. The remainder of the invariant torus survives, to a given order in the perturbation expansion. Invariant tori are thus connected through resonances, meaning that, in contrast with the one-action case, a single trajectory can cover the whole of phase space [47, 29, 15]. We term this characteristic behaviour of two-action Liouvillian maps *resonance-induced* dispersion.

5. Resonance-induced dispersion in two-action flow

As we have seen, two-action Liouvillian maps have particularly interesting behaviour, since there are no barriers to motion and a form of global dispersion is present. To investigate this resonance-induced dispersion further in a fluid dynamical context, in this section we construct a model of a realistic flow showing two-action properties.

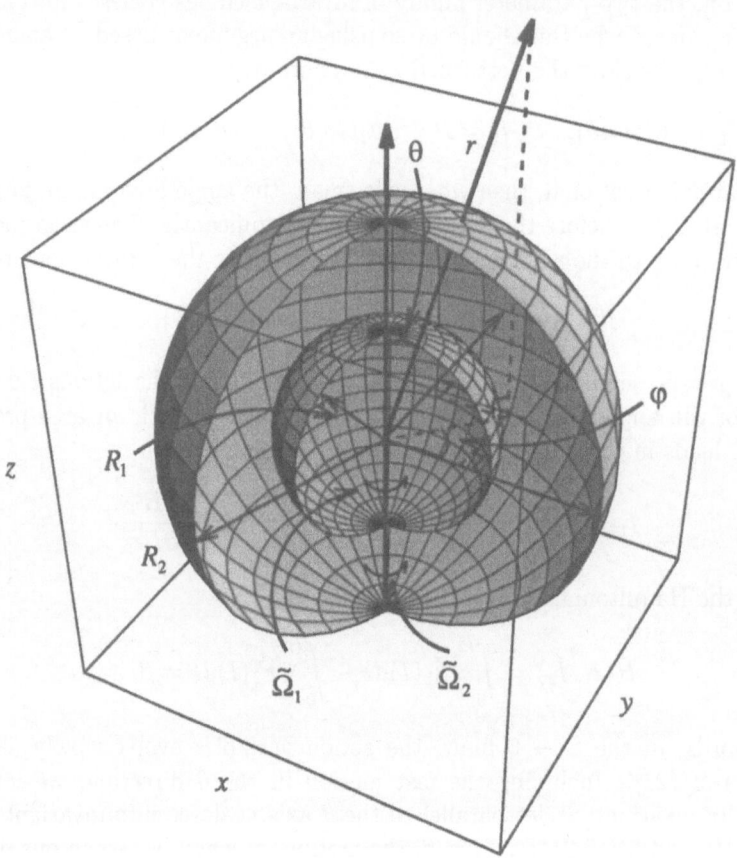

Fig. 3. — Two concentric spheres have radii R_1 and R_2, and constant angular velocities $\tilde{\Omega}_1$ and $\tilde{\Omega}_2$ about a common axis through their centres. The space between them is filled with an incompressible viscous fluid. We term the ratio R_1 to R_2, η, and the ratio $\tilde{\Omega}_1$ to $\tilde{\Omega}_2$, μ. The radial coordinate is rescaled as $r = \tilde{r}/R_2$ in order that it be dimensionless.

5.1. Spherical Couette flow

Spherical Couette flow — steady flow at low Reynolds numbers between coaxially-rotating concentric spheres — is a natural generalization of the Couette flow between two rotating cylinders. In contrast with the latter, however, it possesses both a primary and a secondary flow, and it is dependent on the Reynolds number. References to experimental and numerical studies of spherical Couette flow are given by Koschmieder [35].

Consider concentric spheres with fixed radii R_1 and R_2, and constant angular velocities about a common axis through their centres $\tilde{\Omega}_1$ and $\tilde{\Omega}_2$, as pictured in figure 3. In the spherical polar coordinate system of figure 3, with the radial coordinate rescaled as $r = \tilde{r}/R_2$, the steady flow of a viscous, incompressible fluid contained between the two spheres is assumed to be independent of the longitude ϕ. We may then write the Navier–Stokes equations in terms of a stream function ψ in the meridian plane, and a longitudinal velocity function Ω

$$\frac{1}{R}eD^2\Omega = -\frac{\psi_r\Omega_\theta - \psi_\theta\Omega_r}{r^2\sin^2\theta}, \tag{25}$$

$$\frac{1}{R}eD^4\psi = \frac{2\Omega}{r^3\sin^2\theta}(\Omega_r r\cos\theta - \Omega_\theta\sin\theta) + \frac{2D^2\psi}{r^3\sin^2\theta}(\psi_r r\cos\theta - \psi_\theta\sin\theta)$$
$$- \frac{1}{r^2\sin\theta}(\psi_r(D^2\psi)_\theta - \psi_\theta(D^2\psi)_r), \tag{26}$$

where

$$D^2 = \frac{\partial^2}{\partial r^2} + \frac{1}{r^2}\frac{\partial^2}{\partial\theta^2} - \frac{1}{r^2}\cot\theta\frac{\partial}{\partial\theta}, \tag{27}$$

and the subscripts associated with ψ and Ω denote differentiation.

The Reynolds number is defined as $Re = \tilde{\Omega}_c R_2^2/\nu$, where R_2 is a characteristic length given by the radius of the outer sphere, $\tilde{\Omega}_c$ is a characteristic angular velocity, and ν is the kinematic viscosity. The characteristic angular velocity may be taken as that of the outer sphere $\tilde{\Omega}_2$, or that of the inner sphere $\tilde{\Omega}_1$, depending on which is the dominant influence on the flow in the case being considered. The boundary conditions are firstly that $\psi = \psi_r = 0$ on both spheres, when $r = \eta$ and $r = 1$. Secondly, with $\tilde{\Omega}$ as the physical angular velocity, and choosing $\tilde{\Omega}_2$ as the characteristic angular velocity, the dimensionless angular velocity $\omega = \tilde{\Omega}/\tilde{\Omega}_2 = \Omega/r^2\sin^2\theta$ should be $\omega = \mu$ on the inner sphere and $\omega = 1$ on the outer. Hence the other two parameters, apart from the Reynolds number, are the radius ratio $\eta = R_1/R_2$, and the angular velocity ratio $\mu = \tilde{\Omega}_1/\tilde{\Omega}_2$.

The above problem can be solved analytically by a perturbation method in the Reynolds number. Haberman [30] gave the solution to first order in the Reynolds number. Subsequently Munson & Joseph [41] derived general results through to seventh order in the Reynolds number. Here we need only the first order approximation valid in the low-Reynolds-number limit $Re \lesssim 1$ (with the higher-order solutions one can obtain good approximations for Reynolds numbers in the hundreds). The solution to first order in the Reynolds number is

$$\psi = Re\left(\frac{A_1}{r^2} + A_2 + A_3r^3 + A_4r^5 + \frac{a_2}{4}\left(\frac{a_2}{r} - a_1r^2\right)\right)\sin^2\theta\cos\theta, \tag{28}$$

$$\Omega = \left(a_1r^2 + \frac{a_2}{r}\right)\sin^2\theta, \tag{29}$$

where

$$a_1 = \frac{1 - \mu\eta^3}{1 - \eta^3}, \qquad a_2 = \frac{(\mu - 1)\eta^3}{1 - \eta^3}, \tag{30}$$

and

$$A_1 = \Xi\eta^4\left((2\eta^5 - \eta^4 - 16\eta^3 - 20\eta^2 - 8\eta - 2)\mu\right.$$

$$+ 2\eta^5 + 8\eta^4 + 20\eta^3 + 16\eta^2 + \eta - 2\Big), \tag{31}$$

$$
\begin{aligned}
A_2 &= -\Xi\eta^2\Big(\big(10\eta^7 + 25\eta^6 + 40\eta^5 + 57\eta^4 + 48\eta^3 + 20\eta^2 + 8\eta + 2\big)\,\mu \\
&\quad - 2\eta^7 - 8\eta^6 - 20\eta^5 - 48\eta^4 - 57\eta^3 - 40\eta^2 - 25\eta - 10\Big),
\end{aligned}
\tag{32}
$$

$$
\begin{aligned}
A_3 &= -\Xi\Big(\big(6\eta^6 + 24\eta^5 + 35\eta^4 + 32\eta^3 + 15\eta^2 - 2\eta - 5\big)\,\mu\eta^2 \\
&\quad + 5\eta^6 + 2\eta^5 - 15\eta^4 - 32\eta^3 - 35\eta^2 - 24\eta - 6\Big),
\end{aligned}
\tag{33}
$$

$$
A_4 = \Xi\Big(\big(2\eta^4 + 8\eta^3 + 5\eta^2 - 2\eta - 3\big)\,\mu\eta^2 + 3\eta^4 + 2\eta^3 - 5\eta^2 - 8\eta - 2\Big),
\tag{34}
$$

with

$$
\Xi = \frac{(\mu - 1)\eta^3}{4(4\eta^6 + 16\eta^5 + 40\eta^4 + 55\eta^3 + 40\eta^2 + 16\eta + 4)(\eta - 1)^2(\eta^2 + \eta + 1)^2}.
\tag{35}
$$

The dimensionless flow velocities are given by

$$
u_r = \frac{1}{r^2 \sin\theta}\frac{\partial\psi}{\partial\theta} = \dot{r},
\tag{36}
$$

$$
u_\theta = -\frac{1}{r \sin\theta}\frac{\partial\psi}{\partial r} = r\,\dot\theta,
\tag{37}
$$

$$
u_\phi = \frac{1}{r \sin\theta}\Omega = r \sin\theta\,\dot\phi,
\tag{38}
$$

so that

$$
\dot{r} = \frac{R}{e}r^2\left(\frac{A_1}{r^2} + A_2 + A_3 r^3 + A_4 r^5 + \frac{a_2}{4}\left(\frac{a_2}{r} - a_1 r^2\right)\right)(3\cos^2\theta - 1),
\tag{39}
$$

$$
\dot\theta = -\frac{R}{e}r^2\left(-\frac{2A_1}{r^3} + 3A_3 r^2 + 5A_4 r^4 + \frac{a_2}{4}\left(-\frac{a_2}{r^2} - 2a_1 r\right)\right)\sin\theta\cos\theta,
\tag{40}
$$

$$
\dot\phi = a_1 + \frac{a_2}{r^3}.
\tag{41}
$$

The solution shows a primary flow Ω about the axis of rotation, together with a secondary flow given by ψ in the meridian plane. In general, one of the two spheres dominates the flow pattern, generating a pair of vortices, one on either side of the equator, whose sense of rotation depends on the angular velocity ratio μ. The spheres may corotate or counterrotate, and in the latter case (μ negative) there is a part of the (η, μ) parameter space where neither sphere is dominant and we observe two counterrotating vortices in each hemisphere [55], giving a total of four vortices. Figure 4 shows examples of both the two-vortex and the four-vortex patterns. Since all three velocity components depend only on r and θ, spherical Couette flow is in the category of flows we named $(2+1)$-dimensional in §1. The secondary flow stream function ψ is a one-degree-of-freedom time-independent Hamiltonian of the type (2), so the steady flow is identifiable as an integrable Hamiltonian system. Now consider

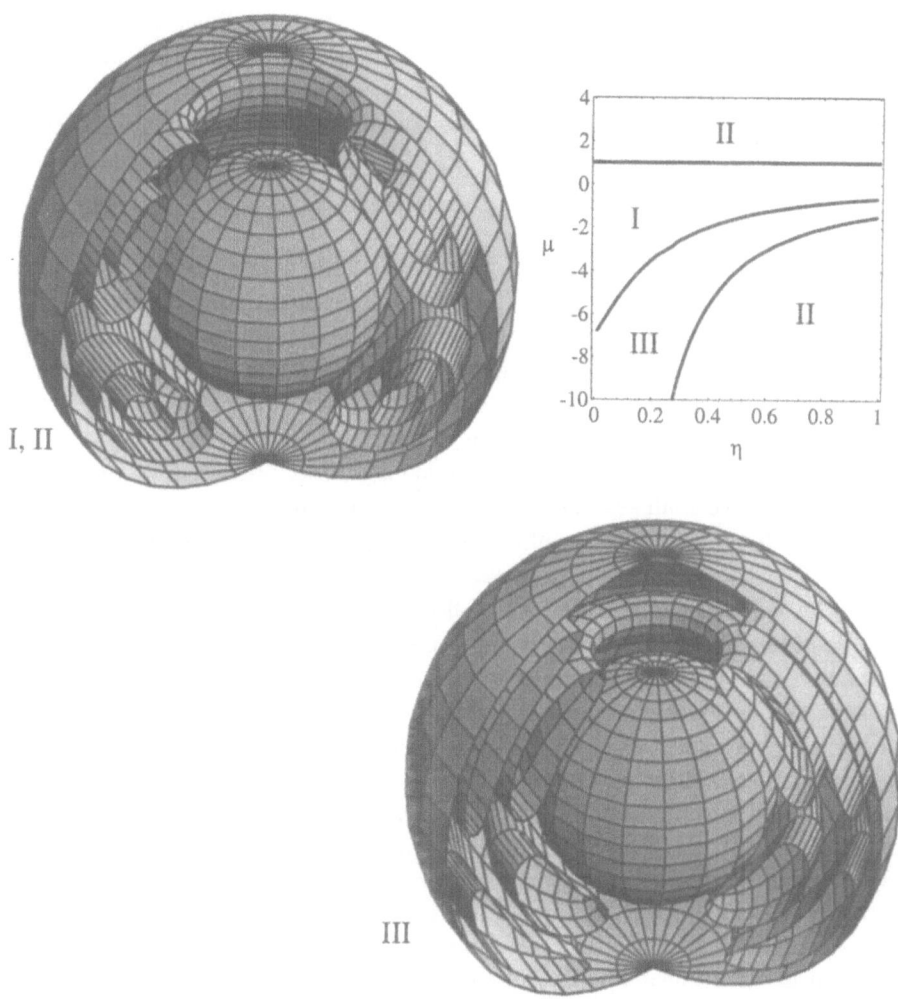

Fig. 4. — Shown here are the regions of parameter space for which the different secondary flow patterns between the spheres exist, and typical examples of the two-vortex and four-vortex flow are represented as cutaway pictures showing the spheres together with some level surfaces of the vortices. When $\mu = 1$ there is rigid body rotation and no secondary flow. Either side of this line two vortices are present. In region I the outer sphere is the dominant influence on the flow, and fluid is transported from the poles to the equator near the outer sphere, and back near the inner sphere. In the two disjoint parts of region II the inner sphere is the dominant influence, and fluid is transported in the opposite direction, from the equator to the poles near the outer sphere, and back near the inner sphere. In between, in the linguiform region III, neither sphere is the dominant influence and there are four vortices. These rotate such that fluid is transported to the equator near the inner and outer spheres, and to the poles at the separatrix between the two vortices of each hemisphere.

the flow in terms of the categorization we discussed in §4. Inserting typical values for
the magnitudes of the variables and parameters (by which we mean typical of those
we use throughout the rest of this paper) into (39–41) shows that at the values of the
Reynolds number $Re \leqslant 1$ for which (39–41) hold, the one-dimensional primary flow
Ω governed by ϕ is hundreds to thousands of times faster than the two-dimensional
secondary flow ψ governed by r and θ. Thus we have two slow action variables r
and θ, and one fast angle variable ϕ; the flow falls into the two-action category.

5.2.   Biaxial unsteady spherical Couette flow

The existence of a secondary flow in spherical Couette flow is crucial for our purposes,
since it enables us to construct a fully three-dimensional flow simply by using two
different axes of rotation for the spheres. Rotating periodically about first one axis,
then the other acts to couple the longitude to the latitude and the radial coordinate,
giving a three-dimensional flow that is also time dependent, figure 5. We call the time
for which we rotate about the first axis, the first semiperiod, T_1, and that about the
second axis, the second semiperiod, T_2, making a total forcing period of $T = T_1 + T_2$.
For simplicity we shall set $T_1 = T_2 = T/2$, and we shall vary T as necessary. In the
following, all our numerical results will be obtained by the numerical integration of
the Navier–Stokes equations for spherical Couette flow given in (39–41).

Biaxial unsteady spherical Couette flow is then a piecewise combination of the
flow solutions about each axis. This piecewise combination of solutions is the same
technique used to obtain the unsteady journal-bearing flow extensively investigated
as a model for chaotic advection in two-dimensional unsteady flows. However, an
important conceptual difference between spherical Couette flow and journal-bearing
flow is that in the latter problem the solution depends linearly on the rotation rates of
the inner and outer cylinders. Thus in journal-bearing flow, particle paths depend
only on the boundary displacements. As long as the cylinders are set in motion
alternately, the details of the modulation of the rotation rate are not important;
provided the angular displacements of the boundaries are the same, so will be the
path of a passive scalar. This is not true of spherical Couette flow, since the stream
function is not a linear function of the rotation rate. Thus in the spheres case we
have to decide the way in which to modulate the rotation. The two extremes are
the discontinuous limit, and the quasistatic limit. In the discontinuous limit we
change the rotation rate as discontinuously as possible, with instantaneous startup
and shutoff of the rotation, as with a square wave. We then model the problem
as a piecewise-linear combination of the two separate steady spherical Couette flow
solutions. This is a reasonable approximation if the fluid inertia is negligible, so that
the fluid stops moving as soon as the sphere stops rotating, which is why we require
low-Reynolds-number flow. In the quasistatic limit, on the other hand, we change the
rotation rate as slowly as possible, with smooth modulation of the rotation in time,
as with a sine wave. The quasistatic approximation then consists of considering the
resulting stream function to be given instantaneously by the steady stream function
for the rotation rate at that instant. We have chosen to use the discontinuous
approximation to model the unsteady flow here. We shall use radius ratio $\eta = 1/2$
throughout this paper, and we can elect to use different angular velocity ratios μ_1
and μ_2 in each semiperiod to alter the character of the flow solution.

The alternative to the above choices for modulating the rotation rates of the

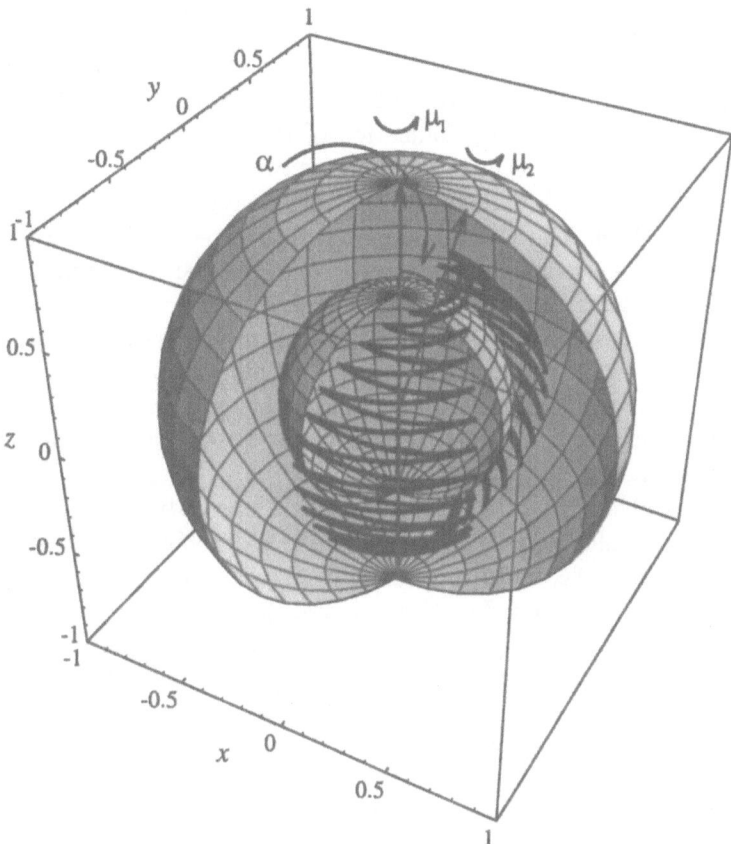

Fig. 5. — A trajectory in biaxial unsteady spherical Couette flow illustrates the piecewise (mathematically speaking, C^0) nature of the solution. The trajectory covers thirty periods. The radius ratio is $\eta = 0.5$, the angular velocity ratios are $\mu_1 = 0$ and $\mu_2 = -3$, and the semiperiods are $T_1 = T_2 = T/2 = 2$. The Reynolds number is $Re = 0.1$ and the axis separation is $\alpha = 10°$. The initial conditions are $r = 0.64$, $\theta = 60°$, and $\phi = 30°$.

spheres would be to model the full developing flow; the so-called spin-up problem discussed by Pearson [45]. This makes the model far more complex, and we feel justified in ignoring the complication since the discrepancy is small at low Reynolds numbers. In the case of either unsteady journal-bearing flow or biaxial unsteady spherical Couette flow, one is neglecting transient effects such as the diffusion of momentum. This neglect requires that the acceleration terms in the Navier–Stokes equation

$$Re\left(Sr\,\frac{\partial u}{\partial t} + u.\nabla u\right) = -\nabla p + \nabla^2 u \qquad (42)$$

be negligible, where $Re = VL/\nu$ is the Reynolds number as before, and $Sr = fL/V$ is the Strouhal number, with L a characteristic length, V a characteristic velocity, and f a characteristic frequency; ν being the kinematic viscosity. To make the acceleration terms negligible, Re and Sr must be small. This last condition is violated by the discontinuous modulation of the rotation rate in time; at the discontinuities, $\partial u/\partial t$ will be infinite. However, we argue that the effects of this violation will be slight at low Reynolds numbers (many of the studies of journal-bearing flow have been made with discontinuous changes in the cylinder rotation rates), and to place us squarely in the low Reynolds number regime we fix $Re = 0.1$ here. Because the dynamical phenomena we observe in this system are structurally stable, that is to say robust against small changes in the system, the small extra displacement suffered by a fluid particle not responding instantaneously when the flow stops and restarts will not lead to a qualitative change in its long term behaviour. This robustness is also a prerequisite for being able to observe these phenomena in a real fluid experiment.

To explore the dynamics of a three-dimensional unsteady flow, we may plot a stroboscopic map, which gives us a three-dimensional volume-preserving map like the Liouvillian maps of §4, but a three-dimensional view of a Liouvillian map often fails to give much information. In order to reveal the information contained within a three-dimensional map, we need to open it up to view. There are various methods to achieve this; we shall use two in this paper. The first is to slice the map open by taking a section through it. We select a thin slice and record the iterates that lie within it. There is a trade off between the thickness of the slice and the quality of the information that is obtained; a thicker slice has more points lying within it, but is fuzzier because of its three dimensionality, a thinner slice gives a sharper image but requires longer time integration of the system to have the same number of points in the slice. In the spheres case we use a slice between $-0.01 < y < 0.01$. The second technique we use is to project the map down onto a plane. Unlike the slice technique, which will always produce useful information, this technique is only useful if the Liouvillian map is near to being integrable. If we are in possession of the action and angle variables, projecting down onto an action–angle plane serves to fold up all the information along the other action or angle variable. In the spheres case, with small axis separation, this can be achieved by plotting a projection of the stroboscopic map onto the action plane. We do this by taking the points of the stroboscopic map expressed in action–angle variables and forgetting about the angle coordinate; plotting the points only in the actions. One can think of the operation geometrically as concertinaing the spheres about the angle variable into one plane.

Apart from the period T, the other new parameter in biaxial unsteady spherical Couette flow is the angle α between the two axes. We have chosen to have the two axes lying in the (x, z)-plane of our coordinate system (see figure 5), since this simplifies the algebra whilst not affecting the dynamics. If $\alpha = 0$, the spheres rotate about a single axis, and we have a $(2 + 1)$-dimensional flow that may be steady, if $\mu_1 = \mu_2$ — which corresponds to the spherical Couette flow described in §5.1 — or unsteady if $\mu_1 \neq \mu_2$. Whilst in the former case the flow is integrable, in unsteady spherical Couette flow the spheres rotate about a single axis with angular

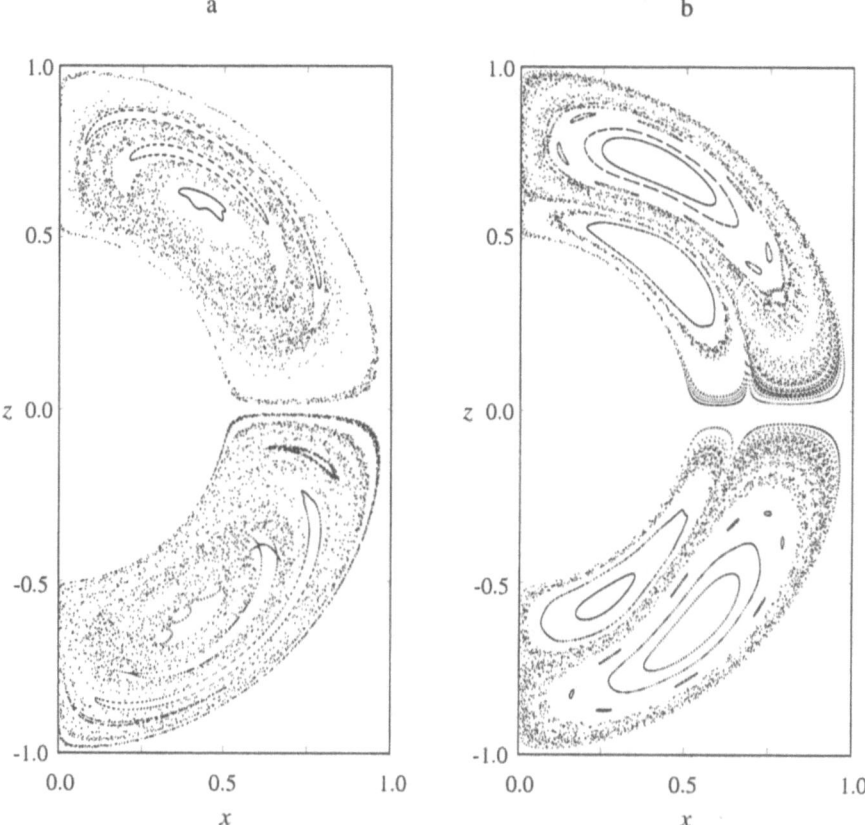

Fig. 6. — Projections of stroboscopic maps onto the plane $y = r \sin \theta \sin \phi = 0$ show Hamiltonian chaos in the flow when $\alpha = 0$. Both projections show iterates resulting from several different initial conditions. The parameter values in both cases are $\eta = 0.5$, $Re = 0.1$ and $T = 12000$. (a) At left: $\mu_1 = 0$, $\mu_2 = -8$, and the flow is a chaotic perturbation of two-vortex flow. (b) At right: $\mu_1 = 0$, $\mu_2 = -3$, and the flow is a chaotic perturbation of four-vortex flow.

velocities modulated with period T, and the secondary flow stream function ψ may be identified with a one-degree-of-freedom time-dependent Hamiltonian, which will generically exhibit chaos. Although the secondary flow in (r, θ) is decoupled from the primary flow in ϕ, it provides input to the primary flow equation (41), which will be a chaotic input when the secondary flow is chaotic[5]. Numerically integrating the flow with $\alpha = 0$ for angular velocity ratios μ_1 and μ_2, where $\mu_1 \neq \mu_2$, we find the

[5] Thus in order to fully understand transport in $(2+1)$-dimensional flows like spherical Couette flow, it is not enough to consider the Hamiltonian dynamics of the secondary flow alone. Nevertheless, it is possible to determine many properties of the flow from the secondary flow dynamics, and since the secondary flow equations (39) and (40) have the Reynolds number Re as an overall multiplying factor, if one is only interested in the secondary flow dynamics, the product $Re\,T$ can be treated as a single parameter.

Hamiltonian chaos we expected, as we show in figure 6. We find that a significant measure of Hamiltonian chaos appears in the flow only for large values of T: for low-Reynolds-number flow $Re = 0.1$, the period T needs to be approximately 10^4 before Hamiltonian chaos becomes apparent in numerical simulations. The increasing amount of Hamiltonian chaos with longer periodicities has also been noted in the investigations of journal-bearing flow.

If we introduce a small coupling between (r, θ) and ϕ, by using two rotation axes with a small angle α between them, we can simultaneously introduce both time dependence and fully-fledged three dimensionality to the problem. In order to analyse this, we take spherical Couette flow (39–41) and derive a Liouvillian map describing the strobed behaviour of biaxial unsteady spherical Couette flow in the limits of low Reynolds numbers and small axis separation. We have no reason to favour one of the axes over the other, so it is natural to use a coordinate system located in the $\alpha/2$-rotated reference frame $(r, \tilde{\theta}, \tilde{\phi})$ oriented midway between the two axes, that collapses back to (r, θ, ϕ) when $\alpha = 0$. In this frame the map can be written

$$
\begin{aligned}
r' &= r + \frac{Re}{2}\left(f_1(r) + f_2(r)\right)\left(3\cos^2\tilde{\theta} - 1\right)T \\
&\quad - \frac{3Re}{2}\left(f_1(r)\cos\tilde{\phi} - f_2(r)\left(2\cos\zeta - \cos\tilde{\phi}\right)\right)\sin\tilde{\theta}\cos\tilde{\theta}\,T\alpha \\
&\quad + O(\alpha^2),
\end{aligned}
\tag{43}
$$

$$
\begin{aligned}
\tilde{\theta}' &= \tilde{\theta} + \frac{Re}{2}\left(g_1(r) + g_2(r)\right)\sin\tilde{\theta}\cos\tilde{\theta}\,T \\
&\quad - \frac{1}{2}\Big(2\cos\zeta - \cos\tilde{\phi} - \cos\xi + \frac{Re}{4}f_1(r)h_2'(r)\left(3\cos^2\tilde{\theta} - 1\right)\sin\xi\,T^2 \\
&\quad - \frac{Re}{2}\left(g_1(r)\cos\tilde{\phi} - g_2(r)\left(2\cos\zeta - \cos\tilde{\phi}\right)\right)\left(2\cos^2\tilde{\theta} - 1\right)T\Big)\alpha \\
&\quad + O(\alpha^2),
\end{aligned}
\tag{44}
$$

$$
\begin{aligned}
\tilde{\phi}' &= \tilde{\phi} + \frac{1}{2}\left(h_1(r) + h_2(r)\right)T + \frac{Re}{4}f_1(r)h_2'(r)\left(3\cos^2\tilde{\theta} - 1\right)T^2 \\
&\quad + \frac{1}{2}\Big(2\mathrm{cosec}\,\zeta - \mathrm{cosec}\,\tilde{\phi} - \mathrm{cosec}\,\xi + \frac{Re}{2}\left((g_1(r) + g_2(r))\,\mathrm{cosec}\,\xi\right. \\
&\quad - 2g_1(r)\mathrm{cosec}\,\zeta + \frac{1}{2}f_1(r)h_2'(r)\left(\left(3\cos^2\tilde{\theta} - 1\right)\mathrm{cosec}\,\xi\cot\xi\right. \\
&\quad \left.\left. - 6\sin^2\tilde{\theta}\cos\tilde{\phi}\right)T\right)T\Big)\cot\tilde{\theta}\,\alpha + O(\alpha^2),
\end{aligned}
\tag{45}
$$

where $h_2'(r) = dh_2(r)/dr$,

$$
\zeta = \tilde{\phi} + \frac{1}{2}h_1(r)T, \qquad \xi = \tilde{\phi} + \frac{1}{2}\left(h_1(r) + h_2(r)\right)T,
\tag{46}
$$

and

$$
f_j(r) = \frac{1}{r^2}\left(\frac{A_{1j}}{r^2} + A_{2j} + A_{3j}r^3 + A_{4j}r^5 + \frac{a_{2j}}{4}\left(\frac{a_{2j}}{r} - a_{1j}r^2\right)\right),
\tag{47}
$$

$$g_j(r) = -\frac{1}{r^2}\left(-\frac{2A_{1j}}{r^3} + 3A_{3j}r^2 + 5A_{4j}r^4 + \frac{a_{2j}}{4}\left(-\frac{a_{2j}}{r^2} - 2a_{1j}r\right)\right),$$

(48)

$$h_j(r) = a_{1j} + \frac{a_{2j}}{r^3},$$

(49)

and j is 1 or 2; a_{ij} and A_{ij} being a_i (30) and A_i (31–34), respectively, evaluated with $\mu = \mu_j$. The terms of order α are those through which r and $\tilde{\theta}$ couple to $\tilde{\phi}$ and perturb the $(2+1)$-dimensional flow into being fully three dimensional. Note that when $\mu_1 = \mu_2$, so that $f_1(r) = f_2(r)$, $g_1(r) = g_2(r)$, and $h_1(r) = h_2(r)$, the two semiperiods are identical and there is cancellation within the the order-α terms which results in the coupling being weaker than otherwise. We emphasize that we shall be using the map (43–45) and results derived from it to analyse the behaviour of biaxial unsteady spherical Couette flow, but that we continue to integrate numerically the flow itself taking the full Navier–Stokes solution of (39–41).

When Re and α are small, The map (43–45) has the same form as the perturbed two-action Liouvillian map (20), with r and $\tilde{\theta}$ as the actions and $\tilde{\phi}$ as the angle variable. As we showed in §4, a map of the action plane $(r, \tilde{\theta})$ can be constructed by averaging (43–44) over the angle $\tilde{\phi}$, giving

$$r' = r + \frac{R}{e}2\left(f_1(r) + f_2(r)\right)\left(3\cos^2\tilde{\theta} - 1\right)T,$$

(50)

$$\tilde{\theta}' = \tilde{\theta} + \frac{R}{e}2\left(g_1(r) + g_2(r)\right)\sin\tilde{\theta}\cos\tilde{\theta}\,T.$$

(51)

Since, as we mentioned above, $Re\,T$ can be treated as a single parameter for the purposes of the secondary flow, taking either of the limits $Re \to 0$ or $T \to 0$ of the Euler map (50–51) has the same effect, and the actions evolve according to the stream function

$$H = \frac{1}{2}r^2\left(f_1(r) + f_2(r)\right)\sin^2\tilde{\theta}\cos\tilde{\theta},$$

(52)

$$= \frac{1}{2}\left(\frac{A_{11} + A_{12}}{r^2} + (A_{21} + A_{22}) + (A_{31} + A_{32})r^3 + (A_{41} + A_{42})r^5\right.$$

$$\left. + \frac{a_{21}}{4}\left(\frac{a_{21}}{r} - a_{11}r^2\right) + \frac{a_{22}}{4}\left(\frac{a_{22}}{r} - a_{12}r^2\right)\right)\sin^2\tilde{\theta}\cos\tilde{\theta}.$$

(53)

The invariant tori in the small-axis-separation limit are then the level surfaces of (53). Note that (53) reduces to (28), up to a constant, when $\mu_1 = \mu_2$. The resonant surfaces in the small-axis-separation limit may be obtained as in (24) by considering the values of the action variables for which the above averaging procedure fails, when the angle variable changes by a rational fraction of 2π each period. We consider (45) in the limits $Re \to 0$ and $\alpha \to 0$, so that the terms of order α and Re disappear and

$$\tilde{\phi}' - \tilde{\phi} = \frac{2\pi k}{n} = \frac{1}{2}\left(h_1(r) + h_2(r)\right)T.$$

(54)

The resonances are then the solutions of (54) where k and n are integers. These occur when

$$r_{k/n} = \left(\frac{(1 - \bar{\mu})\eta^3 nT}{(1 - \bar{\mu}\eta^3)nT - 2\pi k}\right)^{\frac{1}{3}},$$

(55)

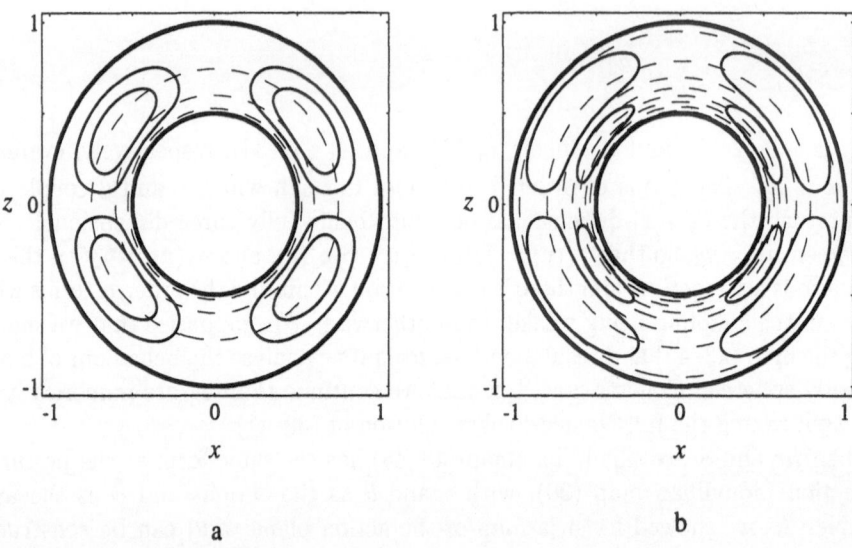

Fig. 7. — Primary resonances (55) — shown dashed — and adiabatic invariant surfaces (53) in biaxial spherical Couette flow are shown for small axis separation in the low-Reynolds-number regime. (a) At left: $\eta = 0.5$, $\mu_1 = 0$, $\mu_2 = -8$, and $T = 4$; a two-vortex flow. The resonances in the spheres are shown where, from the inner sphere outwards, k/n is $-2/1$, $-1/1$, and $0/1$. (b) At right: $\eta = 0.5$, $\mu_1 = 0$, $\mu_2 = -3$, and $T = 20$; a four-vortex flow. The increase in the axis-switching period T from (a) means there are more resonances in this the picture: those where k/n is $-4/1$, $-3/1$, $-2/1$, $-1/1$, $0/1$, $1/1$, $2/1$, and $3/1$ are shown.

which are shells of constant r, whose radius depends on the mean value of μ_1 and μ_2, $\bar{\mu} = (\mu_1 + \mu_2)/2$, as well as on η, T and the order k/n of the resonance. For example, in figure 7 (a) we show the distribution of some of the resonances in a two-vortex flow with $\eta = 1/2$, $\mu_1 = 0$, $\mu_2 = -8$, and $T = 4$. The terms of order α and Re in (45) constitute negligible perturbations of these resonance shells when the axis separation and the Reynolds number are small. The resonances are dense throughout space, but they have a hierarchical structure in which their relative strengths are governed by the Fourier expansion of (43–45), such that in general the resonances with smaller denominators n are stronger; the primary resonances where $n = 1$ have the largest influence, and the effect of the secondary and higher-order resonances $n \geqslant 2$ is far smaller. One can tune the resonances using $\bar{\mu}$ and T so that more or less primary resonances be inside or outside the region $\eta < r < 1$ between the spheres. In figure 7 (b) we show an example with $\eta = 1/2$, $\mu_1 = 0$, $\mu_2 = -3$, and $T = 20$ of a four-vortex flow with more primary resonances than figure 7 (a).

In the projection of the stroboscopic map of the flow in figure 8 (a), we can see that the trajectory is, as predicted, lying on the invariant tori of (53), except for intervals where it jumps from one invariant torus to another. Also shown dashed in figure 8 (a) is the principal resonant surface, and it is apparent that the resonances are the means

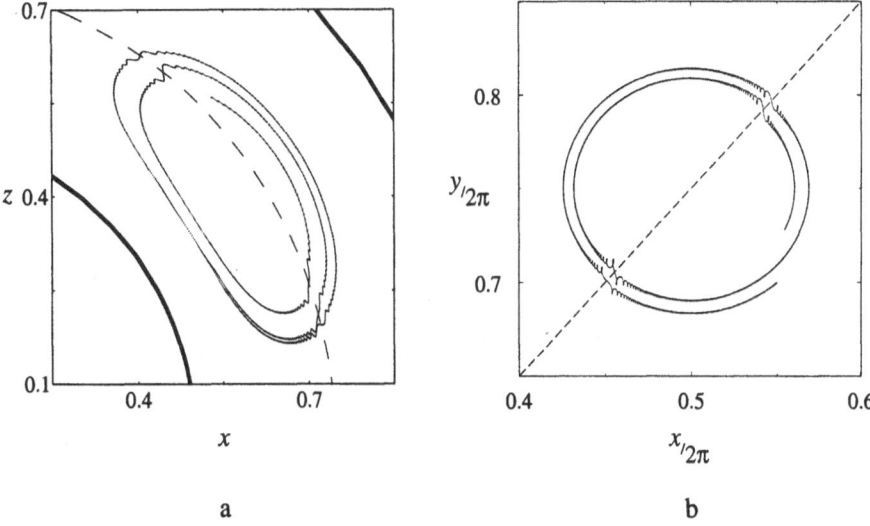

Fig. 8. — (a) At left: a projection of a stroboscopic map onto the plane $y = r \sin \bar{\theta} \sin \tilde{\phi} = 0$ illustrates the jump behaviour typical of resonance-induced dispersion in two-action flows. The parameter values are as in figure 7 (a), with $Re = 0.1$ and $\alpha = 0.1°$. These parameter values are such as to give a two-vortex flow, and to put the primary 0/1 resonance in the middle of the region between the spheres, whilst leaving other primary resonances distant. Notice that because this is a projection, trajectories can appear to cross. (b) At right: a projection of a trajectory in a Liouvillian map shows resonance-induced dispersion. The diagonal dashed line indicates the location of the primary resonance. Close to this line the trajectory oscillates wildly and jumps from one invariant torus to another. The map used is $x' = x + 0.001(\sin z + 2 \cos y)$, $y' = y + 0.001(1.5 \sin x' + 2.5 \cos z)$, $z' = z + 4(\cos y' + \sin x')$.

by which the trajectory jumps from invariant torus to invariant torus. To compare with figure 8 (a), in figure 8 (b) we show an example of the same resonance-induced dispersion behaviour occurring in a two-action Liouvillian map. As we increase the coupling α, the resonances have increasing influence, and the length of time for which trajectories are captured into resonance increases. In figure 9 (a) we show another projection of a stroboscopic map, this time for a larger value of α than in figure 8 (a), but with the other parameters remaining unchanged. Trajectories are now captured by resonances for much longer times, and by this means they can rapidly cross the separatrix between the vortices at the equator and disperse throughout practically the whole region between the spheres. This is shown graphically in figure 10 (a), which illustrates a slice between $-0.01 < y < 0.01$ through the same stroboscopic map as figure 9 (a). The stroboscopic map slice shows the evolution of a single trajectory, and comparing it with the distribution of the resonances shown in figure 7 (a), we observe that the only regions as yet unvisited in figure 10 (a) are close to the poles, and the centres of the vortices that lie inside the resonance 0/1 shell and are not crossed by any primary resonance. Resonance-induced dispersion into

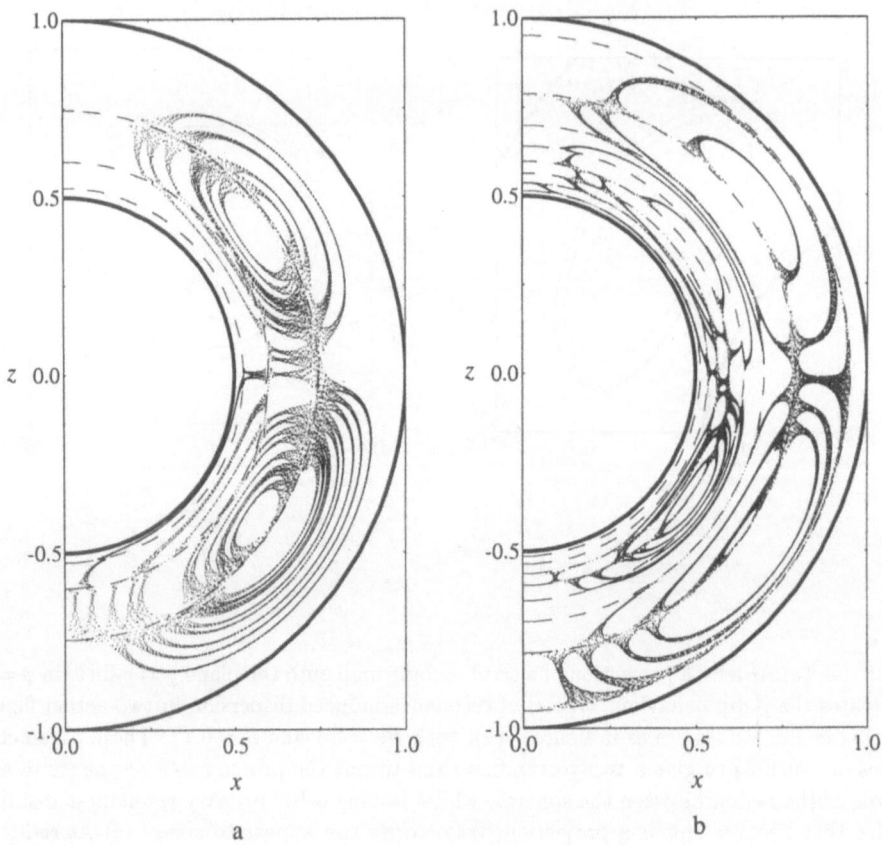

Fig. 9. — Projections of stroboscopic maps onto the plane $y = r \sin \tilde{\theta} \sin \tilde{\phi} = 0$ for $Re = 0.1$ and $\alpha = 1°$. (a) At left: other parameters are as in figure 7 (a), and a single trajectory is depicted in this two-vortex flow. (b) At right: other parameters are as in figure 7 (b), and two trajectories are shown. This is a four-vortex flow; there is no mixing across the separatrix between the inner and outer vortices.

these regions will occur slowly through weak high-order resonances that lie densely between the primary resonances. To compare with this global mixing behaviour in the two-vortex case, in figures 9 (b) and 10 (b) we show that for four-vortex flow, resonance-induced dispersion operates separately within the inner and outer vortices; there is no mixing across the separatrix between them, since this separatrix is concentric with, and therefore not crossed by, any of the resonance shells.

Looking at figure 9 side by side with the $\alpha = 0$ Hamiltonian chaos of figure 6 shows how different resonance-induced dispersion appears in comparison; the absence of KAM tori shows that we are dealing with chaos that is not a small perturbation of an integrable Hamiltonian system. Whilst the Hamiltonian chaos of figure 6 is produced by long-period flow, $T = 12000$, the resonance-induced dispersion of figure 9 is already present when the flow period is much shorter; $T = 4$. At these short periodicities, the measure of Hamiltonian chaos is minute. In the case where

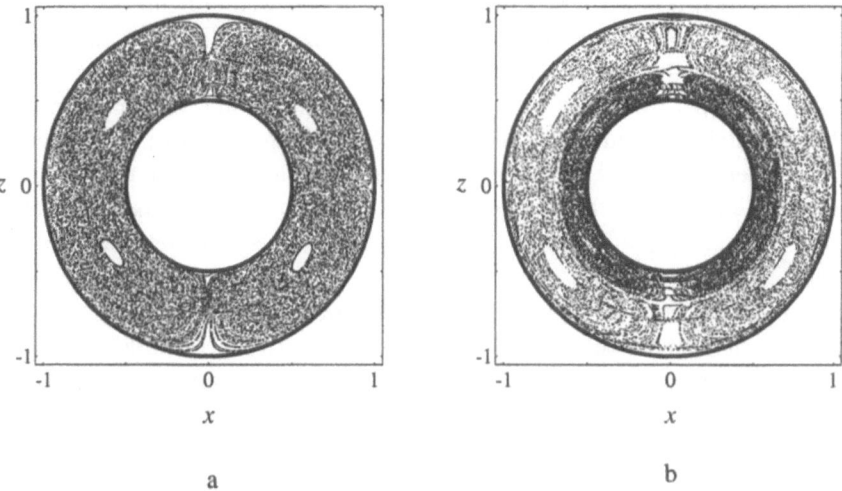

Fig. 10. — Slices of stroboscopic maps between $-0.01 < y < 0.01$. (a) At left: the parameter values and initial conditions are as in figure 9 (a); the resonances are distributed as in figure 7 (a). Forty thousand points (corresponding to many times this number of periods) are shown, all from the same initial condition. This slice illustrates how effective resonance-induced dispersion is at mixing; the only regions where the trajectory has not yet ventured are close to the poles, and the centres of the vortices, which lie inside the 0/1 resonance. In time, the trajectory will disperse into these regions too, through the action of the higher-order resonances that are dense throughout the space. If, by altering $\bar{\mu}$, the 0/1 resonance were tuned to pass right through the centres of the vortices, these regions would be visited as frequently as others. (b) At right: two initial conditions are used to illustrate dispersion in the trajectory with parameter values as in 9 (b); the resonances are distributed as in figure 7 (b). The space is divided into two regions by the separatrix between the two vortices in each hemisphere. There in no mixing between these two regions, but effective mixing occurs separately within each region.

$\mu_1 = \mu_2$ there is no Hamiltonian chaos when $\alpha = 0$. Recall from our discussion of (43–45) that when $\mu_1 = \mu_2$ the coupling between $(r, \tilde{\theta})$ and $\tilde{\phi}$ is weaker than otherwise. This implies that when $\alpha \neq 0$ and $\mu_1 = \mu_2$ all the resonances will have less influence than normal. Furthermore, there is additional cancellation that takes place for $r = r_{0/1}$ when $\mu_1 = \mu_2$; in this case $h_1(r) = h_2(r) = 0$ and the order-α terms cancel completely, so that the 0/1 resonance is not present when $\mu_1 = \mu_2$. This anomalous behaviour when $\mu_1 = \mu_2$ is why we have chosen throughout this paper to use examples where $\mu_1 \neq \mu_2$ for illustrating resonance-induced dispersion.

There are several parameters one can alter to control the rate of resonance-induced dispersion. Tuning the resonances with (55) is one option. The stroboscopic map projections of figures 8 (a) and 9 (a) were made with $\mu_1 = 0$ and $\mu_2 = -8$ so that the 0/1 primary resonance passed right through the centre of the region $\eta < r < 1$ and cut almost all invariant tori of (53), to ensure that resonance-induced dispersion with a primary resonance would occur every time the trajectory wound once around the vortex. Increasing the axis-switching period T would speed up the dispersion by increasing the density of primary resonances, as we showed in figure 7 (b). If, on the other hand, we choose to tune the system by changing $\bar{\mu}$ and T so that all the primary resonances lie outside the interval $\eta < r < 1$, we would still have resonance-induced dispersion, but operating only with the weaker secondary and higher resonances, so that the dispersion would be slower with smaller jumps. Following the discussion above, another possibility is to make μ_1 the same as or different to μ_2, as the resonances are stronger when $\mu_1 \neq \mu_2$. We can also alter the axis separation α, since the resonances gain strength with increasing α, as the comparison of figure 8 (a) with figure 9 (a) shows.

For a different view of resonance-induced dispersion, we can look at the time evolution of the stream function (53). If the adiabatic approximation of (50–51) were exact, this would be constant in time. Instead, if resonance-induced dispersion were occurring, we would expect to see that this graph should be almost flat when the trajectory evolved on a level surface of an invariant torus, but these flat regions would be interspersed with oscillations leading to sudden jumps, as each time the trajectory approached a resonance intersecting the invariant torus it would be transported to another level surface with a different value of H, rather like playing a game of snakes and ladders. Figure 11 (a), for the same parameter values and initial conditions as figure 8 (a), is such a graph, and indeed we observe this snakes-and-ladders behaviour exactly as we have just described — we are seeing resonance-induced dispersion in action. As an example to compare with figure 11 (a), we show in figure 11 (b) a similar graph of resonance-induced dispersion in a two-action Liouvillian map.

We find numerically that the small-axis-separation approximation for biaxial unsteady spherical Couette flow remains good as long as α is less than around ten degrees. The calculations of invariant tori and resonant surfaces we made for the small-axis-separation flow become increasingly poor approximations when the axis separation α is made larger. For α greater than one degree, the noise arising from the fact that the small-axis-separation approximation is just an approximation starts to overwhelm the signal. However, we can recuperate it by applying a running average to the noisy signal from (53), as we show in figure 12. For α more than about ten degrees, this averaging starts to lose its effectiveness, and consequently we are unable perform much analysis for larger axis separations. We believe that the large-axis-separation flow is still of two-action type; the interaction between $\tilde{\theta}$ and $\tilde{\phi}$ will produce one action and one angle variable, r being the other action. However since the large-axis-separation flow is not a small perturbation of a $(2+1)$-dimensional flow, we are unable to obtain analytical expressions for the action and angle variables in the large-axis-separation case in the same way as was possible for small-axis-separation flow.

6.   Chaotic advection in three-dimensional unsteady flow

Following our classification in §4 of three-dimensional unsteady flows by the number of fast angle and slow action variables, biaxial unsteady spherical Couette flow is of the two-action class. Three-dimensional flows often arise from what we have termed $(2 + 1)$-dimensional flows by the coupling of a one-dimensional primary flow to a

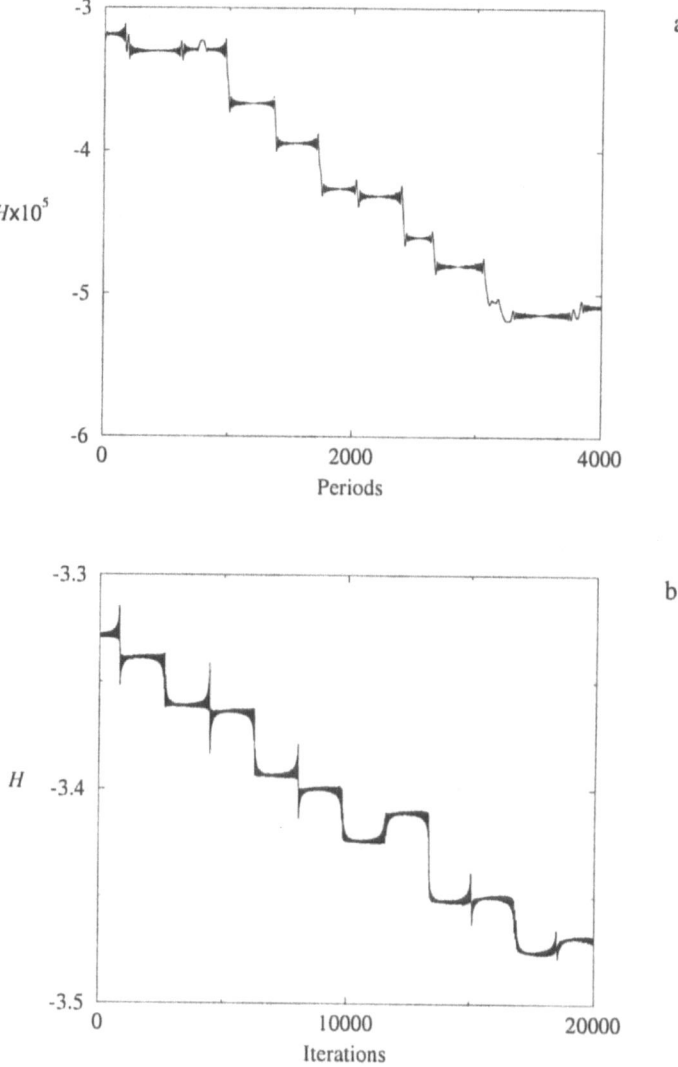

Fig. 11. — (a) Above: resonance-induced dispersion is illustrated here with a graph of the adiabatic invariant (53) against time of the trajectory in figure 8 (a). (b) Below: resonance-induced dispersion in a Liouvillian map is shown using a graph of the adiabatic invariant (23) against time of the trajectory in figure 8 (b). In both (a) and (b) the adiabatic invariant remains nearly constant almost all the time, but jumps chaotically from one invariant torus to another at each intersection with a primary resonance. The general downward trend in the graphs corresponds to the trajectory winding towards a vortex centre, as is visible in figures 8 (a) and (b).

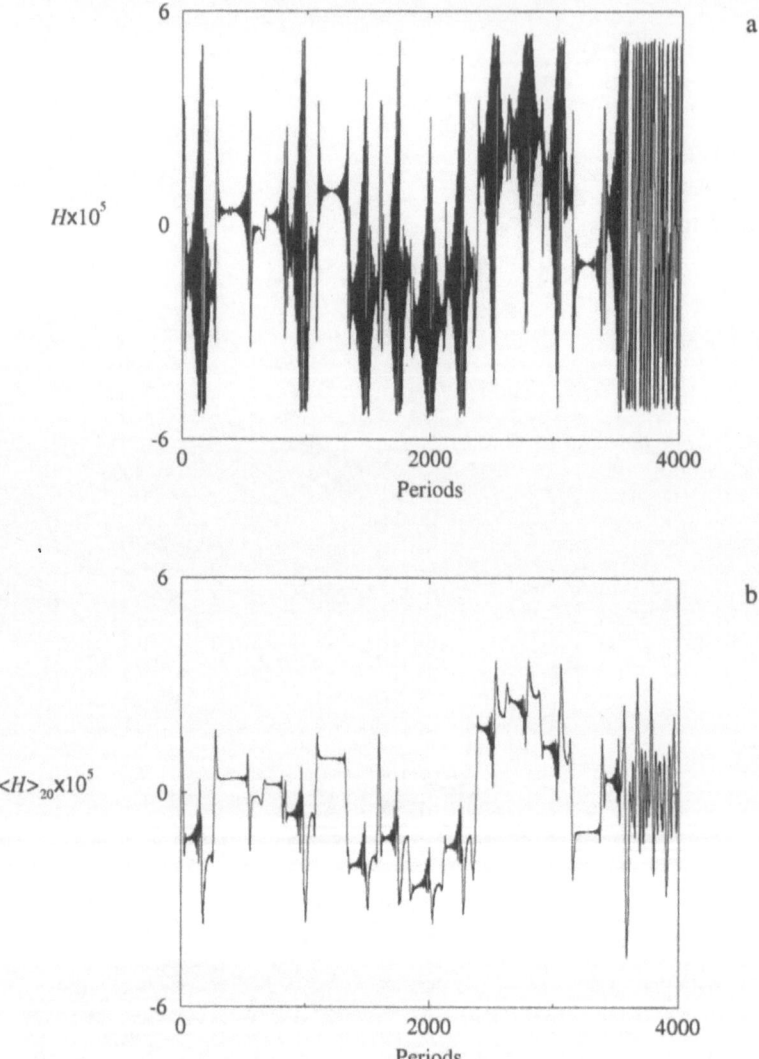

Fig. 12. — Averaging at work; the initial conditions and parameter values here are as in figure 8 (a), except for the axis separation angle α, which is now $10°$. The upper graph shows the noisy adiabatic invariant H (53) against time. The noise arises because H is only approximately the invariant, and the approximation worsens with increasing α. In the lower graph we have performed a twenty-point running average on the trajectory. The running average has smoothed out the noise substantially, so that we can now see the jumps much more clearly. This shows that even in cases where we do not possess the exact adiabatic invariant, it is still possible using averaging to see the characteristic two-action jump behaviour at work.

two-dimensional secondary flow, giving one variable of one type and two variables of the other. Generically this will preclude zero-action and three-action flows, but a large class of three-dimensional unsteady flows are two-action flows, and one-action flows will be equally common [39]. With biaxial unsteady spherical Couette flow the relationship between the primary and secondary flow rates depends on the Reynolds number — the secondary flow is hundreds to thousands of times slower than the primary flow when $Re = 0.1$ — so that although it would be possible in our model to increase the Reynolds number sufficiently to give two fast secondary variables and one slow primary variable, making the flow of one-action type, doing so would make the low-Reynolds-number approximation we started with invalid, landing us squarely in the turbulent regime [13], Other possible candidate flows for obtaining a one-action three-dimensional unsteady flow model that can both be treated analytically and is experimentally realizable should be sought out and investigated.

We have seen in biaxial unsteady spherical Couette flow that the zero-axis-separation limit of $\alpha = 0$ can lead us either to two-dimensional steady flow without chaos, if $\mu_1 = \mu_2$, or to two-dimensional unsteady flow with chaos, if $\mu_1 \neq \mu_2$. The limit we cannot explore with biaxial unsteady spherical Couette flow corresponds to the transition between three-dimensional unsteady flow and three-dimensional steady flow. One way to arrive at a three-dimensional steady flow model using spherical Couette flow would be to displace the spheres so that they would no longer be coaxial, like the cylinders in journal-bearing flow. In this way we would break the ϕ symmetry and obtain a three-dimensional steady flow, spherical-bearing flow, which could then be modulated in time using different μ_1 and μ_2 to give a three-dimensional unsteady flow. The only work we know of on this geometry is that of Wannier [54], who however only treats the the problem of spherical-bearing flow in the limit of a very small gap between the spheres, such that the radial component of the motion can be assumed to be zero. This limit of the Stokes equation leads to the Reynolds equations of lubrication theory, which give a two-dimensional flow in θ and ϕ. Munson [40] solves perturbatively the case where the spheres are displaced along their common axis, so that they are eccentric but remain coaxial. This however retains the ϕ symmetry, so it is another example of a $(2+1)$-dimensional steady flow rather than a three-dimensional steady flow.

Another possibility for investigating the transition between three-dimensional steady and unsteady flow is given by the flow introduced by Bajer & Moffatt [10], who treat the three-dimensional steady flow between two concentric spheres rotating about different axes. This flow is particularly interesting because it exhibits the consequences of the breakdown of the Hamiltonian description of steady three-dimensional flows that occurs at stagnation points of the flow. This phenomenon is similar in origin to the resonance-induced dispersion we have been investigating here in three-dimensional unsteady flows. In both cases averaging fails in certain regions of space. In three-dimensional unsteady flows these regions are generically two-dimensional resonant surfaces, whereas in three-dimensional steady flows they are zero-dimensional stagnation points. However, despite the less important rôle, occasioned by their lower dimensionality, that these regions play in three-dimensional steady flow, in appropriate circumstances the global topology of the flow can be altered by the breakdown of averaging at the stagnation points, leading to what Bajer & Moffatt have termed trans-adiabatic drift [9, 8]. Trans-adiabatic drift can be considered to be the equivalent in three-dimensional steady flows of resonance-induced dispersion. It should thus be particularly interesting to model the addition of time dependence to the flow described above. We should point out that adiabatic descriptions of processes that are valid everywhere except in certain neighbourhoods are ubiquitous in many areas of physics, and lead to graphs of the type of figure 11 being seen in many contexts.

Occupying as they do a position somewhere between symplectic maps of two and four dimensions from two and three-degree-of-freedom Hamiltonian systems, the Liouvillian maps that underlie three-dimensional unsteady flows can exhibit types of behaviour analogous to both their lower and higher-dimensional siblings. An important difference between Hamiltonians of two and three degrees of freedom is that Arnold diffusion[6] exists in the latter but not in the former. With two degrees of freedom, the invariant KAM tori completely separate different regions of phase space. This is no longer true when another degree of freedom is added, and as a result Arnold diffusion can take place throughout the phase space. This occurs, albeit extremely slowly, for arbitrarily small values of the nonlinearity parameter ε. Resonance-induced dispersion in two-action Liouvillian maps, although reminiscent of Arnold diffusion in the four-dimensional volume-preserving maps from three-degree-of-freedom Hamiltonians, is engendered by a different mechanism. Whilst Arnold diffusion occurs through an interconnected web of resonances, in resonance-induced dispersion the available space is covered using the invariant tori, and the resonant surfaces are a means to jump from one torus to another. The consequence of this difference is that the rates of the two processes are radically different; with nonlinearity ε, an estimate of the upper bound on the rate of Arnold diffusion is $O(\exp(-\varepsilon^{-1/2}))$, whilst previous work on Liouvillian maps shows that resonance-induced dispersion has an $O(\varepsilon^2)$ dependence [47].

The presence of resonance-induced dispersion in two-action type flows, and its absence in those of one-action type, is from the mixing and transport point of view the crucial difference between these two classes of flows. In the one-action class of three-dimensional unsteady flows, as with two-dimensional unsteady flows, the existence of KAM-type theorems proves that complete mixing will not generically be achievable, as the invariant tori present barriers to transport. On the other hand, the addition of a third dimension to a two-dimensional unsteady flow leads in the two-action case to much faster dispersion than with two-dimensional flow alone, owing to the presence of resonance-induced dispersion. This should lead to better mixing. What we mean by efficient mixing here is that the density of an initial concentration of passive scalars should rapidly tend to a constant value throughout the region being mixed. This property is related to the Lyapunov exponents of the flow, one of which must be positive to allow the exponential separation of neighbouring trajectories that is necessary for chaos. Even a small coupling between primary and secondary flows is effective in two-action flows in producing a large improvement in dispersion over two-dimensional chaotic advection alone through the action of resonance-induced dispersion, particularly when the flow period is relatively short when there is very little chaotic advection in two dimensions. In practical terms, we have seen that even a wobble from a bad bearing between the two spheres leading to a free play

---

(6) One of the terminological differences between nonlinear dynamics and fluid dynamics is that *diffusion* in nonlinear dynamics refers to a process that in fluid dynamics is called *dispersion* (diffusion in fluid dynamics carrying connotations of molecular diffusion). In previous papers [47, 15, 16] we used the term diffusion in the dynamical systems sense.

of 0.1° is enough to produce a spectacular growth in dispersion rate. Although we have run some of our plots for long times, by adjusting the parameters it is possible to arrive at a flow in which the effects of resonance-induced dispersion are visible after only a handful of periods.

Biaxial unsteady spherical Couette flow forms a bridge between theory and experiment for investigating chaotic advection in three-dimensional unsteady flows; we have an analytical solution for the stream function and the flow can be realized in the laboratory. Journal-bearing flow has performed this function extremely well for two-dimensional unsteady flows, as has twisted-pipe flow in three-dimensional steady flows. We hope that our model of biaxial unsteady spherical Couette flow might spur similar interest in the analysis and investigation of chaotic advection in three-dimensional unsteady flows, and that the challenge of devising suitable apparatus for the experimental observation of resonance-induced dispersion and the other phenomena described here will be taken up.

References

[1] N. Acharya, M. Sen, and H. C. Chang. Heat transfer enhancement in coiled tubes by chaotic mixing. *J. Heat Mass Transfer*, 35:2475, 1992.

[2] H. Aref. Stirring by chaotic advection. *J. Fluid Mech.*, 143:1–21, 1984.

[3] H. Aref and S. Balachandar. Chaotic advection in a Stokes flow. *Phys. Fluids*, 29:3515–21, 1986.

[4] H. Aref, S. W. Jones, S. Mofina, and I. Zawadzki. Vortices, kinematics and chaos. *Physica D*, 37:423–40, 1989.

[5] V. I. Arnold. Sur la topologie des écoulements stationnaires des fluides parfaits. *C. R. Acad. Sci. Paris A*, 261:17–20, 1965.

[6] V. I. Arnold. *Mathematical Methods of Classical Mechanics*. Springer, 1978.

[7] D. K. Arrowsmith and C. M. Place. *An Introduction to Dynamical Systems*. Cambridge University Press, 1990.

[8] K. Bajer. Hamiltonian formulation of the equations of streamlines in three-dimensional steady flows. *Chaos, Solitons & Fractals*, 4:895–911, 1994.

[9] K. Bajer and H. K. Moffatt. On a class of steady confined Stokes flows with chaotic streamlines. *J. Fluid Mech.*, 212:337–63, 1990.

[10] K. Bajer and H. K. Moffatt. Chaos associated with fluid inertia. In H. K. Moffatt, G. M. Zaslavsky, P. Comte, and M. Tabor, editors, *Topological Aspects of the Dynamics of Fluids and Plasmas*, pages 517–34. Kluwer, 1992.

[11] B. Y. Ballal and R. S. Rivlin. Flow of a Newtonian fluid between eccentric rotating cylinders: Inertial effects. *Arch. Ration. Mech. Anal.*, 62:237–294, 1976.

[12] M. V. Berry, N. L. Bazacs, M. Tabor, and A. Voros. Quantum maps. *Ann. Phys.*, 122:26–63, 1979.

[13] K. Bühler. Pattern formation of instabilities in spherical Couette flow. In S. Kai, editor, *Pattern Formation in Complex Dissipative Systems: Fluid Patterns, Liquid Crystals, Chemical Reactions*, pages 298–302. World Scientific, 1992.

[14] J. H. E. Cartwright, M. Feingold, and O. Piro. Dynamically diffusive Lagrangian trajectories in time-periodic three-dimensional flows. In H. Peerhossaini and A. Provenzale, editors, *Heat Transfer Enhancement by Lagrangian Chaos and Turbulence*, pages 101–7. ISITEM, Univ. Nantes, 1994.

[15] J. H. E. Cartwright, M. Feingold, and O. Piro. Passive scalars and three-dimensional Liouvillian maps. *Physica D*, 76:22–33, 1994.

[16] J. H. E. Cartwright, M. Feingold, and O. Piro. Global diffusion in a realistic three-dimensional time-dependent nonturbulent fluid flow. *Phys. Rev. Lett.*, 75:3669–72, 1995.

[17] J. H. E. Cartwright, M. Feingold, and O. Piro. Chaotic advection in three-dimensional unsteady incompressible laminar flow. *J. Fluid Mech.*, 316:259–84, 1996.

[18] J. H. E. Cartwright and O. Piro. The dynamics of Runge–Kutta methods. *Int. J. Bifurcation and Chaos*, 2:427–49, 1992.

[19] J. Chaiken, R. Chevray, M. Tabor, and Q. M. Tan. Experimental study of Lagrangian turbulence in Stokes flow. *Proc. R. Soc. Lond. A*, 408:165–74, 1986.

[20] J. Chaiken, C. K. Chu, M. Tabor, and Q. M. Tan. Lagrangian turbulence and spatial complexity in a Stokes flow. *Phys. Fluids*, 30:687–99, 1987.

[21] C.-Q. Cheng and Y.-S. Sun. Existence of invariant tori in three-dimensional measure preserving mappings. *Celest. Mech.*, 47:275–92, 1990.

[22] W.-L. Chien, H. Rising, and J. M. Ottino. Laminar mixing and chaotic mixing in several cavity flows. *J. Fluid Mech.*, 170:355–77, 1986.

[23] W. R. Dean. Note on the motion of fluid in curved pipes. *Phil. Mag.*, 4:208–23, 1927.

[24] W. R. Dean. The streamline motion of fluid in curved pipes. *Phil. Mag.*, 5:673–93, 1928.

[25] T. Dombre, U. Frisch, J. M. Greene, M. Hénon, A. Mehr, and A. M. Soward. Chaotic streamlines in the ABC flows. *J. Fluid Mech.*, 167:353–91, 1986.

[26] M. Feingold, L. P. Kadanoff, and O. Piro. A way to connect fluid dynamics to dynamical systems: Passive scalars. In A. J. Hurd, D. A. Weitz, and B. B. Mandelbrot, editors, *Fractal Aspects of Materials: Disordered Systems*, pages 203–5. Materials Research Society, 1987.

[27] M. Feingold, L. P. Kadanoff, and O. Piro. Diffusion of passive scalars in fluid flows: Maps in three dimensions. In R. Jullien, L. Peliti, R. Rammal, and N. Boccara, editors, *Universalities in Condensed Matter*, pages 236–41. Les Houches, Springer, 1988.

[28] M. Feingold, L. P. Kadanoff, and O. Piro. Passive scalars, three-dimensional volume-preserving maps, and chaos. *J. Stat. Phys.*, 50:529–65, 1988.

[29] M. Feingold, L. P. Kadanoff, and O. Piro. Transport of passive scalars: KAM surfaces and diffusion in three-dimensional Liouvillian maps. In P. Collet, E. Tirapegui, and D. Villarroel, editors, *Instabilities and Nonequilibrium Structures II*. Reidel, 1989.

[30] W. L. Haberman. Secondary flow about a sphere rotating in a viscous liquid inside a coaxially rotating spherical container. *Phys. Fluids*, 5:625–6, 1962.

[31] M. Hénon. Sur la topologie des lignes de courant dans un cas particulier. *C. R. Acad. Sci. Paris A*, 262:312–4, 1966.

[32] S. W. Jones, O. M. Thomas, and H. Aref. Chaotic advection by laminar flow in a twisted pipe. *J. Fluid Mech.*, 209:335–57, 1989.

[33] D. V. Khakhar, J. G. Franjione, and J. M. Ottino. A case study of chaotic mixing in deterministic flows. The partitioned pipe mixer. *Chem. Engng Sci.*, 42:2909–26, 1987.

[34] D. V. Khakhar, H. Rising, and J. M. Ottino. An analysis of chaotic mixing in two chaotic flows. *J. Fluid Mech.*, 172:419–51, 1986.

[35] E. L. Koschmieder. *Bénard Cells and Taylor Vortices*. Cambridge University Press, 1993.

[36] H. A. Kusch and J. M. Ottino. Experiments on mixing in continuous chaotic flows. *J. Fluid Mech.*, 236:319–48, 1992.

[37] Y. Le Guer, C. Castelain, and H. Peerhossaini. Experimental study of chaotic advection regime in a twisted duct flow. *preprint*, 1996.

[38] R. S. MacKay. Transport in 3D volume-preserving flows. *J. Nonlinear Sci.*, 4:329–54, 1994.

[39] I. Mezić and S. Wiggins. On the integrability and perturbation of three-dimensional fluid flows with symmetry. *J. Nonlinear Sci.*, 4:157–94, 1994.

[40] B. R. Munson. Viscous incompressible flow between eccentric coaxially rotating spheres. *Phys. Fluids*, 17:528–31, 1974.

[41] B. R. Munson and D. D. Joseph. Viscous incompressible flow between concentric rotating spheres. Part 1. Basic flow. *J. Fluid Mech.*, 49:289–303, 1971.

[42] S. E. Newhouse, D. Ruelle, and F. Takens. Occurrence of strange axiom A attractors near quasiperiodic flows on T^m, $m \geqslant 3$. *Commun. Math. Phys.*, 64:35–40, 1978.

[43] J. M. Ottino. *The Kinematics of Mixing: Stretching, Chaos, and Transport*. Cambridge University Press, 1989.

[44] T. S. Parker and L. O. Chua. *Practical Numerical Algorithms for Chaotic Systems*. Springer, 1989.

[45] C. E. Pearson. A numerical study of the time-dependent viscous flow between two rotating spheres. *J. Fluid Mech.*, 28:323–36, 1967.

[46] H. Peerhossaini, C. Castelain, and Y. Le Guer. Heat exchanger design based on chaotic advection. *Exp. Thermal & Fluid Sci.*, 7:333, 1993.

[47] O. Piro and M. Feingold. Diffusion in three-dimensional Liouvillian maps. *Phys. Rev. Lett.*, 61:1799–802, 1988.

[48] D. Ruelle and F. Takens. On the nature of turbulence. *Commun. Math. Phys.*, 20:167–92, 1971. **23**, 343–4.

[49] T. H. Solomon and J. P. Gollub. Chaotic particle transport in time dependent Rayleigh–Bénard convection. *Phys. Rev. A*, 38:6280–6, 1988.

[50] H. A. Stone, A. Nadim, and S. H. Strogatz. Chaotic streaklines inside drops immersed in steady linear flows. *J. Fluid Mech.*, 232:629–46, 1991.

[51] P. D. Swanson and J. M. Ottino. A comparative computational and experimental study of chaotic mixing of viscous fluids. *J. Fluid Mech.*, 213:227–49, 1990.

[52] G. I. Taylor. *Low Reynolds Number Flow.* Educational Services Incorporated, 1960. (16 mm film).

[53] A. Thyagaraja and F. A. Haas. Representation of volume-preserving maps induced by solenoidal vector fields. *Phys. Fluids*, 28:1005–7, 1985.

[54] G. H. Wannier. A contribution to the hydrodynamics of lubrication. *Quart. Appl. Math.*, 8:1–32, 1950.

[55] I. M. Yavorskaya, Y. N. Belyaev, A. A. Monakhov, N. M. Astaf'eva, S. A. Scherbakov, and N. D. Vvedenskaya. Stability, non-uniqueness and transition to turbulence in the flow between two rotating spheres. In *Proceedings of the XV International Congress of Theoretical and Applied Mechanics*, pages 431–43. IUTAM, 1980.

[56] V. A. Zheligovsky. A kinematic magnetic dynamo sustained by a Beltrami flow in a sphere. *Geophys. Astrophys. Fluid Dynamics*, 73:217–54, 1993.

Renormalization Group Method in Chaotic Mixing

GEORGE M. ZASLAVSKY

Courant Institute of Mathematical Sciences, New York University, 251 Mercer St., New York, NY 10012, USA
and
Department of Physics, New York University, 2-4 Washington Place, New York, NY 10003, USA

There exist special values of a control parameter for which a self-similar hierarchy of islands occurs in the phase space of a system with chaotic dynamics. This set of islands impose an anomalous diffusion of particles which can be described by a fractional kinetic equation with fractional derivatives in space and time. The corresponding exponents of the derivatives can be related to the self-similarity properties of the islands using the renormalization group approach. Examples are given for the web-map and standard map.

1. Introduction

From its very beginning chaotic dynamics was considered as a dynamics of trajectories with intrinsic property of mixing in the phase space [1]. From a formal point of view, the mixing property of a system can be expressed by the decomposition of the time-correlation function taken along a trajectory.

Two different approaches in considering chaotic systems are related to a local or a large-scale phase space analysis of the system in time-space. Studying local properties provides the information on Kolmogorov-Sinai entropy, singularities and resonances, etc. Large scale space-time dynamics provides the asymptotics of the distribution function and its moments. One can connect micro-dynamics with macro-dynamics via the derivation of the kinetic equation for the coarse-grained distribution function. Such an equation includes micro-properties through a "collisional term"

Mixing: Chaos and Turbulence, edited by Chaté *et al.*
Kluwer Academic / Plenum Publishers, New York 1999.

or its analog. We say that the kinetic equation describes transport properties or simply, transport, of the system.

Considering a Hamiltonian system of low dimensions one can introduce canonical pairs of action-angle variables $(I\vartheta)$ and look at the kinetic equation for distribution function of $f(I,t)$ which is obtained by an appropriate averaging procedure over phases ϑ. In a typical simplified case, the trajectories in the phase space obey the Gaussian process, and distribution function $f(I,\vartheta)$ satisfies diffusional equation. In this case we say that the diffusion is normal, i.e. asymptotically

$$\langle(\Delta I)^2\rangle \sim t^\mu \tag{1}$$

with $\mu = 1$. Nevertheless many cases are discovered for which $\mu \neq 1$ and the diffusion is anomalous. The case $1 < \mu < 2$ is called superdiffusion and the $0 < \mu < 1$ case subdiffusion. Superdiffusion means a strong acceleration of the mixing process and *vice versa* for subdiffusion. These processes are often related often to the so-called Lévy processes or Lévy flights (see [2, 3, 4, 5] for more details).

In a set of our publications [6, 7, 8, 9, 10] the anomalous properties of transport were connected to the topological properties of the phase space of a system. Namely, it was shown that the transport exponent μ in (1) can be expressed through the characteristic exponents of self-similarity of fine structure of islands' boundary layer in the phase space. A special renormalization method was applied for both space and time [7, 9, 10] to find a kinetic equation. In this article we follow the same scheme and describe some new applications of the method, related to the Poincaré recurrences and exit time distributions. In fact, the main idea of the Renormalization Group for kinetics (RGK) method used in the article is to eliminate the leading term of the asymptotic expansion for the kinetic equation which we call Fractional Fokker-Planck-Kolmogorov (FFPK) equation. Speaking about asymptotics one should keep in mind a possibility to obtain an intermediate asymptotics as well as an absolute one. In this sense an important result was obtained in [11] where an intermediate asymptotics for the anomalous transport was derived for 2D particle advection in the presence of convective cells and small diffusivity. Our consideration assumes the absence of any diffusivity, i.e. the randomness occurs only as a result of chaotic dynamics.

2. A simple motivation for superdiffusion

Here we would like to make a simple comment about the occurrence of an abnormal value for the transport exponent μ in (1). Imagine a trjectory which consists of many different kinds of pieces: pieces which looks like a normal random walk with a characteristic scale ℓ_r of displacement, and pieces of almost regular motion along a direction which do not have a characteristic scale at all and can be arbitrarily long (we call them "flights"). Then the mean square displacement $\langle R^2(t)\rangle$ can be expressed as

$$\langle R^2(t)\rangle \sim c\langle R^2(t)\rangle_n + (1-c)\langle R^2(t)\rangle_{an}$$

where c is a part of the "normal" random walk pieces and is supposed to be close to one. Using (1) we have

$$\langle R^2(t)\rangle \sim cDt + (1-c)\cdot At^\mu \tag{2}$$

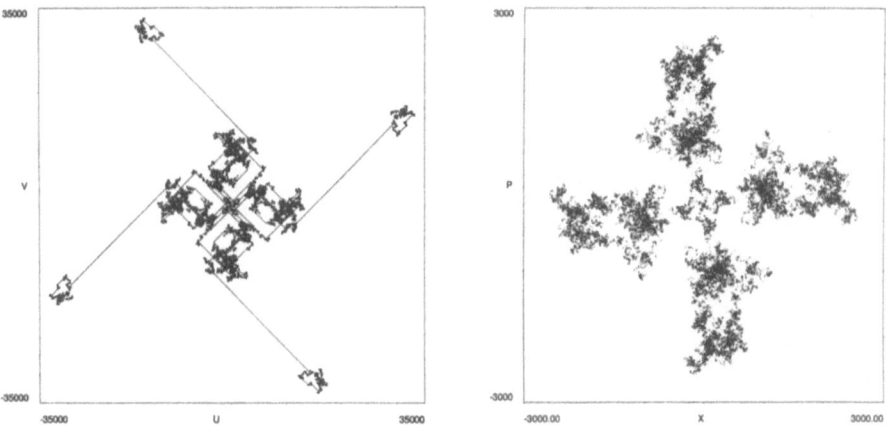

Fig. 1. — Two examples of trajectories for the web-map of the four-fold symmetry: (a) with long flights, $K = 6.349972$; (b) without long flights, $K = 7.2$.

where A is a constant and $\mu > 1$. Although $1 - c \ll c$, the second term becomes a leading term if time t is sufficiently large. This explains how "ballistic" parts of the trajectories induce superdiffusion if the observation time is sufficiently large, and the influence of random forces or friction is sufficiently small. In the case of the existence of different exponents $\mu_1, \mu_2, \ldots$, different intermediate asymptotic may occur. In the limit $t \to \infty$, the largest one survives independently of how small the coefficient is in the expansion-like (2) over different powers of t. Precisely such situation can be observed in experimental and numerical studies.

3.   Numerical and experimental examples

The numerical and experimental results given below can help to understand some of the specific features of the chaotic kinetics responsible for the non-Gaussian behavior.

The first example is related to the web-map of the four-fold symmetry [12]:

$$
\begin{aligned}
u_{n+1} &= v_n \\
v_{n+1} &= -u_n - K \sin v_n .
\end{aligned}
\tag{3}
$$

The map (3) corresponds to a kicked linear oscillator with four kicks per the oscillator period. The variables (u, v) are dimensionless coordinate and velocity of the oscillator, and K is the amplitude of a kick. Equation (3) generates a stochastic web in the phase space, which is a net of finite width channels with unbounded chaotic motion inside of them. A typical random walk-like process is shown in Fig. 1(a). It resembles a Brownian motion in (u, v) space with 4-fold symmetry. Unusual random walk with very long flights is shown in Fig. 1(b) [13]. The difference beween the two figures is only a small change of the parameter K.

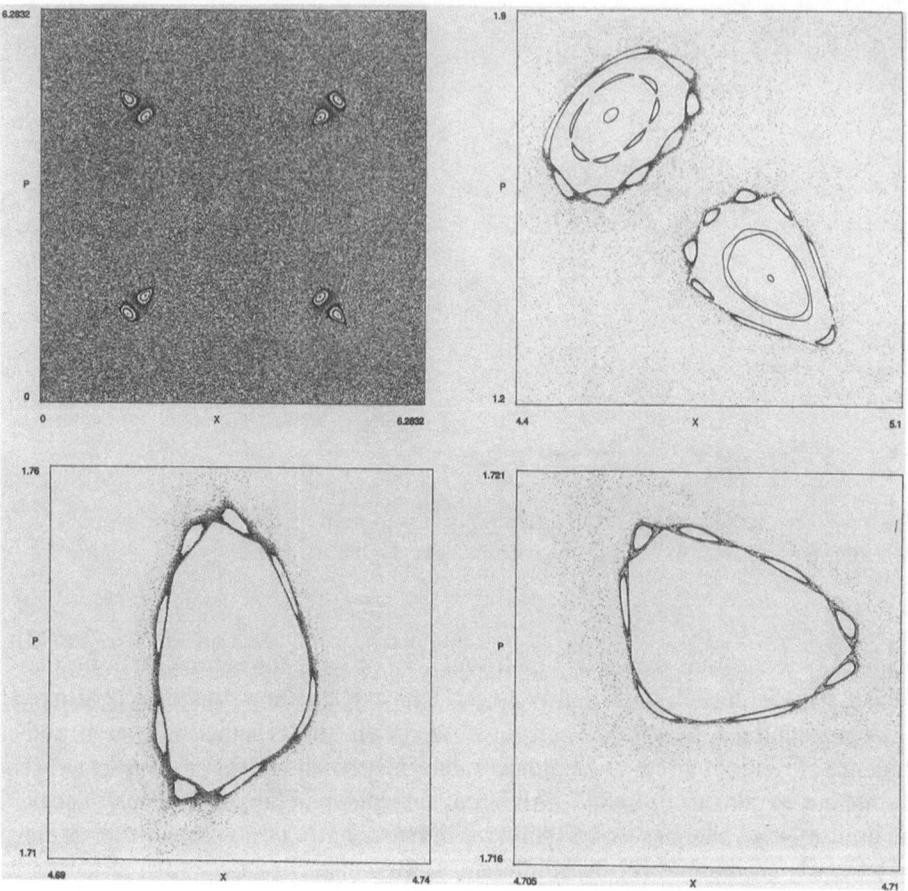

Fig. 2. — The phase space of the web-map for $K = K_{an} = 6.349972$: (a) full phase space with islands that correspond to an accelerator mode; (b) magnification of the accelerator-islands from the bottom-right of (a); (c) magnification of the subisland from the top-right of (b); (d) magnification of the bottom left island of (c).

A detailed analysis [13] shows that the flights are imposed by particle trapping into a singular zone of the phase space, which consists of a self-similar structure of islands-around-islands. Such a topology was observed in [14, 15]. A new property of the singular zone is the possess of exact self-similarity for special values of the parameter K [13, 16]. An example of the self-similarity in the island structure is shown in Fig. 2 [13]. A flight occurs as a result of the particle trapping into a narrow boundary layer near an island in the phase space. The smaller is the island, the narrower is its boundary layer, and the longer is a corresponding flight.

The second example is related to the standard map [17]

$$
\begin{aligned}
p_{n+1} &= p_n - K \sin x_n \\
x_{n+1} &= x_n + p_{n+1} ,
\end{aligned}
\tag{4}
$$

which has accelerated mode solutions as well. Self-similar structure in the phase space was described for $K = 6.908745$ in [16] (see other examples in [18, 19]). The phase portrait is presented in Fig. 3(a), and two subsequent generations of 8-islands chain are in Fig. 3(b,c).

The list of numerical examples can be extended. There are few experimental results on the advected particle diffusion in fluids [20, 21]. In [20], passive particle dynamics with trappings and flights has been observed in a flow with a regular set of vortices generated in the rotating tank. In [21], diffusion of passive particles has been observed in a quasi two-dimensional turbulent flow with randomly distributed vortices.

One can assume from all this data that there exists a situation when the anomalous properties of the diffusional process occur as a result of special singular properties of the space of motion (phase space). The diffusion pattern reveals trappings and flights which are similar to Lévy flights process [22, 23]. The corresponding type of kinetic approach to such processes was called "strange kinetics" [2]. To some extent the strange kinetics is similar to the anomalous kinetics near a critical point of phase transition. But the dynamical origin of flights and trappings makes it necessary to consider singularities in the phase space in order to formalize the kinetic description. In this situation the renormalization group approach (RG) is a natural approximation to the problem. We call it renormalization group for kinetics (RGK).

4. Self-similarity of the singular zone

As it was mentioned above, the phase space of chaotic dynamics is not uniform, and there exist numerous islands embedded into the chaotic sea. The crucial property of the islands is that their boundary is sticky and a particle spends fairly long time being trapped in a narrow boundary layer near an island. The virtual boundary layer can not be a well defined notion. To specify the situation, assume the following structure (see Fig. 4): Near the boundary of an island there is a set of smaller islands and near each of the islands of first generation there is a set of much smaller islands of the second generation and so on. In fact such a self-similar structure of the boundary layer does exist (at least for simplified models) [8, 13, 14, 15, 16, 19]. Furthermore, precisely such a just the structure can be used as a basic element for RGK. For the web-map and standard map the existence of the described self-similar structure of the islands' hierarchy can be proved for special values of the parameter K [13, 16].

To parameterize the self-similarity property we introduce two scaling parameters λ_s and λ_t such that

$$
\begin{aligned}
s_{n+1} &= \lambda_s s_n \\
\Delta t_{n+1} &= \lambda_t \Delta t_n ,
\end{aligned}
\tag{5}
$$

where s_n is the area of a k-th generation island and Δt_n is the rotation time (period) around the n-th generation island. In fact, the properties of the trajectories have

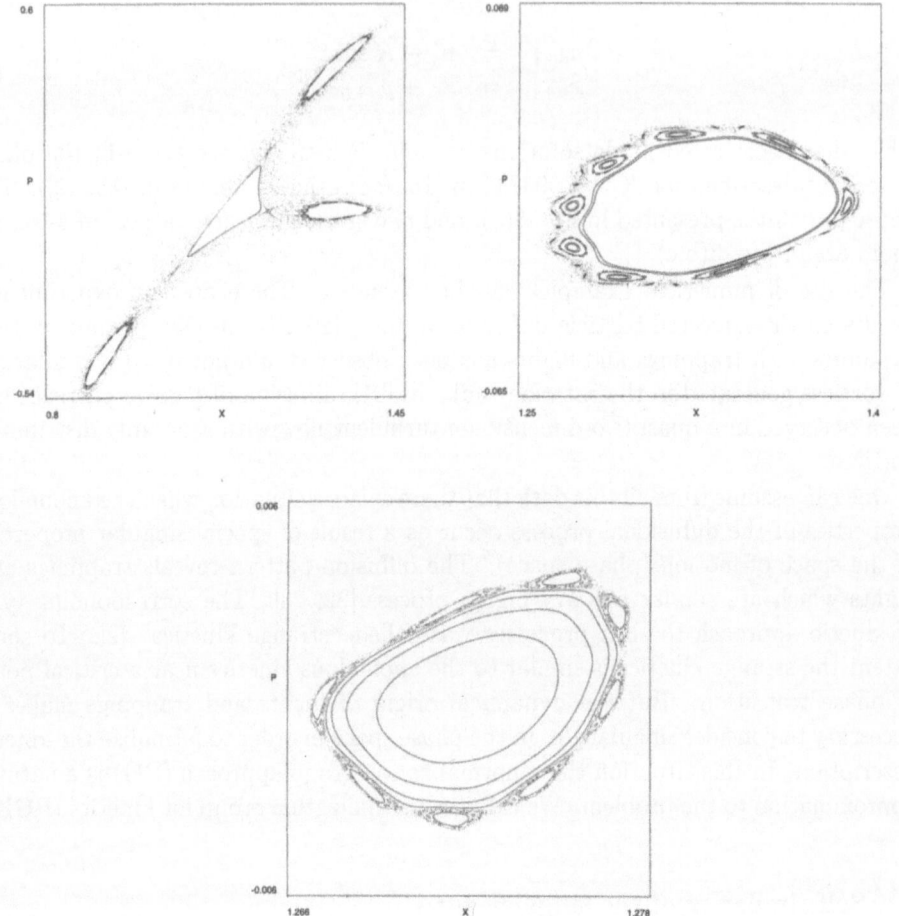

Fig. 3. — Island's hierarchy for the standard map with $K = 6.908745$: (a) the main island and the first 3-island generation; (b) magnification of the right subisland in (a); (c) magnification of the left subisland in (b).

more self-similarity. Namely,

$$
\begin{aligned}
\sigma_{n+1} &= \lambda_\sigma \sigma_n \\
\ell_{n+1} &= \lambda_\ell \ell_n \\
q_{n+1} &= \lambda_q q_n \,,
\end{aligned}
\tag{6}
$$

where σ is Lyapunov exponent, ℓ is the length of a flight, and q_n is the number of islands in the n-th generation. For the situation related to the web-map and standard map

$$
\lambda_s \leq 1 \,, \quad \lambda_t \geq 1 \,, \quad \lambda_\sigma \leq 1 \,, \quad \lambda_\ell \geq 1 \,, \quad \lambda_q \geq 1 \,.
\tag{7}
$$

The islands are the sections of cylinders or tori in the phase space. So the closer a

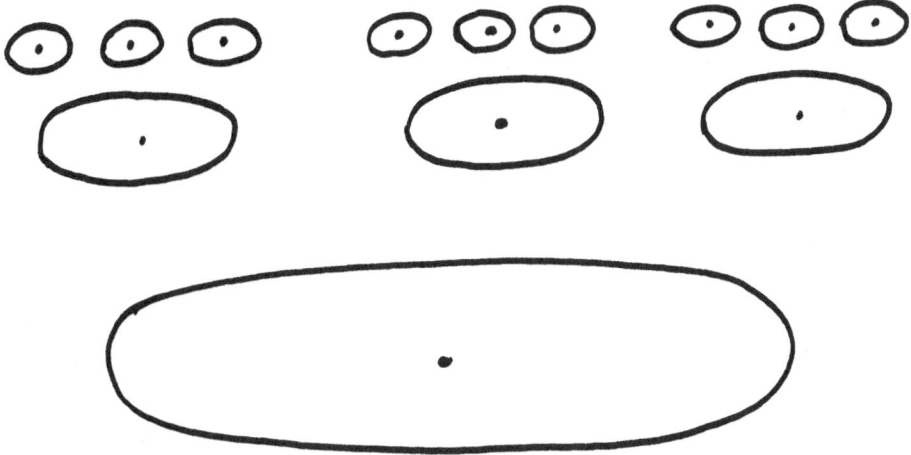

Fig. 4. — Symbolic representation of the islands-set of different generations.

trajectory to the island's boundary, the less Lyapunov exponent σ and the longer the flight length ℓ are.

Not all of the scaling parameters λ, are independent. We assume that only two are significant for kinetics, for example λ_s, λ_t, although a nonuniversal dependence between λ_s and λ_t can exist. It was shown in [9, 10] that for a typical situation

$$\lambda_\ell = 1/\lambda_s^{1/2} \tag{8}$$

and

$$\lambda_t = 1/\lambda_\sigma \tag{9}$$

Moreover, for a simplified situation of the exact self-similarity in the islands' proliferation we have

$$\lambda_t = \lambda_q . \tag{10}$$

The occurrence of islands in the phase portrait is due to the resonances of different order. Appearance of the resonance set of orders $\lambda_q, \lambda_q^2, \lambda_q^3, \ldots$ means a special "tuning" of the system. At the same time there could be different sequences $\{q_1, q_2, \ldots\}$ of islands proliferation with the values of q_j near some effective value q^*. Precisely this situation was observed in [8] for the advection problem in a hexagonal Beltrami flow.

There exists a method [24, 25] for calculating parameters λ; although it has not yet been applied for any example. The method is based on the considering the nonlinear resonance of the corresponding order and iterating it to higher orders.

5. Kinetic exponents

This part of the problem of kinetic description of a particle in the presence of stickiness of islands can be compared to the evaluation of critical exponents in the phase transition theory. In a large time asymptotics the long memory parts of full trajectories are responsible for the kinetics which correspond to long flights. This phenomena has an alalogy to the long distance correltions near a critical point.

Let ℓ be a flight length and the flight is defined as an almost regular part of the trajectory up to some small scale. Introduce a normalized distribution function of flights $P(\ell, t)$

$$\int_{\ell_0}^{\infty} P(\ell, t) d\ell = 1 \tag{11}$$

The lower limit ℓ_0 is not well defined and should correspond to a minimal time interval Δt_0 in (5), which is the time trajectory spends in the main island boundary layer. Consider infinitesimal deviation of the distribution function during a interval Δt

$$\delta_t P(\ell, t) = P(\ell, t + \Delta t) - P(\ell, t) \tag{12}$$

It is important that $P(\ell, t)$ should also depend on some initial value ℓ_0 at the time instant $t = 0$ and that we consider only large scale deviation of ℓ.

Applying the general idea of the kinetic description of motion, which consists expressing the time-deviation of the probability distribution function through its space-deviations, i.e. through the probability distribution of particles from the neighbor domain, one can write

$$\delta_t P(\ell, t) = \sum_{\Delta\ell} \overline{\{P(\ell + \Delta\ell, t) - P(\ell, t)\}} \equiv \delta_\ell P(\ell, t) \tag{13}$$

where sum over $\Delta\ell$ means summation over all such paths within the infinitesimal interval $\Delta\ell$ that gives the infinitesimal evolutional change $\delta_t P$, and the bar means averaging over the paths. As an example, one can take the conventional averaging over the phases. Equation (13) can be rewritten in the form

$$(\Delta t)^\beta \frac{\partial^\beta P}{\partial t^\beta} = \sum_{\Delta\ell} \frac{\partial^\alpha}{\partial \ell^\alpha} ((\overline{\Delta\ell})^\alpha \mathcal{D} P) \tag{14}$$

where $\mathcal{D}$ is some expansion coefficient. Here we have introduced fractional derivatives [26, 27] without specifying the exponents α and β.

In correspondence with (5),(6),(8) the transform

$$\hat{R}_k : \quad \Delta\ell_{k+1} = \lambda_\ell \Delta\ell_k , \qquad \Delta t_{k+1} = \lambda_t \Delta t_k , \tag{15}$$

will be called the RGK-transform. We assume that the kinetics as well as dynamics should satisfy the self-similarity condition (15) if a trajectory jumps from a flight of length $\ell \sim \ell_k$ to the flight of length $\ell \sim \ell_{k\pm1}$. More distant flights can be neglected due to their negligible probability. This assumption is typical for the method of renormalization group while its application differs depending on the problem. In other words we assume that Eq. (14) should be invariant under the transform (15) with restrictions that

$$\Delta\ell \geq \ell_{\min}, \qquad \Delta t \geq \Delta t_0 . \tag{16}$$

The invariance under the RGK-transform (15) can be written in the form

$$\hat{R}_k(\delta_t P) = \hat{R}_k(\delta_\ell P) \tag{17}$$

or after n-time application of (15) we have

$$\hat{R}_k^n(\delta_t P) = \hat{R}_k^n(\delta_\ell P) \tag{18}$$

with restrictions (16). Here we again should note that for a fairly large n one expects that the probability distribution function $P(\ell, t)$ is concentrated at fairly large ℓ and t so that the conditions (16) do not influence the process of increasing n in Eq. (18). Direct application of (18) to (14) means

$$[\hat{R}_k^n(\Delta t)^\beta] \frac{\partial^\beta P}{\partial t^\beta} = \sum_{\Delta t} \frac{\partial^\alpha}{\partial \ell^\alpha} \overline{([\hat{R}_k^n(\Delta \ell)^\alpha] \mathcal{D} \cdot P)} \tag{19}$$

or after substitution of (15) into (19)

$$(\Delta t)^\beta \frac{\partial^\beta P}{\partial t^\beta} = \left(\frac{\lambda_\ell^\alpha}{\lambda_t^\beta}\right)^n \sum_{\Delta \ell} \frac{\partial^\alpha}{\partial \ell^\alpha} \overline{((\Delta \ell)^\alpha \mathcal{D} \cdot P)} \tag{20}$$

This equation can be rewritten in a more convenient form

$$\frac{\partial^\beta P}{\partial t^\beta} = \left(\frac{\lambda_\ell^\alpha}{\lambda_t^\beta}\right)^n \sum_{\Delta t} \frac{\partial^\alpha}{\partial \ell^\alpha} \frac{\overline{(\Delta \ell)^\alpha}}{(\Delta t)^\beta} \mathcal{D} \cdot P \tag{21}$$

Small deviations $\Delta \ell, \Delta t$ are actually not so small, and $\Delta \ell \sim 1/s_0^2$ where s_0 is the area of the largest island. Nevertheless the magnitudes of $\Delta \ell, \Delta t$ are too small in comparison with large time-space asymptotics for which Eq. (21) is applicable. That is the general scheme of kinetic equation derivation to consider a limit $\Delta \ell \to 0$, $\Delta t \to 0$ which actually means the limits $\Delta \ell/\ell \to 0, \Delta t/t \to 0$ with $\ell, t \to \infty$. The main question of the kinetic description is what the conditions are for which the value

$$\mathcal{A} \equiv \sum_{\Delta \ell} \frac{\overline{(\Delta \ell)^\alpha}}{(\Delta t)^\beta} \mathcal{D} \tag{22}$$

is finite in the limit $\Delta t \to 0$. For the Gaussian process $\mathcal{A}$ is finite if $\alpha = 2$ and $\beta = 1$. One can play with exponents (α, β) depending on the situation.

Here we consider a limit $n \to \infty$ in (21). Then Eq. (21) and its initial form (14) makes sense if

$$\lambda_t^\alpha = \lambda_t^\beta \tag{23}$$

or there exists a fixed point

$$\mu_0 \equiv \beta/\alpha = \ln \lambda_\ell / \ln \lambda_t . \tag{24}$$

Conditions (23),(24) do not define α or β, but only their ratio μ_0, which is sufficient to get the transport exponents.

The equation

$$\frac{\partial^\beta P}{\partial t^\beta} = \frac{\partial^\alpha}{\partial \ell^\alpha}(\mathcal{A}P) \tag{25}$$

is called fractional generalization of the Fokker-Planck-Kolmogorov equation (FFPK). More information about it can be found in [9, 10, 16].

k	q_k	T_k	t_k/T_{k-1}	ΔS_k	$\Delta S_k/\Delta S_{k-1}$	δS_k	$\delta S_k/\delta S_{k-1}$
0	1	16.4	–	0.436	–	0.436	–
1	8	131.8	8.04	5.24^{-3}	1.20×10^{-2}	4.19×10^{-2}	0.0961
2	8	1049	7.96	5.30^{-5}	1.01×10^{-2}	3.39×10^{-3}	0.0809
3	8	8420	8.02	5.32^{-7}	1.00×10^{-2}	2.72×10^{-4}	0.0802

Table I. — Parameters of self-similarity for the web-map with $K = 6.349972$.

6. Moments and transport

The equation for the moments of distribution function $P(\ell,t)$ can be derived the directly from (25). Multiplying the equation by ℓ^α and integrating it we obtain

$$\frac{\partial^\beta P}{\partial t^\beta}\langle \ell^\alpha \rangle = \text{const} \tag{26}$$

or

$$\langle \ell^\alpha \rangle = \text{const} \cdot t^\beta \tag{27}$$

The equation (27) does not imply the existence of self-similarity for all moments. Nevertheless it is clear from (25) that if $\mathcal{A} = \text{const}$, then the self-similarity exists and one can find it through iterations. From (25) we have

$$\frac{\partial^\beta P}{\partial t^\beta}\langle \ell^{2\alpha} \rangle = \text{const} \cdot \langle \ell^\alpha \rangle$$

or using (27)

$$\langle \ell^{2\alpha} \rangle = \text{const} \cdot t^{2\beta} \tag{28}$$

Similarly, the itertion of (27),(28) yields

$$\langle \ell^{2\alpha} \rangle = \text{const} \cdot t^{m\beta}$$

with integer m, or approximately

$$\langle \ell^{2m} \rangle \sim t^{m\mu} \tag{29}$$

with

$$\mu = 2\beta/\alpha \tag{30}$$

Using the expressions (24) and (8) we have from (30)

$$\mu = 2\mu_0 = |\ln \lambda_s|/\ln \lambda_t \tag{31}$$

which is the final expression for the anomalous diffusion exponent.

The results of the computer simulation for the web-map (3) are presented in Table 1 for $K = K_{an} = 6.349972$ [13].

The first column is the order of generation k, and q_k is the number of islands for the k-th generation. T_k is a period of rotation very close to the boundary of k-th generation islands, ΔS_k is the area of an island of k-th generation, and

$$\delta S_k = q_k\Delta S_k . \tag{32}$$

The scaling parameters λ_s, λ_t are obtained via the relations

$$T_{k+1} = \lambda_t T_k, \qquad \delta S_{k+1} = \lambda_s \delta S_k. \tag{33}$$

Table 1 and formula (31) give

$$q = \lambda_q = 8, \qquad \mu = 1.21 \ . \tag{34}$$

A direct simulation for the transport exponents obtained from the expression

$$\langle R^2 \rangle = \langle u^2 + v^2 \rangle \sim t^\mu \tag{35}$$

where averaging is performed over a large number of initial conditions, gives $\mu = 1.26$ which is in a good agreement with (34)

7.   Anomalous distribution of the Poincaré cycles

In the previous sections we introduced a construction of a singular zone existence of which is responsible for the anomalous kinetics and transport. This zone is a bounary layer near an island with a hierarchical set of subislands. For a special value of the control parameter k, there occurs a precise hierarchy of islands-around-islands, which induces the anomalous kinetics. There are different critical values of k for different hierarchies. In the following two sections we shall consider how the presence of the islands' hierarchy influences the time dependence of different distributions.

Consider a small domain $\Delta\Gamma$ in the phase space of a system and take an initial condition $(u_0, v_0) \in \Delta\Gamma$. Poincaré recurrences is the phenomenon of the trajectory passing through the domain $\Delta\Gamma$ an infinite number of times. For the set of time instants $\{t_j\}$, $j = 0, 1, \ldots$ of a trajectory's passages through the domain, one can introduce Poincaré cycles

$$\{\tau_j\} = \{t_{j+1} - t_j\} \ , \qquad j = 0, 1, \ldots \tag{36}$$

Let us introduce a normalized probability distribution function of the Poincaré cycles $P(\tau; \Delta\Gamma)$

$$\frac{1}{\Delta\Gamma} \int_0^\infty P(\tau; \Delta\Gamma) d\tau = 1 \tag{37}$$

For example, for a periodic orbit with period T_0, $P(\tau; \Delta\Gamma)/\Delta\Gamma$ is simply $\delta(\tau - T_0)$. In a generic case, $P/\Delta\Gamma$ depends on $\Delta\Gamma$. We assume the existence of a limit

$$P(\tau) = \lim_{\Delta\Gamma \to 0} P(\tau; \Delta\Gamma)/\Delta\Gamma \tag{38}$$

It is known that for some simple system with perfect mixing properties, $P(\tau)$ has a Poissonian distribution [28]:

$$P(\tau) = \frac{1}{\langle \tau \rangle} \exp(-\tau/\langle \tau \rangle) \tag{39}$$

Nevertheless, it was observed in a numerical simulation of trajectories in the stochastic layer [29] that $P(\tau)$ has a power-like tail for large τ. In our article [30], for a

Hamiltonian flow, it was shown that $P(\tau)$ has a Poissonian type (39) when the parameters of the system correspond to normal (i.e. Gaussian) particle kinetics, and that $P(\tau)$ has a power-like asymptotics

$$P(\tau) \sim \tau^{-\gamma}, \qquad \tau \to \infty \tag{40}$$

when the kinetics is anomalous. Similar results for the escape-time were obtained in [31] where the standard map was considered. The connection between the exponents γ and β is proposed below.

8. Relations between different time-scales and time-distribution functions

Let

$$P_e(t; \Delta\Gamma) = \int_0^t \psi(\tau; \Delta\Gamma)d\tau \tag{41}$$

be the probability of exit from the domain $\Delta\Gamma$ during the time interval t, and $\psi(t; \Delta\Gamma)$ the corresponding probability density to escape from the $\Delta\Gamma$ in a time not less than t. Then the survivial probability is

$$\Psi(t; \Delta\Gamma) = 1 - P_e(t; \Delta\Gamma) = 1 - \int_0^t \psi(\tau; \Delta\Gamma)d\tau = \int_t^\infty \psi(\tau; \Delta\Gamma)d\tau \tag{42}$$

where the normalized condition and (41) are used:

$$P_e(t \to \infty; \Delta\Gamma) = 1 \tag{43}$$

One can consider the function $\psi(\tau; \Delta\Gamma)$ also as the probability density of particles to be trapped in the domain $\Delta\Gamma$ for the time span not less than τ. The mean time of trapping is

$$t_s(\Delta\Gamma) = \int_0^\infty \tau\psi(\tau; \Delta\Gamma)d\tau \tag{44}$$

It is important to stress that all the introduced functions ψ, Ψ, P_e and magnitude t_s are local characteristics of the particle motion, i.e. they are defined for an infinitesimal domain $\Delta\Gamma$ and can depend on the shape and location of $\Delta\Gamma$. In other words, ψ, Ψ, P_e, and t_s depend only on the properties of the system's trajectories inside the $\Delta\Gamma$. The situation is opposite for the distribution of Poincaré cycles $P(\tau)$, which depends on full, infinitely long trajectories and does not depend on $\Delta\Gamma$ for sufficiently small $\Delta\Gamma$. This means that $P(\tau)$ is a global characteristic of the system. Nevertheless, a connection between $P(\tau)$ and $P_e(\tau; \Delta\Gamma)$ can be established for specific cases.

Let $\psi(\tau; \Delta\Gamma)$ possesses an asymptotic property

$$\psi(\tau; \Delta\Gamma) \sim t^{-(1+\beta')}, \qquad t \to \infty, \ \beta' > 0 \tag{45}$$

Then from the definition (42)

$$\Psi(\tau; \Delta\Gamma) \sim t^{-\beta'}, \qquad t \to \infty, \ \beta' > 0 \tag{46}$$

and from (44)

$$t_s(\Delta\Gamma) \sim \int_{const}^{\infty} \tau^{-\beta'} d\tau = \begin{cases} \langle\tau\rangle < \infty, & \beta' > 1 \\ \lim_{t\to\infty} \ln t = \infty, & \beta' = 1 \\ \infty & \beta' < 1 \end{cases} \qquad (47)$$

It follows from (47) that only for $\beta' > 1$ the mean trapping time is finite, otherwise, for $\beta \leq 1$, the mean trapping time is infinite and the domain $\Delta\Gamma$ works like a real trap in the sense of the time average. It is this property that was used in [32] for a speculation on Maxwell's Demon.

Assume that the phase space is uniform with good mixing properties. Then one can expect the Poissonian distribution (39) for $P(\tau)$ to be the same as the one found in [30, 32]. For the nonuniform phase space there are domains of halt in the phase space with islands-around-islands, which impose the power-like laws (45),(46) for $\psi(\tau; \Delta\Gamma_s)$, and $\Psi(\tau; \Delta\Gamma_s)$, where $\Delta\Gamma_s$ is the sticky boundary layer.

Now let us take a small domain $\Delta\Gamma$ fairly far from the domain $\Delta\Gamma_s$. A trajectory, after departure from $\Delta\Gamma$, will occasionally reach the domain $\Delta\Gamma_s$ and then repeatedly come back to $\Delta\Gamma$. For fairly short times, $P(\tau)$ resembles the Poissonian distribution, and for fairly long times, $\tau \gg \tau_0$ the property of stickiness of the domain $\Delta\Gamma_s$ imposes the anomalous asymptotics of the (45),(46) kind. It follows that

$$P(\tau) \sim \tau^{-(1+\beta')}, \qquad \tau \gg \tau_0 \qquad (48)$$

where τ_0 can be estimated as a characteristic time required for a particle to reach $\Delta\Gamma_s$ if it starts at $\Delta\Gamma$. In (46) the exponent β' depends on the choice of $\Delta\Gamma_s$. The exponent γ in (40) for the Poincaré recurrences does not depend on the choice of $\Delta\Gamma$. Then, the existence of singular sticky domains imposes its dominant asymptotics which reveals itself in the expression (40) for the anomalous, non-Poissonian, distribution.

Now let us recall that in the $\Delta\Gamma_s$, the trapping time should be scaled in correspondence to the self-similarity law (5),(6). The probability to spend a time t in $\Delta\Gamma_s$ is the probability to have a flight of the t duration. It will be shown in the next section that

$$\beta' = 1 + |\ln \lambda_s|/\ln \lambda_T = 1 + \mu, \qquad (49)$$

which after the substitution to (48) gives

$$P(\tau) \sim \tau^{(2+\mu)} \qquad (50)$$

Comparing this expression to (40), we finally obtain

$$\gamma = 2 + \mu \qquad (51)$$

The following few figures provide an illustration for the web-map model (3). In Fig. 5(a), we plot the $\log_{10}\psi(t)$ versus t (number of steps) for two values of K: triangulars for $K = K_{an} = 6.349972$, and crosses for $K = K_{no} = 6.25$. For the time $t < \tau_0 \sim 15 \cdot 10^3$, $\psi(\tau)$ follows the Poissonian law (39) for both cases, while for $t > \tau_0$, the case of the anomalous diffusion ($K = K_{an}$) provides the power-like tail. The last case is presented in more details in Fig. 5(b) where the data are obtained from 10,000 trajectories of the duration $\sim 10^6$ each. The corresponding slope in Fig. 5(b) gives $\beta' \sim 2.2$ which is in a good agreement with the value $1 + \mu = 2.21$ that follows from Table 1.

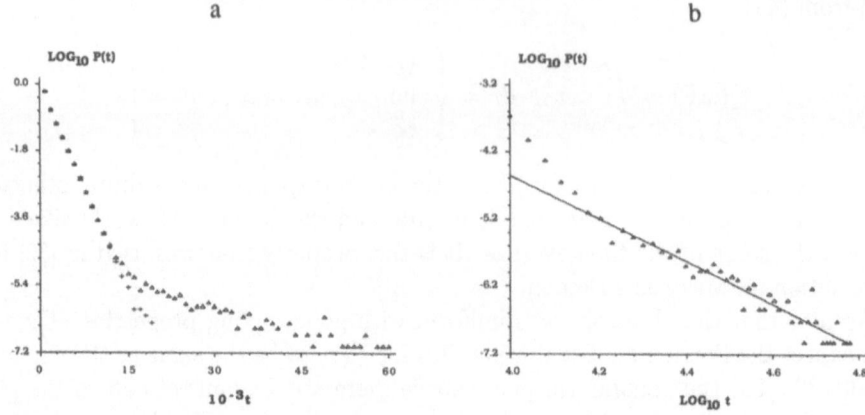

Fig. 5. — Logarithm of the probability distribution function $P(t)$ of the Poincaré cycles for the web-map: (a) case of the normal diffusion (crosses for $K = 6.25$) and case of the anomalous one (triangles for $K = 6.349972$); (b) magnification of the power-like tail from (a).

9. Renormalization formulas for the exit time distribution

It was mentioned above that the exit time probability distribution $P_e(t; \Delta\Gamma)$ depends on the location of the domain $\Delta\Gamma$. If, for example, $\Delta\Gamma$ is taken in a domain $\Delta\Gamma_S$ of a singular zone around an island, then $\dot{P}_e$ should be dependent on t similarly to $P(t)$ in (48), and P_e should satisfy the self-similarity condition.

In order to formulate this property, let us consider different domains $\Delta\Gamma_q$, in the BIC of the corresponding generation order $q = 0, 1, \ldots$. This means that

$$\Delta\Gamma_{q_{s+1}} \sim \lambda_s \Delta\Gamma_{q_s} \tag{52}$$

Then the exit time distribution $P_e(t; \Delta\Gamma_{q_i})$ and the corresponding survival distribution $\Psi(t; \Delta\Gamma_q)$ satisfy the scaling equation that follows from (51) as a result of the renormalization of time:

$$\Psi(t - t_{s+1}; \Delta\Gamma_{q_{s+1}}) = 1 - P_e(t - t_s; \Delta\Gamma_{q_{s+1}}) = \Psi(t - t_s; \Delta\Gamma_{q_s}) = (1 - P_e(t - t_s; \Delta\Gamma_{q_s})) \tag{53}$$

where each of the functions $P_e(t - t_i; \Delta\Gamma_{q_i})$ is normalized per corresponding domain $\Delta\Gamma_{q_i}$ and $t_s \sim \lambda_T^s$. A similar equation can be also written for the density distribution

$$\psi(t - t_{s+1}; \Delta\Gamma_{q_{s+1}}) = \psi(t - t_s; \Delta\Gamma_{q_s}) \tag{54}$$

Both equations (53) and (54) are valid only for the tail part of distribution functions. In the logarithmic scale, using (41) and (50), we obtain a shift formula

$$\ln \psi(\ln(t - t_s) - \ln \lambda_T; \Delta\Gamma_{q_{s+1}}) = \ln \psi(\ln(t - t_s); \Delta\Gamma_{q_s}) \tag{55}$$

Precisely such a shift has been observed in our simulations (see Fig. 6), in which the value of the shift is of the order $\ln \lambda_T = \ln 8 \approx 2.2$.

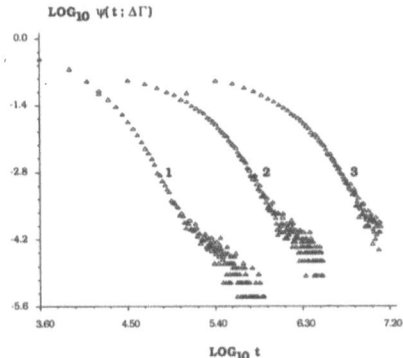

Fig. 6. — Exit time distributions for domains taken near the boundary of islands of the first, second, and third generations (curves 1, 2, and 3 correspondingly), obtained after averaging over $9 \cdot 10^4$ trajectories of the web-map.

Let us now show that the equation (53) or (54) gives us the possibility to obtain the exponent β' for the distribution (12). For simplicity introduce the domain

$$\delta t_s = ((1 + \delta^-) \lambda_T^{s-1}, (1 + \delta^+) \lambda_T^s) , \tag{56}$$

where $|\delta^{\pm}| < \lambda_T$ are some numbers. We will consider

$$t - t_s \in \delta t_s \tag{57}$$

with $t_s \sim \lambda_T^{s-1}$ in order to specify the area of consideration of functions $\Psi(t - t_s; \Delta\Gamma_{q_s})$. Now from the defintions of $\Psi(t - t_s; \Delta\Gamma_{q_s})$ and $\Psi(t - t_s; \Delta\Gamma_{q_{s+1}})$ follows

$$
\begin{aligned}
\Psi(t - t_{s+1}; \Delta\Gamma_{q_{s+1}}) &= \frac{\text{const.}}{(t - t_{s+1})^{\beta'}} \\
&= \Psi\left(\lambda_T\left(\frac{t}{\lambda_T} - t_s\right); \Delta\Gamma_{q_s} \cdot \frac{\Delta\Gamma_{q_{s+1}}}{\Delta\Gamma_{q_s}}\right) \\
&= \Psi\left(\lambda_T\left(\frac{t}{\lambda_T} - t_s\right); \lambda_s \Delta\Gamma_{q_s}\right) \\
&= \frac{\lambda_T}{\lambda_s} \frac{1}{\lambda_T^{\beta'}} \Psi(t - t_{q_s}; \Delta\Gamma_{q_s})
\end{aligned}
\tag{58}
$$

where the argument of Ψ in the last expression satisfies $t - t_s \in \delta t_s$. Applying (53) to (58) we get the equation $\lambda_s \lambda_T^{\beta'-1} = 1$ or

$$\beta' = 1 + |\ln \lambda_s| / \ln \lambda_T = 1 + \mu \tag{59}$$

This is precisely the result that was introduced in (49).

Summarizing the results of this section we can write for the exit probability density function (see (45))

$$\psi(t; \Delta\Gamma) \sim t^{-(2+\mu)} \tag{60}$$

for the Poincaré cycles distribution density (see (40))

$$P(t) \sim t^{-(2+\mu)} , \tag{61}$$

and for the integral waiting probability (see (46))

$$\Psi(t; \Delta\Gamma) \sim t^{-(1+\mu)} \tag{62}$$

Thus, we have completed the evaluation of different time distributions in the case of the existence of a singular zone in the phase space of a system.

Acknowledgments

This work was supported by the U.S. Department of Energy, Grant No. DE-FG02-92ER54184 and the U.S. Department of Navy, Grant No. N00014-96-1-0055.

References

[1] N.S. Krylov, *Papers on the Foundation of Statistical Physics* (Princeton Univ., Princeton, NJ, 1979).

[2] M.F. Shlesinger, G.M. Zaslavsky, and J. Klafter, Nature **363**, 31 (1993).

[3] J.P. Bouchaud and A. Georges, Phys. Rep. **195**, 127 (1990).

[4] J. Klafter, M.F. Shlesinger, G. Zumofen, Physics Today **49**, 33 (1996).

[5] "Lévy Flights and Related Topics in Physics", Eds. M. Shlesinger, G. Zaslavsky, and U. Frisch (Springer, Heidelberg, 1995).

[6] V.V. Afanas'ev, R.Z. Sagdeev and G.M. Zaslavsky, Chaos **1**, 143 (1991).

[7] G.M. Zaslavsky, in: "Topological Aspects of the Dynamics of Fluids and Plasmas", Eds. H.K. Moffatt, G.M. Zaslavsky, P. Comte, and M. Tabor (Kluwer, Boston, 1992) p. 481.

[8] G.M. Zaslavsky, D. Stevens, H. Weitzner, Phys. Rev. E **48**, 1683 (1993).

[9] G.M. Zaslavsky, Chaos **4** 25 (1994).

[10] G.M. Zaslavsky, Physica D **76** 110 (1994).

[11] W. Young, A. Pumir, and Y. Pomeau, Phys. Fluids A **1**, 462 (1989).

[12] G.M. Zaslavsky, M. Yu. Zakharov, R.Z. Sagdeev, D.A. Usikov and A.A. Chernikov, Sov. Phys.-JETP **64** 294 (1986).

[13] G.M. Zaslavsky, B.A. Niyazov, Phys. Reports (to appear).

[14] J.D. Meiss, Phys. Rev. A **34**, 2375 (1986).

[15] J.D. Meiss, Rev. Mod. Phys. **64**, 795 (1992).

[16] G.M. Zaslavsky, M. Edelman, and B. Nyazov, Chaos **7**, No. 1 (1997).

[17] B.V. Chirikov, Phys. Reports **52**, 264 (1979).

[18] J. Klafter, G. Zumofen, and M.F. Shlesinger, in *Lévy Flights and Related Topics in Physics*, Eds. M.F. Shlesinger, G.M. Zaslavsky, U. Frisch, Springer, 1995, p. 196.

[19] S. Benkadda, S. Kassibrakis, R.B. White, and G.M. Zaslavsky, (not published).

[20] T.H. Solomon, E.R. Weeks and H.L. Swinney, Phys. Rev. Lett. **71**, 3975 (1993);

[21] O. Cardoso, B. Gluckmann, O. Parcollet, and P. Tabeling, Phys. Fluids **8**, 209 (1996).

[22] P. Lévy, Theorie de l'Addition des Variables Aletoires (Gauthier-Villiers, Paris, 1937).

[23] E.W. Montroll and M.F. Shlesinger, in "Studies in Statistical Mechanics", Eds. J. Lebowitz and E. Montroll (North-Holland, Amsterdam, 1984), vol. 11, p. 1.

[24] D. Escande, Phys. Rep. **121** (1985) 163.

[25] G. Zaslavsky, *Chaos in Dynamic Systems*, (Harwood Acad. Publ., NY, 1995).

[26] S.G. Samko, a.A. Kilbas, and O.I. Marichev, Fractional Integrals and Derivatives and Their Applications, Nauka i Tekhnika, Minsk, 1987. (Translation by Harwood Academic Publishers.)

[27] K.S. Miller and B. Ross, *An Introduction to the Fractional Differential Equations* (John Wiley & Sons, NY, 1993).

[28] R. Ishizaki, T. Horito, T. Kobayashi, and M. Mori, Progr. Theor. Phys. **85**, 1013 (1991).

[29] B.V. Chirikov, D.L. Shepeliansky, Physica D **13**, 394 (1984).

[30] G.M. Zaslavsky and M.K. Tippett, Phys. Rev. Lett. **67** 3251 (1991).

[31] S. Benkadda, Y. Elskens, and B. Ragot, Phys. Rev. Lett. **72** 2859 (1994).

[32] G.M. Zaslavsky, Chaos **5** 653 (1995).

Fluctuations and Mixing of a Passive Scalar in Turbulent Flow

BORIS I. SHRAIMAN[1] AND ERIC D. SIGGIA[2]

[1] *Bell Laboratories, Lucent Technologies,*
700 Mountain Avenue, Murray Hill, NJ 07974, USA
[2] *Laboratory of Atomic & Solid State Physics, Cornell University,*
Ithaca, NY 14853-2501, USA

The article reviews the aspects of the intermittency phenomena exhibited by the advected passive scalar and summarizes the recent theoretical ideas concerning the statistics of large fluctuations and the origin of anomalous scaling of multipoint correlators.

1. Introduction

This article will review some of the recent ideas pertaining to the statistics of passive scalar advected by turbulent flow. The passive scalar could be the smoke or a fluorescent dye or temperature, in the latter case with the precaution that the buoyancy must be negligible compared to inertial stresses. The typical observables are the time series of temperature [1] measured at one or several points or snapshots of the fluorescent dye distribution [2]. The phenomenology of the passive scalar is quite similar to that of the turbulent velocity fluctuations. Like the statistics of the turbulent velocity fluctuations the statistics of the passive scalar is strongly non-gaussian [3, 44], and exhibits deviations away from Kolmogorov's scaling: [5, 6, 7] the intermittency phenomenon which manifests itself in an excess of large fluctuations on small scales. The interesting task thus is to understand the nature and statistics of large fluctuations lying at the origin of the anomalous (i.e. non-Kolmogorov) scaling. This phenomenology has been recently and excellently reviewed by Sreenivasan [1], Sreenivasan and Antonia [8] and Frisch [9].

Mixing: Chaos and Turbulence, edited by Chaté *et al.*
Kluwer Academic / Plenum Publishers, New York 1999.

The remarkable fact about the passive scalar is that its non-trivial statistical properties are not merely a shadow of the non-trivial statistics of the turbulent velocity field. Even a "simple" synthetic, Gaussian, random velocity field results in interesting intermittency phenomena for the scalar, as demonstrated by the numerical simulations of Holzer and Siggia, [10] Pumir [11] and by Chen [12]. The study of the passive scalar advected by Gaussian random (white in time) velocity field was initiated by Kraichnan [13, 14] in 1968. Yet only recently did the appreciation that the relatively simple problem of advection by *random* velocity offers a glimpse of intermittency stimulate a flurry of activity. [15, 16, 17, 18, 19]

Below, we will first (Sec. 2) review the phenomenology of the passive scalar and delineate the manifestations of the non-Kolmogorov behavior emphasizing the possible breakdown of Kolmogorov's notion of small scale isotropy and universality due to rare large fluctuations. Section 3 will outline the analysis of the rare event statistics in the simplest case, that of advection by a single scale random flow, arriving at the conclusion that in the presence of an external gradient the probability distribution function (PDF) of the scalar has exponential tails. Section 4 will address the more difficult problem of random flow with multiple spatial scales (i.e. with Kolmogorov-like power spectrum) and review the recent ideas of Kraichnan [15], Gawedsky and Kupiainen [19], Chertkov et al, [18] and of the present authors [16, 17] on the origin of anomalous scaling. Finally, Section 5 will address the problem of advection by a velocity field with proper temporal correlations and discuss the expected behavior of the N-point scalar correlation functions in terms of the their conjectured approximate $SL(N-1, R)$ symmetry. [20] It will argue that the small scale statistics, in contradiction with K41, exhibits persistent anisotropy and discuss the prospects for confronting the theoretical predictions (for the simplest non-trivial case of the 3d order function) with experiment. Section 6 is the conclusion.

2. Phenomenology of scalar fluctuations

The dynamics of the passive scalar is governed by the following deceptively simple equation:

$$\partial_t \Theta + (\vec{u} \cdot \nabla)\Theta = \kappa \nabla^2 \Theta \qquad (1)$$

where $\vec{v}(\vec{r}, t)$ is an incompressible velocity field. In an experiment, v, is the turbulent velocity field, although for the modeling purposes we will contemplate a synthetic random field. In the regime of interest, molecular diffusivity κ is small compared to the diffusion by random advection VL, where V is the root-mean-square velocity and L the correlation length, so that Peclet number $Pe \equiv \frac{VL}{\kappa} \gg 1$. To achieve statistical stationarity the scalar must be continuously forced which in an experimental situation is typically accomplished via boundary conditions which impose a large scale gradient. [21, 22, 23] Thus we shall assume $< \Theta(r) > = \vec{g} \cdot \vec{r}$. Note that the homogeneity is recovered for the fluctuating field $\theta \equiv \Theta(r) - < \Theta(r) >$ which obeys Eq. (1) but with a "source" $-\vec{v} \cdot \vec{g}$ added to the right hand side.

The phenomenology of the passive scalar is quite similar to that of the turbulent velocity fluctuations. A common object of study is the structure function, $S_n(r) = < [\theta(r) - \theta(0)]^n >$ constructed in analogy with the moments of the (longitudinal) velocity difference $S_n^{(v)}(r) = < [v(r) - v(0)]^n >$. Equivalently one may look at the probability distribution function (PDF) of scalar (or velocity) differences measured at points separated by r: $P(\Delta_r\theta) = \langle \delta(\Delta_r\theta - (\theta(r) - \theta(0))) \rangle$ where $\langle ... \rangle$ denotes time and space average and presumes statistical stationarity and homogeneity of the flow. These objects are particularly interesting in view of the Kolmogorov's (K41) theory of turbulence [5], which has been transcribed into the context of the passive scalar by Obukhov [6] and Corrsin. [7] Kolmogorov's theory predicts the existence of a scaling range $L > r > \eta$, where the large scale, L, is set by the forcing and the boundary conditions and the small scale cutoff, $\eta = Re^{-3/4}L$, is controlled by the viscosity via the Reynolds number $Re \equiv VL/\nu$. The K41 assumption that the only dimensionful parameter in the scaling range is the rate of energy dissipation, ϵ, results in a prediction for the scaling of the structure function: $S_n^{(v)}(r) \sim (\epsilon^{1/3}r^{1/3})^n$. Kolmogorov's reasoning applied to the passive scalar replaces the notion of cascading energy by the cascade and dissipation of the scalar variance which is assumed to be passed down from scale to scale over the time of the eddy turnover. [6, 7] The θ-dissipation rate $\langle \Delta_r\theta^2 \Delta_r v \rangle /r = \epsilon_\theta$ is assumed to be independent of the scale r leading to $S_{2n}(r) \sim (\epsilon_\theta\epsilon^{-1/3}r^{2/3})^n$ provided K41 scaling of the velocity field. More generally $S_{2n}(r) \sim r^{n(2-\zeta_v)}$ where ζ_v is the scaling exponent of $\langle \Delta_r v^2 \rangle$. If $\kappa/\nu \approx 1$ the scaling range for the scalar is expected to coincide with the inertial range of the velocity field. Often discussed is the Batchelor regime for $\kappa \ll \nu$ which corresponds to length scales smaller than the viscous cutoff, η but larger than the diffusion scale $\eta\sqrt{\kappa/\nu}$. In that range $\Delta_r v \sim r$ predicting $S_2(r) \sim r^0$ or $\ln(r)$ according to Batchelor. [24] In the Batchelor regime, the turbulent velocity acts on the small scale fluctuations simply as random strain.

The distinction between the even and odd moments of the scalar structure function is due to parity which, unless broken by the large scale boundary conditions, would render the odd moments equal to zero. The K41 type reasoning [25, 26] dictates that odd moments are proportional to the large scale anisotropy:

$$S_{2n+1}(r) \sim \langle(\Theta(r) - \Theta(0))\rangle S_{2n}(r) \sim \vec{G} \cdot \vec{r}(\epsilon_\theta\epsilon^{-1/3}r^{2/3})^n \qquad (2)$$

The odd moments provide a measure of anisotropy surviving on a given scale r, e.g. skewness $s_3 \equiv S_3(r)/S_2(r)^{3/2}$ which is predicted to decay on small scales as $(r/L)^{2/3}$. Finally, as a consequence of the predicted decay of large scale anisotropy and of the linear dependence of scaling exponents on n the PDFs of $\Delta_r v$ and $\Delta_r \Theta$ normalized by their root-mean-squares are predicted to be *universal*, i.e. independent of r in the inertial range and Reynolds number (provided it is large) and independent of the large scale boundary conditions.

The experimental studies of velocity fluctuations, while supporting K41 theory on the level of the second moment (or the power spectrum) exhibit anomalous behavior of the higher order moments (see Sreenivasan [8] for a comprehensive review). The velocity power spectrum follows K41 5/3 law and appears to approach isotropy [27] on small scales as predicted (the relative magnitude of the anisotropic contribution decays with wavenumber as $k^{-2/3}$), yet the moments > 3 show anomalous scaling. Alternatively, the velocity difference PDF depends on r in a way not fully parametrized by its variance. The shape of the PDF evolves from Gaussian for $r \sim L$ to stretch- exponential for $r \sim \eta$ strongly suggesting that the anomalous scaling with r is related to the appearance of long tails of the PDF. [3, 4]

The observed behavior of the passive scalar is in a close apposition with that of turbulent velocity. Although the scaling of the scalar power spectrum is not as well established experimentally, the observed exponents appear to converge [1] towards the Kolmogorov-Obukhov-Corrsin prediction [28] with increasing Re. (Curiously, a considerable scaling range, extending to lower than expected frequencies, opens up at moderate Re.) Higher moments of the structure function exhibit anomalous scaling and the PDF evolves with r from Gaussian on large scale to stretched exponential in the dissipation range $r \sim \eta$.

One interesting property of the passive scalar is that in the presence of the large scale gradient [22, 23] an exponential tail of the PDF appears even for a single point measurement which is equivalent to the difference measured at points separated by $r > L$, the correlation length. The physics underlying this phenomenon will be discussed in detail in Sec. 3.

Another interesting and well documented passive scalar phenomenon is the anomalously large skewness. The experiment of Mestayer [21] (1984) measured the derivative skewness, $s_d = < (\partial_x \theta)^3 > / < (\partial_x \theta)^2 >^{3/2}$, in the heated boundary layer flow and found it to be of $O(1)$ apparently independent of the Reynolds number (in the range explored). The presence of the vertical gradient of temperature $< \partial_z \Theta > \neq 0$ and of the vertical gradient of longitudinal velocity, $< \partial_z v_x > \neq 0$ break $x \to -x$ parity and allow non-zero s_d. The surprise however is that the skewness is much larger then what is expected on the basis of K41 which predicts that $s_d \sim s_3(r = \eta) \sim (\eta/L)^{2/3} \sim Re^{-1/2}$. Physically the skewness corresponds to the asymmetry in the rate of upward and downward drifts of the temperature: the time series exhibits a conspicuous sequence of gentle "ramps" terminating in sharp "cliffs." [1, 21] Remarkably, this behavior is not a consequence of the peculiar structure of velocity fluctuations in the boundary layer. It was also observed in the numerical simulations of Holzer and Siggia [10] utilizing a Gaussian random field. In a snapshot of the scalar configuration the ramp-and-cliff structure is identified as a gradient sheet formed transiently along the unstable manifold of a hyperbolic point in the flow (see Fig. 1). The surprising aspect is that such structures can be formed with relatively high probability even in a strongly fluctuating multi-scale flow! The first steps towards understanding this phenomenon will be discussed later in this article.

The anomalous skewness is interesting for several reasons. It indicates that the effect of large scale anisotropy does not decay with scale as predicted by K41 and persists down to the dissipation range. Thus the statistics of small scale fluctuations after all does depend on the boundary conditions and, contrary to the dogma of Kolmogorov's theory, is not universal. The relation of skewness to the gradient sheets goes along well with the intuition that the violations of the K41 scaling are due to rare large fluctuations which couple large and small scales directly, short-circuiting the cascade. What we learn from skewness, is that these large fluctuations are not-so-rare and dominate already the 3d moment. Scalar gradient sheets themselves, like vortex filaments which are recognized as the dominant intermittent events for the velocity field [29, 30], defy classification as large or small structures: their core is on dissipation scale, while the scalar jump and velocity circulation respectively are closer to the large scale values. While it is plausible that these events underl y the appearance of the anomalous scaling, the relation between the scaling of the statistical quantities like the structure function and the spatial structure for the intermittent events remains elusive.

Fig. 1. — The gray level plot of a snapshot of the $\theta(\vec{r})$ field advected by a synthetic multiscale Gaussian random velocity field (from Holzer and Siggia [10]) at Pe = 1544 on 512^2
lattice. The large scale gradient is in the horizontal direction. Note the appearance of the
sharp transitions from "white" to "black", which are asymmetric with respect to left-right
reflection. The corresponding skewness: $s_d \approx 1$.

3. Exponential tails of the scalar pdf

An interesting feature of the passive scalar phenomenology is the appearance of an
exponential tail of the PDF of the scalar measured at a given point, provided that
there is a large scale gradient. Below we shall analyze this effect for the simplest
case of a single scale random flow: Gaussian random incompressible $\vec{v}(\vec{r}, t)$ with the
correlation length L and correlation time $\tau_c = L/V$, where V is the RMS velocity.

The two main aspects of passive scalar physics are 1) transport, 2) mixing. The
first describes the large distance behavior of the scalar field coarse-grained over the
scale of L and τ. This coarse-grained field, $< \Theta(r, t) >$ on the scale much larger
than L obeys an effective diffusion equation:

$$\partial_t < \Theta >= D_{\text{eff}} \nabla^2 < \Theta > \qquad (3)$$

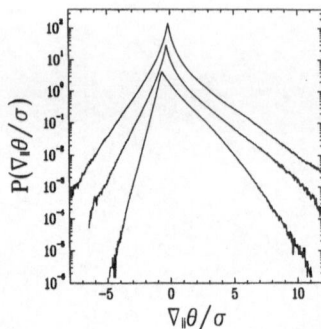

Fig. 2. — Schematic drawing of a one dimensional cut through the scalar field along the direction of the gradient.

where $D_{\text{eff}} \equiv \int dt < v(\vec{r}(t), t)v(0,0) >$ is the Taylor diffusivity [31] defined as an integral over the velocity correlator along the Lagrangian trajectory of a material point. This statement presumes ergodicity of Lagrangian trajectories and is intuitively obvious but hard to prove rigorously. (We shall not be interested in situations such as a time independent velocity field in 2D, where this ergodicity fails.) Note that D_{eff} does not at all depend on the molecular diffusivity κ!

The molecular diffusivity governs the mixing of the scalar. Without κ all the moments of Θ would be conserved and coarse-grained $< \Theta^n >$ would obey Eq. (3). With $\kappa \neq 0$, in the absence of the large scale gradient, $< (\Theta - < \Theta >)^n > \rightarrow 0$ corresponding to molecular mixing. Consider the evolution of a blob of "ink" of size $l(0) \sim L$. Because of the strain and shear in the flow the blob will evolve tendrils, along the unstable manifolds of locally hyperbolic flow, and form "jelly-rolls" in the vorticity dominated regions as shown in Fig. 2. The smallest dimension of the distorted region will decrease exponentially with time $l(t) \sim Lexp(-\gamma t)$, with γ of the order of the average rate of strain in the flow. Eventually, when $l^2(t) < \kappa/\gamma$, the molecular diffusion on time-scale of the eddy turnover, given by the inverse of the rate of strain γ, will take over and wipe out the "tendril". This molecular mixing will occur on average after the mixing time $t = t_* \equiv \gamma^{-1}ln(L^2\gamma/\kappa) \sim \gamma^{-1}lnPe$.

Let us now consider the evolution of $\theta(r,t)$ in the absence of molecular diffusion, $\kappa = 0$, starting from a prepared linear profile at $t = 0$: $\theta(x,0) = gx$ (let's for simplicity consider d=1). The probability of finding value θ' at position x and time t is equal to the probability that a Lagrangian point initially $(t = 0)$ at position $x' = \theta'/g$ reaches the point x at time t:

$$P(\theta', x, t) \sim \exp\left[-\frac{(\theta' - \theta(x,0))^2}{g^2 D_{\text{eff}} t}\right] \qquad (4)$$

The distribution is Gaussian with the variance increasing in time as the material points from progressively farther away reach the point of observation. Clearly, since in the $\kappa = 0$ limit the material points of the flow are permanently labeled by their initial value of θ, the PDF cannot reach a steady state.

The effect of $\kappa \neq 0$ is to allow the scalar field carried by a small parcel of fluid to equilibrate with its environment. The crude mean-field type estimate would be to assume that the "lifetime" of a given θ value of the fluid parcel is the mixing time t_* and substitute the latter in lieu of t in the exponent in Eq. (4). A material point can random walk a distance $\sqrt{D_{\text{eff}} t_*}$ before losing the memory of the scalar concentration at its place of origin: the resulting scalar PDF is Gaussian with variance $g^2 D_{\text{eff}} t_*$.

It is clear at this stage that the tails of the distribution arise from the relatively rare events where a material point traverses a long distance without dissipating its θ-content, i.e. without mixing with its environment. Indeed, the different Lagrangian trajectories will have different histories of strain along them, hence different γ and different "mixing times". The above mean-field type argument can be generalized by introducing such a distribution of lifetimes, $\mu(\tau)$:

$$P(\theta, \vec{r}) \sim \int d\tau \mu(\tau) \exp\left[\frac{-(\theta - < \theta(\vec{r}) >)^2}{g^2 D_{\text{eff}} \tau}\right]. \tag{5}$$

The average lifetime is the mixing time $t_* \equiv \int d\tau \tau \mu(\tau)$ however it is easy to see that if $\mu(\tau)$ has an exponential tail, e.g. $exp(-\tau/t_*)$, then $P(\theta)$ also has an exponential tail.

The qualitative argument leading to the expression for the PDF given by Eq. (5) is supported by a more formal computation using the Lagrangian path integral representation of the Green's function (for Eq. (1) given in [16] which also calculates the asymptotics of $\mu(\tau)$. Here we will only give a back-of-the-envelope argument. Observe that according to Eq. (3), the mixing time is inversely proportional to the rate of strain, γ. Let us measure time and the rate of strain in the units of the correlation time of the flow ξ/V (the average eddy turnover time). The lifetime τ can be achieved by suppressing the rate of strain: $\gamma < \tau^{-1}$. The latter will occur with probability $\sim \tau^{-1}$ per correlation time and with probability $\sim (\tau^{-1})^\tau$ over the required length of the Lagrangian trajectory, which traverses τ correlation volumes. Hence we get $\mu(\tau) \sim exp(-\tau ln(\tau))$ valid for large τ so that $\mu(\tau)$ is close to having an exponential tail.

Because of the uniformity of the bound imposed on γ along the trajectory, this simple argument underestimates the tail of $\mu(\tau)$. A better calculation [16] allows for the fluctuations of the rate of strain along the trajectory which gets rid of the ln in the exponent. This calculation is based on a semiclassical approximation to the path integral representation of the Green's function:

$$\mathcal{G}(r', t'|r, t) \equiv \int_{r(t)=r}^{r(t')=r'} Dr \exp\left(-\frac{4}{\kappa} \int_t^{t'} dt'' [\dot{r}_a - v_a(r, t'')]^2\right) \tag{6}$$

which solves Eq. (1) with the right hand side replaced by a δ-function source. In the large Peclet number limit, it is sufficient to consider the behavior of classical trajectories, which extremize the classical action appearing in the exponent of (6) and share the endpoint with the Lagrangian trajectory, $\dot{r} = \vec{v}(\vec{r}(t), t)$, which corresponds to the absolute minimum, zero value, of the action. The mixing time can be shown to correspond to the spreading of these classical trajectories over distance L, which is, predictably, controlled by the cumulative action of strain along the Lagrangian trajectory. The latter can be evaluated in terms of a product of random matrices and its statistics can be calculated in detail.

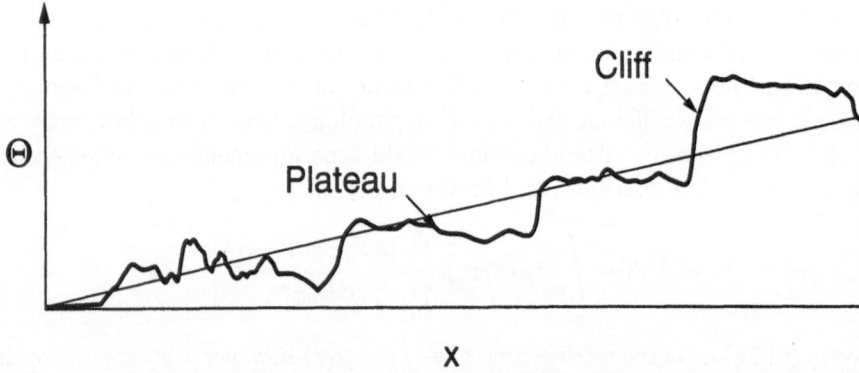

Fig. 3. — Probability distribution function of the longitudinal scalar gradient, $\vec{G} \cdot \vec{\partial}\Theta$ for advection by single scale random velocity field (from Holzer and Siggia [10]). Pe number increases from bottom to top (Pe= 62.8, 314, 1257). Note 1) the central cusp shifting to $\partial\Theta = 0$ at high Pe; 2) the asymmetry; 3) the appearance of an exponentail tail.

The semiclassical method can be extended to the analysis of the PDF of the scalar gradient. The result is interesting. The PDF tails (in the single scale flow considered) are again exponential. More surprising is the behavior of the PDF for small gradients (smaller than the variance $< (\nabla\Theta)^2 > \sim Pe < \nabla\Theta >^2$). The peak of the PDF of $\hat{g} \cdot \nabla\Theta$ does not lie at $< \nabla\Theta >= \vec{g}$ but is at zero instead, provided $Pe \gg 1$. The correct average longitudinal gradient appears via overall asymmetry of the PDF under reflection, see Fig. 3. This asymmetry also contributes an an anomalously large derivative skewness. The cuspiness of the PDF at $\nabla\Theta = 0$ points to an excess of well mixed regions and can be interpreted in terms of the "cliff and plateau", Fig. 2, picture where the mean gradient is expelled from the well mixed plateau regions [32] and concentrated into the "cliffs" corresponding to sheets of scalar gradient. The latter should have a width on the dissipation length-scale $\eta = LPe^{-1/2}$ and a height distributed with exponential probability thus accounting for the exponential tails of the gradient PDF. This interpretation, while supported by the numerical simulations of Holzer and Siggia [10] and Pumir [11] of course goes beyond the information available from a single point PDF.

Two problems arise at this point. The first is the extension of the analysis to the more interesting case of the multi-scale velocity field, which would include the case of turbulent flow with Kolmogorov power spectrum. The large Pe number approximation formally is no longer helpful, since the effect of strain has to be compared to the effect of eddy diffusivity which is nominally of the same order, leading to the effective $Pe \sim O(1)$. The second, we would like to go beyond the single point statistics. To that effect in the next two sections we will take up an alternative approach to scalar statistics based on the so called Hopf equation.

The application of the semiclassical approximation (more fashionably referred to as "instanton calculus") to the study of the rare event statistics has been pursued in the work of Chertkov et al [33], Falkovich *et al* [34] and Chertkov [35]. The latter work is particularly interesting as a way of establishing a link between the semiclassical and the Hopf equation methods, of which more below.

4. The δ−correlated velocity model of passive scalar advection

4.1. Formulation

Kraichnan has initiated [13] the investigation of Passive Scalar advection by a Gaussian random velocity field, δ-correlated in time. Such a field is specified by the velocity correlation function:

$$C_{ab}(r - r', t - t') = < v_a(r, t) v_b(r', t') > = \delta(t - t') \left[D_{\text{eff}} \delta_{ab} - D_{ab}^{(\gamma)}(r - r') \right] \quad (7)$$

with :

$$D_{ab}^{(\gamma)}(r) = A \left((d + 1 - \gamma) \delta_{ab} - (2 - \gamma) \frac{r^a r^b}{|r|^2} \right) |r|^{2-\gamma} \quad (8)$$

where γ is related to the velocity power spectrum exponent ζ_v via $\gamma = 1 - \zeta_v/2$. This relation arises from taking the white noise limit, i.e. correlation time $\tau \to 0$, in such a way as to keep the "diffusivity" $D_{ab}^{(\gamma)}(r) = \int dt[C_{ab}(r) - C_{ab}(0)] \sim \tau(r) < \Delta_r v^2 >$ finite. If, $< \Delta_r v^2 > \sim r^{\zeta_v}$, the eddy turnover time is $\tau(r) \sim r^{1-\zeta_v/2}$ yielding $D^{(\gamma)}(r) \sim r^{1+\zeta_v/2}$. Thus the K41 case, $\zeta_v = 2/3$ corresponds also to $\gamma = 2/3$. The projection operator in Eq. (8) insures incompressibility $\partial_a D_{ab}^{(\gamma)} = 0$. Finally, we shall pick dimensionful units so as to make $A = 1$ in Eq. (8).

Kraichnan proceeded to derive and solve [14] the equation for the 2-point equal time correlator of Θ. This equation arises from the stationarity condition

$$\partial_t < \Theta(r_1 t) \, \Theta(r_2 t) > = 0 \quad (9)$$

and is the simplest example of a "Hopf equation." [36] Using the equation of motion Eq. (3):

$$
\begin{aligned}
0 &= < -v_a(1) \, \partial_1^a \, \Theta(1) \, \Theta(2) - \Theta(1) \, v_a(2) \, \partial_2^a \, \Theta(2) > \\
&\quad + \kappa < \nabla^2 \, \Theta(1) \, \Theta(2) + \Theta(1) \nabla^2 \Theta(2) > \\
&= -[D_{ab}^{(\gamma)}(0) \, (\partial_1^a \, \partial_1^b + \partial_2^a \, \partial_2^b) + 2D_{ab}^{(\gamma)}(1 - 2) \, \partial_1^a \, \partial_2^b] < \Theta(1)\Theta(2) > \\
&\quad + \kappa < \nabla^2 \, \Theta(1) \, \Theta(2) + \Theta(1) \nabla^2 \Theta(2) >
\end{aligned}
\quad (10)
$$

where the second line is obtained by expressing $\Theta(r, t) = \Theta(r, t - \Delta t) + \int_{t-\Delta t}^{t} dt \, \vec{v} \cdot \vec{\nabla} \Theta$ and performing the Gaussian average over white velocity. We have dropped explicit reference to t and use arguments 1,2 as the shorthand for $r_{1,2}$. Unlike Kraichnan [14] who studied the case where Θ is forced by introducing a Gaussian source into Eq. (1) we shall adopt the more natural "gradient forcing": i.e. let us force the Θ field via its larger scale gradient $< \vec{\nabla}\Theta >= \vec{g}$. The translational invariance is recovered for the fluctuation field $\theta(r,t) \equiv \Theta(r,t) - \vec{g} \cdot \vec{r}$ which obeys

$$[2D_{ab}^{(\gamma)}(r) + \delta_{ab}\kappa] \, \partial_r^a \, \partial_r^b < \theta(r)\theta(0) >= g_a \, g_b[D_{ab}^{(\gamma)}(0) - D_{ab}^{(\gamma)}(r)] \qquad (11)$$

obtained by using the translational invariance which insures the proper "subtraction" of the $D_{ab}^{(\gamma)}(0)$ term on the left hand side. The $\mathcal{L}_\gamma \equiv D_{ab}^{(\gamma)}(r) \, \partial_r^a \, \partial_r^b$ operator appearing in Eq. (11) is nothing but the Richardson diffusion [36] operator which governs the evolution of the distance between two material points of the fluid as they follow their Lagrangian trajectories. The probability of their separation being r is governed by $\partial_t P(r,t) = \partial_r^a \, D_{ab}^{(\gamma)}(r) \, \partial_r^b \, P(r,t)$ from which, by dimensional analysis for $D^{(\gamma)}(r) \sim r^{2-\gamma}$ one obtains $< r^2(t) >\sim t^{2/\gamma}$. Using K41 scaling $\gamma = 2/3$ one obtains $< r^2 >\sim \epsilon t^3$, the Richardson law.

The scaling of the two point function is obtained readily from Eq. (11) by observing that the right hand side goes to a constant for small r, hence for $D^{(\gamma)}(r) \sim r^{2-\gamma}$, $< \Theta(r)\,\Theta(0) >\sim r^\gamma$. This result was obtained in [14] and it agrees with Kolmogorov theory as applied to the passive scalar by Obukhov [6] and Corrsin [7].

4.2. Kraichnan's 94 closure Ansatz [15]

More recently Kraichnan attempted the same approach to computing all even moments of the structure function. The stationarity condition

$$\partial_t < [\Theta(r) - \Theta(0)]^{2n} >= 0 \qquad (12)$$

leads to

$$\mathcal{L}_\gamma < [\Theta(r) - \Theta(0)]^{2n} > +2n\kappa < [\nabla^2\Theta(r) - \nabla^2\Theta(0)] \, [\Theta(r) - \Theta(0)]^{2n-1} >= 0 \quad (13)$$

The molecular diffusivity term here cannot be neglected since

$$\kappa\nabla^2\Theta(r) \, \Theta^n(r) = -\kappa n[\nabla\Theta(r)]^2 \, \Theta^{n-1}(r) + \kappa\vec{\nabla}(\vec{\nabla}\Theta \, \Theta^n) \qquad (14)$$

and the $\kappa[\nabla\Theta(r)]^2$ term on the right hand side is recognized as the scalar dissipation $\epsilon_\theta(r)$ which is finite even as $\kappa \to 0$. The correlation functions of $\nabla^2\Theta$ fields with Θ at other points are not expressible in terms of the structure function and to "close" Eq. (13). Kraichnan has proposed an Ansatz [15] based on an assumption about the conditional probability $P(\nabla^2\Theta(r) - \nabla^2\Theta(0) \,|\, \Delta_r\Theta)$ of the type first introduced by Sinai and Yakhot [37] and studied by Ching and Pope [38]. Specifically:

$$\int dy y \, P(y|\Delta_r\Theta) = C\kappa^{-1}r^{-\gamma}\Delta_r\Theta \qquad (15)$$

resulting in

$$< \kappa \, (\nabla^2\Theta(r) - \nabla^2 \, \Theta(0)) \, (\Theta(r) - \Theta(0))^{2n-1} >= C_n r^{-\gamma} < (\Delta_r\Theta)^{2n} > \qquad (16)$$

with the proportionality constant $C_n = C$ fixed by the case of $n = 1$ to be $C = 1$. Eq. (13) then closes as a homogeneous equation for $< (\Delta_r \Theta)^{2n} >$ with a scaling solution $< [\Delta_r \Theta]^{2n} > \sim r^{\zeta_{2n}}$. The exponent is

$$\zeta_{2n} = \frac{1}{2} \sqrt{4 \, nd \, \gamma + (d - \gamma)^2} - \frac{1}{2} (d - \gamma) \tag{17}$$

which by construction (i.e. the choice of constant C) agrees with Kolmogorov-Obukhov-Corrsin for the second moment, $\zeta_2 = \gamma$ but predicts an anomalous $n^{1/2}$ asymptotic behavior for large n.

It is important to note that Kraichnan's Ansatz actually may be split into two propositions. The first is the proportionality of the correlator of the dissipation with the scalar fields (l.h.s. of (16) to the correlator of the scalar fields only (r.h.s. of (16)). The second is the notion that the proportionality constant C_n in Eq. (16) is *independent* of n. It has been argued by Fairhall et al [39] that the linear relation between the correlators can be obtained by the local analysis of the Hopf equation for a multipoint correlator (which will be discussed in the next sections) with points approaching each other. However the stronger statement that the proportionality constant, C_n, is n independent obtained from the strong form of the "linear" Ansatz given by Eq. (15), does not follow from that argument. In fact the linear relation between the two correlators in Eq. (16) is just a restatement of Eq. (13) assuming that the scaling solution exists. The support of the non-trivial, strong form, i.e. Eq. (15), of the Ansatz is mostly empirical. [40]

4.3. Hopf equation for multi-point correlators

There, however, is an alternative which avoids the closure difficulty inherent in Eq. (13). Consider instead the stationarity condition for N-point function

$$\partial_t < \Theta \, (r_1 \, t) \, \Theta(r_2 \, t)...\Theta \, (r_N t) >= 0 \tag{18}$$

A generalization of Kraichnan's derivation [13] of the 2-point equation leads to the multipoint Hopf equation [16]

$$\sum_{i \neq j}^{N} \left(D_{ab}(r_i - r_j) + \kappa \delta_{ab} \right) \partial_{r_i}^a \partial_{r_j}^b < \theta(r_1)...\theta(r_N) >=$$
$$\sum_{i \neq j}^{N} g_a g_b \left(D_0 \delta_{ab} - D_{ab}(r_i - r_j) \right) < \theta... >_{ij}^{N-2} \tag{19}$$
$$-2 \sum_{i \neq j}^{N} g_a D_{ab}(r_i - r_j) \partial_j^b < \theta... >_i^{N-1}$$

(with implicit summation over repeating indices).

The right hand side involves correlators of $N - 1$ and $N - 2$ points which couple to the N-point function via external gradient g. An analogous equation but with a different inhomogeneous term on the right hand side can be derived for the "gaussian forcing" case. [18, 19, 41] Remarkably Eq. (19) is a perfectly self-contained PDE hierarchy which can be solved, in principle, systematically starting with $N = 2$. We restrict ourselves to the inertial range of scales, where r (the distance between any pair of points) is large enough so that the molecular diffusivity can be neglected: $r \gg \eta \equiv (\kappa/D_0)^{1/\gamma}$. The circumscribed domain of validity of Eq. (19) means that it must be supplemented by appropriate boundary conditions at large, $r \sim L$, and small scales $r \approx \eta$. The small scale conditions are better thought of as matching conditions allowing a smooth crossover from the inertial to dissipative behavior. The large scale boundary conditions are not universal and are impractical to impose in detail, however their mere existence will require construction of a solution sufficiently general to accommodate them in principle.

Thus, the closure problem does not arise however instead of an ODE entailed in Kraichnan's Ansatz, Eq. (13-16), one is now faced with an inhomogeneous PDE in $(N-1) \times d$ dimensions.

4.4. The origin of anomalous scaling

Quite generally one observes that the solution of the Hopf equation (19) consists of the homogeneous and inhomogeneous contributions and can be written schematically:

$$
\begin{aligned}
\langle \Theta_1...\Theta_N \rangle &= cst - I_N(r_{ij}) - H_N(r_{ij}) \\
&= cst - \left(\frac{R}{L}\right)^{\zeta_N} \hat{I}\left(\frac{r_{ij}}{R}\right) \\
&\quad - \sum_{\lambda_\alpha > 0} \left(\frac{R}{L}\right)^{\lambda_\alpha} \psi_\alpha\left(\frac{r_{ij}}{R}\right) - \sum_{\lambda_\beta < 0} \left(\frac{R}{\eta}\right)^{\lambda_\beta} \psi_\beta\left(\frac{r_{ij}}{R}\right)
\end{aligned}
\tag{20}
$$

where we have made the scaling exponents explicit by normalizing r_{ij} by a common length scale R, i.e. $R^2 = \sum_{\{ij\}} r_{ij}^2$. The homogeneous solution H_N in general consists of a superposition of all existing zero modes, ψ_α, of the Hopf operator (the left hand side of Eq. (19)) which are excited via boundary conditions or matching with the solution outside of the scaling domain described by Eq. (19). The zero modes with the positive scaling index, λ_α, are excited at the integral scale $R = L$, at which they are of the same order of magnitude with inhomogeneous term. The zero modes ψ_β with the negative scaling index λ_β are required in general to satisfy matching with dissipative scales and are therefore normalized with η so that they are of $o(1)$ at the dissipative cutoff, but decay in the inertial range, $R \gg \eta$.

Now, in the inertial range, $r \ll L$, it is the mode with the smallest (positive) scaling exponent λ which dominates. The scaling exponent of the inhomogeneous solution obeys the dimensional analysis and by examining Eq. (19) can be seen to obey $\zeta_N = \zeta_{N-2} + \gamma$ for even N and $\zeta_N = \zeta_{N-1} + 1$ for odd N. In the absence of the homogeneous solution the result of such a construction would necessarily agree with K41 theory. On the other hand the scaling indices λ_α of the zero modes cannot be determined by dimensional analysis, and require solving $\mathcal{L}_\gamma(R^\lambda \psi_\alpha) = 0$. The scaling indices enter this equation as eigenvalues. Existence of the $\lambda_\alpha < \zeta_N$ would indicate anomalous scaling. This realization of the role of the zero modes of the Hopf operator in determining the anomalous scaling was advanced independently in [17, 18, 19].

It is straightforward to construct all zero modes of for the N=2 case [18] (the modes are classified by the angular momentum) and demonstrate that all $\lambda_\alpha > \zeta_2 = \gamma$. Hence the zero modes are irrelevant confirming the results of Kraichnan, [13, 14] and Obukhov [6] - Corrsin [7] prediction for the power spectrum. [41] On the other hand it appears that for $N \geq 3$ relevant λ_α's and the anomalous scaling exist.

4.5. Perturbative approaches to anomalous scaling

The simplest calculation of anomalous scaling for δ-correlated model pertains to the case of $\gamma = 2 - \epsilon$, which corresponds to near diffusion limit, and is due to Gawedski and Kupiainen [19] who computed the anomalous scaling of the 4-point function $< \Theta(r_1)...\Theta(r_4) >$. Let us look for the zero mode of $\mathcal{L}_\epsilon$. The operator may be expanded in ϵ:

$$
\begin{aligned}
\mathcal{L}_{2-\epsilon} &= \sum_{ij} |r_{ij}|^\epsilon \left[\delta_{ab} \left(1 + \frac{\epsilon}{d-1} \right) - \frac{\epsilon}{d-1} \frac{r_{ij}^a r_{ij}^b}{r_{ij}^2} \right] \partial_i^a \partial_j^b \\
&= \sum_{ij} \partial_i \cdot \partial_j + \epsilon \sum_{ij} \ln r_{ij} \partial_i \cdot \partial_j + ... \\
&= \mathcal{L}_2 + \epsilon \mathcal{L}'
\end{aligned}
\tag{21}
$$

The first step is to solve $\mathcal{L}_2 \Phi^{(0)} = 0$ which is easy because $\mathcal{L}_2$ may be turned into a Laplacian in $3 \times d$ dimensions by a linear coordinate transformation. Its zero modes are harmonic polynomials, $\Phi^{(0)} = R^\ell \Psi_\ell(\hat{\Omega})$, written here in polar coordinates with radius R and angles parameterized by the unit vector $\hat{\Omega}$. The symmetry of the 4-point correlator under permutation of points restricts the set of relevant polynomials. In addition, it can be shown that none of the symmetric polynomials with $l < 4$ contribute to the structure function, i.e. they vanish for the superposition of configurations: $< \Theta^4(0) > -4 < \Theta(r)\Theta^3(0) > +3 < \Theta^2(r)\Theta^2(0) >$. The first interesting case is $\ell = 4$, hence $\lambda_4 = 4 + O(\epsilon)$. More generally one seeks a perturbative solution in the form $R^{4+\epsilon\Delta} [\Psi_4(\hat{\Omega}) + \epsilon \, \delta\Psi(\hat{\Omega})]$ and determines the scaling index correction Δ via the Fredholm solvability condition [42]

$$
\int d\hat{\Omega} \ \Psi_4^*(\hat{\Omega}) \ [\mathcal{L}'(R^4 \Psi_4(\hat{\Omega}) + \Delta \mathcal{L}_2(R^4 \ln R)\Psi_4(\hat{\Omega})] = 0
\tag{22}
$$

which expresses the fact that the perturbation must be orthogonal to the zero modes of the unperturbed operator $\mathcal{L}_2$ so that the latter could be inverted in order to calculate $\delta\Psi(\hat{\Omega})$.

The result for $d = 3$ is $\lambda_4 = 4 - \frac{14}{5}\epsilon$ which implies for the 4-point function

$$
< \Theta_1\Theta_2\Theta_3\Theta_4 >= cst - \left(\frac{R}{L}\right)^{2(2-\epsilon)} \left\{ F_{IH}\left(\frac{r_{ij}}{R}\right) + \left(\frac{R}{L}\right)^{-\frac{4}{5}\epsilon} F_H\left(\frac{r_{ij}}{R}\right) \right\}
\tag{23}
$$

where the factor outside the bracket corresponds to K41 dimensional scaling obeyed by the inhomogeneous solution (denoted F_{IH}). The second term in the bracket arising from the zero mode is anomalous and dominates in the inertial range. This calculation has been subsequently extended to other moments yielding, for even moments [43]: $\zeta_{2n} = \gamma n - 2(2 - \gamma)n(n - 1)/(d + 2) + O((2 - \gamma)^2)$ and for the 3d moment [44]: $\zeta_3 = (d + 4)/(d + 2)$.

An important aspect of the $2 - \gamma \to 0$ limit is that the multi-point correlators factorize into products of two-point correlators reflecting Gaussianity of the statistics. The same is true in the large d limit, which thus provides another point of departure for the perturbation theory which was employed by Chertkov et al [18, 45]. Since in large d most pairs of vectors are orthogonal, the "inter-particle" distances r_{ij} are, to the leading order in $1/d$ independent of each other. Provided, $d \gg N$, the operator $\mathcal{L}_\gamma$ can be rewritten in terms of $|r_{ij}|$ length only: $\sum_{k=1}^{N(N-1)/2} s_k^{3-d-\gamma} \partial_{s_k} s^{d-1} \partial_{s_k}$ where s_k stand for the lengths. The zero modes, to the leading order, factorize pairwise, e.g. for $N = 4$: $\Psi = |r_{12}|^\gamma |r_{34}|^\gamma + (permutations)$ and the logarithmic terms found in the next order are interpreted as a correction to the scaling exponent. The result of Chertkov and Falkovich [45] for the even moments is: $\zeta_{2n} = \gamma n - 2n(n-1)(2-\gamma)/d$, provided $n(2-\gamma)/d \ll 1$. An exciting aspect of this large d expansion is that there is a chance that it may transcend its present context. Indeed, while the possibility that large d limit of turbulence may be close to Gaussian has been surmised many years ago independently by Siggia and Migdal [46], the calculation of Chertkov et al was the first one to demonstrate this intuition to be correct at least in the context of the passive scalar.

The anomalous scaling exponents for the $N = 2n$ point functions calculated by Chertkov et [45] and by Bernard et al. [43] agree in the large d limit and both predict the correction to Komogorov's scaling increase quadratically with n, i.e. $2(2-\gamma)d^{-1}n(n-1)$. Such a dependence is generic for nearly Gaussian behavior and is shared by the log-normal theory of Kolmogorov and Obukhov [9, 36] but is not expected to hold in the large n limit (lest the exponents violate monotonicity as a function of n). Indeed, the domain of validity of the two perturbative calculations is limited to $n \ll d$ and $n(2 - \gamma) \ll 1$ respectively.

Comparing the above results with the ζ_{2n} predicted by the Kraichnan's 94 Ansatz (Eq. (17)) one notes the disagreement. For the $\delta-$correlated model at hand, the $2 - \gamma$ and $1/d$ expansions appear well controlled and must be given precedence over an Ansatz, no matter how plausible the latter. The problem with the Ansatz lies not in the assumed relation of the dissipation correlator with the θ correlator in Eqs. (15-16), which after all follows from the exact Eq. (13), provided that a scaling solution exists; but with the exact value of the proportionality constant, C_n, which in order to agree with the perturbation theory must depend on n. To the extent that the constancy of C_n has empirical (and numerical) support [12, 40, 47] one may expect it to be approximately true. The question then is how good and uniformly valid an approximation it is 1) as a function of γ for the δ-correlated model, 2) for the broader class of models with more physical velocity correlations.

Setting aside the disagreement between the Eq. (17) and the calculated exponents in the perturbative limits, one may inquire about the validity of the $\zeta_{2n} \sim \sqrt{n}$ asymptotics predicted by the Ansatz. [15, 47] Chertkov [35] has argued for the asymptotic n-independence of the exponents, however, this issue remains open at present.

Figure 4 illustrates the regions of the γ, d parameter space accessible to perturbation theory. In addition to the two different expansions mentioned above, it is possible to carry out an expansion near the Batchelor limit, $\gamma \ll 1$ [48]. Unlike the other two limits the $\gamma \to 0$ limit is not Gaussian in the sense that the relevant zero modes do not (even approximately) factorize pairwise. Yet the Batchelor limit is completely integrable [17, 49]. The perturbation theory about it however is singular and is therefore quite involved. It will be discussed briefly in the next section.

4.6.   Local expansion of the Hopf equation

Before leaving the subject of the δ-correlated model let us briefly discuss the the behavior of a multipoint correlator in the limit when a pair of points, say 1,2, ap-

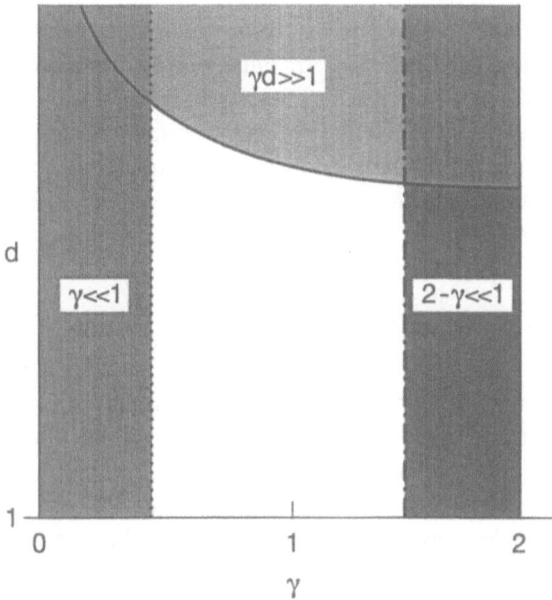

Fig. 4. — The domains of different perturbative expansions for the Kraichnan's $\delta-$ correlated velocity model in the γ, d. plane.

proach each other. When $\rho \equiv |r_{12}|$ becomes much smaller than all other interpoint distances, $|r_{ij}| \sim R$ one could develop an expansion in $\rho/R \ll 1$, which intuitively one expects to start as:

$$\langle \Theta_1 \Theta_2 ... \Theta_N \rangle \approx \langle \Theta^2(r_1) ... \Theta(r_N) \rangle + \langle (\Theta(r_1) - \Theta(r_2))^2 \rangle \langle \mathcal{O}(r_1) ... \Theta(r_N) \rangle \qquad (24)$$

This form along with the definition of the "operator" $\mathcal{O}(r)$ follow from the direct expansion of Eq. (19) which is obtained by observing that $\mathcal{L}_\gamma$ contains a term $\mathcal{L}'_\gamma \equiv D_{ab}^{(\gamma)}(\rho)\partial_\rho^a\partial_\rho^b$ which scales as $\rho^{-\gamma}$ and the rest, call it $\mathcal{L}''_\gamma$ which is of $o(1)$. Hence, the solution in the $\rho \ll R$ limit has the form $F_1(r_1, r_3, ...) + \rho^\gamma F_2(r_1, r_3, ...) + ...$ The singular $\mathcal{L}'_\gamma$ term annihilates F_1, since it is ρ independent, but yields a constant (call it a_γ) acting on ρ^γ resulting in the relation $a_\gamma F_2 = -\mathcal{L}''_\gamma F_1$. Since $< (\Theta_1 - \Theta_2)^2 > = cst \rho^\gamma$ the composit local field $\mathcal{O}$ (whose correlator with Θ's up to a constant factor is given by F_2) is identified by acting on both sides of Eq.4.17 with $\mathcal{L}'_\gamma$, which yields $\mathcal{O}(r) \sim \lim_{\rho \to 0} D_{ab}^{(\gamma)}(\rho)\partial^a\Theta(r)\partial^b\Theta(r+\rho)$. Thus, up to a constant factor, $\mathcal{O}$ is just the local scalar dissipation operator. Strictly speaking as $\rho \to 0$ this inertial range definition sensible for $R \gg \rho \gg \eta$ must be replaced by the one appropriate in the dissipation range, $\kappa\partial^a\Theta(r)\partial^a\Theta(r)$, to which it crosses over at $(\rho/L)^{2-\gamma} \approx \kappa/D_{\text{eff}}$.

Note, that this *local* analysis determines neither the multipoint correlator nor its scaling exponent both of which require a *global* solution of the multivariate PDE. However it does provide a useful relation between the scaling of the nth order dissipation correlator and and the $2n$ correlator of Θ. This was first noted by Chertkov et al [18] and Fairhall et al [39] by considering a 4-point function with two distinct pairs "fusing" which yields: $\zeta_4 = 2\gamma + \Delta$ where Δ is the (non-positive) scaling exponent of the dissipation-dissipation correlator. Fairhall et al [39] also employed the local analysis (sometimes referred to as "fusion rules"), to "derive" the scaling laws proposed by Kraichnan (i.e. Eq. (16)). It must be noted however, that to the extent that the stationarity equation for the structure function (Eq. (13)) is in itself a limit of the Hopf equation (Eq. (19)) (with coincident points and proper subtractions) the local analysis does not provide any additional information to complete the closure. In computing the scaling exponents the present authors see no alternative to solving the multipoint PDE!

5. Advection by flows with finite correlation time

Kraichnan's δ-correlated velocity model is most appealing for its analytical tractability - in the very least at allows an exact derivation of the Hopf equation. It is clearly a good source of insight into the origin of non-Kolmogorov phenomena. Yet, from the physical point of view, the $\tau(r) \to 0$ limit does not appear appropriate and the predictions of the model may not be directly comparable with the experiments. It is therefore desirable to have a broader class of models which could provide a good approximation to the physical reality even at the expense of few parameters. Below we shall consider a class of such phenomenological models constructed on the level of the Hopf equation. Were the notions of universality characteristic of the scaling behavior near the thermodynamic critical points [50] applicable, the anomalous exponents would be independent of the details of the phenomenological models and would depend but on a handful of "relevant" parameters, such as the scaling exponent of the random velocity field. There would be hope that the experimental situation may be in the same universality class as the Kraichnan model. Regrettably our results lead to the conclusion that anomalous scaling exponents of the scalar depends on the details of velocity statistics. Hence one should not expect an a priori prediction of the experimental exponents. What could then be a meaningful basis for comparison with experiment? It is this question, along with the derivation and the analysis of the phenomenological Hopf equation, which will be the goal of this Section.

Let us consider the turbulent flow with the Lagrangian correlation time $\tau(R) \sim R^{2/3}$ corresponding to the eddy turnover on scale R. Consider a bunch of material points, $\vec{r}_i$, separated by distances of the order of R. Their evolution over a correlation time, $\tau(R)$, can be decomposed into three components: 1) the translation of the center of mass due to effectively uniform velocity arising from scales $\gg R$ which contribute only through their action on the mean temperature gradient, 2) coherent straining and shearing of the point configuration due to the action of the velocity from scales $\sim R$, 3) incoherent random motion of the points due to scales $\ll R$:

$$r_i'^a - r_i^a = \tau v_<^a + \tau \mathbf{m_{ab}} r_i^b + \int_0^\tau dt\, v_>^a(r_i, t) \qquad (25)$$

with the random strain-vorticity matrix $\mathbf{m_{ab}} \equiv \partial_b v_a$ representing the effect of the $\sim R$ scales. K41 scaling implies that the magnitude of $\tau m \sim R^{2/3} R^{-2/3} \sim O(1)$ and that of the last term in Eq. (25) scales with R, from which one concludes that this mapping is scale invariant. This is the consequence of the Navier Stokes equations possessing no intrinsic scales except for viscosity so that the action of v_i over one Lagrangian correlation time amounts to a change in the r_i configuration which scales with the r_{ij} differences themselves.

Since the mapping steps are statistically independent, the random fields, which are assumed to be Gaussian, can be averaged freely. If in addition one takes the right hand side of Eq. (25) to be small, the evolution of the $\Theta(\vec{r}_i)$ correlator implied by the mapping becomes governed by a differential equation (instead of the integral equation which is valid more generally). The stationarity dictates the Hopf equation:

$$(\mathcal{L}_0 + \mathcal{L}_D) < \theta(1)... \theta(N) >= F \qquad (26)$$

where the RHS denotes the inhomogeneous forcing term similar to that in Eq. (19) apart from an extra multiplicative factor of $R^{\frac{2}{3}}$ which arises from $\tau(R)$. The advection by random strain is represented by Richardson operator $\mathcal{L}_0$ defined by Eq. (20) with exponent $\gamma = 0$ corresponding to the Batchelor limit. $\mathcal{L}_D$ (D - for dissipation) represents effective diffusivity due to small scales and is most plausibly taken as $\alpha R^{\frac{2}{3}}\mathcal{L}_{\frac{2}{3}}$, where $\mathcal{L}_{\frac{2}{3}}$ is just the white velocity Hopf operator with 2/3 scaling exponent corresponding to the Kolmogorov spectrum and the multiplicative $R^{\frac{2}{3}}$ factor arises from $\tau(R)$ and ensures that $\mathcal{L}_D$ has the same scaling dimension as $\mathcal{L}_0$). Finally α parameterizing the relative strength of random strain and "eddy diffusivity" will be used as a formal expansion parameter.

One may note that Eq. (26) corresponds to an alternative way of taking the white noise limit where instead of $\tau < \Delta v^2 >$ it is $\tau^2 < \Delta v^2 >$ that is kept finite. The underlying motion is more ballistic then diffusive. This results in the overall scale invariance of the evolution operator on the LHS of Eq. (26) to be contrasted with scaling dimension $-\gamma$ of the Hopf operator for the δ-correlated model. Nevertheless, in the limit of $\gamma \to 0$ the δ-correlated model upon expansion in γ reduces to the form of Eq. (26) with some particular form of $\mathcal{L}_D$. The general analysis discussed below will be applicable to that case as well. [48]

As in the case of the Kraichnan model one can show that the scaling exponents of the inhomogeneous solutions are governed by naive dimensional analysis. The general considerations concerning the relation of anomalous scaling and the homogeneous solutions holds as well. Hence in order to determine the anomalous scaling we need to compute the scaling exponents, λ, of the zero modes of the $\mathcal{L}_0 + \mathcal{L}_D$ operator.

Let us start with the zero modes of $\mathcal{L}_0$. For the $(n+1)$ point correlation function of r_i eliminate the center of mass by defining n reduced vectors, $\rho_1 = (r_1 - r_2)/\sqrt{2}$, $\rho_2 = (r_1 + r_2 - 2r_3)/\sqrt{6}$, $\rho_3 = (r_1 + r_2 + r_3 - 3r_4)/\sqrt{12}$, etc, with the linear combinations chosen as to make $\rho^2 - i$ independent. Since the coherent part of the velocity field is modeled as a single matrix acting on all ρ_i, one observes that this dynamics, and hence $\mathcal{L}_0$ are invariant under all volume preserving linear transformations $\vec{\rho}_i \to g_{ij}\vec{\rho}_j$ acting on the "pseudo-space" label of ρ's. These form the $SL(n, R)$ group of linear volume preserving transformation acting on n-component vector. In addition $\mathcal{L}_0$, is invariant under spatial rotations, $SO(d)$, and a dilation $\wedge = \rho_i^a \partial_i^a$. Since $\mathcal{L}_0$ is second order in derivatives, the above mentioned invariances imply that it can be expressed entirely in terms of their Casimir operators (i.e. angular momentum squared for $SO(d)$ and an appropriate generalization thereof for $SL(n)$. [17, 51]

These group theoretic considerations allow one to immediately determine all the eigenmodes of $\mathcal{L}_0$. In the simplest case of the skewness in two dimensions, they can be labeled by the angular momentum quantum number ℓ, the eigenvalue of the $SL(n = 2)$ Casimir, ν, which labels the irreducible representations, the quantum number q - the "pseudo-space" angular momentum corresponding to the rotation subgroup of $SL(2)$ and finally the dilation exponent λ. Demanding $\mathcal{L}_0 \Psi_\alpha = 0$ and imposing boundary conditions leads to $\lambda = 1$, $\nu = \frac{1}{2}$, $\ell = 1$, and arbitrary q (plus other less relevant λ's). The rest of the analysis will address the deviation of λ away from 1.

Before considering $\mathcal{L}_D$ we note another important physical property of $\mathcal{L}_0$ namely its reducibility; acting on any function behaving as $|r_{ij}|^x$, $x > 0$; the limit of co-incident points, say $r_{12} \to 0$, reduces $\mathcal{L}_0(n)$ to $\mathcal{L}_0(n-1)$ acting on the remaining variables. This is exactly what one would expect for the advection without dissipation which conserves all powers of Θ so that the field Θ^2 behaves exactly like Θ. This reducibility must not hold for the dissipative operator $\mathcal{L}_D$. Whenever two points, say points 1 and 2, approach so that $\rho = r_{12}$ becomes much smaller than other $r_{ij} \sim R$ a local expansion of the Hopf equation can be developed along the lines discussed already in the previous Section. Since locally the dominant part of $\mathcal{L}_D \sim R^{\frac{2}{3}} \rho^{\frac{4}{3}} \partial_\rho{}^2$ scales like $\rho^{-\frac{2}{3}}$ the local solution must have the form

$$\left(\frac{R}{L}\right)^\lambda \left[f_1 + \left(\frac{\rho}{R}\right)^{\frac{2}{3}} f_2 + \ldots \right] \tag{27}$$

where $f_{1,2}$ are functions of the r_{ij} distances excluding the small $r_{12} = \rho$. The non-analytic dependence on the short distance ρ has the Kolmogorov scaling and hence we call the point in configuration space where any of the $r_{ij} \to 0$ a 'Kolmogorov point'. The action of $\mathcal{L}_0$ on the first term is balanced by the action of $\mathcal{L}_D$ on the second. In the limit of $\rho \to 0$ the Hopf equation takes the form:

$$\mathcal{L}_0 < \theta^2(1)\theta(2).. > + \alpha R^{\frac{2}{3}} < \epsilon(1)\theta(2).. > = 0 \tag{28}$$

The new field ϵ is the local dissipation rate, expressible as $\mathcal{L}_{\frac{2}{3}}(\rho)\theta(\rho)\theta(0)$ in the inertial range and matching to $\kappa(\nabla\theta)^2$ in the dissipation range. Of course Eq. (27) is not a closed equation, and corresponds closely to the equation studied by Kraichnan [15] for the structure function. The relation between the scaling dimensions of the $< \theta^2(1)\theta(2)... >$ and $< \epsilon(1)\theta(2).. >$ correlators is explicit in Eq. (28). This equation provides a useful lower bound on the scaling index which follows from the fact that $< \epsilon\theta.. >$ must have a positive scaling dimension (since it can be bounded from above by $\sup_r |\theta(r)|$ and $< \epsilon >$ which are both set by the large scales). Thus the exponent for $< \theta\theta... >$ must exceed $\frac{2}{3}$.

So far we discussed the singularity of the θ correlator with nearly "fused" points as the property of the local solution of the Hopf equation. Alternatively since the form of the local two point singularity is related to dissipation and is fixed by the velocity scaling exponent the same way as for the $< (\theta(\rho) - \theta(0))^2 >$ function (e.g. ρ^γ in the case of the Kraichnan model) it can be held as known and used as a condition to be imposed on the $\mathcal{L}_D$ operator in the phenomenological Hopf equation. The simplest possible dissipation operator, $R^2 \vec{\partial}_i \cdot \vec{\partial}_i$, is too strong in that sense and would force analyticity of the correlator at the "Kolmogorov points". However, the variant with the same scaling properties $\mathcal{L}_D = \alpha R^{\frac{2}{3}} \beta(\rho)\nabla_\rho^2$ where β has a scaling dimension $\frac{4}{3}$ and vanishes as $r_{ij}^{\frac{4}{3}}$ when any $r_{ij} \to 0$ would allow the correct local singularity

(Eq. (27)). For algebraic simplicity we have used this form of dissipation in some of the calculations.

The zero modes of $\mathcal{L}_0 + \mathcal{L}_D$ can be found perturbatively in α. There are two subtleties. The first is that the zero modes of $\mathcal{L}_0$ that we are expanding about are infinitely degenerate. The degeneracy is the consequence of the $SL(N-1)$ symmetry. This symmetry is broken by $\mathcal{L}_D$ which therefore also lifts the degeneracy of the zero modes. The second subtlety is that the dissipative term is actually a *singular* perturbation which was already evident from its dominance near the Kolmogorov points, where pairs of r_i coalesce. It turns out to be dominant also for "collinear" configurations where the points fall on the same line.

The perturbative calculation in $\mathcal{L}_D$ thus requires dealing with the non-trivial crossover from the $\mathcal{L}_0$ dominated region to the $\mathcal{L}_D$ dominated one, which for the skewness in 2 or 3 dimensions occurs for $|\vec{\rho}_1 \wedge \vec{\rho}_2|/R^2 \sim \alpha^{\frac{1}{2}}$. Note that $|\vec{\rho}_1 \wedge \vec{\rho}_2|$ is the area of the triangle defined by the N=3 measurement points and its zero value corresponds to the collinear configuration. This calculation has been described in [20, 52]. Its result, aside from the scaling exponent $\lambda(\alpha) = 1 + o(\alpha)$, is the explicit form of the correlator for a general configuration of points. Away from collinearity the latter can be represented as a linear superposition of the $\mathcal{L}_0$ zero modes, ψ_α,

$$< \theta(1)\theta(2)\theta(3) >= \sum_q a_q \psi_{\nu=\frac{3}{2},\ell=1,q}(\rho) \tag{29}$$

The coefficient a_q is related to this correlator with points collinear.

We have thus investigated several models of $\mathcal{L}_D$ including the one corresponding to the $\gamma \ll 1$ expansion of the Kraichnan's δ−correlated model. [17, 20, 48] The results for the exponent and the correlation function are model dependent, which is clear the moment one observes that the two are respectively the eigenvalue and an eigenfunction of a linear operator appearing in the Hopf equation. This operator is not universal, depends on the details of the random velocity ensemble above and beyond the scaling exponent of the velocity correlator, and the spatial dimension. The restriction of the parameter space to the latter two in the Kraichnan δ− correlated model is a consequence of the "white noise" limit. Of course our phenomenological modeling takes the freedom of the parameters to another extreme. More conservatively, Chertkov et al [53] have confirmed the lack of universality by computing the effect of short but non-zero correlation time on the exponents within the large d limit.

One result which does not depend on the details of the model except the for the notion of the dominant effect of strain (i.e. $\alpha \ll 1$ assumption) is that the scaling exponent for the 3d moment is $\zeta_3 \approx 1$ which is smaller then 5/3 predicted via K41. Evaluating the normalized 3d moment: $< (\Delta_r\theta)^3 > / < (\Delta_r\theta)^2 >^{3/2}$ at $r = \eta \sim Re^{-3/4}$ one obtains an estimate of the derivative skewness $s_d \sim o(1)$ independent of Re (since $< (\Delta_r\theta)^2 > \sim r^{2/3}$). This result is consistent with the experimental observation [21] and provides a degree of credence to the underlying assumption.

The comparison to the experiment is not limited to the scaling exponents. In addition to the dependence of the correlator on the scale, R, one could measure its dependence on the point configuration. Most generally the 3-point function is represented by a sum:

$$< \theta(1)\theta(2)\theta(3) > = \sum_{\alpha} \left(\frac{R}{L}\right)^{\lambda_\alpha} \Psi_\alpha(w, \chi) \tag{30}$$

where α labels terms with different scaling and the configuration dependence is parametrized by the reduced variables w, χ where $w \equiv 2|\hat{\rho}_1 \wedge \hat{\rho}_2|/R^2$ is the scaled area of the triangle and $\chi \equiv .5 \, sin^{-1}\left(\hat{\rho}_1 \cdot \hat{\rho}_2/2R^2\sqrt{1-w^2}\right)$ The meaning of angle χ is best understood for collinear configurations, $w = 0$, where it corresponds to moving one point continuously from coincidence with the second ($\rho_1 = 0$) to coincidence with the 3^d ($\rho_1 = \sqrt{3}\rho_2$). This choice of reduced variables is natural from the point of view of SL symmetry. The angle χ parametrizes pseudo-space rotations and corresponds to the SL angular momentum quantum number q mentioned earlier.

The theory and experiment could confront each other via direct comparison of the zero mode computed analytically for different solved models of dissipation with the measured configuration dependence. A weaker, but less model dependent comparison would involve the relation between the correlator in the collinear configuration, i.e. $\Psi(w = 0, \chi)$, with the correlator for the general configuration via Eq. (29), which emerges from the singular perturbation theory and is the consequence of the underlying SL symmetry. If the approximate symmetry holds then the collinear correlator predicts the non-collinear one. Alternatively, the approximate symmetry which should be good away from collinearity, can be tested directly as it predicts a certain linear relation between correlators $\Psi_\alpha(\xi w, \chi) = \int d\chi' g_\xi^{(\alpha)}(\chi' - \chi)\Psi_\alpha(w, \chi')$ with the transformation kernel $g_\xi^{(\alpha)}$ fixed by group theory [51, 20, 52].

The possible existence of several terms in the α sum, Eq. (30) with close scaling exponents makes the comparison with experiment more difficult. Aside from fitting the predicted zero modes to the data, one could attempt to use the multipoint measurements to disentangle the crossover: e.g. by picking a linear superposition of correlators in different configurations (different w and χ) which maximally extends the range of clean scaling. Clearly, if the eigenfunctions Ψ_α were known, one would be able to pick out the leading mode by an appropriate linear transformation.

6. Conclusion

The important question is: 1) what is the significance of the lack of universality of the randomly advected scalar problem and 2) how can one make contact between the theory and experiment? It may be argued that the non-universality in the random advection problem is merely the result of our ignorance of the correct turbulent velocity statistics: provided that *real* turbulent velocity field has universal statistics there is a unique correct Hopf operator which has a unique and universal set of zero modes. The problem then is entirely technical - we do not have a luxury of being able to choose a simple model in the *same universality class* with physical reality. The situation however may be worse. Independent of the exact exponents, the anomalous skewness that we find in the scalar indicates persistence of anisotropy and the dependence of the statistics of small scale fluctuations on the non-universal large scale boundary conditions. Were the same to hold for the velocity fluctuations themselves the notion of universality would have to be abandoned. Measurements of velocity intermittency, the derivative flatness, show considerable scatter which is somewhat masked by the tendency for there to be not much overlap in the Re range accessible through different geometries. Thus, more pertinent is the second question - comparing the theory and experiment. The accurate measurement of the scaling exponents is notoriously difficult and we would like to find other points of contact between theory and experiment. We would also like to extract from the present analysis of random advection something which can be directly compared to reality. Our proposal is to measure the full configuration dependence of the multi-point correlators (starting with N=3) and check the accuracy of the $SL(n-1)$ symmetry. This will 1) confirm or falsify the notion that the effect of the large scale strain/vorticity dominates the effect of small scale eddy diffusivity, 2) provide a framework for interpretation of the multi-point correlators which contain more information than just the overall scaling index.

We close with a few remarks about the velocity. Recently very convincing evidence has been given for the small scale isotropy of the two point velocity correlations. [23] One should not conclude that higher order correlations are isotropic. Motivated by analogy with the scalar skewness Pumir and Shraiman studied homogeneous shear $(< \vec{v} > \alpha \, y\hat{x})$ numerically and found that both ω_z and $\partial_y v_x$ had a normalized third moment $\sim O(1)$ and Re independent. [24, 25] Kolmogorov theory predicts again $Re^{-\frac{1}{2}}$. Should the trend seen only for $R_\lambda \stackrel{<}{\sim} 100$ numerically persist to much higher Re, we will again have a clear violation of Kolmogorov theory.

Acknowledgments

The authors are indebted to L.P.Kadanoff for his comments and suggestions with regard to the scope of the present review. EDS was supported by the NSF under contract DMR9121654. BIS wishes to thank the organizers of the Cargèse 1996 Summer School on "Mixing: Chaos and Turbulence."

References

[1] K.R. Sreenivasan, Proc. R. Soc. London **A 434**, 165 (1991).

[2] H. J. Catrakis and P. Dimotakis, J Fluid Mech **317**, 369 (1996); P.L. Miller and P. Dimotakis, J Fluid Mech **308** 129 (1996).

[3] R. A. Antonia and C. W. Van Atta, J. Fluid Mech. **84**, 561 (1980).

[4] F. Anselmet, Y. Gagne, E.J. Hopfinger and R.A. Antonia, J. Fluid Mech. **140**, 63 (1984).

[5] A.N. Kolmogorov, Dokl. Akad. Nauk SSSR **30**, 301 (1941).

[6] A. M. Obukhov: *Structure of the Temperature Field in a Turbulent Flow*. Izv. Akad. Nauk SSSR, Geogr.

[7] S. Corrsin, NACA R & M 58B1 (1958).

[8] K. R. Sreenivasan and R. Antonia, Annual Review of Fluid Mechanics, *29*, 435, (1997).

[9] U. Frisch, *Turbulence: the Legacy of A.N. Kolmogorov*, (Cambridge University Press, Cambridge, 1995).

[10] M. Holzer and E.D. Siggia, Phys. Fluids **6**, 1820 (1994).

[11] A. Pumir, Phys. Fluids **6**, 2118 (1994).

[12] R. H. Kraichnan, V. Yakhot, and S. Chen, Phys. Rev. Lett. **75**, 240, (1995).

[13] R. H. Kraichnan, Phys. Fluids **11**, 945 (1968).

[14] R. H. Kraichnan, J. Fluid Mech. **64**, 737 (1974).

[15] R. H. Kraichnan, Phys. Rev. Lett. **72**, 1016 (1994).

[16] B. I. Shraiman and E. D. Siggia, Phys. Rev. **E49**, 2912 (1994).

[17] B.I. Shraiman and E.D. Siggia, C. R. Acad. Sci. Paris, **321**, Sér. IIb, 279 (1995).

[18] M. Chertkov, G. Falkovich, I. Kolokolov, and V. Lebedev Phys. Rev. **E 52** , 4924 (1995).

[19] K. Gawędzki and A. Kupiainen, Phys. Rev. Lett. **75**, 3834 (1995).

[20] B. I. Shraiman and E. D. Siggia, Phys. Rev. Lett, **77**, 2463, (1996).

[21] P.G. Mestayer, J. Fluid Mech. **125**, 475, (1982).

[22] J. Gollub et al, Phys. Rev. Lett. **67**, 3507, (1991).

[23] Jayesh and Z. Warhaft, Phys. Rev. Lett. **67**, 3503, (1991).

[24] G. K. Batchelor, J. Fluid Mech. **5**, 113 (1959).

[25] J.L. Lumley, Phys. Fluids **10**, 855 (1967).

[26] S.Tavoulares and S. Corrsin, J. Fluid Mech. **104**, 311, (1981).

[27] S. Saddoughi and S.V. Veeravalli, J. Fluid Mech. **268**, 333, (1994).

[28] The Batchelor k^{-1} spectrum is even more controversial (see [1, 2] and B.Williams and J. Gollub, to be published) although the numerical simulations [10, 11] suggest that it only appears at very high Pe, the existence of this long crossover covering most experimentally relevant situations is interesting and requires an explanation in itself.

[29] E.D. Siggia, J. Fluid Mech. **107**, 375, (1981).

[30] S. Douady, Y. Couder, and M.E. Brachet, Phys. Rev. Lett. **67**, 983, (1991).

[31] G.I. Taylor, Proc. London Math. Soc., ser 2, **20**, 196,(1921).

[32] The appearance of a cusp in the center is ubiquitous and is observed experimentally both in the PDFs of $\Delta_r\Theta$ and $\Delta_r v$. The stretch exponentials employed in fitting these PDFs (see P. Kailasnath, K. Sreenivasan, and G. Stolovitzky, Phys. Rev. Lett. **68**, 2766, (1992); P. Tabeling, G. Zocci, F. Belin, J. Maurer, and H. Willaime, Phys. Rev. E, **53**, 1613, (1996) and E. Ching, Phys. Rev A44, 3622 (1991)) are to a large degree controlled by these cusps rather than the tails of the distribution. This interesting phenomenon deserves further investigation.

[33] M. Chertkov , G. Falkovich, I. Kolokolov, and V. Lebedev, Phys. Rev. E **51**, 3974, (1995).

[34] G. Falkovich, I. Kolokolov, V. Lebedev, and A. Migdal, Phys. Rev. E **54**,4896,(1996).

[35] M. Chertkov, Phys. Rev. E **55**, (1997).

[36] A.S. Monin and A.M. Yaglom, *Statistical Fluid Mechanics*, (MIT Press, Cambridge, 1971).

[37] Y. Sinai and V. Yakhot, Phys Rev Lett, **63**, 1962, (1989).

[38] E.S.C.Ching, Phys. Rev. Lett. **70**, 283 (1993); S.B. Pope and E.S.C.Ching, Phys. Fluids A **5**, 1529 (1993).

[39] A. Fairhall, O.Gat, V.Lvov, and I. Procaccia, Phys Rev. E, **53**, 3518, (1996).

[40] E. Ching, V. L'vov, A. Podivilov, and I. Procaccia, chao-dyn/9608008.

[41] The irrelevance of zero modes and non-anomalous behavior should not be taken for granted: e.g. M. Vergassola have demonstrated the anomalous scaling of the 2-point function in the passive vector model (see Phys.Rev.E **53**, R3021 (1996).

[42] R. Courant and D. Hilbert, *Mathematical Methods of Physics*, (Wiley Inter-science, New York, 1953).

[43] D. Bernard, K. Gawedski, and A. Kupiainen, Phys. Rev. E **54**, 2564,(1996).

[44] A. Pumir, Europh. Lett. **34**, 25 (1996).

[45] M. Chertkov and G. Falkovich, PRL **76**, 2706, (1996).

[46] J-D. Fournier, U. Frisch, and H. A. Rose, J Phys A **11**, 187, (1978).

[47] R.H. Kraichnan, preprint (1996).

[48] A. Pumir, B.I. Shraiman, and E.D. Siggia, PRE, Rapid Comm., to appear (1996).

[49] E.Balkovsky, M. Chertkov, I. Kolokolov, and V. Lebedev, JETP Lett. **61**, 1012, (1995).

[50] L.P. Kadanoff, "Scaling, Universality and Operator Algebras," in *Phase transitions and Critical Phenomena*, Vol. 5A, Domb and Green eds., (Academic Press, Boston, 1976).

[51] G. Vilenkin, in *Representations of groups and special functions* Kluwer Acad. Publ., 1991.

[52] B.I. Shraiman and E.D. Siggia, to be published.

[53] M. Chertkov, G. Falkovich, and V. Lebedev, Phys. Rev. Lett., **76**, 3707, (1996).

Cascade Models of Turbulence and Mixing

Leo P. Kadanoff

*The Research Institutes, The University of Chicago,
5640 South Ellis Ave , Chicago Illinois 60637, USA*

This note describes two kinds of work on turbulence. First it describes a simplified model turbulent energy-cascades called the GOY model. Second it mentions work on a model of mixing in fluids derived as an extension of this very same GOY model. In addition to a brief historical discussion, I include a some mention of our own work carried on at the University of Chicago by Jane Wang, Detlef Lohse, Roberto Benzi, Norbert Schörghofer, and Scott Wunsch. Our own studies are in large measure the outgrowth of a paper by M. H. Jensen, G. Paladin, and A. Vulpiani[1]. I mention this connection with some sadness because I recall Paladin's recent death in a mountain accident.

1. Turbulence and mixing

Turbulence is one of *those* problems. Interesting. Vexing. Long-standing. Unsolved.

Mixing is also in much the same situation. In both cases, we know all the basic equations involved, but we really do not understand the physics in any full way. Instead we have models and metaphors which are believed to provide a partial understanding of what is going on, but both the relevance of the models, and the conclusions to be drawn from them, are quite controversial.

Today I shall focus on turbulence and the closely associated problem of mixing in a turbulent environment. I shall be interested in the conceptual and mathematical models which describe these phenomena, and most specifically on how one draws physical conclusions from such models.

The basic physics of turbulence is easy to describe: Imagine a fluid which initially has some sort of simple flow with a non-uniform velocity. In the situation

Mixing: Chaos and Turbulence, edited by Chaté *et al.*
Kluwer Academic / Plenum Publishers, New York 1999.

in which there is a velocities small in comparison to the speed of sound the flow is approximately incompressible. It is then described by the Navier-Stokes equation:

$$\partial_t \mathbf{U}(\mathbf{r}, t) + \mathbf{U} \cdot \nabla)\mathbf{U} = \nu \nabla^2 \mathbf{U} - \nabla P(\mathbf{r}, t) + \mathbf{F}(\mathbf{r}, t)$$
$$\nabla \cdot \mathbf{U} = 0 \tag{1}$$

in units in which the density is equal to one. Here the velocity $\mathbf{U}$ is carried from place to place by the non-linear term $\mathbf{U} \cdot \nabla)\mathbf{U}$. The diffusion term with its associated viscosity, ν, tends to smooth out the velocity field while $\mathbf{F}$ is an external force which describes the stirring process which puts the fluid into motion. The incompressibility condition is that the divergence of $\mathbf{U}$ is zero and is enforced by having the system pick an appropriate spatial dependence of the pressure, P. To get an alternative form of the Navier Stokes equation take the spatial Fourier transform of equation (1). Use the symbols $\mathbf{u}$, p, and $\mathbf{f}$ for the spatial Fourier transforms of velocity, pressure, and force. The resulting equation reads

$$\partial_t \mathbf{u}(\mathbf{k}) + i \int d\mathbf{q}(\mathbf{u}(\mathbf{k} - \mathbf{q}) \cdot i\mathbf{q})\mathbf{u}(\mathbf{q}) = \nu \kappa^2 \mathbf{u}(\mathbf{k}) - i\mathbf{k}p(\mathbf{k}) + \mathbf{f}(\mathbf{k})$$
$$\mathbf{k} \cdot \mathbf{u}(\mathbf{k}) = 0. \tag{2}$$

To describe the mixing imagine some quantity suspended in the flow. This quantity might for example be heat. The density of the quantity, perhaps the temperature $T(\mathbf{r}, t)$ obeys the flow equation

$$\left[\partial_t + \mathbf{U} \cdot \nabla - D\nabla^2\right] T(\mathbf{r}, t) = 0. \tag{3}$$

Here D is the diffusion coefficient for the passive scalar, while the term in $\mathbf{U}$ describes how the velocity field moves the heat about.

The nonlinear processes built into the Navier-Stokes equations produce the interesting structures in turbulent flow. As time goes on, the interactions among the fluid elements, represented by the $\mathbf{U} \cdot \nabla \mathbf{U}$ term, will tend to produce more and more complex structures, with more and more fine details. This complexity will continue to grow until, at a very fine scale, it is limited by viscous damping which tends to smooth everything out.[2] Of course, the $\mathbf{U} \cdot \nabla T$ term term will also produce complex structures. The difference is that this convective mixing equation, equation (3), has no non-linear character. Because of this lack of feedback, the structures in T can be quite different from the structures in $\mathbf{U}$. The parameter for describing the degree of complexity in velocity is the Reynolds number, $\mathrm{Re} = LU/\nu$, where U is the typical non-uniformity in velocity, L is the size of a region in which the non-uniformity occurs and ν is the kinematic viscosity of the fluid in question. A similar parameter in T is LU/D. In everyday systems, fine scales are produced precisely because these parameters tend to be quite large, as a consequence of the small size of the diffusion coefficients. For example, the kinematic viscosity of water is of order $10^{-6}\mathrm{m}^2/\mathrm{sec}$.

Conservation laws are always important. Since the diffusion parameters ν and D are small, it is reasonable to ask what quantities are conserved in the absence of the diffusion. For the Navier-Stokes equation, the conserved quantities include:

- the momentum $= \int d\mathbf{r}\rho\mathbf{u}(\mathbf{r}, t)$
- the kinetic energy $= \int d\mathbf{r}\frac{\rho}{2}\mathbf{u}(\mathbf{r}, t)^2$, and
- the helicity $= \int d\mathbf{r}\mathbf{u}(\mathbf{r}, t) \cdot (\nabla \times \mathbf{u}(\mathbf{r}, t))$.

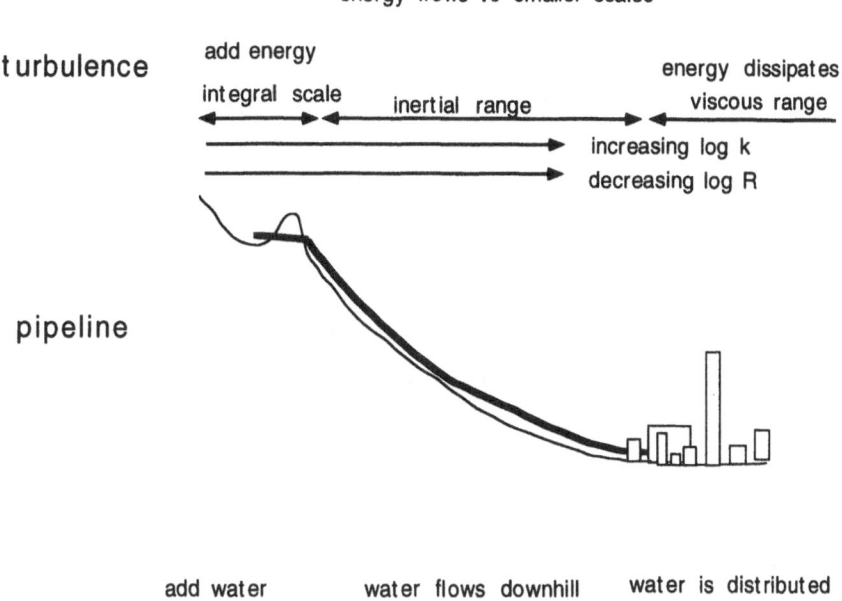

energy flows to smaller scales

turbulence

add energy

integral scale   inertial range   energy dissipates
                                       viscous range

increasing log k
decreasing log R

pipeline

add water   water flows downhill   water is distributed

Fig. 1. — A pipeline model of turbulence

The mixing equation moves temperature from place to place but, in the absence of diffusion, does not change the distribution of temperature-values over the entire system. For this reason, all moments of the temperature

- $\int d\mathbf{r}[T(\mathbf{r}, t)]^p$

are equally conserved.

It is usually considered that the conservation of kinetic energy is particularly crucial in defining the qualitative nature of turbulent processes. The figure shows a cartoon view of these processes, a view first introduced by Richardson[3] and then given more mathematical form by Kolmogorov in his wonderful 1941 paper[4]. The idea is best described in Fourier transform language in which the wave vector $\mathbf{k}$ serves to describe the inverse of the different spatial scales. Some large scale process, perhaps stirring, is used to feed energy into the fluid at small $k = |\mathbf{k}|$. The range of scales at which this addition of energy occurs is called the integral range. Then the non-linear coupling of velocities causes the energy to cascade toward larger wave vectors. The range of scales dominated by the cascade is called the inertial range. This range can be quite large since it is proportional to a power of the Reynolds number. For even larger k, one passes into the viscous range. Here, viscous process dissipate the energy into heat and smooth out all details. The problem of turbulence is the problem of predicting and understanding what happens when the inertial range is large. The mixing process can be understood in a similar way as a question about what happens to some additive thrown into the pipe at the top.

2. Turbulent models

A quantitative analysis of the energy flow was introduced by Kolmogorov in two papers written in 1941 and 1962, and called K41 and K62. (See [4]). The key idea is that all the energy which is added on the largest scale flows unchanged through the inertial range toward higher k. The flow is described by an energy current, ε. By dimensional analysis Kolmogorov found that the current of energy is proportional to $k[u(k)]^3$. The conservation of energy implies that in the inertial range, the time average of this current must be independent of k. In the 1941 paper it was assumed that the fluctuations in energy current are not very large. In that case, the velocity has a typical value which is given an order of magnitude statement

$$u(\mathbf{k}) \sim \left(\frac{\varepsilon}{k}\right)^{1/3} \tag{4}$$

where ε is the time-averaged energy current or equivalently the average rate of energy dissipation. There are two crucial assumptions in this K41 paper:

a. The energy flux, ε, is assumed to be roughly constant, and thus essentially equal to the energy input per unit time.

b. Further, it is assumed that far into the inertial range, the system remembers as little as it possibly can about the conditions at which energy is added or dissipated. It must remember the value of ε since that determines the total energy flow all scales in the inertial range. In K41 it is assumed that this ε determines the flow and that the system does not know anything about either the value of L or of the dissipation length scale η.

This argument, and many subsequent arguments, sees the energy flow as a kind of pipeline. (See again the figure.) Energy is added to the pipe at the large scales, flows through the very long pipe, and then flows out via dissipation at the far end. The 1941 analysis assumes that the flow only depends upon the average ε. The 1962 analysis includes fluctuation in ε, but it assumes that the flow through the pipe at a given point only depends upon processes upstream of that point. Thus for example, inertial range behavior would not be influenced by the value of the viscosity since the viscosity appears only in the dissipative range, far downstream. Since the fluctuations in different regions of the pipe would be independent multiplicative effects, one might expect to get fluctuations which are much larger than Gaussian.

But the pipeline analogy makes the assumption that influence only flows downstream seem doubtful. Certainly a choking of the outlet would tend to diminish the flow.

In the first decades after the 1941 paper, scholars of turbulence focused upon the other basic assumption of the K41 theory, the assumption of the constancy of energy dissipation. Partially in response to a criticism by Landau, Kolmogorov changed his own point of view. In K62, he argued that, because of fluctuations, the energy input should be function of space and time which can vary over many, many orders of magnitude. These huge fluctuations were termed intermittency. While still assuming that influence only flows downstream, Benoit Mandelbrot[5] and others developed richer cascade models which showed how a highly fluctuating behavior might arise and be reflected in a complex fractal structure of $\mathbf{u}(\mathbf{k})$.

The mixing defined by equation (3) can be described in much the same terms. The K41 ideas suggest that there should be an entropy current proportional to

$k\mathbf{u}(\mathbf{k})[T(\mathbf{k})]^2$ and that this current should not dependent upon k in the entire range of k^{-1} between the forcing scale and the dissipation scale. The consequence once again is a $k^{-1/3}$ behavior, but now for $T(k)$. However, once again, one might ask whether fluctuations will produce very important deviations from the simple estimated behavior.

Theory cannot for the moment distinguish between the weak fluctuations of K41 or the intermittency of K62. To settle questions of this kind we should look to experiment. The experimental evidence suggests the existence of intermittency, but is not yet conclusive. We might also look to simulations. Turbulence problems can easily be set up on the computer since the turbulent system may be correctly described by a well-known partial differential equation, the Navier-Stokes equation. Unfortunately, computers are too slow to allow simulations to cover a sufficient range of scales. How can we learn more? One thing is to set up models which correctly realize in specific form the previous theoretical ideas and then see how consistent these ideas might be. One effort in this direction has been the development of a one-dimensional model which includes in a very rough way cascades from shell to shell in momentum space.

3. GOY model

The model is connected with the names of Gledzer, Ohkitani, and Yamada (inelegantly called GOY)[6, 7]. It is among the simplest models which can correctly realize a cascade through orders of magnitude of wave vector, k. To derive it, look at the Fourier transform version the Navier-Stokes equation, as given by equation (2). Here, the system is described by the Fourier coefficients of the velocity, $u(k)$. Imagine a cascade which goes from a k of order the size of the system, to one in a shell which contains k's a factor of λ larger, and then a higher shell with k's of order $\lambda^2 L$, and so forth through $\lambda^n L$, until finally we reach $\lambda^N L$, a wave vector which sits far into the dissipative range. Thus we have k-values given by $k_n = k_0 \lambda^n$. The real system involves many, many Fourier coefficients, $\mathbf{u}(\mathbf{k})$, with the number per shell sharply increasing with the shell-number, n. This model tries to describe the flow by keeping precisely one Fourier coefficient for each shell, so it is described by $u_n(t)$, for $n = 1, 2, \ldots, N$ with each u_n being a complex number. The equations of motion are picked to mimic Navier-Stokes, with non-linearity, viscous damping and energy conservation all included. Thus the model equation has the form

$$\frac{\mathrm{d}}{\mathrm{d}t} u_n(t) = C_n + D_n + F_n, \quad n = 1, 2, 3, \ldots, N \tag{5}$$

where will we choose a number of shells, N, to be some reasonable number like 25. The dissipation term is just like the one in equation (2), namely

$$D_n = -\nu k_n^2 u_n \tag{6}$$

We will take the integral scale length, L, to be equal to one and then insure a large inertial range by picking the viscosity, ν, to be much smaller than one. The forcing term is taken to be at large scales

$$F_n = \delta_{n,1} f \tag{7}$$

Finally, one needs a term to represent the cascade which sends energy to higher and higher wave vectors. To mimic the structure of the Navier-Stokes equation, this cascade term should be of first order in wave vector and second order in velocity. The Richardson-Kolmogorov picture sees a step by step process in which information is carried from one wave vector to neighboring ones, so we should couple nearby shells. One structure which will do this is the GOY form:

$$C_n = (ak_n u_{n+1} u_{n+2} + bk_{n-1} u_{n+1} u_{n-1} + ck_{n-2} u_{n-2} u_{n-1})^* \qquad (8)$$

with a, b, and c being undetermined coefficients.

There is a lot in this model-building which is undetermined. There is no particularly good reason for picking these specific three terms to make up the cascade term (8). We did it in our work[8, 9] mostly because the previous authors had done so and we wished to build upon their ideas and results. Previous authors, K. Ohkitani and M. Yamada, [7] and M. H. Jensen, G. Paladin, and A. Vulpiani [1] and also R. Benzi, L. Biferale, and G. Parisi[10] used it because it gave a pattern of u_n versus time which looked rather like the fluctuating velocity fields $U(r, t)$ of turbulence.

Notice that in the absence of forcing and dissipation we have two conserved quantities of the form

$$C = \sum_n |u_n|^2 z^n \qquad (9)$$

which are conserved so long as the coefficients in equation (8) obey

$$a + bz + cz^2 = 0 . \qquad (10)$$

A reasonable definition of a conserved kinetic energy is a sum of velocities squared of the form:

$$E = \frac{1}{2} \sum_n |u_n|^2 \qquad (11)$$

which then requires that the coefficients obey

$$a + b + c = 0 \qquad (12)$$

Given this condition the energy obeys the conservation law

$$\frac{\mathrm{d}}{\mathrm{d}t} E_n = (J_n - J_{n-1}) + \delta_{n,1} \mathrm{Re}(f u_1^*) - 2\nu k_n^2 |u_n|^2$$

The first term on the right hand side is the changed by the energy cascade. It is of the form of the divergence of an energy current, where the current has the form

$$J_n = a\Delta_n - c\Delta_{n-1} \ \text{ with } \ \Delta_n = \mathrm{Re}(k_n u_n u_{n+1} u_{n+2})$$

The other terms represent the addition of energy at the forcing scale and its dissipation by viscosity.

We are now ready to start to understand the model. We can, without loss of generality, pick

$$a = 1 = f \qquad (13)$$

since these conditions only sets the scale for time and for velocity. Then the model contains three parameters. The most physical one is ν, which is essentially the inverse Reynolds number and is taken to be very small. Next, λ sets the scale for discretization. If the model is a good one, results should be independent of λ. Finally the one remaining parameter is b, an artificial parameter which essentially defines different turbulence models. Given equation (12), the parameter b determines the ratio of the energy flux toward higher wave number to the flux in the opposite sense. Thus b might be considered to be a free parameter in the theory. But, in [7] we argue that fixing a particular relation of b to λ,

$$b = -1 + \lambda^{-1} , \qquad (14)$$

makes the most sense physically because that makes the second conserved quantity in the GOY model scale with distance in the same way as the helicity.

But the connection (14) between b and λ seems a bit contrived. If we put this connection aside, the model involves two parameters which have no direct counterpart in Navier-Stokes: the shell width, λ, and b. We study the model for in the range $-1 < b < 0$. Because the model equations involve many fewer variables than those in the Navier-Stokes equations, its qualitative properties can be established with the aid of simulations. We can thus ask, do the size of the fluctuations in the model agree with K41 or with K62? The question is sharply posed and it may be sharply answered. The answer is yes. Yes? ....Yes!

In one range of parameters (b close to zero) the system shows a static behavior, no fluctuations, and exactly the K41 scaling of equation (4). In contrast, in another range (more negative b) the static solution is unstable against fluctuations. The model solution then shows a time-dependent state with fluctuations in u_n which become stronger for higher values of n in the inertial range. The net result is something more like K62 behavior. (In fact if the relation between λ and b is chosen in accord with equation (14) there is a very close quantitative relation between the intermittency observed in experiments and simulations on turbulent systems and the intermittency of the GOY model. (This point was developed in detail in [7].) In the GOY model, we can easily understand how we might find very large fluctuations in the magnitude of u_n for large n. The small n behavior is fixed by the forcing. To understand larger n, notice that the model permits independent fluctuations in the velocity ratio between neighboring shells. This idea was pointed out by R. Benzi, L. Biferale, and G. Parisi[10]. When one multiplies out a set of many independent random multipliers, the result can have truly huge fluctuations. Thus the model serves as a partial justification for both K41 and the later, more fluctuating, theories.

However, K41 and K62 both suggest that only the value of ε matters. In K41, ε is constant throughout the inertial range; in K62, ε fluctuates and varies increasingly as one passes to smaller and smaller distances. But ε is still the only piece of information effectively passed along through the inertial range. As a consequence of this information flow from large to small distances, the both theories suggest that the inertial range behavior should be independent of ν. The GOY model disagrees with these conclusions. In the static region of the model-behavior, not one but three pieces of information are passed through the inertial range. These pieces of information are the values of ([8]):

$$u_{3q}/\lambda^q \quad u_{3q+1}/\lambda^q \text{ and } u_{3q+2}/\lambda^q .$$

The product of these three gives the energy flux. However the remaining two quantities represent information carried faithfully through the inertial range, but not predicted by the Kolmogorov arguments. Because more information is passed along, there is more possibility for communication between the integral and the dissipative scales. In response to some prompting from Z.S. She, a group of us (Norbert Schörghofer, Jane Wang, Detlef Lohse, Roberto Benzi, and I) showed that in the static situation the energy flux down the pipe does certainly depend upon ν, even in the limit of high Reynolds number. In contrast to the early theories (*both* K41 and K62) here the flow down the energy pipeline does depend upon an interaction between inlet and outlet conditions, induced by correlations carried up and down the pipe. The mechanism for the correlation in the GOY model is quite unphysical in that it makes direct use of the shell thickness, λ. However this result points us toward the possibility that both inlet and outlet might play a role in determining the flow in real turbulence. If so, there will be some Reynolds number dependence of inertial range behavior. Thus, a vastly oversimplified model led to insights which might well play back to a deeper understanding of a truly complex phenomenon.

4. Mixing

The ν-dependence described above only holds in the static case. I would guess that when fluctuations are important, the extra pieces of information carried through the inertial range are destroyed by an averaging process. Nonetheless there are two issues raised by this argument which might indeed be relevant for real turbulence:

- How much information is carried through the inertial range.

- Can ends of this range be replaced by simple boundary conditions. In particular can the viscosity be replaced by a boundary condition in which the velocity is required to be zero at a certain large value of k. If that is true, probably the exact form of the dissipation is irrelevant to the structure of inertial range behavior.

Somewhat similar questions hold for mixing. Turbulent stirring mixes the temperature distribution and produces small-scale structure from large. There is a kind of inertial range in which one is far away from the full smoothing produced by the diffusion term in equation (3). To understand this range, it is reasonable to ask about how much information is transferred through this scaling regime. Maybe the existence of many conservation laws is relevant for this discussion. One could also ask whether the dissipation can effectively be replaced by a boundary condition at the start of the dissipative range. As part of the same discussion one could ask also whether the value of the diffusion coefficient matters.

Recent calculations have focused upon the special case in which fluctuations in the temperature field are extremely rapid time-variations in a Gaussian velocity field. Many recent authors agree that the resulting temperature field shows extremely large fluctuations roughly like that of the K62 theory. In fact, Robert H. Kraichnan, Victor Yakhot, and Shiyi Chen[11] claim that they have found an apparently exact solution to the this multiscaling problem, based upon old theoretical work of Robert Kraichnan[12]. It is further claimed that this result will be robust under a small changes in the way the flow is set up. In contrast, other authors[13, 14, 15] do look at the problem slightly differently and fail to find the claimed results and/or robustness. There is a real disagreement.

One possible approach to discussing this disagreement and also to gaining a better understanding of what makes the very large fluctuations it to deal with model problems which have roughly the right nature. One can extend the GOY model to handle temperature mixing. In fact this was done in the early paper of M. H. Jensen, G. Paladin, and A. Vulpiani[1], More recently, the calculation has been extended to the rapid-time-variation situation by A. Wirth and L. Biferale[16] who wrote down an equation of the form

$$\left(\frac{\mathrm{d}}{\mathrm{d}t} + Dk_n^2\right)\theta_n(t) = f(t)\delta_{n,1} + C_n^* \quad \text{with}$$

$$
\begin{aligned}
C_n &= k_n(\theta_{n+1}u_{n+2} + u_{n+1}^*\theta_{n+2}) + k_{n-1}(\theta_{n+1}u_{n-1} - u_{n+1}\theta_{n-1}) \\
&\quad -k_{n-2}(\theta_{n-2}u_{n-1}^* + u_{n-2}\theta_{n-1})
\end{aligned}
$$

where they take the u_n's to be a prescribed velocity field varying rapidly in time. They report a result more complex than simple Kolmogorov scaling, even though they have one conservation law, for $\sum_n |\theta_n|^2$. It will be interesting to see if this model shed any light on the nature of the mixing problem.

At Chicago, we are working on a slightly different model involving a flow through two one-dimensional pipes, labeled by an index σ which takes on the values plus or minus one. The equations are:

$$\left[\frac{\partial}{\partial_t} - D\frac{\partial^2}{\partial x^2} + \sigma u(x,t)\frac{\partial}{\partial_x}\right]T_\sigma(x,t) + \frac{v(x,t)}{2}[T_+(x,t) - T_-(x,t)] \tag{15}$$

with $\partial_x u = v$ being a kind of incompressibility condition. This model has two conservation laws. Does it have large fluctuations? If we pick a simple scaling behavior for v, e.g. Gaussian with zero time correlation range, and scaling properties in x, then what will we find for T?

We can do a different version of the model which has an infinite number of conservation laws in the zero diffusion limit. Pick x to be a discrete index, $x = 1, 2, \ldots, N$. Then allow two kinds of processes, first diffusion

$$T_\sigma(x, t+\tau) = T_\sigma(x,t) + D\tau \left[T_\sigma(x+1,t) + T_\sigma(x-1,t) - T_\sigma(x,t)\right].$$

The second process is advection which involves the composition of a series of swirls or vortices. The basic swirl has a new temperature which obeys

$$
\begin{aligned}
T_\sigma(x, t+\tau) &= T_{-\sigma}(x, t) \\
T_\sigma(x+1, t+\tau) &= T_\sigma(x, t) \\
T_{-\sigma}(x+1, t+\tau) &= T_\sigma(x+1, t) \\
T_{-\sigma}(x, t+\tau) &= T_{-\sigma}(x+1, t)
\end{aligned}
$$

One gets larger swirls by putting together several of these smaller swirls. Notice that in the absence of diffusion, the probability distribution for temperature is invariant in time. The combination of swirl plus diffusion can produce many of the characteristic effects of passive scalars in an incompressible flow.

Models are fun, and sometimes even instructive.

Acknowledgments

This talk is based upon two recent articles: Leo P. Kadanoff, A model of turbulence, Physics Today, (September 1995) p 11. and Leo P. Kadanoff, Turbulent Excursions, Nature 382 p.116 (1996). The modeling work described here was done in collaboration with Jane Wang, Detlef Lohse, Roberto Benzi, Norbert Schörghofer, and Scott Wunsch. That work was supported in part by the DOE and the ONR. This work is similarly supported.

References

[1] M. H. Jensen, G. Paladin, and A. Vulpiani, Phys. Rev. A **43**, 798 (1991). See also T. Bohr, M.H. Jensen, G. Paladin, A. Vulpiani, *Dynamical Systems Approach to Turbulence*, Cambridge Nonlinear Science Series (in press).

[2] See, for example, the extremely complex small structures in the picture by Harry Catrakis and Paul Dimotakis in Physics Today, (September 1995) p 11.

[3] See the writeup in Benoit Mandelbrot, *The Fractal Geometry of Nature*, (Freeman and Company, New York, 1983) p.401ff.

[4] This paper is called K41 in the field. It and Kolmogorov's other work on turbulence are elegantly described in U. Frisch, *Turbulence*, (Cambridge University Press, 1995).

[5] See the Frisch book, op. cit., for details.

[6] E. B. Gledzer, Sov. Phys. Dokl. **18**, 216 (1973).

[7] K. Ohkitani and M. Yamada, Prog. Theor. Phys. **81**, 329 (1989); J. Phys. Soc. Jpn. **56**, 4210 (1987); Prog. Theor. Phys. **79**, 1265 (1988).

[8] L. Kadanoff, D. Lohse, J. Wang, R. Benzi, Physics Fluids **7** 617 (1995); L. Kadanoff, Physics Today **48**, 11 (1995).

[9] N. Schörghofer, L. Kadanoff, and D. Lohse, Physica D **88** 40 (1995); N. Schörghofer, L. Kadanoff, and D. Lohse, Physica D, in press.

[10] R. Benzi, L. Biferale, and G. Parisi, Physica D **65**, 163 (1993).

[11] R. H. Kraichnan, V. Yakhot, and Shiyi Chen, Phys. Rev. Lett. **75** 240 (1995).

[12] R. H. Kraichnan, Phys. Fluids **11**, 945 (1968).

[13] B. I. Shraiman and E. D. Siggia, C.R. Acad. Sci. Paris, **321** Série II b, 279 (1995).

[14] M. Chertkov, G. Falkovich, I. Kolokolov, and V. Lebedev, Phys. Rev. E **52** 4924 (1995).

[15] A. Gawedzki and A. Kupiainen, Phys. Rev. Lett. **75**, 3608 (1995).

[16] A. Wirth and L. Biferale, preprint 1996.

INDEX